Strongly Correlated Fermions and Bosons in Low-Dimensional Disordered Systems

NATO Science Series

A Series presenting the results of scientific meetings supported under the NATO Science Programme.

The Series is published by IOS Press, Amsterdam, and Kluwer Academic Publishers in conjunction with the NATO Scientific Affairs Division

Sub-Series

I. Life and Behavioural Sciences	IOS Press
II. Mathematics, Physics and Chemistry	Kluwer Academic Publishers
III. Computer and Systems Science	IOS Press
IV. Earth and Environmental Sciences	Kluwer Academic Publishers
V. Science and Technology Policy	IOS Press

The NATO Science Series continues the series of books published formerly as the NATO ASI Series.

The NATO Science Programme offers support for collaboration in civil science between scientists of countries of the Euro-Atlantic Partnership Council. The types of scientific meeting generally supported are "Advanced Study Institutes" and "Advanced Research Workshops", although other types of meeting are supported from time to time. The NATO Science Series collects together the results of these meetings. The meetings are co-organized bij scientists from NATO countries and scientists from NATO's Partner countries – countries of the CIS and Central and Eastern Europe.

Advanced Study Institutes are high-level tutorial courses offering in-depth study of latest advances in a field.
Advanced Research Workshops are expert meetings aimed at critical assessment of a field, and identification of directions for future action.

As a consequence of the restructuring of the NATO Science Programme in 1999, the NATO Science Series has been re-organised and there are currently Five Sub-series as noted above. Please consult the following web sites for information on previous volumes published in the Series, as well as details of earlier Sub-series.

http://www.nato.int/science
http://www.wkap.nl
http://www.iospress.nl
http://www.wtv-books.de/nato-pco.htm

Series II: Mathematics, Physics and Chemistry – Vol. 72

Strongly Correlated Fermions and Bosons in Low-Dimensional Disordered Systems

edited by

I.V. Lerner

School of Physics and Astronomy,
University of Birmingham,
Edgbaston, Birmingham, U.K.

B.L. Althsuler

Physics Department,
Princeton University,
New York, U.S.A. and
NEC Research Princeton,
New York, U.S.A.

V.I. Fal'ko

Physics Department,
Lancaster University,
Lancaster, U.K.

and

T. Giamarchi

Laboratoire de Physique des Solides,
CNRS-UMR 8502,
Orsay, France

Kluwer Academic Publishers

Dordrecht / Boston / London

Published in cooperation with NATO Scientific Affairs Division

Proceedings of the NATO Advanced Study Institute on
Strongly Correlated Fermions and Bosons in Low-Dimensional Disordered Systems
Windsor, United Kingdom
13–26 August 2001

A C.I.P. Catalogue record for this book is available from the Library of Congress.

ISBN 1-4020-0748-5

Published by Kluwer Academic Publishers,
P.O. Box 17, 3300 AA Dordrecht, The Netherlands.

Sold and distributed in North, Central and South America
by Kluwer Academic Publishers,
101 Philip Drive, Norwell, MA 02061, U.S.A.

In all other countries, sold and distributed
by Kluwer Academic Publishers,
P.O. Box 322, 3300 AH Dordrecht, The Netherlands.

Printed on acid-free paper

Printed in the Netherlands.

Contents

PART THREE / TWO-DIMENSIONAL SYSTEMS

Preface

The physics of strongly correlated fermions and bosons in a disordered environment and confined geometries is at the focus of intense experimental and theoretical research efforts. Advances in material technology and in low temperature techniques during the last few years led to the discoveries of new physical phenomena including Bose condensation of atomic gases and a possible metal-insulator transition in two-dimensional high mobility electron structures. Situations were the electronic system is so dominated by interactions that the old concepts of a Fermi liquid do not necessarily make a good starting point are now routinely achieved.

This is particularly true in the theory of low dimensional systems such as carbon nanotubes, or in two dimensional electron gases in high mobility devices where the electrons can form a variety of new structures. In many of these systems disorder is an unavoidable complication and lead to a host of rich physical phenomena. This has pushed the forefront of fundamental research in condensed matter towards the edge where the interplay between many-body correlations and quantum interference enhanced by disorder has become the key to the understanding of novel phenomena.

To describe these new physical phenomena and systems, the theorist has to face new challenges. Indeed, in many of these systems studies of transport properties (volt-ampere characteristics, noise, spin-current, mesoscopic fluctuations etc.) are crucial, simultaneously being the main tool for investigations of the fundamental properties of new materials and providing a way towards their development into devices that can be directly used for applications. The need to deal with those complications makes theorists to develop new and considerably modify old techniques. As a result, together with the incredibly enriched experimental situation, the theoretical toolkit has also been substantially improved and developed over the recent decade.

This situation makes educating a new generation of young theorists one of the prime targets of researchers active in the field. The structure of NATO Advanced Study Institutes is ideally suited for presenting the new exciting developments together. This volume is the product of a NATO ASI entitled 'Field Theory of Strongly Correlated Bosons and Fermions in Low-Dimensional Disordered Systems', held at the Cumberland Lodge of St. Catherine's Foundation (Windsor Great Park, Windsor - UK) in August 2001.

Lectures and review articles presented in this book make together a comprehensive course in the field theory techniques and their applications to recently discovered new physical phenomena. Topics included in this volume cover many exciting surprises found in novel nanostructures and low-dimensional devices of submicron size, including quantum intereference and electron-electron interaction effects in the transport in metals, semiconductors and superconductors, theories of the Luttinger liquid of electrons in carbon nanotubes, Wigner crystal in two dimensional electronic systems, the Kondo effect in quantum dots in the Coulomb blockade regime, quantum chaos in ballistic and disordered microstructures, mesoscopic effects in superconductors. Theoretical papers are complemented by four experimental reviews of recent achievements in the physics of the Kondo effect in zero-dimensional devices, transport in one-dimensional nanotubes, magnetooptics of skyrmions and composite fermions in the quantum Hall regime, and a possible metal-insulator transition in two-dimensional electron and hole gases.

Acknowledgements. We are thankful to NATO Science Programme for providing an excellent opportunity for bringing together theorists and experimentalists working in the field of strongly correlated disordered systems. We gratefully acknowledge an additional support provided for the organization of this ASI by EPSRC, USARDSG-UK, the CMT group and TMP group (IOP, UK), NEC Research Institute (Princeton) and the University of Birmingham. As the organisers of this NATO ASI, we would like to thank all the lecturers of this summer school for their efforts and, on behalf of all the participants, express our gratitude to the staff of the Cumberland Lodge and to the St. Catherine's Foundation for creating a warm and hospitable atmosphere during the school.

Igor Lerner

Boris Altshuler

Thierry Giamarchi

Vladimir Fal'ko

THE KONDO SCREENING CLOUD

IAN AFFLECK
Physics Department, Boston University, Boston, MA02215, USA

1. The Kondo effect and the screening cloud

A single impurity in a metal is described by the Kondo (or s-d) model:

$$H = \sum_{\vec{k}\sigma} \psi^\dagger_{\vec{k}\sigma} \psi_{\vec{k}\sigma} \epsilon_k + J \vec{S}_{\text{imp}} \cdot \vec{S}_{\text{el}}(r=0). \quad (1)$$

Here $\vec{S}_{\text{imp}}$ is the impurity spin operator (with S=1/2) and $\vec{S}_{\text{el}}$ is the electron spin density at position $\vec{r}$. After expanding the electron field, $\psi(\vec{r})$, in spherical harmonics and keeping only the s-wave and linearizing the dispersion relation we obtain a relativistic quantum field theory, defined on a half-line with the impurity at the origin [1]. (See fig. 1.)

The Hamiltonian reduces to:

$$H = iv_F \int_0^\infty dx \left[\psi^\dagger_L \frac{d}{dx} \psi_L - \psi^\dagger_R \frac{d}{dx} \psi_R \right] + 2\pi v_F \lambda \vec{S}_{\text{imp}} \cdot \vec{S}_{\text{el}}(0). \quad (2)$$

Here λ is the dimensionless Kondo coupling constant, $J\nu$, where ν is the density of states. To study the problem at low energies, we may apply the renormalization group, integrating out high energy Fourier modes of the electron operators, reducing the band-with, D:

$$d\lambda/d\ln D \approx -\lambda^2 + \ldots \quad (3)$$

with solution:

$$\lambda_{\text{eff}}(D) \approx \frac{\lambda_0}{1 - \lambda_0 \ln(D_0/D)} + \ldots \quad (4)$$

Figure 1.

I.V. Lerner et al. (eds.),
Strongly Correlated Fermions and Bosons in Low-Dimensional Disordered Systems, 1–12.

Figure 2.

The effective coupling becomes O(1) at an energy scale T_K:

$$T_K \approx De^{-1/\lambda_0}. \tag{5}$$

Here λ_0 is the bare Kondo coupling and D_0 is of order the Fermi energy. After reducing the bandwidth the effective Hamiltonian has a *wave-vector* cut off:

$$|k - k_F| < T_K/v_F \equiv \xi_K. \tag{6}$$

This defines a characteristic *length scale* for the Kondo effect; it is typically around .1 to 1 micron.

At low energies, $T << T_K$, λ_{eff} seems to get large. This strong coupling physics is easiest to understand in a tight-binding model.

$$H = -t\sum_{j=0}^{\infty}(\psi_j^\dagger \psi_{j+1} + h.c.) + JS_{\text{imp}} \cdot \vec{S}_{\text{el}}(0). \tag{7}$$

(See fig. 2.) For $J >> t$, we simply find the groundstate of the last term in this Hamiltonian. This has 1 electron at site $j = 0$ forming a singlet with the impurity, $|\phi_0 >= (| \uparrow\downarrow> -| \downarrow\uparrow>)/\sqrt{2}$. The other electrons are free except that they must not go to $j = 0$ since they would break the singlet. Effectively they feel an infinite repulsion at $j = 0$, corresponding to a $\pi/2$ phase shift. For finite (small) λ_0, this description only holds at low energies and small $|k - k_F|$. Only long wavelength probes see this simple $\pi/2$ phase shift. This is the basis of Nozières' local Fermi liquid theory of the Kondo effect [2]. The short distance physics is more complicated, involving the singlet formation. Heuristically, we may think of an electron in a wave-function which is spread out over this large distance, ξ_K, which is forming a singlet with the impurity. (More accurately, we should think of the singlet as being formed by a linear superposition of an electron and a hole since there is not neccessarily any local modification of the charge density around the impurity.)

2. Non-observation of the screening cloud in conventional Kondo experiments

This long length scale is surprisingly difficult to observe, perhaps mainly because it so large.

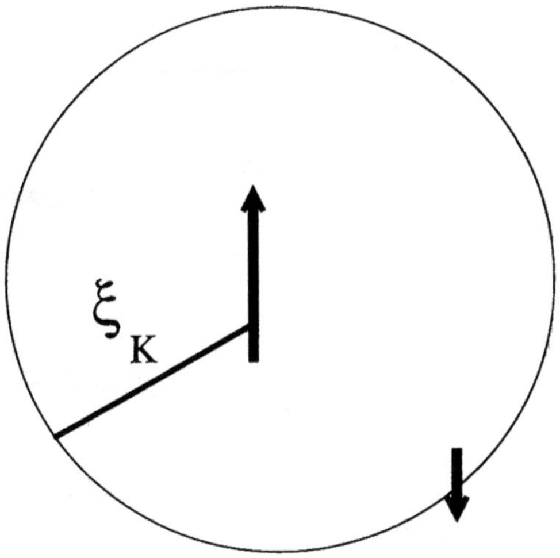

Figure 3.

An important point to realise is that single impurity Kondo behaviour can be observed even when the average inter-impurity separation, R_{imp}, is much greater than ξ_K. This happens because the screening cloud wave-functions from different impurities are nearly orthogonal even when they are strongly overlapping in space [3]. The condition for single impurity Kondo behavior is probably

$$R_{\text{imp}} >> \xi_K^{1/3} k_F^{-2/3}. \tag{8}$$

This follows from Nozières "exhaustion principle" which states that the number of electrons within $|k - k_F| < \xi_K^{-1}$ should be greater than the number of impurities. This, perhaps optimistic, estimate assumes that if enough states are available the individual screening clouds will manage to be almost orthogonal. However, in a one-dimensional system, the condition is simply that $R_{\text{imp}} >> \xi_K$.

The Knight shift as a function of distance from an impurity can be measured by nuclear magnetic resonance [4]. At $T = 0$ this takes the form [5]:

$$\chi(r) = \chi_0 + \frac{\cos(2k_F r)}{r^2} f\left(\frac{r}{\xi_K}\right), \tag{9}$$

where χ_0 is the susceptibility of the pure system and f is some universal scaling function. The rapid oscillations and the power law pre-factor make this scaling behavior difficult to observe. Typically the signal gets far too small long before r is as large as ξ_K. Furthermore, at such long distances the Knight shift will be given by a superposition of contributions from many impurities.

The charge density around a Kondo impurity varies only over short length scales of $O(1/k_F)$. For instance, in a particle-hole symmetric model, such as

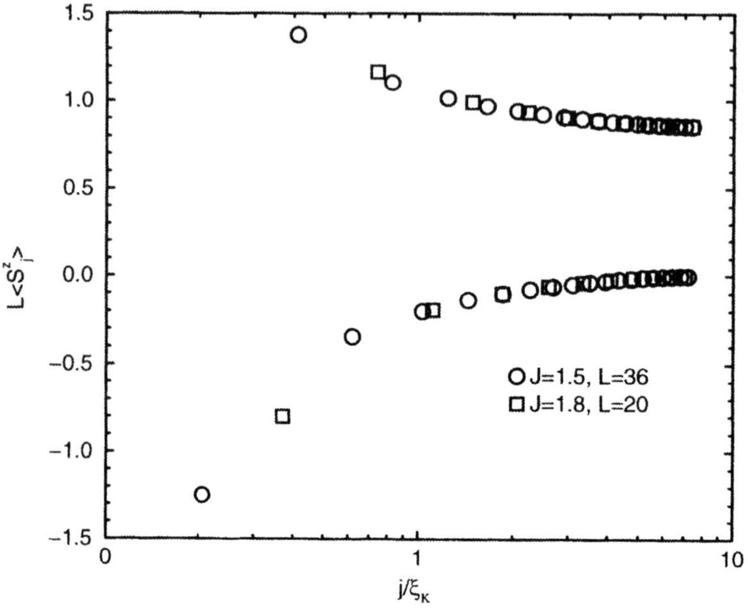

Figure 4. L times the expectation value of the z-component of the electron spin, $< S_j^z >$, as a function of $j/\xi_K(J)$. Two systems are shown: $J = 1.8$, $\xi_K = 2.7$, $L = 20$ and $J = 1.5$, $\xi_K = 4.85$, $L = 36$ Thus in both cases we have $L/\xi_K \approx 7.4$. Clearly the data collapses onto a universal curve.

a tight-binding model at half-filling, the charge density can easily be proven to be completely uniform. This is connected with "spin-charge separation" in this effectively one-dimensional problem. The Kondo effect takes place purely in the spin sector. Furthermore, the energy and r-dependent density of states, probed by scanning tunnelling microscopy, only varies on short length scales [6]. For the case of a δ-function Kondo interaction, the electron self-energy can be easily shown to have the form:

$$\Sigma(r, \omega) \propto G_0(r, \omega) T(\omega) G_0(r, \omega). \tag{10}$$

Here $G_0(r, \omega)$ is the free electron Green's function, which has trivial dependence on r. $T(\omega)$ has non-trivial dependence on ω and varies on the characteristic energy scale T_K. However, the r-dependence of Σ, and hence the density of states, *is* trivial.

This stubborn refusal of the Kondo length scale to show up in experiments might make one wonder if it really exists. E. Sørensen and I demonstrated [7] that it does exist by doing density matrix renormalization group simulations on large finite chains. The system we studied is the one sketched in fig. 2, with an even

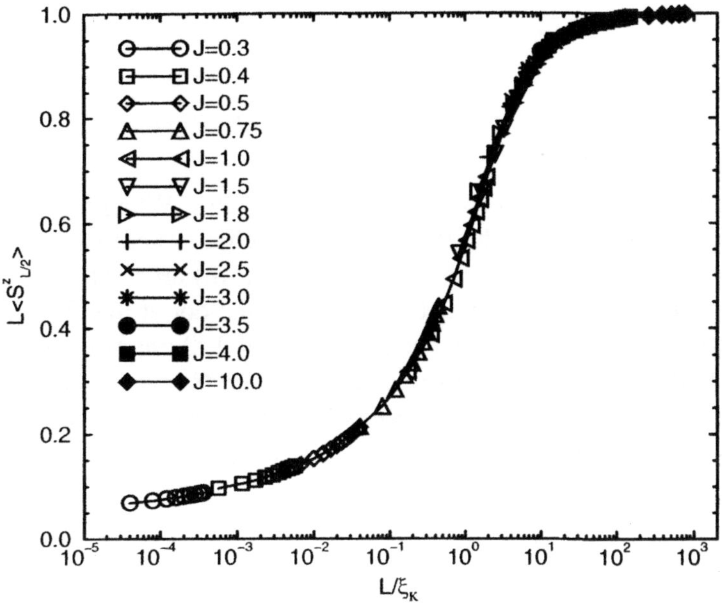

Figure 5. Logaritmic plot of $L < S^z_{L/2} >$ as a function of chain length L/ξ_K for a range of different coupling constants. The initial point corresponds in all cases to $L = 4$. The solid lines are guides to the eye. The strong coupling limit corresponds to $L < S^z_{L/2} >\approx 1$

number of sites L and open boundary conditions and the electron density set at 1/2-filling. In this case the groundstate has spin S=1/2. We measured $< S^z_j >$ in the groundstate with $S^z_T = +1/2$. This takes the value corresponding to a non-interacting chain with 1 site excised far from the screening cloud ($j >> \xi_K$). This is $O(1/L)$ and oscillates. At shorter distances $j \leq \xi_K < S^z_j >$ exhibits more complicated behavior. We showed that the data appears to collapse onto a single scaling curve when plotted vs. j/ξ_K. (See figs. 4 and 5.) The values of ξ_K obtained from this data collapse had the expected exponential dependence on J.

The question remains whether such behaviour could ever be observed experimentally.

3. Kondo effect in transmission through a quantum dot

Quantum dots and wires can be made by first forming a two-dimensional electron gas in a semi-conductor heterostructure layer and then further confining the electrons by etching and gate voltages varying at the sub-micron scale. Typical dots have been formed with radius around .1 micron containing around 50 electrons. The number of electrons on the dot can be varied in unit steps by varying a

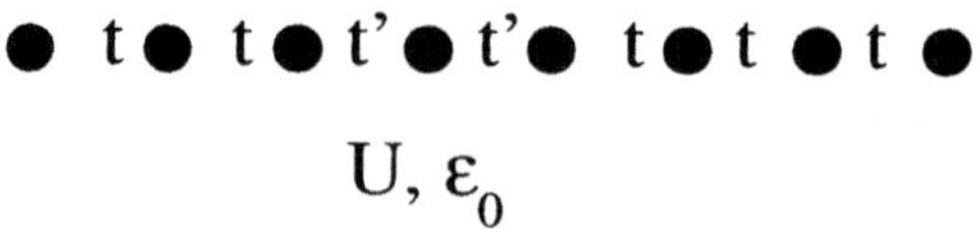

Figure 6.

gate voltage. Such a system is a single electron transistor. When the tunnelling amplitude between the dots and the leads is weak such a system can exhibit the "Coulomb blockade". The energy cost to add (or remove) an electron to the dot becomes sigificant and inhibits conductance through the dot at low T. When the number of electrons on the dot is *odd* it acts like an $S = 1/2$ impurity. Now a type of Kondo effect takes place which corresponds to *perfect transmission* rather than perfect reflection [8–10]. A simplified model which captures the essential physics is the one-dimensional Anderson model [12, 13]. See fig. 6.

$$H = -t\left[\sum_{j\leq -2} + \sum_{j\geq 1}\right](\psi_j^\dagger \psi_{j+1} + h.c.) - t'[\psi_0^\dagger(\psi_{-1} + \psi_1) + h.c.] + \epsilon_0 \psi^\dagger \psi_0 + U n_{0\uparrow} n_{0\downarrow}. \quad (11)$$

The conductance is determined by the transmission amplitude through the dot at the Fermi energy. In the non-interacting case ($U = 0$) there is a transmission resonance when $\epsilon_0 = 2(t'^2 - t^2)\cos k_F$, with a width of $O(t')$. In the Kondo limit, $t' << -\epsilon_0, U + \epsilon_0$, the dot acts like an S=1/2 impurity with a Kondo interaction:

$$H_{\text{int}} = J(\psi_{-1}^\dagger + \psi_1^\dagger)\frac{\vec{\sigma}}{2}(\psi_{-1} + \psi_1) \cdot \vec{S}_{\text{imp}}. \quad (12)$$

Here $\vec{S}_{\text{imp}}$ represents the electron spin on the dot and

$$J = 2t'^2[-1/\epsilon_0 + 1/(U + \epsilon_0)]. \quad (13)$$

As usual, we expect J to renormalize to large values as the band-width is reduced. Again it is useful to consider the behaviour for large bare coupling, J. Now the screening electron goes into the symmetric orbital on sites 1 and (-1). Electrons are transmitted through the dot, in this limit, by passing through the anti-symmetric orbital:

$$\psi_a \equiv (\psi_1 - \psi_{-1})/\sqrt{2}. \quad (14)$$

The effective low energy Hamiltonian is obtained by taking $J \to \infty$ and projecting out the symmetric orbital that screens the impurity.

$$H - -t\left[\sum_{j\leq -2} + \sum_{j\geq 1}\right](\psi_j^\dagger \psi_{j+1} + h.c.) - (t/\sqrt{2})[-\psi_{-2}^\dagger \psi_a + \psi_a^\dagger \psi_2 + + h.c.]. \quad (15)$$

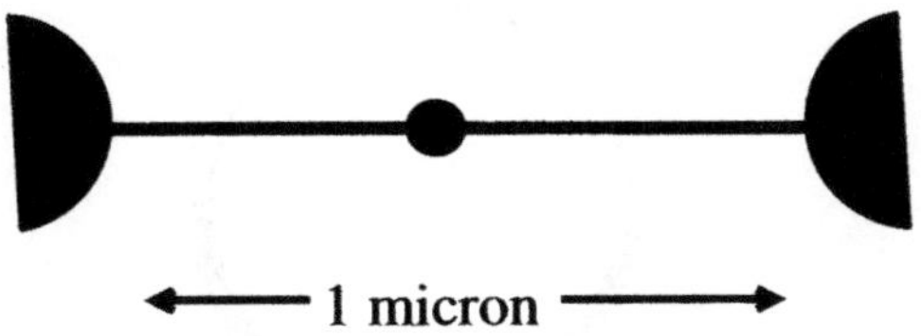

Figure 7.

This non-interacting Hamiltonian exhibits perfect conductance at half-filling.

In general additional potential scattering terms are induced in the Kondo Hamiltonian of Eq. (12) which are also of $O(J)$. This generally occurs unless the Hamiltonian has particle-hole symmetry. (For the Anderson model of Eq. (11), exact particle-hole symmetry requires half-filling and $\epsilon_0 = 0$.) If these were added to the large-J effective Hamiltonian they would in general move it off resonance and reduce the conductance. However, another important result about the Kondo model is that such particle-hole symmetry breaking is strictly marginal. These terms do not grow large at low energies but remain of order the bare Kondo coupling. Thus for small bare Kondo coupling the resonance remains pinned at essentially the Fermi energy. Consequently there is a plateau in the transmission as a function of gate voltage, ϵ_0, over which the conductance is close to the idea value $2e^2/h$, with a width of $O(U)$. Thus, near perfect conductance occurs at low T over the entire range of gate voltage where the occupancy of the dot is an odd integer. Such low temperature conductance plateaux have been observed by the Delft group [11].

4. Observing the screening cloud in a persistent current experiment

It is convenient to consider a narrow quantum wire connected to the quantum dot which is eventually connected to macroscopic leads. (See fig. 7.) Now the screening could will live in the leads. It is likely to be considerably larger than the quantum dot and perhaps about equal to the length of the quantum wire leads in some experiments (eg. those of the Delft group). However, these quantum wires are generally connected to macroscopic leads, as sketched in fig. 7. The screening cloud can also exist in the macroscopic section of the leads. Thus it is not obvious how the size of the screening cloud will manifest itself in most experimental setups.

One simple possibility (from a theoretical point of view) is to study a closed ring containing a quantum dot. (See fig. 8.) Now the screening cloud is trapped on the ring and can't escape into any macroscopic leads. The experimental difficulty is to measure the current through the dot! This could perhaps be done by applying a magnetic field to the ring and measuring the resulting persistent current. When the cloud size, $\xi_K << L$, the ring circumference, the perfect transmission through

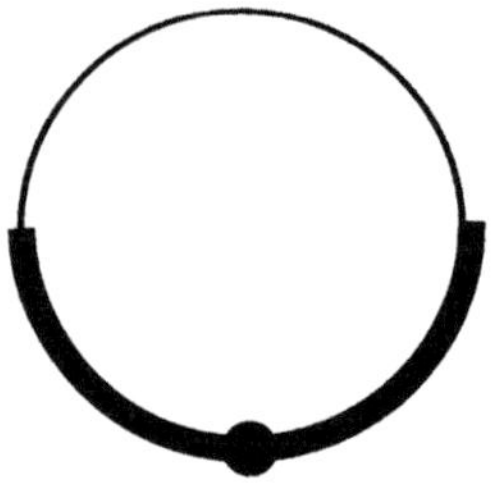

Figure 8.

the dot, due to the Kondo effect, implies that the persistent current should be the same as for an ideal ring with no dot. On the other hand, when $\xi_K >> L$, the Kondo effect doesn't take place; the infrared divergence of the Kondo coupling, λ, is cut off by the finite size of the ring. In this limit the persistent current can be calculated perturbatively in λ.

Analytical expressions for low order perturbation theory can be derived in the tight binding model, for $L >> 1$, by using continuum limit approximations: linearizing the dispersion relation, etc [14]. To include magnetic flux we simply give the hopping terms to the dot phases $\pm\alpha/2$:

$$H_{\text{int}} \rightarrow J(e^{i\alpha/2}\psi^\dagger_{-1} + e^{-i\alpha/2}\psi^\dagger_1)\frac{\vec{\sigma}}{2}(e^{-i\alpha/2}\psi_{-1} + e^{i\alpha/2}\psi_1) \cdot \vec{S}_{\text{imp}}. \qquad (16)$$

The corresponding flux is $\Phi = (c/2)\alpha$. The persistent current (at $T = 0$) is calculated from the flux-dependence of the groundstate energy:

$$j = -cdE/d\Phi. \qquad (17)$$

When the total number of electrons (including the electron on the dot), N, is even, there is 1 unpaired electron at the Fermi surface which forms a singlet with the impurity spin. This leads to a contribution of E of first order in λ. When N is odd, E is second order in λ since the groundstate of the $(N-1)$ non-interacting electrons is a spin singlet. To calculate next order corrections in λ we use the continuum propogator:

$$G(\tau, x) \approx \sum_{n=0}^{\infty} e^{-(v_F\tau+ix)\pi n/L} = 1/(1-e^{-v_F\tau+ix)\pi/L}). \qquad (18)$$

This leads to the following expressions, for N even or odd:

$$\begin{aligned} j_e(\alpha) &\approx \frac{3\pi v_F e}{4L}\{\sin\tilde{\alpha}[\lambda + \lambda^2 \ln Lc] + (1/4 + \ln 2)\lambda^2 \sin 2\tilde{\alpha}\} + \ldots \\ j_O(\alpha) &\approx \frac{3\pi v_F e}{16L}\sin 2\alpha[\lambda^2 + 2\lambda^3 \ln Lc'] + \ldots. \end{aligned} \qquad (19)$$

Here $\tilde{\alpha}$ is α for $N/2$ even or $\alpha + \pi$ for $N/2$ odd. c and c' are dimensionless constants which we have not bothered to compute. Note that Lj depends on λ and L only through the renormalized coupling constant at scale L:

$$\lambda_{\text{eff}}(L) \approx \lambda + \lambda^2 \ln L + \cdots. \tag{20}$$

This is expected to be exactly true, in all orders of perturbation theory due to standard scaling arguments. Perturbation theory breaks down when $\xi_K \leq L$ but we expect Lj to be given by some universal scaling functions of ξ_K/L. These need to be calculated from large scale numerical simulations, or perhaps using integrability methods. The current in the two limits $L << \xi_K$ and $L >> \xi_K$ are plotted in fig. 9. In the latter case we simply use the result for an ideal ring with no quantum dot. We expect the universal scaling functions to give a smooth cross over between these two limits.

Our results disagree with those of several other groups [16–19].

There are many effects left out of this simple model which are probably important to a complete understanding of potential experiments in this field. One of these is the presence of several channels in the quantum wire. The Delft experiments suggest about 5 partially filled 1D bands. If we also include more structure in the quantum dot then it might be appropriate to consider a multi-channel Kondo model with all the Kondo couplings, λ_i, different. The corresponding RG equations, to third order, are:[20]

$$-d\lambda_i/d\ln D = \lambda_i^2 - (1/2)\lambda_i \sum_j \lambda_j^2 + \ldots \tag{21}$$

For unequal bare couplings, these equations predict that the largest one gets larger and the rest shrink towards zero under renormalization. The physical picture is that one channel screens the quantum dot and has perfect conductance at $T \to 0$, but the other channels have zero conductance at $T \to 0$. Thus the basic picture of the Kondo effect is not changed. Assymetric leads (with different hopping parameters t'_L and t'_R on the two sides of the quantum dot) reduce the conductance and hence the maximum current but we still expect a cross over behaviour at $L \approx \xi_K$. Luttinger liquid interactions in the quantum wire destroys the plateau in the conductance vs. gate voltage, leaving perfect conductance only at one resonant value of the gate voltage [21] (or none if the interactions are strong enough [22]), at least in a one channel model. Non-magnetic disorder in the leads reduces the persistent current even in the limit $L >> \xi_K$.

5. Side-coupled dot

Another interesting model has the quantum dot "side-coupled" to the quantum wire ring, as shown in fig. 10. The simplified Anderson Hamiltonian now becomes:

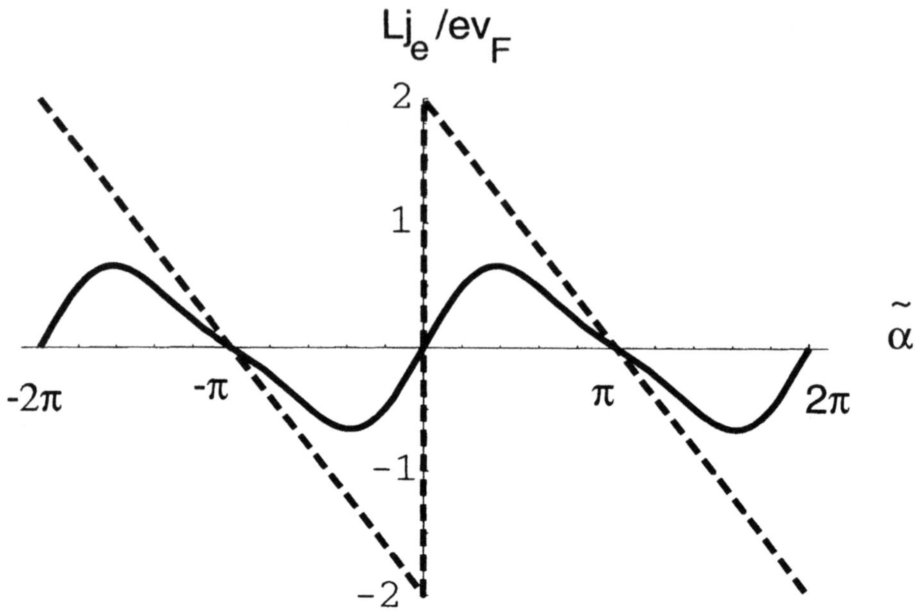

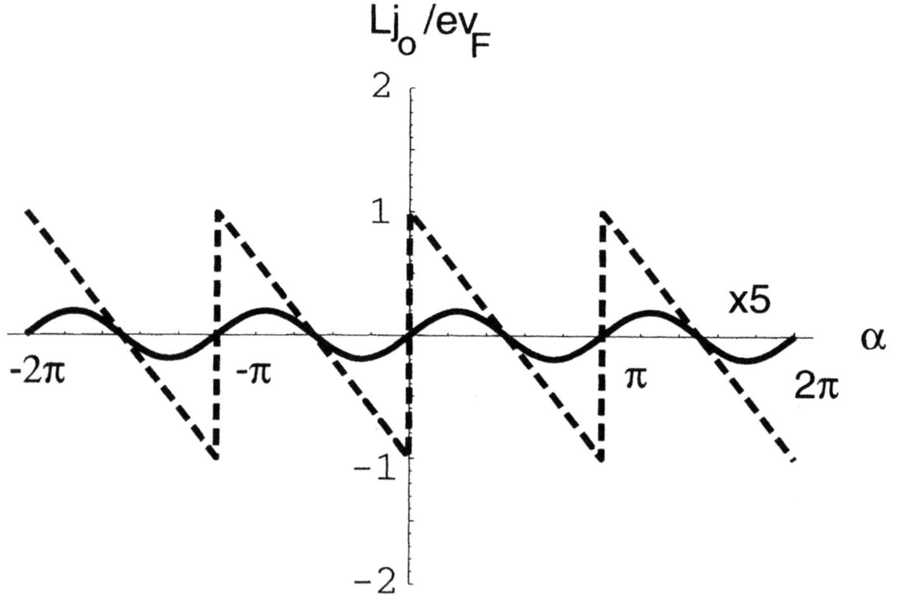

Figure 9. Persistent current vs. flux for an even or odd number of electrons for $\xi_K/L \approx 50$ (solid line) and for $\xi_K/L << 1$ (dashed line). j_o is multiplied $\times 5$ at $\xi_K/L = 50$ for visibility. The solid lines are obtained from Eq. (19) using the effective coupling $\lambda(L) \approx 1/\ln(\xi_K/L)$.

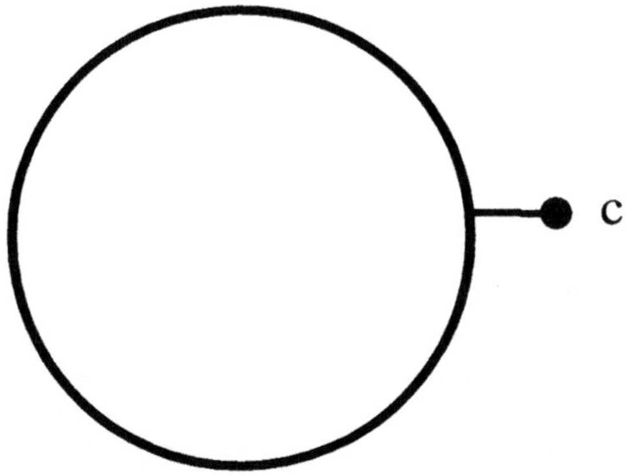

Figure 10.

$$H = -t \sum_{j=0}^{L-1}](\psi_j^\dagger \psi_{j+1} + h.c.) - t'[\psi_0^\dagger c + h.c.] + \epsilon_0 c^\dagger c + U n_{c\uparrow} n_{c\downarrow}. \quad (22)$$

Now when $t' = 0$ there is a perfect current. The growth of the Kondo coupling under renormalization drives the current to 0 for a large ring. The formation of the screening cloud now interferes with the current. Again this is clear in the large J limit. A single electron sits at site 0 and screens the impurity. Other electrons must stay away from 0 corresponding to a $\pi/2$ phase shift in the even channel. The odd channel wave-functions vanish at 0 so there is no conduction route for electrons in the large J limit. Again we predict scaling behaviour in ξ_K/L. The difference beween the two cases may be understood from the formula for the transmission through an impurity with even and odd phase shifts:

$$|T|^2 = \sin^2(\delta_e - \delta_o), \quad (23)$$

where δ_e and δ_o are the phase shirts at the Fermi energy in the even and odd channel. When the effective even phase shift goes from near 0 to near $\pi/2$ for a long ring, T can switch between values near 0 and 1. Which occurs in which limit depends on δ_0.

6. Conclusions

The Kondo effect due to spin impurites in metals or due to a quantum dot always involves a large screening cloud whenever the dimensionless Kondo coupling is small. This has never been observed experimentally. In mesoscopic experiments the screening cloud may "escape" into macroscopic leads, in general. This can be avoided in a closed ring experiment where a clear experimental signal emerges from the dependence of the persistent current on the ratio ξ_K/L. How screening manifests iteself in transmission experiment with open leads appears much more subtle [23–25].

References

1. For a review of the mapping to one dimension see Appendix A of I. Affleck and A.W.W. Ludwig, Nucl. Phys. **B360**, 641 (1991).
2. P. Nozières in Proc. 14th Int. Conf. on Low temperature physics, ed. M. Krusius and M. Vuorio, Vol. 5 (North-Holland, Amsterday, 1975) p, 339; J. Low Temp. Phys. **17**, 31 (1974).
3. See, for example, V. Barzykin and I. Affleck, Phys. Rev. **B61**, 6170 (2000).
4. J.P. Boyce and C.P. Slichter, Phys. Rev. Lett. **32**, 61 (1974); Phys. Rev. **B13**, 379 (1976).
5. V. Barzykin and I. Affleck, Phys. Rev. Lett. **76**, 4959 (1996); Phys. Rev. **B57**, 432 (1998).
6. O. Ujsaghy, J. Kroha, L. Szunyogh and A. Zawadowski, Phys. Rev. Lett. **85**, 2557 (2000).
7. E.S. Sørensen and I. Affleck, Phys. Rev. **B51**, 16115 (1995).
8. D. Goldhaber-Gordon, H. Shtrikman, D. Mahalu, D. Abusch-Magder, U. Meirav and M.A. Kaster, Nature **391**, 156 (1998).
9. S.M. Cronewett, T.H. Oosterkamp, L.P. Kouwenhoven, Science **281**, 540 (1998).
10. F. Simmel, R.H. Blick, U.P. Kotthaus, W. Wegsheider, M. Blichler, Phys. Rev. Lett. **83**, 804 (1999).
11. W.G. van der Wiel, S. De Franceschi, T. Fujisawa, J.M. Elzerman, S. Tarucha and L.P. Kouwenhoven, Science, **289**, 2105 (2000).
12. L.I. Glazman and M.E. Raikh, JETP Lett. **47**, 452 (1988).
13. T.K. Ng and P.A. Lee, Phys. Rev. Lett. **61**, 1768 (1988).
14. I. Affleck and P. Simon, Phys. Rev. Lett. **86**, 2854 (2001).
15. P. Simon and I. Affleck, Phys. Rev. **B64**, 100506 (2001).
16. A.A. Zvyagin and P. Schlottmann, Phys. Rev. **B54**, 15191 (1997).
17. H.-P. Eckle, H. Johannesson and C.A. Stafford, Phys. Rev. Lett. **87**, 016602 (2001).
18. S. Y. Cho, K. Kang, C.K. Kim and C.-M. Ryu, Phys. Rev. **B64**, 0333314 (2001).
19. E.V. Anda, C. Busser, G. Chiappe and M.A. Davidovich, cond-mat/0106055.
20. P. Nozières and A. Blandin, J. Phys, (Paris) **41**, 193 (1974).
21. C.L. Kane and M.P.A. Fisher, Phys. Rev. **B46**, 15233 (1992).
22. E.H. Kim, cond-mat/0106575.
23. W.B. Thimm, J. Kroha and J. von Delft, Phys. Rev. Lett. **82**, 2143 (2001).
24. P. Cornaglia and C. Balseiro, private communication.
25. P. Simon and I. Affleck, in progress.

QUANTUM INTERFEROMETRY WITH ELECTRONS: OUTSTANDING CHALLENGES

YUVAL GEFEN
Department of Condensed Matter Physics, The Weizmann Institute of Science, Rehovot 76100, Israel

1. Introduction and perspectives

Much of the effort in studying mesoscopic systems was directed towards the analysis of thermodynamic or transport properties *per se*. However, measurements (and, subsequently, the analysis) of such quantities can teach us a lot about the *phase* of the wavefunctions involved. Evidently, one needs to explicitly incorporate quantum interference into such measurements to allow for the analysis of the quantum phase. Earlier interferometry experiments in mesoscopic conductors focused on various aspects of Aharonov-Bohm (AB) [1] oscillations vis-a-vis transport [2–5] or thermodynamics (persistent currents and orbital magnetism [6–9]), initially interpreting the data within the framework of single electron physics. This indeed led to an impressive number of novel and, at times, unexpected effects. Over the past few years it has become clear, though, that the physically motivated, yet naive, picture of independent electrons does not suffice to account for the important details that have emerged in the course of the experimental work. Moreover, further theoretical analysis suggested that the presence of electron-electron interaction may indeed give rise to novel important physics.

Thus the new generation of interference experiments [10–14] focused on setups where the role of e-e interaction is emphasized and can be controlled. (Here we leave aside the very interesting issue of thermodynamics of mesoscopic systems). The obvious choice is to incorporate quantum dots (QDs) [15–17] in the interferometers. For QDs which are "closed" (i.e. unconnected to external leads) the electron-electron interaction may be modeled by a 0-mode capacitive energy term, that is an interaction term which does not have any spatial dependence. Higher, space-dependent, interaction modes are smaller in powers of the inverse diemsnionless conductance of the uncoupled dot [18–21], $1/g$. Fig. 1 depicts schematically the conductance through a QD as the applied gate voltage, V_G, is varied. Coulomb peaks appear at values of V_G for which the energy of the

I. V. Lerner et al. (eds.),
Strongly Correlated Fermions and Bosons in Low-Dimensional Disordered Systems, 13–41.

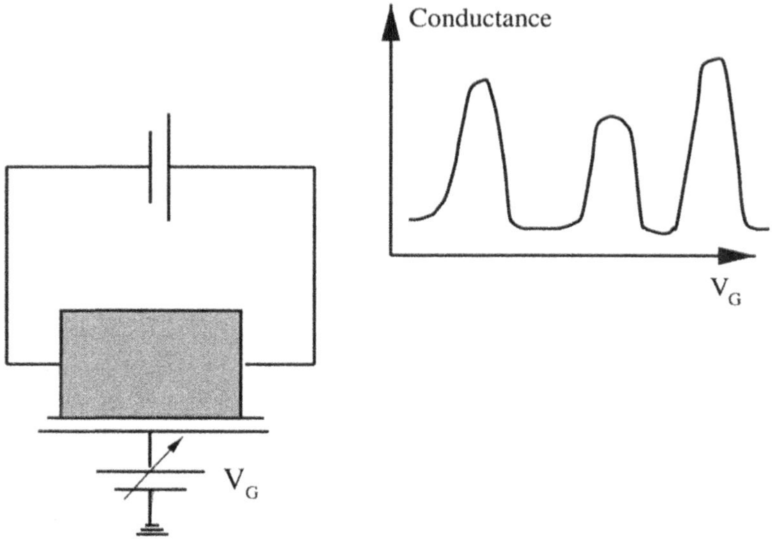

Figure 1. A schematic equivalent circuit of a quantum dot. Inset: The linear conductance through the dot as function of the gate voltage V_G (schematic).

entire system (i. e., the QD and the reservoirs) is (nearly) insensitive to the removal/addition of an electron from the leads. Hereafter we denote the width of the Coulomb peaks by $\tilde{\Gamma}$. Such a quantum dot can now be incorporated into an AB interferometer. The two obvious parameters to vary are the enclosed AB flux, Φ, and the gate voltage on the QD. The former is parametrized by $\varphi \equiv 2\pi\Phi/\Phi_0$, where $\Phi_0 \equiv e/hc$ is the flux quantum. Evidently, one may study the dependence on other important parameters, such as the temperature k_BT, the dimensionless conductance of the uncoupled dot g, the coupling strength of the QD to the leads Γ etc.

The quantum dot interferometry experiments yielded a rich set of results. Most of them came initially as a surpirse, but further analysis provided satisfactory explanations to the observed effects within single particle theories or otherwise the orthodox theory of Coulomb blockade. Certain effects, though, turned out to be more puzzling, initiating an extensive theoretical effort which is yet to prove fruitful. Some of the puzzling results even went unnoticed. I shall not attempt to present here a comprehensive review of the physics of coherence and interferometry on mesoscopic scales, nor shall I review all the relevant (and important) literature. Instead, I provide an updated compendium of the experimental results, the dilemmas they present, and some theoretical perspectives concerning the attempts to address this physics. In fact, the present overview can be regarded as a shopping list of the present challenges in this field. Due to the limited scope of this presentation I will not address the very low temperature limit, where *Kondo physics* comes to play. Interesting effects are expected in that context, cf. e.g.

Refs. [12, 13, 22–25]. I also note that the magnetic fields discussed here are sufficiently weak to ignore spin-related Zeeman splitting. Likewise, the systems considered are always small (or narrow) enough for the quantum Hall effect not to show up.

The outline of this paper is the following. In the next section I review some basic facts concerning AB interferometry in electronic systems, including a reference to symmetries at equilibrium (Onsager) and away from equilibrium, a brief discussion of partial coherence, and a simple demonstration of the breakdown of the Landauer formula in the presence of interaction. In Section 3 I discuss the phenomenon of phase locking (and a scenario for its breakdown). Section 4 addresses the issue of the transmission phase correlations. In Section 5 we discuss the asymmetric features of the coherent AB amplitude. Finally, in Section 6 I comment on the dilemmas involving the width of the Coulomb peaks as well as that of the phase lapses.

This short overview is largely based on past and present collaborations. Parts of this manuscript consist of results obtained recently in collaboration with D, E. Feldman, J. König, Y. Oreg and A. Silva.

2. Aharonov-Bohm interferometry in electronic systems: some basics

Earlier on in the development of the field of *Mesoscopics* it became clear that AB interferometry is a most useful tool. This, of course, has to do with the fact that electrons are charged particles, and therefore the electric current is (minimally) coupled to the vector potential, **A**, generated by a Aharonov-Bohm flux. The vector field **A** influences the phase of the electron and thus affects the outcome of interference experiments. The first intriguing effect that has been noticed in that context was that while the periodicity (say, of the conductance) under the AB flux should, in principle, be $\delta\varphi = 1$ [26, 27], the AB periodicity of a flux threading a (dirty) conducting cylinder was found to be $\delta\varphi = \frac{1}{2}$ [28], a result which was confirmed in experiment [2]. While the latter periodicity was certainly in line with the Byers-Yang theorem [29], stating that equilibrium (hence linear-response) observables should be periodic under $\varphi \to \varphi + 1$, the effect of period halving came as a surprise. It was later understood [30] that period halving was the result of ensemble averaging [31, 32], and that in the absence of such averaging the fundamental period should be 1 rather than $\frac{1}{2}$. This indeed was found to be in line with experimental observations [3, 4]. Other exciting, interferometry-related, effects followed (such as persistent currents, orbital magnetism, flux-dependent "universal" conductance fluctuations).

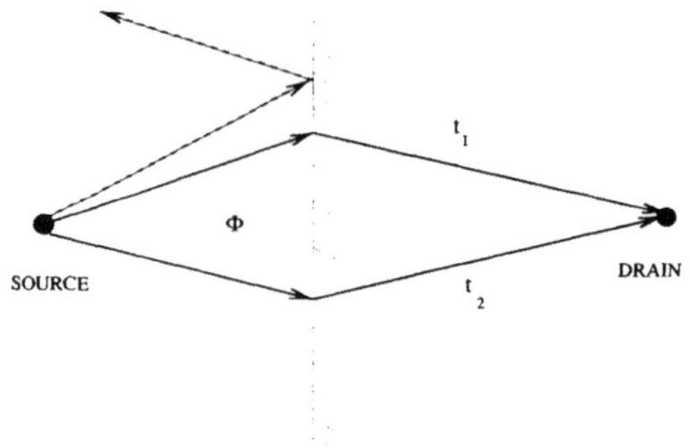

Figure 2. Setup for a two-slit interferometry.The dashed line represents a trajectory of an electron absorbed by a "far-gate".

2.1. PHASE LOCKING

One concept that emerged from those early studies was the phenomenon of *phase locking* [33–35]. To understand this effect let us first consider a two-slit experiment, depicted in Fig. 2, for which phase locking is not satisfied.

A single electron physics is assumed. The electron is emitted from the source and may be absorbed by either the drain, the source, or be "lost" (i.e. absorbed by any of the other "far gates"). This motivates later reference to this setup as *open geometry*. The partial amplitudes for the electron to be transmitted through slits 1 or 2 (and eventually be absorbed by the drain) are t_1 and t_2 respectively, with

$$t_m = |t_m| \exp(i\alpha_m) \ , \tag{1}$$

with $m = 1, 2$. In the presence of an AB flux the partial transmission amplitudes assume additional flux induced, gauge dependent phases,

$$\alpha_m \to \alpha_m + 2\pi\varphi_m \ . \tag{2}$$

The relative phase of the two trajectories described by these two partial amplitudes is

$$\alpha + 2\pi\varphi \equiv \alpha_1 - \alpha_2 + 2\pi\varphi_1 - 2\pi\varphi_2 \ , \tag{3}$$

where $\varphi \equiv \varphi_1 - \varphi_2$ is the gauge invariant AB phase. The total transmission amplitude, $t_{total} = t_1 + t_2$. Employing the Landauer formula [36, 33], the total transmission *probability* is given by

$$T_{tr} = |t_{total}|^2 , \tag{4}$$

where t_{total} is the total transmission amplitude. For our geometry it follows that

$$T_{tr} = |t_1|^2 + |t_2|^2 + 2|t_1 t_2| \cos(\alpha + 2\pi\varphi) \ . \tag{5}$$

The flux sensitive interference term $2|t_1 t_2|\cos(\alpha + 2\pi\varphi)$ is evidently periodic in φ (flux) with a period 1. The asymmetry of the AB signal with respect to $\varphi = 0$ is due to the appearance of the orbital phase α. It is quite suggestive to refer to a situation where this phase shift disappears (e.g. due to some underlying symmetry, see below) as *phase locking*.

2.2. PARTIAL COHERENCE AND VISIBILITY

The above expression for the total transmission probability through the double-slit configuration was derived under the conditions of full coherence. The conductance is related to the transmission probability through

$$\mathcal{G} = \frac{e^2}{h} T_{tr} \ . \tag{6}$$

This relation holds for interacting systems as well. To check whether there is some degree of cohernece in the system is a relatively easy task. We only need to note that there are certain values of the flux for which the total transmission is smaller than the sum of the individual transmissions through each channel, i.e., $T_{tr} < |t_1|^2 + |t_2|^2$. To assert that there is *full coherence* in the system is a more demanding task. Consider the expression for the total transmission, Eq.5. For $|t_1| \neq |t_2|$ the AB amplitude is smaller than the flux-averaged signal. Referring to the conductance we can write the above inequality as

$$\langle \mathcal{G} \rangle_\varphi = \frac{e^2}{h}(|t_1|^2 + |t_2|^2) > \mathcal{G}^{AB} = 2\frac{e^2}{h}|t_1 t_2| > 0 \ , \tag{7}$$

with $\langle \mathcal{G} \rangle_\varphi$ being the flux-averaged conductance, $\mathcal{G}^{AB}$ is the amplitude of the (periodic) flux dependent term. It is therefore convenient to define the *visibility*, $\mathcal{V}$, of such an interferometer

$$\mathcal{V} \equiv \mathcal{G}^{AB} / \langle \mathcal{G} \rangle_\varphi \ . \tag{8}$$

There could be 3 different reasons for the visibility to be smaller than one:
(i) The transmission is fully coherent, yet the transmissions through the two interferometer arms are asymmetric–one arm transmits better than the other. This is the scenario outlined above. Evidently when the two interferometer's arms are symmetric, $|t_1| = |t_2|$, $\mathcal{V} = 1$.
(ii) There are several transmission channels through each arm, each carrying its own orbital (and possibly AB) phase. It follows that the conditions for destructive interference are different among the different channels, and may not be satisfied simultaneously.
(iii) The transmitted electrons are coupled to other degrees of freedom [37, 38] which give rise to dephasing, setting the stage for *partial coherence*.

The first two scenarios are fundamentally different from the third one, as they correspond to full coherence (although the observed visibility may be smaller than unity). Indeed, full coherence implies that *in principle* it is possible to tune or modify the parameters of one of the arms ("the reference arm"), and vary the flux such that full destructive interference ($\mathcal{V} = 1$) is obtained.

From the theoretical point of view there are two main approaches for incorporating dephasing processes in interferometry devices. The first one [39, 40] is phenomenoligical. One starts with writing down a scattering theory formalism for the problem at hand (*a-la* Landauer), and then adding "dephasing reservoirs" which absorb and emit electrons, usually without modifying the current. What a dephasing reservoir does is to erase any phase memory of the electrons that go through. Operationally one describes the scattering into/out of the dephasing reservoir by assigning some complex amplitude to such a process, with a phase θ. Once an observable (e.g. the transmission probability through the device) is calculated, an average over θ is taken. Various generalizations of this aproach are possible, e.g. the introduction of numerous, weakly coupled reservoirs along the transmission line which mimicks gradual, continuous dephasing processes [39, 40]. There are certain caveats with this procedure: The dephasing agent here is completely classical. This means that in the process of dephasing energy may be pumped into to electronic system [41]. Also, subtle quantum effects and correlations are ignored in this approach. Finally, the averaging over the phase θ does not commute with the time reversal operation; performing this averaging and then time reversing the problem may present problems with unitarity [40]. The second approach to the introduction of dephasing is microscopic One can classify the dephasing agents according to whether the coupling term in the Hamiltonian does or does not commute with the Hamiltonian of the uncoupled reservoir [42]. The physics that emerges is elaborate and not yet fully exhausted.

2.3. TWO-TERMINAL VS. MULTI-TERMINAL GEOMETRIES

As we have seen above, owing to the different orbital phases of the two interferometer's arms, the AB signal is, in general, asymmetric with respect to $\varphi = 0$, i.e., no phase locking takes place. This is the case for the open geometry depicted in Fig. 2, and similarly for a multiple-terminal setup, Fig. 3.

In contradistinction, for a two-terminal geometry one expects phase locking to take place. We note that in such a geometry an electron leaving the source may, eventually, be either reflected back to the source or transmitted to the drain (unlike in a multi-terminal geometry or in an open geometry where the electron may end up in one of the other "gate" terminals). The total transmission and the reflection probabilities satisfies then

$$T_{tr} + R_{ref} = 1 \ . \tag{9}$$

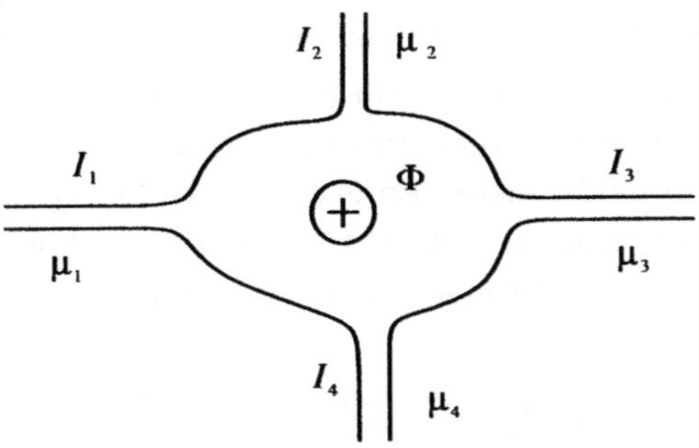

Figure 3. A 4-terminal AB interferometer.

Employing Eqs. 6 and 9 the flux dependence of $\mathcal{G}$ can be fully deduced from R_{ref}, the latter obtained by taking the square modulus of the sum of the self-returning amplitudes. To accomplish this task we pretend that those amplitudes can be evaluated through an infinite set of semiclassical trajectories which we denote by $\{A_j^{(n)}\}$, [43]. This notation is rather symbolic, pretending that there is a countable set of relevant partial amplitudes. We classify the various partial amplitudes according to their winding number, n, and a running index j within each class of n. Each of these partial amplitudes includes an orbital phase which we denote by $\alpha_j^{(n)}$. Furthermore, we need to assign an AB phase (which depends on the winding number only), $\varphi^{(n)} \equiv 2\pi n\varphi$. Thus

$$A_j^{(n)} = |A_j^{(n)}| \exp(i\alpha_j^{(n)}) \exp(i\varphi^{(n)}) \; . \tag{10}$$

Let $A_j^{(n)}$ and $A_j^{(-n)}$ represent two trajectories which are mutually time reversed. It follows that

$$\begin{aligned} A_j^{(n)} &= A_j^{(-n)} \; , \\ \varphi^{(n)} &= -\varphi^{(-n)} \; . \end{aligned} \tag{11}$$

The reflection probability is then given by

$$R_{ref} = r \times r^* = \left[\sum_{j,n} A_j^{(n)}\right] \times \left[\sum_{j,n} A_j^{(n)}\right]^* \; . \tag{12}$$

Performing this multiplication we obtain several distinct types of terms:

(i) **"diagonal terms"**, arising from the multiplication of $A_j^{(n)}$ by $\left[A_k^{(n)}\right]^*$ (same winding number, j, k are general). These sample specific terms amount to a flux independent contribution. In particular, the product of the partial amplitude $A_j^{(n)}$ with its complex conjugate rids of the (sample specific) orbital phase. Adding together the diagonal contributions of $A_j^{(n)}$ and $A_j^{(-n)}$ one obtains

$$|A_j^{(n)}|^2 + |A_j^{(-n)*}|^2 = 2|A_j^{(n)}|^2 \; . \tag{13}$$

(ii) **"time reversed" terms**, arising from the product of $A_j^{(n)}$ and $\left[A_k^{(-n)}\right]^*$ with $j = k$ (we will be concerned with time-reversed pairs; $j \neq k$ pairs possess sample specific orbital phases; such terms contribute to the statistical fluctuations, in much the same way as the cross-terms of type (iii)). By Eq.11 the orbital phase of a time-reversed pair cancels out, and one is left with a flux-dependent phase only. Adding up two related reversed pairs one obtains

$$A_j^{(n)} \times A_j^{(-n)*} + A_j^{(-n)} \times A_j^{(n)*} = 2|A_j^{(n)}|^2 \cos(4\pi n\varphi) \ . \tag{14}$$

(iii) **"cross terms"**. These are all the rest. Their magnitude is sample specific, their (orbital) phase is sample specific (hence strongly fluctuating), and in general they are flux dependent.

Detailed analysis of these terms may yield a wide spectrum of effects, basically the entire single-electron mesoscopics in a nutshell, including the effects of period-halving, negative magneto-resistance at weak magnetic fields, conductance fluctuations and more. This is beyond the scope of the present analysis. What we would like to stress here is the emergence of phase locking. This can be easily seen for the ensemble averaged problem. The terms which survive ensemble averaging are given by Eqs.(13, 14) (summation over j is implied); all the other, strongly fluctuating, terms average to zero. Symmetry with respect to $\varphi = 0$ is evident. Phase locking, though, is a more robust phenomenon, valid on the level of sample-specific observables. We only need to note that when performing the multiplication implied by Eq.12 we add up the following terms together: $A_k^{(n)} \times A_j^{(m)*} + A_k^{(n)*} \times A_j^{(m)} + A_k^{(-n)} \times A_j^{(-m)*} + A_k^{(-n)*} \times A_j^{(-m)*} = 4|A_j^{(m)} A_k^{(n)}| \cos(\alpha_j^{(m)} - \alpha_k^{(n)}) \cos((n-m)\varphi)$. Phase locking is then manifest.

Dephasing or inelastic relaxation processes may suppress the coherent transmission and reflection amplitudes (giving rise to an incoherent background)– more generally they will suppress the related single-particle Green function. It is important to notice, though, that such processes do not destroy the phase locking symmetry. A mechanism for breaking down this symmetry is discussed in Section 3.

Careful examination of our analysis of the noninteracting system reveals that phase locking is satisfied within any infinitesimal energy window (we have been dealing with partial amplitudes each possessing a certain energy). In that case the validity of phase locking (for the two-terminal geometry) is independent of the electron energy distribution function, and it extends beyond linear response. This is not the case in general, where electron-electron interactions render the notion of a "sharply defined single electron energy" inadequate. One may still prove the validity of phase locking assuming that the reservoirs at the boundaries of the system (e.g. the source and the drain) are described by a gas of non-interacting electrons . This proof, originally proposed by Büttiker [33], relies on the general

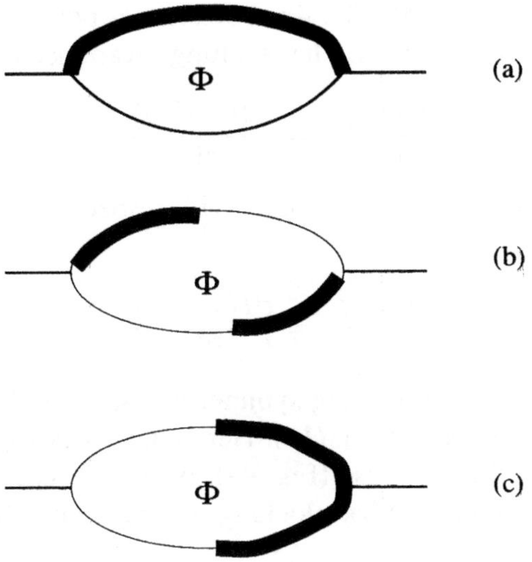

Figure 4. Two terminal AB interferometers with different spatial symmetry. Different lines represent different lead geometries/impurity configurations. (a) possesses mirror symmetry with respect to a vertical axis, (b) possesses a point symmetry: invariance under rotation at angle π, and (c) has a mirror symmetry with respect to a horizontal axis. The chemical potentials of the left/right reservoirs are indicated. In cases (b) and (c) phase locking is preserved even at finite bias.

Onsager relations. The fact that a finite external voltage bias can destroy phase locking has been nicely demonstrated by Bruder, Fazio and Schoeller [44].

It should be noted, though, that even away from equilibrium, when linear response does not apply, there may still be certain symmetries governing the behavior of the conductance. König and Gefen [37] noted the connection between the spatial symmetries of the underlying setup and the symmetries of the transport coefficients. In Fig. 4 we depict three different cases in which the system has a distinct spatial symmetries [37]. The general relation for all two-terminal setups

$$\frac{\partial I(V,\varphi)}{\partial V} = \frac{\partial I(-V,-\varphi)}{\partial V}, \tag{15}$$

where V is the applied bias, yields as a direct consequence the Onsager relation

$$\left.\frac{\partial I(\varphi)}{\partial V}\right|_{V=0} = \left.\frac{\partial I(-\varphi)}{\partial V}\right|_{V=0} \tag{16}$$

which leads to phase locking in linear response. Fig. 4a represents a system with a mirror symmetry with respect to its vertical axis. One clearly can reverse the direction of the bias *and* the sign of the AB-flux leaving the magnitude of the current unchanged. The resulting equation is

$$\frac{\partial I(V,\varphi)}{\partial V} = \frac{\partial I(-V,-\varphi)}{\partial V} \tag{17}$$

which coincides with Eq (15). Fig. 4b represents a point symmetry: rotation at angle π with respect to the center. The resulting invariance is expressed through

$$\frac{\partial I(V,\varphi)}{\partial V}=\frac{\partial I(-V,\varphi)}{\partial V}\,. \tag{18}$$

Finally, Fig. 4c depicts a mirror symmetry with respect to a horizontal axis, leading to the equation

$$\frac{\partial I(V,\varphi)}{\partial V}=\frac{\partial I(V,-\varphi)}{\partial V}\,. \tag{19}$$

In the two latter cases phase locking symmetry is satisfied [37]. It either follows directly or after making use of Eq. (15). Here phase locking is a consequence of spatial symmetry. In the first case (Fig. 4a), or in the absence of any particular spatial symmetry, breaking of phase locking occurs at finite bias voltages.

2.4. WATCH THE LANDAUER FORMULA FOR INTERACTING SYSTEMS

The Landauer formula provides a convenient framework to study the conductance of specific setups, relating the transmission amplitude through the system to the total transmission probability, cf. Eqs. (4,6). Things are not as simple when it comes to an interacting system. In that case any given electron interacts with other degrees of freedom (e.g. other electrons), and its energy is not conserved– one needs to resort to a many-particle, continuous energy-channel scheme (cf. [38, 35]. The Landauer formula has indeed been generalized employing Green's function techniques for interacting systems [45–47]. To demonstrate the failure of the naive Landauer formula, and to relate it to the concepts of partial coherence and decoherence, we present here a toy model which captures these themes [37]. Let us consider a single-level QD with level energy ϵ, measured from the Fermi energy of the leads. The Hamiltonian

$$H=H_L+H_R+H_D+H_T \tag{20}$$

consists of $H_r=\sum_{k\sigma}\epsilon_{kr}a^{\dagger}_{k\sigma r}a_{k\sigma r}$ for the left and right lead, $r=L/R$. The isolated dot is described by $H_D=\epsilon\sum_{\sigma}n_{\sigma}+Un_{\uparrow}n_{\downarrow}$, where $n_{\sigma}=c^{\dagger}_{\sigma}c_{\sigma}$, and $H_T=\sum_{k\sigma r}(t_r a^{\dagger}_{k\sigma r}c_{\sigma}+\text{H.c.})$ models tunneling between dot and leads (we neglect the energy dependence of the tunnel matrix elements $t_{L/R}$). Due to tunneling the dot level acquires a finite line width $\Gamma=\Gamma_L+\Gamma_R$ with $\Gamma_{L/R}=2\pi|t_{L/R}|^2N_{L/R}$ where $N_{L/R}$ is the density of states in the leads. The electron-electron interaction is accounted for by the charging energy $U=2E_C$ for double occupancy. To keep the discussion simple we choose $U=\infty$ for the QD.

As was discussed above, a contribution to the transport through a QD is identified as fully coherent if by adding a reference trajectory fully destructive interference can be achieved. Interaction of the dot electrons with an external bath

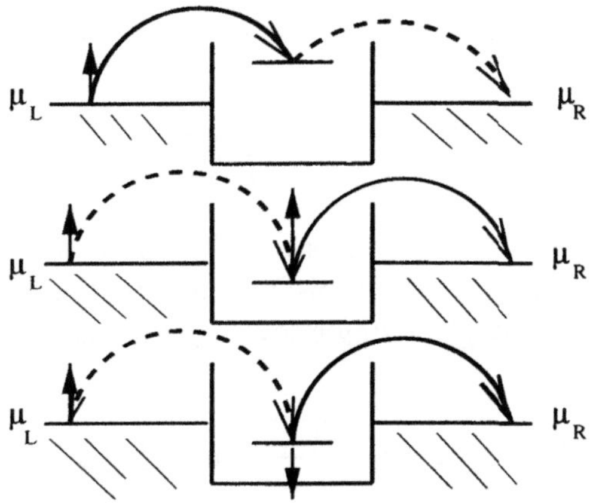

Figure 5. Cotunneling processes for $U = \infty$. The solid line indicates the process that happens first, the dashed line – the process that occurs afterwards. Double occupancy of the QD in the initial, intermediate, or final state is prohibited due to the infinite charging energy.

(e.g. phonons) destroys coherence since interference with a reference trajectory is no longer possible: the transmitted electron has changed its state or, equivalently [48], a trace in the environment is left. One possible mechanism for suppressing coherence in *interacting* QDs is flipping the spin of both the transmitted electron and the QD [49].

Away from resonance, $|\epsilon| \gg k_B T, \Gamma$, transport is dominated by cotunneling. There are three different types of cotunneling processes (for $U = \infty$):
(i) an electron enters the QD, leading to a virtual occupancy, and then leaves it to the other side.
(ii) an electron leaves the QD, and an electron with the same spin enters.
(iii) an electron leaves the QD, and an electron with opposite spin enters.
These three processes are shown schematically in Fig. 5.

Note that double occupancy of the dot (even in a virtual state) is forbidden since we have assumed $U = \infty$. All processes are elastic in the sense that the energy of the QD has not changed between its initial and final state. In particular, process (iii) is elastic in the sense that the energy of the QD has not changed. It is incoherent, though, since the spin in the QD has flipped (and it is therefore possible to determine that the electron under study went through the interferometer's arm with the QD and not through the free reference arm). Such a process will then contribute to the total current but not to the flux-sensitive component thereof, independent of the specific details of the AB interferometer. The observation that energy exchange is not necessary for dephasing [50] and that the latter can take place through, e.g., a spin flip of an external degree of freedom, has been made early on [48]. In our case the electrons in the QD itself (and their spin) serve as the "dephasing bath" [38].

To evaluate the transmission probability, hence the conductance, at finite temperatures, one needs to sum over contributions from different energies. This results in

$$\mathcal{G}_{\text{linear-response}} \equiv \frac{\partial I}{\partial V}\bigg|_{V=0} = -\frac{e^2}{h}\sum_{\sigma}\int d\omega\, T_{tr}^{\sigma}(\omega) f'(\omega)\,. \tag{21}$$

Here $T_{tr}^{\sigma}(\omega)$ denotes the transmission probability for an incoming electron of spin σ and frequency ω while f is the Fermi-Dirac distribution function.

The transmission through a single-level QD appearing in Eq. (21) can be obtained [45–47] from

$$T_{tr(\text{dot})}^{\sigma}(\omega) = -\frac{2\Gamma_L\Gamma_R}{\Gamma}\,\text{Im}\, G_{\sigma}^{\text{ret}}(\omega)\,. \tag{22}$$

Here the dot's Green function, given by

$$G_{\sigma}^{\text{ret}}(\omega) = 1/(\omega - \epsilon + i\Gamma/2)\ , \tag{23}$$

is the Fourier transform of $-i\Theta(t)\langle\left\{c_{\sigma}(t), c_{\sigma}^{\dagger}(0)\right\}\rangle$. For cotunneling, the transmission probabilities of electrons with energy ω near the Fermi level of the leads can also be obtained by calculating the transition rate in second-order perturbation theory and multiplying it with the probabilities P_{χ} to find the system in the corresponding initial state χ. For an incoming electron with spin up the transmission probabilities are $P_{\chi}\Gamma_L\Gamma_R\text{Re}[1/(\omega - \epsilon + i0^{+})^2]$. Here P_{χ} are the probabilities to find the system in the corresponding initial states $\chi = 0, \uparrow, \downarrow$ for case (i), (ii), and (iii), respectively. Since $P_0 + P_{\uparrow} + P_{\downarrow} = 1$ and $P_0 + P_{\sigma} = 1/[1 + f(\epsilon)]$ in equilibrium, we find

$$T_{tr(\text{dot})}^{\sigma}(\omega) = T_{tr(\text{dot,coh})}^{\sigma}(\omega) + T_{tr(\text{dot,incoh})}^{\sigma}(\omega) \tag{24}$$

with [51]

$$\begin{aligned} T_{tr(\text{dot})}^{\sigma}(\omega) &= \text{Re}\,\frac{\Gamma_L\Gamma_R}{(\omega - \epsilon + i0^{+})^2}\,, \\ T_{tr\text{dot,con}}^{\sigma}(\omega) &= \frac{T_{tr(\text{dot})}^{\sigma}(\omega)}{1 + f(\epsilon)}\,. \end{aligned} \tag{25}$$

The (evidently coherent) transmission amplitude through the QD can be defined in the following way

$$t_{(\text{dot})}^{\sigma}(\omega) = i\sqrt{\Gamma_L\Gamma_R}\, G_{\sigma}^{\text{ret}}(\omega)\,. \tag{26}$$

We now show that Eq. (26) is *not* a good building block for calculating the transmission *probability* for interacting QDs. Employing Eq. (23) we obtain

$$t_{\sigma}^{\text{dot}}(\omega) = \frac{i(P_0 + P_{\uparrow})\sqrt{\Gamma_L\Gamma_R}}{(\omega - \epsilon + i0^{+})}\,, \tag{27}$$

from which it follows that

$$|t_\sigma^{\text{dot}}(\omega)|^2 = \frac{T^\sigma_{tr(\text{dot})}(\omega)}{[1+f(\epsilon)]^2} \ . \tag{28}$$

The latter equation does not yield the total transmission through the dot, nor does it represent the contribution to the transmission probability arising from the coherent part of the transmission.

We have thus demonstrated that the the presence of electron-electron interactions gives rise to spin-flip processes, i.e. to incoherent transmission channels, hence to the breakdown of Eq. (4); there is no direct physical meaning of the expression $|t_\sigma^{\text{dot}}(\omega)|^2|$.

3. How to break phase locking

The phenomenon of phase locking has been found [26, 27] and recognized [53] in the early days of Mesoscopics. Experiments on small normal systems (simply and multiply connected) revealed magnetoconductance which was asymmetric under reversal of the direction of the magnetic field [54]. It has been then realized theoretically that the magnetoresistance can be asymmetric in the presence of an AB flux. Following some controversy Büttiker has proposed a multi-terminal effective circuit framework which captures the essential symmetries of the problem. This approach is, in principle, generalizable to a many-body interacting system coupled to external terminals of independent electron gas.

Below we provide a brief review of Büttiker's approach [33] (to be contrasted with the approach outlined e.g. in Section 5). We then make a few comments concerning the breakdown of the two-terminal phase locking mentioned earlier. Consider the four-terminal system depicted in Fig. 3. The four terminals are connected to voltage sources ("electron reservoirs") of chemical potentials $\mu_i, i = 1, ..., 4$. The reservoirs serve both as a source and a sink of carriers and energy. They possess the following properties: At zero temperature they feed the leads with carriers up to the chemical potential μ_i. At finite temperatures they feed the leads at all energies, weighted by the Fermi-Dirac function of the corresponding temperature and chemical potential. Each carrier coming from the lead and reaching the reservoir is absorbed by the reservoir irrespectively of its phase and energy. We first assume that the terminals connecting the respective reservoirs to the system are strictly one-dimensional. This means that at each terminal there are two running states at the Fermi energy, one with positive velocity (away from the reservoir) and the other with negative velocity. At this point we ignore interactions (e.g. electron-electron) or any inelastic processes within the system (inelastic relaxation and dephasing processes take place in the reservoirs). The electrons are scattered elastically in the system. We next assign scattering *probabilities* $\{T_{ij}(\varphi)\}$ for carriers outgoing from terminal j to be transmitted into terminal

i; we also use the notation $\{R_{ii}(\varphi)\}$ to denote reflection probabilities from i to i. It is clear that

$$T_{ij}(\varphi) = T_{ji}(-\varphi) \, ,$$
$$R_{ii}(\varphi) = R_{ii}(\varphi) \, . \tag{29}$$

These relations can be easily verified employing equations akin to Eqs. (10) and (12). As we are interested in linear response, the differences among the various $\{\mu_i\}$ are small, rendering the transmission and reflection probabilities $\{T_{ij}(\varphi)\}$ and $\{R_{ii}(\varphi)\}$ energy independent.

Straightforward algebra leads to the following expression for the current in the i-*th* lead

$$I_i = \frac{e}{h}\left[(1 - R_{ii})\mu_i - \sum_{j \neq i} T_{ij}\mu_j\right] \, . \tag{30}$$

Particle conservation, $R_{ii} + \sum_{j \neq i} T_{ij} = 1$, implies that Eq. (30) is independent of the choice of the reference potential ($\mu = 0$).

Let us first consider an arrangement [33, 55] where the currents satisfy $I_1 = -I_3$ and $I_2 = -I_4$. Inserting into Eq. (30) we obtain

$$I_1 = \chi_{11}(V_1 - V_3) - \chi_{12}(V_2 - V_4) \, , \tag{31}$$
$$I_2 = -\chi_{21}(V_1 - V_3) + \chi_{22}(V_2 - V_4) \, , \tag{32}$$

where $V_i = \mu_i/e$. Büttiker then found the following expressions for the generalized conductances of Eq. (31):

$$\chi_{11} = \frac{e^2}{h}\left[(1 - R_{11})S - (T_{14} + T_{12})(T_{14} + T_{21})\right]/S \, , \tag{33}$$
$$\chi_{12} = \frac{e^2}{h}(T_{12}T_{34} - T_{14}T_{32})/S \, , \tag{34}$$
$$\chi_{21} = \frac{e^2}{h}(T_{21}T_{43} - T_{23}T_{41})/S \, , \tag{35}$$
$$\chi_{22} = \frac{e^2}{h}\left[(1 - R_{22})S - (T_{21} + T_{23})(T_{32} + T_{12})\right]/S \, , \tag{36}$$

where

$$S = T_{12} + T_{14} + T_{32} + T_{34} = T_{21} + T_{41} + T_{23} + T_{43} \, . \tag{37}$$

The last equality is obtained by connecting terminals 1 and 3 together, and similarly terminals 2 and 4. One then obtains a two-terminal geometry, for which it is clear that the transmission $(1, 3) \to (2, 4)$ is equal to the transmission in the reverse direction, $(2,4) \to (1,3)$. Employing Eq. (29) one obtains

$$\chi_{11}(\varphi) = \chi_{11}(-\varphi)\,, \tag{38}$$

$$\chi_{22}(\varphi) = \chi_{22}(-\varphi)\,, \tag{39}$$

$$\chi_{12}(\varphi) = \chi_{21}(-\varphi)\,. \tag{40}$$

Eq. (38) establishes the Onsager relation for this circuit.

We next select the current source and drain to be terminals 1 and 3 respectively, and assume that 2 and 4 are potentiometer terminals, implying that the voltages V_2, V_4 are measures under the condition $I_2 = I_4 = 0$. One can now define *a* four-terminal conductance

$$\mathcal{G}_{13,24} = \frac{1}{2}\frac{I_1 - I_3}{V_2 - V_4}\,. \tag{41}$$

(We have used $I_1 = \frac{1}{2}(I_1 - I_3)$ to cast $\mathcal{G}_{13,24}$ in a symmetric form). From Eqs. (31), (33) and (38) one obtains

$$\mathcal{G}_{13,24} = \frac{\chi_{11}\chi_{22} - \chi_{12}\chi_{21}}{\chi_{21}}\,. \tag{42}$$

Since χ_{21} is not symmetric in φ, $\mathcal{G}_{13,24}$ turns out to be asymmetric as well (although the Onsager relations are clearly satisfied [33]).

We now face the following paradox. Consider a two-terminal AB interferometer. Let us now embed a "conducting island" in each of the interferometer's arms. Since this is a two-terminal setup, we expect phase-locking to hold. On the other hand, once we make the embedded islands sufficiently large, we may expect that eventually they would represent electron reservoirs (as if we have added two extra terminals to the circuit), leading perhaps to the breakdown of phase locking? If this is indeed the case – what is the characteristic island size where this breakdown can be observed? We first call attention to the fact that, as was discussed above, the mere introduction of inelastic or phase breaking processes cannot lead to the breakdown of phase locking. To address these questions we note [56] that in the context of four- (or multi-) terminal geometries it is possible to define other quantities whose dimension is conductance. One quantity of interest is

$$\mathcal{G}_{13,13} = \frac{1}{2}\frac{I_1 - I_3}{V_1 - V_3}\,. \tag{43}$$

Similarly we define

$$\mathcal{G}'_{13,13} = \frac{I_1}{V_1 - V_3}\,, \tag{44}$$

and

$$\mathcal{G}''_{13,13} = -\frac{I_3}{V_1 - V_3}\,. \tag{45}$$

It is clear that as long as $I_2 = I_4 = 0$, $\mathcal{G}_{13,13} = \mathcal{G}'_{13,13} = \mathcal{G}''_{13,13}$, and phase locking follows immediately (we note that in this case the only efffect of the extra terminals is to modify the effective elastic and inelastic scattering rate in the two-terminal interferometer).

We now modify V_i slightly into $V_i' = V_i + \delta V_i, i = 2, 4$, such that there is small current flowing out of terminal 2 while the current through terminal 4 is still zero. Then $I_3 = -(I_1 + I_2) \neq -I_1$. and the definition of conductance $G_{13,13}$ becomes ambiguous: $G_{13,13} = 1/2(I_1 - I_3)/(V_1 - V_3) \neq G'_{13,13} = I_1/(V_1 - V_3) \neq G''_{13,13} = -I_3/(V_1 - V_3)$. A simple calculation yields

$$G_{13,13} = (\chi_{11}\chi_{22} - \chi_{12}\chi_{21})/\chi_{22} - \chi_{12}I_2/[\chi_{22}(V_1 - V_3)] = A(\Phi) + B(\Phi)I_2/(V_1 - V_3) \ . \tag{46}$$

By Eq. (38) the first term in Eq. (46) is invariant under the transformation $\Phi \to -\Phi$. However, the second term which is nonzero in the presence of I_2 is not invariant under this transformation. The condition for the extremum of the conductance (as function of the flux) is $A'(\Phi) + I_2 B'(\Phi)/(V_1 - V_3) = 0$. Since $A'(0) = 0$, we estimate the derivative $A'(\Phi)$ as $A''(0)\Phi$ at small Φ. For small I_2 this gives for the extremum $\Phi_{\text{extr}} = -I_2 B'(0)/[A''(0)(V_1 - V_3)]$. The flux Φ_{extr} is nothing else but $-\alpha\Phi_0/2\pi$, where α is the orbital phase introduced above.

The above observations can be used for a systematic study of a "gradual breaking" of phase locking, characteristic of two-terminal geometries. A study of the breakdown of phase locking in connection the loss of unitarity may also be found in Ref. [57].

We finally note that, in a sense, phase-locking may be broken even in a two-terminal geometry if the applied magnetic flux is not purely of a Aharonov-Bohm type. Once the magnetic flux penetrates into the arms, different semi-classical trajectories of the electron's path will enclose different amount of flux, and the strict periodicity in Φ is broken. It follows that while $\mathcal{G}(\varphi) = \mathcal{G}(-\varphi)$ still holds, the relation $\mathcal{G}(\varphi = n + \delta) = \mathcal{G}(\varphi = n - \delta)$ is broken.

4. The dilemma of the transmission phase

The discussion in Section (2) shows that in general there is a coherent component of the electron transmitted through a QD. It is therefore legitimate to ask what the *phase* associated with the transmitted amplitude is. This is a particularly interesting issue since the role of e-e interactions in a low (zero) dimensional system, i.e. a QD, is expected to be enhanced. This question has indeed been taken up by experimentalists. It is clear that to obtain information about quantum phase one needs to resort to interferometry experiments. Typically the setup of such experiments consists of a 2DEG AB interferometer. The latter includes a G_aA_s/AlG_aA_s heterostructure QD embedded in one of its arms, and a free arm which serves as reference. The QD is manipulated by varying its gate voltage. In addition one controls the AB flux, the strength of the dot-lead coupling and the temperature. The first measurement (Ref.[10]), employing a two-terminal setup, produced only limited information, having to do with the phenomenon of *phase locking*. Later open-geometry experiments (Ref.[11]) yielded the flux dependent component of the conductance of an AB interferometer $\mathcal{G}^{AB}(\varphi)$, as function of the parameter V_G (the gate voltage). This can be written as

$$\mathcal{G}^{AB}(\varphi, V_G) = A(V_G) g^{AB}(2\pi\varphi + \alpha(V_G)), \tag{47}$$

where g^{AB} is a periodic function of φ. The prefactor $A(V_G)$ is expected to be large (small) for values of V_G that correspond to Coulomb peaks (conductance valleys). Recalling Eq. (5) for non-interacting electrons (in the double-slit geometry alluded to above only the first harmonic in the AB flux appears), it is tempting to draw an analogy between $\alpha(V_G)$ of Eq. (4) and the orbital phase of Eq. (3). If the orbital part of the transmission phase through the reference arm, α_1, is insensitive to the gate voltage, one may be motivated to refer to $\alpha(V_G)$ (up to the constant α_1), as the transmission phase through the QD. In doing so we stress again that for interacting electrons the transmission probability is *not* given by the square of the transmission amplitude, Eq. (4).

Thus, in a strict sense, the naive interpretation of $\alpha(V_G)$ outlined above is wrong. One, however, notes that the flux dependent part of the transmission *probability* is given by [37]

$$T_2^{AB}(\omega) = 2\sqrt{\Gamma_L \Gamma_R} |t^{ref}| \cos(2\pi\varphi) \mathrm{Re} G^{ret}(\omega) \tag{48}$$

for a two-terminal geometry and

$$T_{open}^{AB}(\omega) = 2\sqrt{\Gamma_L \Gamma_R} |t^{ref}| \mathrm{Re} |e^{-i(2\pi\varphi+\alpha)} G^{ret}(\omega)| \tag{49}$$

for an open geometry. Here Γ_L, Γ_R are the dot-lead couplings (on the left and right side respectively); t^{ref} is the transmission amplitude through the reference arm alone); G^{ref} is the retarded Green function of the coupled dot and ω is the incident's electron energy measured from the Fermi energy. The expression for $T_{open}^{AB}(\omega)$ is analogous to the interference term in Eq. (5) (roughly speaking, the left-right Green function through the dot replaces the transmission amplitude through the dot, t_2). It is this fact that justifies referring to $\alpha(V_G)$ as the transmission phase.

Typical parameters for the interferometry circuit are $U \approx 500\mu EV$, $k_B T \approx 10\mu eV$ and $\varepsilon_F \approx 10meV$. The number of electrons in the dot $N_{el} \approx 200 - 500$. The mean level spacing $\Delta \approx \varepsilon_F / N_{el} \approx 50 - 150\mu eV$, hence $T \leq \Delta$. The dot-lead coupling $\Gamma \approx 1\mu eV < T$. This set of parameters refers to the large QD experiments of Refs.[10] and [11]. In recent experiments [12], [13] smaller QDs were used in order to facilitate probing of Kondo physics. Our present review does not include discussion of this limit.

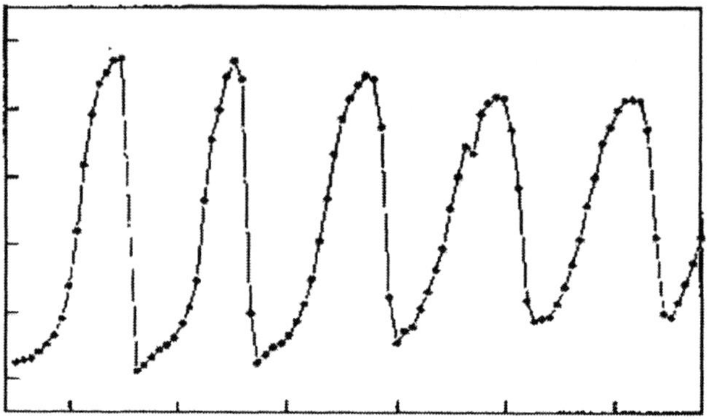

Figure 6. The evolution of the transmission phase $\alpha(V_G)$ from Ref. [11]

The evolution of $\alpha(V_G)$ is shown in Fig. 6 as function of the gate voltage. The latter is swept across consecutive Coulomb resonances. The phase α increases by π as V_G is swept across a Coulomb peak [deviations from π are, presumably, due to the fact that the induced resonances are not entirely independent: the ratio Δ/Γ is not sufficiently large]. This comes as no surprise and can easily be accounted for if one represents each individual resonance by a Breit-Wigner Lorentzian. The width of the π-step in α at resonance is of the order of the resonance width. At low temperatures the latter is denoted by $\tilde{\Gamma}$ (in Section 6 we comment on the differences between the bare (golden rule) level width Γ and $\tilde{\Gamma}$), and remains so as long as $k_3T << \tilde{\Gamma}$. In the conductance valley (i.e., between consecutive Coulomb peaks), the phase appears to drop rather sharply by π, rendering the phase evolution over a single period (in V_G) of Coulomb oscillations 0. These valley-to-valley correlations in the transmission phase have been observed repeatedly in a number of measurements, spanning up to 12 consecutive peaks in a single measurements.

This remarkable result soon attracted the attention of theorists, stimulating a large number of papers attempting to explain this phenomenon. I shall not try to present a comprehensive overview of these works. Instead I will mention the main approaches and state to what extent the problem still remains open.

The first point to note in this context is that these transmission phase correlations cannot be accounted for by independent electron theories [58]. Indeed, for a non-interacting "one-dimensional QD" (a segment of a one-dimensional wire of length L bounded by two potential barriers, whose respective transmission and reflection amplitudes from left (right) are t_1, r_1, t_2, r_2 $(t_1', r_1', t_2', 2_2')$) the transmission amplitude for an incoming electron of wave-number k is

$$t_{1dQD} = \frac{t_1 t_2 e^{ikL}}{1 - r_1' r_2 e^{i2kL}}. \tag{50}$$

It is easy to determine that as one varies the incoming electron's energy (or, alternatively, the base potential of the one-dimensional QD) the phase of t varies by π across a transmission resonance (i.e., across the energy of a quasi-bound state). This, however, is not accompanied by any phase lapse between resonances. The overall change of the transmission phase over a single period (as we increase the energy of the incoming electrons from, say, just below the n*-th* resonance to just below the $(n+1)$*-st* resonance) is therefore π, in stark contradistinction with the experimental data. This change-by-π-per-period is intimately connected with the fact that the sign $(p(n))$ of the product of the coupling matrix elements of the n-th wave function of the dot to the left and to the right leads alternates with n [58] (We consider here a system which possesses time reversal symmetry (e.g., no magnetic field)). The single particle wave functions of the uncoupled dot can thus be chosen to be real; by an appropriate gauge the dot-lead matrix elements can also be made real. The sign of the latter [58], [59] is that of the derivative of the component of the wave function normal to the dot-lead interface, cf. Ref [60]. For a two- or three-dimensional non-interacting dots $p(n)$ is geometry dependent and, in general, does not show any robust n-independence to account for the experimental data. Furthermore, for chaotically shaped or diffusively disordered QDs, the signs of the derivative of consecutive single-particle wave functions (near the lead), hence $p(n)$, are (to leading order) uncorrelated (unlike spectral properties). It is thus evident that we need to go beyond the independent particle framework.

Some of the early attempts to resolve the transmission-phase correlation effect [61] (see also [24]) relied on the Friedel sum rule [62, 63] which provides for a relation between phase and charge. One needs, though, to call attention to the fact that the Friedel sum rule deals with the *scattering matrix* (rather than the transmission matrix). It relates the total charge displaced in the field of a fixed impurity (e.g., the charge added to a quantum dot) to the scattering by that impurity of a free electron at the Fermi momentum k_F. The number of displaced electrons, N_D, is given by

$$N_D = \frac{1}{\pi} \sum_{l,m_l,m_s} \delta_{l,m_l,m_s}, \tag{51}$$

where the sum runs over the scattering phases δ with $(l, m_l), m_s$ being the angular momentum and the spin quantum numbers. More generally one can write

$$N_D = \frac{1}{2\pi i} T_r \ln S(\mu) \tag{52}$$

where $S(\mu)$ is the scattering matrix for single-particle-like excitations at the chemical potential μ. Consider for a moment the scattering of non-interacting electrons in one-dimension [64]. (We assume that the scattering is spin independent, hence suppress the spin index). This two-channel problem is described in terms of a 2×2

matrix

$$S = \begin{pmatrix} r & t' \\ t & r' \end{pmatrix} \tag{53}$$

whose eigenvalues are $e^{i\theta_l}, l = 1, 2$ (primed quantities refer to reflection and transmission coefficients for electrons impinging on the scatterer from the right). The one-dimensional version of the Friedel sum rule asserts that [64]

$$n(\varepsilon) = \frac{1}{2\pi} \frac{\delta(\theta_1 + \theta_2)}{\delta\varepsilon} , \tag{54}$$

where $n(\varepsilon)$ is the density of states contained in the scatterer. It can also be shown that for the 1d case the transmission amplitude, parameterized as $t = |t|e^{i\alpha}$, leads to the relation

$$\frac{\theta_1 + \theta_2}{2} = \alpha + \frac{\pi}{2} \qquad (1d) . \tag{55}$$

Therefore, specifically for the 1d case, the Friedel sum rule can be expressed, e.g., through Eq. (55), with the transmission phase α replacing the scattering phases. This, however, is not a general theorem concerning the transmission phase. Furthermore, there is no reason why the total charge accumulated at the scatterer (the QD and the leads near it) over a Coulomb period should be an integer (let alone 0, as is required if the total charge of the scattering phase were to be 0).

The theoretical effort addressing the correlations in the transmission phase could be divided, in large part, into two approaches.

The first school of thought maintains that there are one or few dot levels which are particularly strongly coupled to the leads. Such a strongly coupled level will dominate a number of consecutive transmission peaks. This would imply that successive resonances are dominated by the same tunneling matrix elements. In other words, it is practically the same level which keeps repeating at consecutive resonances, leading to transmission phase correlations over successive Coulomb periods.

As a specific example one may invoke a model QD whose Hamiltonian is made primarily of an integrable part (its eigenstates are products of longitudinal and transverse modes). The subset of states possessing high longitudinal quantum numbers defines the strongly coupled levels. The location of the gates is chosen in such a way that the energies $\{E^0_\alpha\}$ of the strongly coupled states, $\{F_\alpha\}$, are weakly dependent on the gate voltage. We now switch on a small non-integrable term of the potential. This leads to avoided level crossing (of the original, "bare" levels), as function of V_G. As is demonstrated in Fig.(7) an actual single-particle level, ψ_n, (plotted as function of V_G), is now made piecewise of strongly (flat) and weakly (steep) coupled bare states (F's and S's). In particular, a given F_α will be equal to ψ_n for a certain window of V_G, to ψ_{n+} for the next interval of V_G etc. It

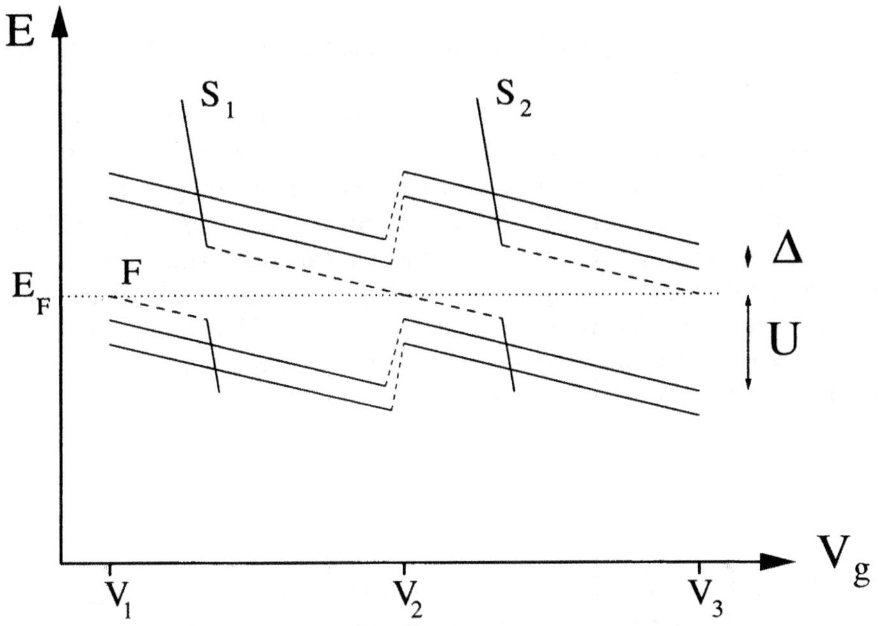

Figure 7. Avoided crossings with the steep bare levels S_1 and S_2 cause the flat- and strongly coupled- bare level F (dashed) to stay close to the Fermi energy ε_F. Resonances dominated by this level occur at gate voltages V_1, V_2, V_3. There is a gap of magnitude U between the last occupied and the first empty level. From Ref. [65]

so happens that as the levels $n, n+1, n+2, \ldots$ cross successively into the Fermi sea, what used to be the (bare) level F_α will keep "floating" over the Fermi level, dominating successive Coulomb resonances. While this picture [66, 67] provides a correlation-generating mechanism, it has a couple of nagging weaknesses. Firstly, for the level F_α to keep "hovering" just above the Fermi energy, an (approximate) commensurability condition is required between intervals (in V_G) of consecutive avoided crossings and intervals (in V_G) over which an additional electron is added to the QD (Coulomb period). This imposes rather stringent constraints on the geometry and the confining potential. Secondly, this picture assumes that the dot's levels (varying as function of V_G) are made piecewise of the original bare levels. The latter, other than at value of V_G that correspond to avoided crossing, do not mix, implying that the effect of the non-integrable term in the Hamiltonian is weak. Some of the QDs studied in the experiments of the Weizmann group might be indeed almost integrable, and the above approach may be suited to describe the pertinent physics. On the whole, though, this approach lacks the flavor of being generic, i.e., pertaining to chaotic or diffusive QDs where level mixing is strong.

As for the first reservation mentioned above the good news is that both finite temperature [65] or quantum [68] fluctuations render the "hovering effect" (i.e., the dominance of a single level S_α over a number of consecutive Coulomb peaks) more robust.

The second school of thought addressing the phase correlation effect takes the opposite point of view. Rather than a single, particularly strongly coupled level, dominating the transmission phase over a wide interval of V_G, here we rely on the fact that the coherent transmission in the "conductance valley" (i.e., between Coulomb peaks) is mostly due to the process of elastic cotunneling. A parametrically large number of levels (of order U/Δ) participate, each making a small (in magnitude) and random (in magnitude and phase) contribution to the transmission amplitude through the QD. Shifting the gate voltage to the next valley, these are almost the very same dot's levels that contribute, leading (with a high probability) to the same phase of the transmission amplitude. This is the mechanism behind the valley-to-valley correlations. Detailed analysis of the phase evolution following this picture is presented in Ref. [69]

5. Asymmetry of the interference signal

Let us consider transport through the QD away from resonance. At temperatures higher than the Kondo temperature this is dominated by cotunneling (second order in Γ). ¿From the discussion of Section 2 [37] it turns out that such cotunneling effects give rise to an asymmetry of the AB amplitude measured on either side of a Coulomb conductance peak. To see this consider, for example, the two level QD modelled by the Hamiltonian of Eq. 20. We tune the gate voltage such that the Fermi energy is a distance $\epsilon(\ll U)$ above (or below) the Coulomb resonance separating the $N_{el} = 0$ from the $N_{el} = 1$ valley (N_{el} is the mean number of electrons on the dot). For $\epsilon \gg \Gamma, k_B T$ the dominating cotunneling process on the $N_{el} = 0$ side of the conductance peak is the first (coherent) process depicted in Fig. 5, while on the $N_{el} = 1$ side cotunneling is dominated by the other two processes depicted in Fig. 5. These two processes are of equal probability $\sim \Gamma_L \Gamma_R \mathrm{Re}[1/(\omega - \epsilon + i0^+)^2]$. Only one of them is coherent (the second in that figure) while the other contributes to the current through the QD but not to the flux dependent conductance $\mathcal{G}^{AB}(\varphi)$. If we compare two values of V_G on either side of the conductance peak for which the flux-averaged conductance is the same, we expect greater visibility on the $N_{el} = 0$ side, hence (cf. Eq. 8) a larger AB amplitude on that side. This would imply that the the AB amplitude is asymmetric with respect to the total conductance (as a function of V_G).

Our detailed analysis [37] reveals that this asymmetry exists both near resonance (going to first order in Γ) and in second order. Considering an AB interferometer with a QD in one of the arms, and tuning the transmission of the reference arms to $|t_{\mathrm{ref}}| = \sqrt{\Gamma_L \Gamma_R}/|\epsilon|$ (to maximize the visibility) one finds for the total conductance

$$\left.\frac{\partial I^{\rm tot}}{\partial V}\right|_{V=0} = \frac{4e^2}{h}\frac{\Gamma_L\Gamma_R}{\epsilon^2}\left[1 - \frac{\epsilon}{|\epsilon|}\cos\varphi\right] \tag{56}$$

for the noninteracting case ($U = 0$) and

$$\left.\frac{\partial I^{\rm tot}}{\partial V}\right|_{V=0} = 4\frac{e^2}{h}\frac{\Gamma_L\Gamma_R}{\epsilon^2}\left[1 - \frac{\epsilon}{|\epsilon|}\frac{\cos\varphi}{1+f(\epsilon)}\right] \tag{57}$$

for $U = \infty$. This shows that cotunneling in the noninteracting case is fully coherent (we can tune both $t_{\rm ref}$ and φ such that the total transmission probability – hence the conductance– vanishes. In the interacting case spin-flip processes are present which spoil coherence. This is described by the asymmetry factor $1/[1 + f(\epsilon)]$, in accordance with our intuitive picture (in the above expressions the conductance peak is at $\epsilon = 0$.)

The above considerations can be generalized for a multilevel QD. The asymmetry factor can be used to obtain information concerning the total spin of the QD in a parameter regime away from the Kondo limit.

In Fig.8 and Fig.9 we present unpublished data concerning the dependence on gate voltage of both the total conductance through the AB interferometer and and magnitude of the flux-modulated amplitude. The curves presented in Fig.8 agree qualitatively with the above discussion: for the small QD data it is possible to identify and distinguish between Coulomb blockade valleys (where the total number of electrons is presumably even) and Kondo valleys (odd number). The latter can be identified through the very low temperature behavior of the conductance, not shown here [12, 13]. The theoretical prediction is that the peaks of the AB curve are shifted asymmetrically (with respect to the conductance peaks) away from the Kondo valleys, which is indeed suggested by Fig.8. Surprisingly enough this seems not to be the case for the large dot curve, Fig.9, where the peaks of the AB amplitude appear to be all shifted to the right of the conductance peaks.

One might note the shape of the conductance (and the AB) peaks is asymmetric as well. It was proposed that this is a manifestation of the Fano resonance effect, cf. Refs. [70–73, 24, 74].

6. On the width of the resonance and the phase lapses

The evolution of the transmission phase discussed above presents us with further dilemmas which have been pretty much ignored till now. These concern with the widths of both the phase change by π at the Coulomb peaks and the phase change (again by π) at the phase lapses. These widths are measured on the scale of (the change of) the gate voltage V_G. The issue is yet far from being resolved. Here we shall present the problems and add a few comments.

Let us first consider the range of V_G (near resonance) over which the phase change by π takes place. This is also the width of the Coulomb peak. We first

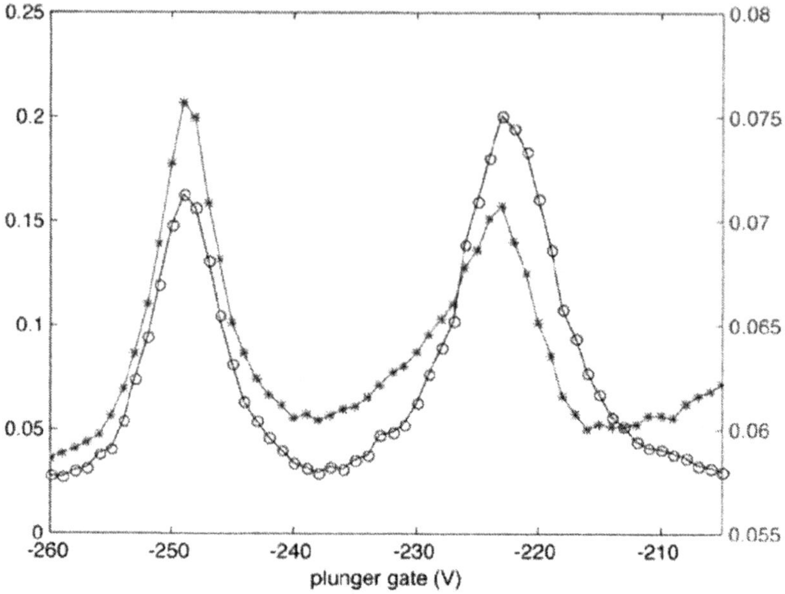

Figure 8. Total conductance(circles) and the AB amplitude (dots) as function of V_G for a single (small) QD interferometer. Data are courtesy of Yang Ji and M. Heiblum

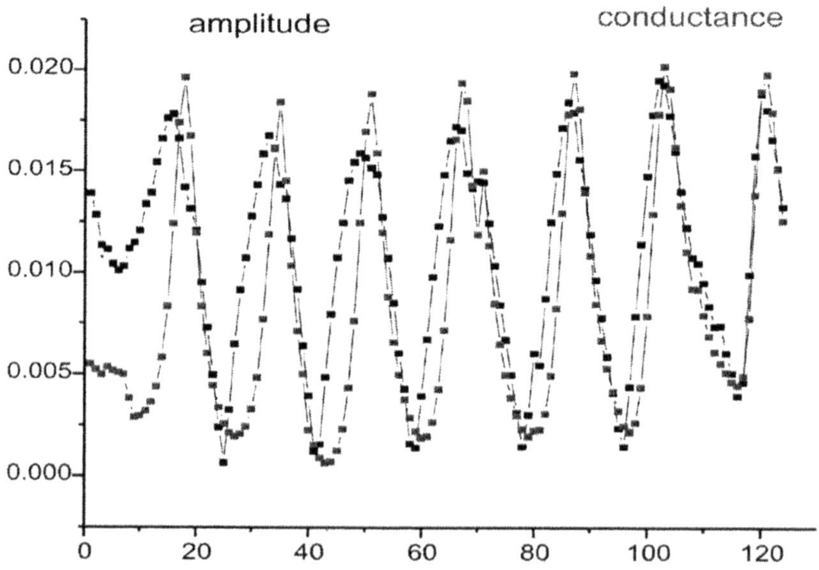

Figure 9. Total conductance (grey) and AB amplitude (black) as function of V_G for a large QD interferometer. Data are courtesy of R. Schuster, Yang Ji and M. Heiblum

consider "metallic dots", meaning that $k_B T > \Delta$. It is commonly accepted that (at least for a multichannel dot-level coupling) the physics of the Coulomb peaks is a function of the dimensionless dot-level conductance, g_{DL}. We will assume that the couplings of the QD to the left and to the right leads are of comparable strengths. We also note that $g_{DL} \sim \frac{\Gamma}{\Delta}$, where Γ is the (bare) golden-rule width of a dot's level. One should be careful distinguishing Γ from $\tilde{\Gamma}$, the latter being the width of the Coulomb peak. In the weak coupling limit [16] $\tilde{\Gamma} \sim \max(\Gamma, k_B T)$ (for a metallic dot $\tilde{\Gamma}$ is dominated by $k_B T$ in this limit)[75]. To get the flavor of the dilemma involved, let us now replace the actual Coulomb peaks by a periodic (in V_G) sequence of identical Lorentzians

$$\frac{d(conductance)}{d(energy)} = \frac{e^2}{\pi h} \sum_m \frac{\tilde{\Gamma}}{(V_G - mU)^2 + \tilde{\Gamma}^2} \tag{58}$$

The Fourier transform of the above expression (with respect to the gate voltage) is

$$\frac{e^2}{2\pi h} \sum_n e^{-\frac{2\pi n \tilde{\Gamma}}{U}} e^{i2\pi n V_G/U}. \tag{59}$$

This behavior is characteristic of mesoscopic systems. For observables which are periodic in some parameter, higher harmonics are suppressed faster (as function of width, inelastic rate, dephasing rate etc.)

Turning now our attention to the strong coupling limit one expects that the periodic modulation (of the conductance, the derivative of the particle number etc.) is all but suppressed. Only exponentially small modulation survives (of order $e^{-\frac{\pi^2}{8} g_{DL}}$) [76, 77]. In the language of the above Fourier expansion this amounts to the first harmonic being

$$e^{-\frac{\pi}{4}\frac{\Gamma}{\Delta}} e^{i2\pi n V_G/U}, \tag{60}$$

(higher harmonics will be suppressed even further) [78]. The fact that the (exponentially small) first harmonic dominates implies that the width of the (exponentially small) "Coulomb peak" is $\sim U$. Comparing Eqs. (59) and (60) (at $\Gamma/\Delta \sim 1$) reveals that in the vicinity of the weak-to-strong-coupling crossover, $\tilde{\Gamma}$ changes dramatically from $\tilde{\Gamma} \sim k_B T$ (or Γ) to $\tilde{\Gamma} \sim U$ over a rather small interval of V_G. We also note that Eqs. (59) and (60) cannot be reconciled within a single parameter scaling theory [79]. We stress that at this point our considerations are rather qualitative. We expect a similar fast crossover of $\tilde{\Gamma}$ (from the weak to the strong coupling limit) with other quantities as well, e.g. $d\langle N_{el}\rangle/d_{V_G}$, where $\langle N_{el}\rangle$ is the expectation value of the number of electrons on the QD. We also expect a fast crossover of $\tilde{\Gamma}$ in the discrete level limit ($k_B T < \Delta$) as well.

Let us now turn our attention to the width of the phase lapses that occur in the "Coulomb valleys", between Coulomb peaks. As has been discussed earlier, a universally accepted theory for such phase lapses is not yet available. Here we focus on another interesting observation – it appears that the experimentally observed typical width of these phase lapses, Γ_{PL}, is significantly smaller than that of the Coulomb peak, $\tilde{\Gamma}$. Presently this is a qualitative observation, yet to be backed up by a detailed study.

It is indeed a challenge to find a mechanism which provides for Γ_{PL} which is parametrically smaller that $\tilde{\Gamma}$. In order to develop a feel as to what the difficulty is let us consider a toy model which exhibits a phase lapse. This is a spinless two-level QD coupled to two leads with the Hamiltonian (cf. Eq. 20)

$$\mathcal{H} = \sum \varepsilon_k c_{k,i}^+ c_{k,i} + \sum_{j=1,2} \tilde{\varepsilon}_j d_j^+ d_j + \sum_{k,i,j} V \left(c_{k,i}^+ d_j + h.c. \right). \tag{61}$$

Here the operators $c_{k,i}$ refer to the electronic states in the leads ($i = L, R$) and the operators d_1, d_2 are associated with the QD states. We note that each of the dot's level will acquire a width Γ due to its coupling to the leads (this width may be further renormalized due to higher order dot-lead tunnelling processes, c.f. Ref. [80]). We have intentionally chosen all dot-lead matrix elements to have the same sign. It can be shown that this indeed leads to a phase lapse [81, 82]. Moreover, the original analysis, treating the tunnelling into/from each dot level independentally, results in the phase lapse having a width of $\Gamma = 2\pi V^2 \rho$ (ρ is the density of status in the leads). However, our recent analysis of Eq. (61) [83] shows that the tunnelling-induced coupling between the QD's levels must not be neglected. Since the toy model at hand is of non-interacting electrons, its analysis is straightforward. The single electron Greens function is given by

$$G(w) = \left[w - \tilde{\mathcal{H}} \right]^{-1} \tag{62}$$

where, in terms of the variables $\tilde{\varepsilon} = -(\varepsilon_1 + \varepsilon_2)/2$ and $\delta_G = (\varepsilon_1 - \varepsilon_2)/2$ the effective Hamiltonian of the QD (in the Hilbert space of levels 1,2) is

$$\tilde{\mathcal{H}} = \begin{pmatrix} -\tilde{\varepsilon} + \delta\varepsilon - i\Gamma & -i\Gamma \\ -i\Gamma & -\tilde{\varepsilon} - \delta\varepsilon - i\Gamma \end{pmatrix}. \tag{63}$$

One can readily find

$$G(w) = \frac{1}{D(w)} \begin{pmatrix} w + \tilde{\varepsilon} + \delta\varepsilon + i\Gamma & -i\Gamma \\ -i\Gamma & w + \tilde{\varepsilon} - \delta\varepsilon + i\Gamma \end{pmatrix} \tag{64}$$

when the determinant $D(w) = (w + \tilde{\varepsilon})^2 - (\delta\varepsilon)^2 + 2i\Gamma(w + \tilde{\varepsilon})$. The transmission amplitude from left to right can be written as

$$t(w) = \Gamma \sum_{i,j=1,2} G_{i,j}(w) = \frac{2\Gamma}{D(w)}(w + \tilde{\varepsilon}). \tag{65}$$

Varying the gate voltage amounts to varying $\tilde{\varepsilon}$ (leaving all other parameters unchanged). It is clear from Eq. (65) that one can tune the gate voltage (hence $\tilde{\varepsilon}$)

to obtain an *exact* zero of the transmission amplitude between two peaks. This occurs for $w = -\tilde{\varepsilon}$. Sweeping V_G around this point results in a sign change of $t(w)$, hence a zero-width phase lapse [83]. This phase lapse acquires a finite width at finite temperatures, or when the hopping matrix elements assume non-trivial relative phases. The above discussion (presented here for a non-interacting QD) demonstrates that the physics responsible for the width of phase lapses is quite different from that applicable at resonances, and may indeed give rise to parametrically narrow Γ_{PL}.

Acknowledgments. I acknowledge useful discussions with Y. Imry, A. Kamenev and H. A. Weidenmüller. This overview employs results and relies on insights obtained in the course of my present collaboration with D. E. Feldman, J. König, Y. Oreg and A. Silva. I am indebted to my colleagues Yang Ji, M. Heiblum and R. Schuster for discussions concerning the experiments and for the permission to present unpublished data. This work was supported by the U.S.-Israel Binational Science Foundation, by the GIF, by the Israel Science Foundation and by the Minerva Foundation.

References

1. Y. Aharonov and D. Bohm, *Phys. Rev.* **115**, 485 (1959).
2. D. Yu. Sharvin and Yu. V. Sharvin, *JETP Lett.* **34**, 272 (1981).
3. R. A. Webb, S. Washburn, C. P. Umbach and R. B. Laibowitz in *Proceedings of the Third International Conference on Superconducting Quantum Devices*, p. 561, Ed. by H. D. Hahlbuhm and H. Lübbig, Berlin, de Gruyter, (1985).
4. R. A. Webb, , S. Washburn, C. P. Umbach and R. B. Laibowitz, *Phys. Rev. Lett.* **54**, 2696 (1985).
5. A. D. Benoit, S. Washburn, C. P. Umbach, R. B. Laibowitz and R. A. Webb, *Phys. Rev. Lett.* **57**, 1765 (1986).
6. L. P. Levy, G. Dolan, J. Dunsmuir and H. Bouchiat, *Phys. Rev. Lett.* **64**, 2074 (1990).
7. V. Chandrasekar, R. A. Webb, M. J. Brady, M. B. Ketchen, W. J. Gallagher and A. Kleinasser, *Phys. Rev. Lett.* **67**, 3578 (1991).
8. D. Mailly, C. Chapelier and A. Benoit, *Phys. Rev. Lett.* **70**, 2020 (1993).
9. P. Mohanty and R. A. Webb, *Phys. Rev. Lett.* **84**, 4481 (2000).
10. A. Yacoby, M. Heiblum, D. Mahalu and H. Shtrikman, *Phys. Rev. Lett.* **74**, 4047 (1995).
11. R. Schuster, E. Buks, M. Heiblum, D. Mahalu, V. Umansky and H. Shtrikman, *Nature* **385**, 417 (1997).
12. Y. Ji, M. Heiblum, D. Sprinzak, D. Mahalu and H. Shtrikman, *Science* **290**, 779 (2000).
13. Y. Ji, M. Heiblum and H. Shtrikman, cond-mat/0106469.
14. W. G. van der Wiel, S. de Franceschi, T. Fujisawa, J. M. Elzerman, S. Tarucha, L. P. Kouwenhoven, *Science* **289**, 2105 (2000).
15. L. P. Kouwenhoven, C. M. Marcus, P. L. Mceuen, S. Tarucha, R. M. Westervelt and N. S. Wingreen in *Mesoscopic Electron Transport*, Ed. by L. L. Sohn, L. P. Kouwenhoven and G. Schoen, Dordrecht, NATO Series, Kluwer (1997).
16. Y. Alhassid, *Rev. Mod. Phys.* **72**, 895 (2000).
17. I. L. Aleiner, P. W. Brouwer and L. I. Glazman, cond-mat/0103008.
18. A. Kamenev and Y. Gefen, unpublished (1995).
19. Ya. M. Blanter, *Phys. Rev. B* **54**, 12807 (1996).

20. B. L. Altshuler, Y. Gefen, A. Kamenev and L. S. Levitov, *Phys. Rev. Lett.* **78**, 2803 (1997).
21. Ya. M. Blanter and A. D. Mirlin, *Phys. Rev. E* **55**, 6514 (1997).
22. U. Gerland, J. v. Delft, T. A. Costi and Y. Oreg, *Phys. Rev. Lett.* **84**, 3710 (2000).
23. E. V. Anda, C. Busser, G. Chiappe and M. A. Davidovich, cond-mat/0106055.
24. W. Hofstetter, J. König, H. Schoeller, *Phys. Rev. Lett.* **87**, 156803 (2001).
25. I. Affleck and P. Simon, cond-mat/0012002.
26. Y. Gefen, Y. Imry and M. Ya. Azbel, *Phys. Rev. Lett.* **52**,129 (1984).
27. Y. Gefen, Y. Imry and M. Ya. Azbel, *Surf. Sci.* **142**, 203 (1984).
28. B. L. Altshuler, A. G. Aronov and B. Z. Spivak, *JETP Lett.* **33**, 94 (1981).
29. N. Byers and C. N. Yang, *Phys. rev. Lett.* **7**, 46 (1961).
30. Y. Gefen, unpublished (1984).
31. M. Murat, Y. Gefen and Y. Imry, *Phys. Rev. B* **34**, 659 (1986).
32. A. D. Stone and Y. Imry, *Phys. Rev. Lett.* **56**, 189 (1986).
33. M. Büttiker, *Phys. Rev. Lett.* **57**, 1761 (1986).
34. Note that the phase locking symmetry is present in the results of ref. [26].
35. Y. Imry *Introduction to Mesoscopic Physics*. New York, Oxford University Press, 1997.
36. R. Landauer, *Phil. Mag.* **21**, 863 (1970).
37. J. König and Y. Gefen, *Phys. Rev. Lett.* **86**, 3855 (2001); cond-mat/0107450.
38. P. A. Mello, Y. Imry, and B. Shapiro, *Phys. Rev. B* **61**, 16570 (2000).
39. M. Büttiker, *Phys. Rev. B* **32**, 1846 (1985).
40. H. F. Cheung, Y. Gefen and E. K. Riedel, *IBM J. Res. Div.* **32**, 359 (1988).
41. Y. Gefen and G. Schön, *Phys. Rev. B* **30**, 7323, (1984).
42. E. Shimshoni and Y. Gefen, *Ann. Phys. (NY)* **210**, 16 (1991).
43. The approach outlined here applies more broadly than the range of validity of semiclassics, where we write down the individual amplitudes within each class of winding number. Rather than doing so we can group all the contributions within such a class into one complex amplitude $t^{(n)}$. The latter quantity (the total transmission amplitude associated with a given winding number) has a wider range of applicability than semiclassics. Also note that the partial amplitudes $\{A_j^{(n)}\}$ may represent processes which include backscattering of point-like impurities, where the validity of semiclasics needs to be reconsidered, cf. S. Chakravarty and A. Schmid, *Phys. repts.* **140**, 193 (1986); N. Argaman, Y. Imry and U. Smilansky *Phys. rev. B* **47**, 4440 (1993).
44. C. Bruder, R. Fazio and H. Schoeller, *Phys. Rev. Lett.* **76**, 114 (1996).
45. Y. Meir and N. Wingreen, *Phys. Rev. Lett.* **68**, 2512 (1992).
46. J. König, H. Schoeller, and G. Schön, *Phys. Rev. Lett.* **76**, 1715 (1996); J. König, J. Schmid, H. Schoeller, and G. Schön, *Phys. Rev. B* **54**, 16820 (1996).
47. H. Schoeller, in *Mesoscopic Electron Transport*, eds. L.L. Sohn *et al.* (Kluwer 1997); J. König, *Quantum Fluctuations in the Single-Electron Transistor* (Shaker 1999).
48. A. Stern, Y. Aharonov, and Y. Imry, *Phys. Rev. A* **41**,3436 (1990).
49. In ref.[37] it is shown why, in the absence of electron-electron interaction in the dot, the amplitude of spin-flip processes vanishes.
50. B.L. Altshuler, A.G. Aronov, D.E. Khmelnitsky, *J. of Phys. C: Solid State Physics* **15**, 7367 (1982).
51. The regularization $+i0^+$ is put here by hand, but can be derived within a complete cotunneling theory [52].
52. J. König, H. Schoeller, and G. Schön, *Phys. Rev. Lett.* **78**, 4482 (1997); *Phys. Rev. B* **58**, 7882 (1998).
53. Büttiker, M., Y. Imry and M. Ya. Azbel: 1984 *Phys. Rev. A* **30**, 1982.
54. See, for example, Umbach, C. P., S. Washburn, R. B. Laibowitz and R. A. Webb, *Phys. Rev. B* **38**, 4848 (1984), Ref. [5] and references therein.

55. H. B. G. Casimir, *Rev. Mod. Phys.* **17**, 343 (1945).
56. D. E. Feldman and Y. Gefen, unpublished.
57. O. Entin-Wohlman, A. Aharony, Y. Imry, Y. Levinson and A. Schiller, cond-mat/0108064.
58. R. Berkovits, Y. Gefen and O. Entin-Wohlman, *Phil. Mag.* **B77**, 1123 (1998).
59. C. B. Duke, *Solid State Phys. Suppl.* **10**, Chap. 18, 207 (1969).
60. J. Bardeen, *Phys. Rev. Lett.* **6**, 57 (1961).
61. A. L. Yeyati and M. Büttiker, *Phys. Rev.* **B52**, R14360 (1995).
62. J. Friedel, *Phil. Mag.* **43**, 153 (1952).
63. J. S. Langer and V. Ambegaokar, *Phys. Rev.* **121**, 1090 (1961).
64. P. W. Anderson and P. A. Lee, *Supp. Prog. Theoretical Physics*, **69**, 212 (1980).
65. R. Baltin, Y. Gefen, G. Hackenbroich and H. A. Weidenmüller, *Eur. Phys. J.* **B10**, 119 (1999).
66. G. Hackenbroich, W. D. Heiss and H. A. Weidenmüller, *Phys. Rev. Lett.* **79**, 127 (1997).
67. Y. Oreg and Y. Gefen, *Phys. Rev. B* **B55**, 13726 (1997).
68. P. G. Silvestrov and Y. Imry, *Phys. Rev. Lett.* **85**, 2565 (2000).
69. R. Baltin and Y. Gefen, *Phys. Rev. Lett.* **83**, 5094 (1999).
70. U. Fano, *Phys. Rev.* **124**, 1866 (1961).
71. P. S. Deo and A. M. Jayannavar, *Mod. Phys. Lett.* **10**, 787 (1996); H. Xu and W. Sheng, *Phys. Rev.* **B 57**, 11903 (1998); C.-M. Ryu and S. Y. Cho, *Phys. Rev.* **B 58**, 3572 (1998); H.-W. Lee, *Phys. Rev. Lett.* **82**, 2358 (1999).
72. J. Göres, D. Goldhaber-Gordon, S. Heemeyer, M. A. Kastner, H. Shtrikman, D. Mahalu and U. Meirav, *Phys. Rev. B* **62**, 2188 (2000).
73. B. R. Bulka and P. Stefanski, *Phys. Rev. Lett.* **86**, 5128 (2001).
74. O. Entin-Wohlman, A. Aharony, Y. Imry and Y. Levinson, cond-mat/0109328.
75. When one goes beyond first order "sequential tunnelling" analysis, the "width of the Coulomb peak" is not a well-defined quantity. For example, 2nd order cotunnelling gives rise to power law tails of the conductance peaks. This underscores the problematics of employing naive sealing to study the width. I am grateful to Y. Nazarov for his comments on this point.
76. Yu. V. Nazarov, *Phys. Rev. Lett.* **82**, 1245 (1999).
77. A. Kamenev, *Phys. Rev. Lett.* **85**, 4160 (2000).
78. Note that while the height and the width of the Coulomb peak are strongly renormalized in the strong coupling limit, this is not the case with the peak-to-peak distance U. The latter might be modified by trivial factors, involving the ratio of self-capacitance/dot-gate capacitance.
79. J. König, Y. Gefen, A. Silva and Y. Oreg, to be published.
80. J. König, Y. Gefen and G. Schön, *Phys. Rev. Lett.* **81**, 4468 (1998).
81. G. Hackenbroich and H.A. Weidenmüller, *Europhys. Lett.* **38**, 129 (1997).
82. Y. Oreg and Y. Gefen, *Phys. Rev. B* **55**, 13726 (1997).
83. A. Silva, Y. Oreg and Y. Gefen, Submitted for publication.

PHOTON ASSISTED TUNNELING IN QUANTUM DOTS

W.G. VAN DER WIEL, T.H. OOSTERKAMP,
S. DE FRANCESCHI, C.J.P.M. HARMANS
AND L.P. KOUWENHOVEN
Department of Applied Physics and DIMES, Delft University of Technology, PO Box 5046, 2600 GA Delft, The Netherlands

We review experiments on single-electron transport through single quantum dots in the presence of a microwave signal. In the case of a small dot with well-resolved discrete energy states, the applied high-frequency signal allows for inelastic tunnel events that involve the exchange of photons with the microwave field. These photon assisted tunneling (PAT) processes give rise to sideband resonances in addition to the main resonance. Photon absorption can also lead to tunneling via excited states instead of tunneling via the ground state of the quantum dot. The manipulation of quantum dots by a microwave field is an important ingredient for the possible application of quantum dots as solid state quantum bits, which forms a motivation for this review.

1. Introduction

Transport properties of quantum dots at zero frequency have been extensively studied and by now many aspects are well understood [1]. In some studies finite, but still low frequency signals were applied to a gate electrode nearby the quantum dot. Capacitance spectroscopy on quantum dots has been performed at kHz frequencies [2]. At MHz frequencies, quantum dots can be operated as turnstiles or pumps [3]. These frequencies, f, are low in the sense that the photon energy, hf, is much smaller than the thermal energy, k_BT, and thus the discrete photon character cannot be discerned. For sufficiently high frequencies, such that $hf \gg k_BT$, the interaction of the electromagnetic field with the electrons confined in a quantum dot would be analogous to light spectroscopy studies on atoms. However, different quantum dots are not microscopically equal. The response of an ensemble of quantum dots to light excitation is therefore strongly averaged. Despite this averaging, excitation studies on arrays of quantum dots by far-infrared light (i.e. the THz regime) have revealed the spectrum of collective modes [4], i.e. the sloshing modes of the whole electron puddle in the external potential. Excitations within the electron puddle are difficult to create since, according to

I.V. Lerner et al. (eds.),
Strongly Correlated Fermions and Bosons in Low-Dimensional Disordered Systems, 43–68.

the generalized Kohn theorem [4], the dipole field of far-infrared radiation does not couple to the relative coordinates of electrons confined in a parabolic quantum dot. This problem is circumvented by inelastic light scattering experiments which have been able to detect electronic excitations in arrays of quantum dots that can be related to a discrete single-particle spectrum [5]. Recently, this technique has also probed excitons in a single quantum dot [6].

In this review, we discuss electron transport experiments on single quantum dots, that are irradiated by a microwave signal. In contrast to light transmission, luminescence, or inelastic light scattering measurements (see for a review Ref. [7]), we measure the dc current in response to a microwave signal. Current can flow through a quantum dot when a discrete energy state is aligned to the Fermi energies of the leads. This resonant current is carried by elastic tunneling of electrons between the leads and the dot. An additional time-varying potential $\widetilde{V}\cos(2\pi ft)$ can induce *inelastic* tunnel events when electrons exchange photons of energy hf with the oscillating field. This inelastic tunneling with discrete energy exchange is known as photon assisted tunneling (PAT). Microwave studies have a long tradition in the field of superconductor-insulator-superconductor tunnel junctions [8], for which the theory was first described by Tien and Gordon in 1963 [9]. Despite many proposals and a long search [10], it took thirty years before PAT was also observed in a non-superconducting system. In 1993 PAT features were seen in the current-voltage characteristics of a GaAs/AlGaAs superlattice under THz irradiation from a free-electron laser [11]. Starting in 1994, PAT was also found in experiments on single-electron transport through semiconductor quantum dots [12-16]. The quantum dots in Ref. [13] were rather large and effectively had a continuous density of states. Here, we focus on PAT processes through quantum dots with well-resolved, discrete energy states. For these small dots, the resonant tunneling peak in the current develops photon sideband resonances when we apply microwaves [16]. The energy separation between main peak and sidebands can be used as a spectroscopic measurement of the energy levels in the dot.

The relevant ac regime for quantum dots is at much lower frequency than visible light. We list the important frequency scales in Table I. The single-particle level spacing $\Delta\varepsilon$ is 0.05 - 0.5 meV for typical dots and the charging energy, e^2/C, is usually 0.2 - 2 meV. To observe effects from a finite $\Delta\varepsilon$ and e^2/C, these energies should exceed the thermal broadening $\sim 4k_BT$. Other characteristic frequencies of the dot are related to the transport times. Γ is the typical rate to tunnel on or off the dot, which can be arbitrarily small for opaque tunnel barriers. This frequency is set by the transmission coefficient of the barriers and should be kept smaller than $\Delta\varepsilon$ otherwise the level broadening exceeds the spacing between the single-particle states. The final time scale is the tunneling time; i.e. the actual time spent during tunneling through the barrier. This time is quite short ($\sim$2 ps) for typical barriers (calculated within the Büttiker-Landauer framework [17]). To access these time scales, ac signals can be applied, and the effects on the dc transport can be measured.

TABLE I. A list of the important energy/frequency scales for transport through quantum dots. For $f = 10$ GHz the photon energy, hf, is 40 μeV.

Quantity	Equivalent frequency	Typical frequencies
Thermal broadening	$\sim 4k_BT/h$	10 GHz (at 100 mK)
Tunneling rate on/off the dot	Γ	0 - 100 GHz
Level spacing (or inverse traversal time)	$\Delta\varepsilon/h$	10 - 100 GHz
Charging energy	e^2/hC	40 - 400 GHz
Tunneling time	$1/\tau_{tunnel}$	200 GHz - 1 THz

If $f \ll \Gamma$ each electron sees an essentially static potential and we are in the adiabatic regime [2, 18]. If $f \gg \Gamma$, each electron experiences many cycles of the ac signal while it is on the dot; i.e. the non-adiabatic regime. If $hf \ll 4k_BT$, single photon processes are masked by thermal fluctuations, and a classical description is appropriate [3]. Thus, the discreteness of the photon energy can be observed in the non-adiabatic, high-frequency regime: $hf \gg h\Gamma, 4k_BT$. This is the quantum, or time-dependent regime for which it is essential to solve the time-dependent Schrödinger equation for the tunneling electron.

This review is divided into two parts. First, in section 2, we discuss the theory of PAT through a single junction (2.1) and calculations of PAT through a single quantum dot using a master equation approach (2.2). Numerical PAT calculations are presented in sections 2.3 and 2.4. Second, in section 3, we discuss experimental results of PAT measurements on single lateral quantum dots. The sample geometry is discussed in section 3.1, the sensitivity of pumping to the applied microwave frequency in 3.2. PAT in the low ($hf < \Delta\varepsilon$) and high frequency ($hf > \Delta\varepsilon$) regime are discussed in 3.3 and 3.4, respectively. In sections 3.5 through 3.7 results are given of the dependence of PAT on microwave frequency, magnetic field and microwave power, respectively. We conclude and discuss the results in section 3.8.

2. Theory

2.1. PAT THROUGH A SINGLE JUNCTION

First, we briefly outline photon assisted transport through a single tunnel junction separating two metallic leads. An oscillating potential difference across a junction, $\widetilde{V}\cos(2\pi ft)$, where $\widetilde{V}$ is the ac amplitude, may be included in the Hamiltonian of one of the leads as: $H = H_0 + H_{ac} = H_0 + e\widetilde{V}\cos(2\pi ft)$, where the unperturbed Hamiltonian, H_0, describes the leads without microwaves. The effect of the oscillating potential is that the time-dependent part of the electron wave function in this lead, when expanded into a power series, contains energy components at E, $E \pm hf$, $E \pm 2hf$, ..., etc. These are called sidebands. The expansion can be done as follows [9]:

$$\begin{aligned}\psi(r,t) &= \varphi(r)\exp\left(-i\int dt[E + e\widetilde{V}\cos(2\pi ft)]/\hbar\right) \\ &= \varphi(r)\exp(-iEt/\hbar)\sum_{n=-\infty}^{\infty} J_n(e\widetilde{V}/hf)\exp(-in2\pi ft) \\ &= \varphi(r)\left(\sum_{n=-\infty}^{\infty} J_n(e\widetilde{V}/hf)\exp(-i[E+nhf]t/\hbar)\right) \qquad (1)\end{aligned}$$

$\varphi(r)$ is the eigenfunction satisfying $H_0\varphi(r) = E\varphi(r)$ and forms the spatial part of the wave function $\psi(r,t)$. $J_n(\alpha)$ is the nth order Bessel function of the first kind (see Fig. 1a) evaluated at $\alpha = e\widetilde{V}/hf$.

Let ρ_l and ρ_r be the unperturbed densities of states of the left and right leads. The tunnel current through a junction without a microwave field is then given by

$$I(V_{SD}) = c\int_{-\infty}^{\infty} dE[f_l(E - eV_{SD}) - f_r(E)]\rho_l(E - eV_{SD})\rho_r(E) \qquad (2)$$

where V_{SD} is the source-drain voltage, $f(E)$ is the Fermi function, and c is a constant proportional to the tunnel conductance.

From Eq. (1), we can write an effective density of states in one of the leads (we choose the right lead) given by:

$$\widetilde{\rho}_r(E) = \sum_{n=-\infty}^{\infty} \rho_r(E + nhf)J_n^2(e\widetilde{V}/hf) \qquad (3)$$

If tunneling is a weak perturbation, the dc current in the presence of microwaves,

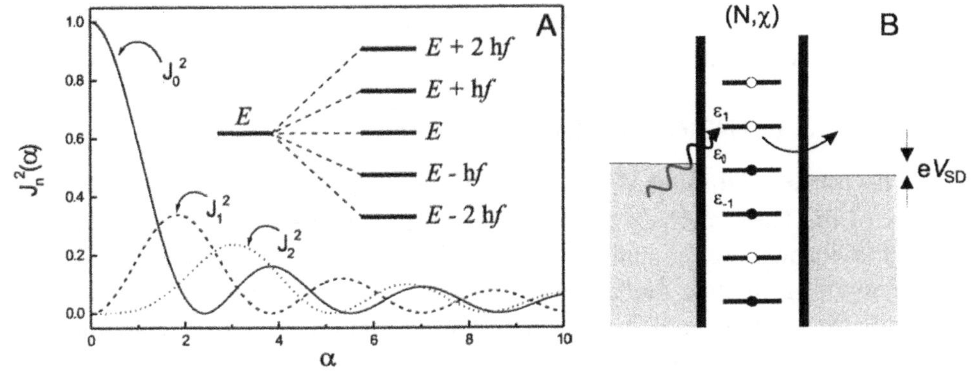

Figure 1. (a) Squared Bessel functions of the first kind $J_n^2(\alpha)$, for n = 0, 1 and 2. The inset shows the development of sidebands of the original energy as a consequence of the microwave field. The population probability $P(n)$ of the different sidebands is given by $P(n) = J_n^2(e\tilde{V}/hf)$. A positive or negative n corresponds to the absorption or emission, respectively, of n photons during the tunnel process. Elastic tunneling corresponds to $n = 0$. (b) Schematic energy diagram of a single dot containing N electrons distributed over the available single-particle levels ε_j in a particular configuration χ. By absorbtion of a photon, an electron can tunnel into ε_1 so that the electron number changes into $N + 1$ and the configuration into χ'.

$\tilde{I}$, is given by [9]:

$$\tilde{I}(V_{SD}) = c \sum_{n=-\infty}^{\infty} J_n^2(e\tilde{V}/hf) \times$$
$$\int_{-\infty}^{\infty} dE[f_l(E - eV_{SD}) - f_r(E + nhf)]\rho_l(E - eV_{SD})\rho_r(E + nhf)$$
$$= \sum_{n=-\infty}^{\infty} J_n^2(e\tilde{V}/hf) I(V_{SD} + nhf/e) \tag{4}$$

We stress that tunneling is assumed to be a weak perturbation, implying that the sidebands are only well-defined for $f \gg \Gamma$. Since there is no electric field in the scattering-free leads, mixing of electron states [19], or photon absorption is absent in the leads. A positive or negative n corresponds to the absorption or emission, respectively, of n photons during the tunnel process. Elastic tunneling corresponds to $n = 0$. *For a single junction the dc current in the presence of microwaves is thus described, simply in terms of the dc current without microwaves.* From the normalization $\sum_{n=-\infty}^{\infty} J_n^2(\alpha) = 1$, it follows that the integrated current does not change due to the oscillating potential: $\int \tilde{I}(V_{SD})dV_{SD} = \int I(V_{SD})dV_{SD}$. We emphasize that although the oscillating field is entirely classical, the interaction with an electron, described by the Schrödinger equation, is only via exchange of *discrete* energy quanta. Equation (4) is only valid for single junctions where the tunneling takes place via a single hop. An extension of Eq. (4) to describe a double junction system is discussed next.

2.2. MASTER EQUATION FOR PAT THROUGH A QUANTUM DOT

Electron transport through double barrier structures is resonant when the Fermi energy of the leads aligns with a discrete energy state between the two barriers. Transport through semiconductor quantum wells are usually well described by non-interacting electron models. Also, their transport properties in the presence of an oscillating signal can, in a first-order approximation, be described by the time-dependent, non-interacting Schrödinger equation. The result of such calculations is that, next to the main resonance, extra peaks appear at distances corresponding to the photon energy, hf [20, 21].

In quantum dots electrons are confined in all directions. The total number of electrons, and the total charge, is thus a discrete value. This makes it essential to include the Coulomb interactions when describing transport. The standard model is known as the single-electron tunneling, or the Coulomb blockade model [22]. This model takes into account that at low voltages and low temperatures only one electron can tunnel at a time. The necessary energy to add an extra electron to a quantum dot consists of the charging energy $E_c = e^2/C$ for a single electron, and a discrete energy difference, $\Delta\varepsilon$, arising from the quantum-mechanical confinement. In practice, a quantum dot has discrete energy states if $\Delta\varepsilon$ exceeds the thermal energy k_BT [1]. Assuming sequential tunneling of single electrons, the current can be calculated with a master equation [23, 24].

PAT through small systems in which Coulomb blockade is important was considered first by Likharev and Devyatov [25], Hadicke and Krech [26], and Bruder and Schoeller [27]. A direct inclusion of the Tien and Gordon equations [9] in a master equation that takes into account Coulomb blockade [22] can be made by writing the tunnel rate through each barrier in the presence of microwaves $\widetilde{\Gamma}(E)$ in terms of the rates without microwaves $\Gamma(E)$ [12]:

$$\widetilde{\Gamma}(E) = \sum_{n=-\infty}^{+\infty} J_n^2(\alpha)\Gamma(E + nhf) \tag{5}$$

Equation 5 has a direct link to studies of the effects of fluctuations in the electromagnetic environment on single-electron tunneling [28]. If the spectral density of the environment is characterized by the probability function $P(hf)$, then the rate including the environment $\Gamma_{env}(E)$, can be written in terms of the rate without the environment $\Gamma(E)$ as [29]:

$$\Gamma_{env}(E) = \int_{-\infty}^{\infty} d(hf)P(hf)\Gamma(E + hf) \tag{6}$$

Whereas Eq. (5) describes a monochromatic environment, the fluctuations in general are broad band in frequency, as described in Eq. (6). Examples of environments that have been studied experimentally, are the impedance in the leads [28], blackbody radiation [30], and phonons [31]. We note, however, that Eqs. (5,6) are valid only for systems with a continuous density of states (i.e. $\Delta\varepsilon \ll k_BT$) and immediate relaxation to the ground state after each tunnel event.

In the case of quantum dots with large level separation (i.e. $\Delta\varepsilon \gg k_BT$), one needs to keep track of the occupation probabilities of each discrete state. This increases the amount of bookkeeping, but has the advantage that intra-dot relaxation and excitation processes can be included. In our model [32] for PAT through small dots, we assume $E_c \gg \Delta\varepsilon$, k_BT, eV_{SD}, nhf, such that we only need to consider two charge states (i.e. the electron number is either N or $N+1$) [33]. We neglect level broadening due to a finite lifetime of the electrons on the dot.

A charge state (Fig. 1b) is described by the electron number N, together with the particular occupation of the electrons in the available single-particle levels $\{\varepsilon_j\}$. If N electrons are distributed over k levels, the number of distinct dot configurations, χ, is given by ${}^{N}C_k$. The probability, $P_{N,\chi}$, for state (N,χ) is calculated from a set of master equations given by:

$$\begin{aligned} \dot{P}_{N,\chi} &= \sum_{\chi'} P_{N+1,\chi'}(\Gamma^{out}_{l,j_{\chi'}} + \Gamma^{out}_{r,j_{\chi'}}) \\ &\quad - P_{N,\chi} \sum_{j=empty} (\Gamma^{in}_{l,j} + \Gamma^{in}_{r,j}) \\ &\quad + \sum_{\chi''\neq\chi} P_{N,\chi''}\Gamma_{\chi''\to\chi} - P_{N,\chi} \sum_{\chi'''\neq\chi} \Gamma_{\chi\to\chi'''} \end{aligned} \tag{7}$$

and the equivalent forms for $\dot{P}_{N+1,\chi'}$. To find a stationary solution, these equations are all set to zero ($\dot{P}=0$) and solved with the boundary condition:

$$\sum_{\chi} P_{N,\chi} + \sum_{\chi'} P_{N+1,\chi'} = 1 \tag{8}$$

For $N=2$ distributed over five different single-particle levels $\{\varepsilon_j\}$, there are ten different configurations, χ, yielding ten equations for $\dot{P}_{N,\chi}$ and also ten equations for $\dot{P}_{N+1,\chi'}$.

The first and second term in Eq. (7) correspond to a change in the occupation probability of a certain distribution due to tunneling (the number of electrons on the dot changes). In the first term an electron tunnels out of the dot. Only those rates are taken into account that correspond to an electron tunneling out of state $j_{\chi'}$ that leave the dot in the distribution (N, χ). In the second term, an electron tunnels onto the dot. One needs to sum over all the states j that are empty when the dot is in configuration χ, because all these events cause a transition from state (N,χ) to a state $(N+1,\chi')$. $\Gamma^{in}_{l,j}$ and $\Gamma^{out}_{l,j}$ are the tunnel rates through the left barrier in and out of single-particle level j on the dot:

$$\Gamma^{in}_{l/r,j}(\varepsilon_j) = \Gamma_{l/r,j}\sum_n J^2_n(\alpha_{l/r})f(\varepsilon_j - \frac{C_g}{C}eV_g - nhf + \eta_{l/r}eV_{SD};T_{l/r}) \quad (9)$$

$$\Gamma^{out}_{l/r,j}(\varepsilon_j) = \Gamma_{l/r,j}\sum_n J^2_n(\alpha_{l/r})[1 - f(\varepsilon_j - \frac{C_g}{C}eV_g - nhf + \eta_{l/r}eV_{SD};T_{l/r})]$$

where $\Gamma_{l/r,j}$ is the tunnel rate through the left or right barrier of energy level j. $\alpha_{l/r}$ is the parameter describing the microwave field at the left or right barrier, C_g is the gate capacitance, C is the total dot capacitance, $T_{l/r}$ is the temperature of the left or right lead. $\eta_{l/r}$ is a parameter describing the asymmetry of the dc voltage drop across the two barriers. We assume that the tunnel rates, $\Gamma_{l/r,j}$, do depend on the level index j, but that they are independent of energy.

In the last two terms of Eq. (7), the number of electrons on the dot is fixed, while only the distribution of the electrons over the states changes. This includes effects from relaxation (i.e. intra-dot transitions, $\chi'' \to \chi$ and $\chi \to \chi'''$, where the total energy decreases) or excitation inside the dot (i.e. intra-dot transitions, $\chi'' \to \chi$ and $\chi \to \chi'''$, where the total energy increases). Below, we take excitation rates equal to zero (i.e. no intra-dot absorption) but, allow for non-zero relaxation rates.

An expression for the dc current can be found by calculating the net tunnel rate through one of the barriers. Using the probabilities $P_{N,\chi}$ and the tunnel rates through the left barrier, this leads to:

$$I = e\sum_{\chi}\sum_{j=empty} P_{N,\chi}\Gamma^{in}_{l,j} - e\sum_{\chi'}\sum_{j=full} P_{N+1,\chi'}\Gamma^{out}_{l,j} \quad (10)$$

In the numerical calculations in the next section we take equal ac amplitudes dropping across the left and right barriers; i.e. $\alpha_l = \alpha_r = \alpha$.

2.3. NUMERICAL RESULTS

Figure 2 shows calculations without relaxation between the states in the dot. The inset shows the case for transport through only a single level; $\Delta\varepsilon \gg hf$. Next to the main resonance, side-peaks develop at multiples of hf/e when the microwave power is increased via the parameter α. The broadening of the resonances is due to a finite temperature. In the main figure transport can also occur via excited states; here $\Delta\varepsilon < hf$. Not only side-peaks develop, but also peaks at other gate voltages. These peaks arise due to the interplay between the discrete single-particle states and the photon energy. Their locations are given by $(m\Delta\varepsilon + nhf)/e$ where $m = 0, \pm1, \pm2, \ldots$ and n is the photon number. Similar simulation results have been reported by Bruder and Schoeller [27].

Figure 3 shows an expansion for the curve with $\alpha = 1$. We have assigned the excited states and the particular PAT processes. The highest occupied single-particle level of the $N + 1$ ground state is denoted by ε_j with $j = 0$; positive

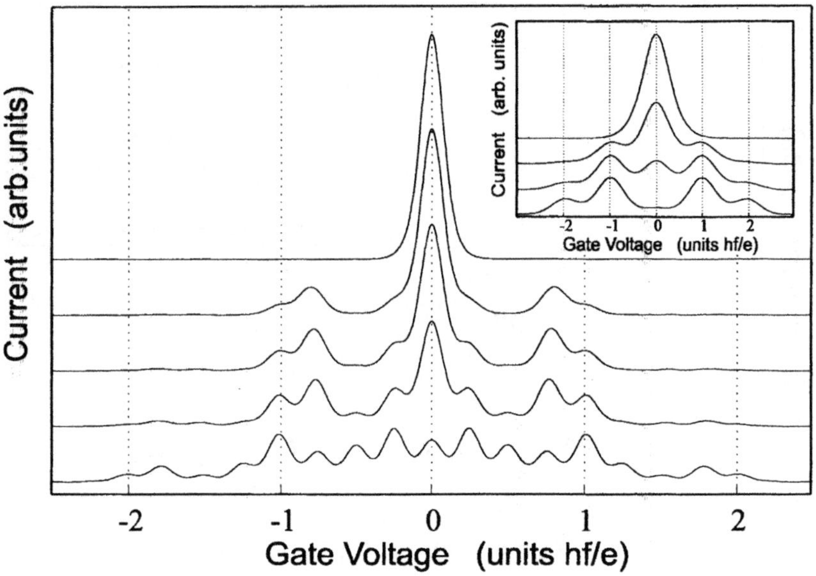

Figure 2. Calculation without relaxation. Curves are offset for clarity. The parameters for the data in the inset are $\Delta\varepsilon = 3hf$, $hf = 5k_BT$, and from top to bottom $\alpha = 0, 1, 1.5, 2$. The parameters for the main figure are $\Delta\varepsilon = 0.75hf$, $hf = 20k_BT$ and from top to bottom $\alpha = 0, 0.5, 0.75, 1, 1.5$.

j's are excited states above ε_0 and negative j's are below ε_0 (see Fig. 1b). The inset shows the effect of relaxation. Upon increasing the relaxation rate, the peaks corresponding to tunneling through excited states decrease, while peaks increase when tunneling occurs through the ground state.

To explain these numerical results, we show energy diagrams in Fig. 4, assuming that only the highest two single-particle levels contribute to the current for the transition between N and $N + 1$ electrons on the dot. For small dc bias voltage and no ac voltages a current resonance occurs when the topmost energy state (i.e. the electrochemical potential) of the quantum dot lines up with the Fermi levels of the leads (see the diagram ε_0). When high-frequency voltages drop across the two barriers, additional current peaks appear. We distinguish two mechanisms. The first mechanism gives photon induced current peaks when the *separation* between the ground state ε_0 and the Fermi levels of the leads *matches* the photon energy (or multiples, nhf), as depicted in the diagrams labeled by $\varepsilon_0 + hf$ and $\varepsilon_0 - hf$. The minus and plus signs correspond to being before or beyond the main resonance. Note that also the case of $\varepsilon_0 - hf$ involves photon absorption. Following the literature on the tunneling time, we call these current peaks: *sidebands* [17]. The second mechanism leads to photon peaks when an excited state is in resonance with the Fermi levels of the leads (see diagram ε_1). Without PAT, transport through

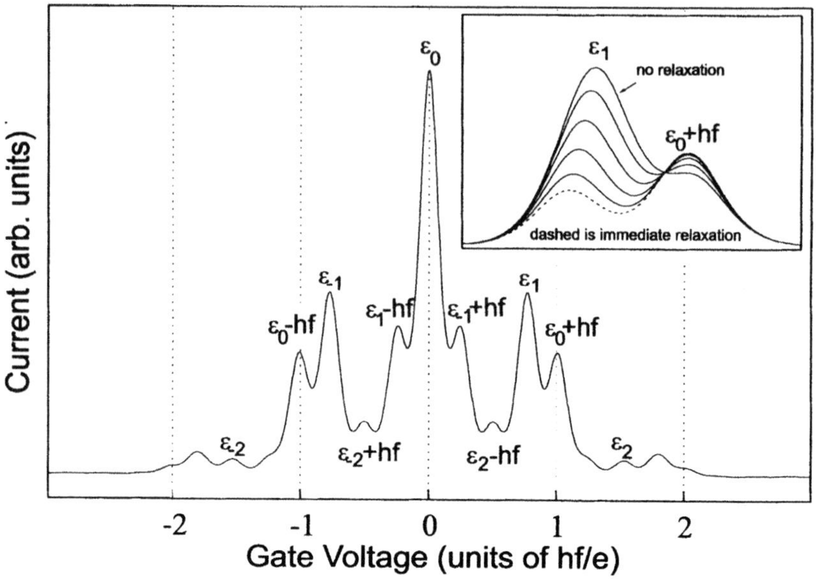

Figure 3. Expansion for the curve $\alpha = 1$ from Fig. 2. The inset shows the effect of an increasing relaxation rate. The relaxation rates divided by the tunnel rate are 0, 0.1, 0.35, 1, 3.5, and infinite.

the excited state, ε_1, is blocked since Coulomb blockade prevents having electrons in both the ground state and the excited state simultaneously. The electron in the ground state cannot escape from the dot, because its energy is lower than the Fermi levels in the leads. PAT, however, can empty the ground state ε_0 when the electron absorbs enough energy and leaves the dot. This process is analogous to photo-ionization. Now, the $(N+1)^{\text{th}}$ electron can tunnel resonantly via the excited state ε_1 as long as the state ε_0 stays empty. Note that for this second mechanism nhf has to *exceed*, but not necessarily *match* the energy splitting $\Delta\varepsilon = \varepsilon_1 - \varepsilon_0$. It is clear from these diagrams that relaxation from ε_1 to ε_0 decreases the height of this resonant peak. More photon peaks are generated when these two mechanisms are combined as in the diagrams labeled by $\varepsilon_1 + hf$ and $\varepsilon_1 - hf$. We thus see that PAT can populate the excited states by tunneling between dot and leads. So, even without intra-dot transitions, we can perform photon spectroscopy on discrete quantum dot states.

2.4. PHOTON-ASSISTED PUMPING

It is important to note that in the diagrams of Fig. 4 only processes with tunneling from or to states in the leads close to the Fermi levels contribute to the net current. Tunnel processes that start with an electron in one of the leads from further below the Fermi level are cancelled by an electron from the other lead. However, this

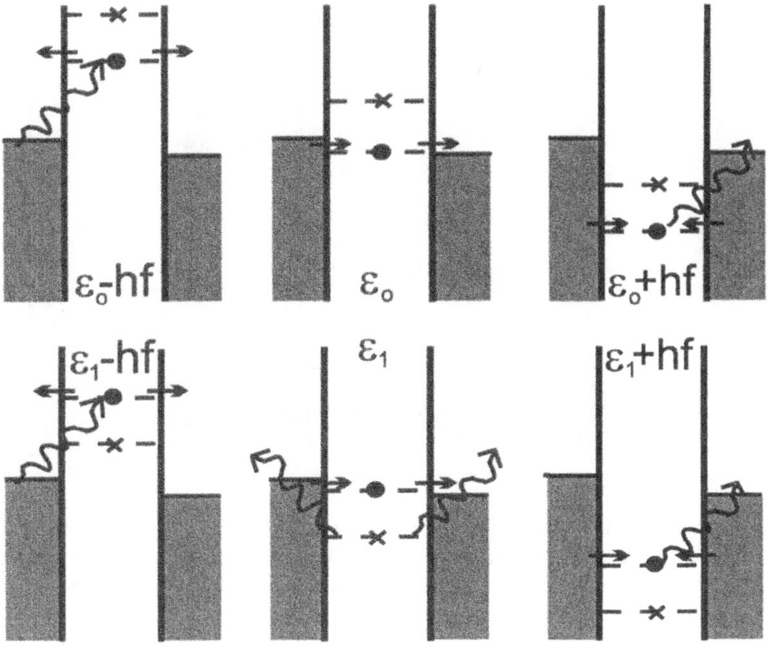

Figure 4. Diagrams depicting the tunneling events which dominantly contribute to the current through a quantum dot at different gate voltages. A small dc bias raises the left Fermi level with respect to the right Fermi level. ε_0 and ε_1 denote the ground state and the first excited state of the $(N+1)$ electron system. When the $(N+1)^{th}$ electron tunnels to one of the two reservoirs, the energy states of the dot drop by the charging energy E_c. The corresponding diagrams for N electrons are not shown.

is only true when the ac voltage drop is the same for both barriers. When the ac voltage drops across the two barriers are unequal, the dot acts as an electron pump [3, 12, 27]. The resulting pumped current makes the resonances discussed above less clear. For this reason we discuss this pumping mechanism in more detail here before proceeding further. Figure 5 shows a calculation of the pumped current as a function of the gate voltage that occurs when the ac voltage drop over one barrier is 5% smaller than over the other barrier. We have taken zero dc bias voltage. To illustrate the origin of the pumped current, the insets show the extreme case, when all the ac voltage drop is across the left barrier. In this case photon absorption occurs only at the left barrier. At negative gate voltage, when the ground-state level of the dot is *above* the Fermi level of the leads, an electron can *enter* the dot from the left lead only (bottom left inset to Fig. 5). Once the electron is in the dot, it can tunnel out through both tunnel barriers. Only tunneling to the right lead contributes to the net current. Therefore, the net (particle) current is to the right. When the ground-state level of the dot is *below* the Fermi level of the leads,

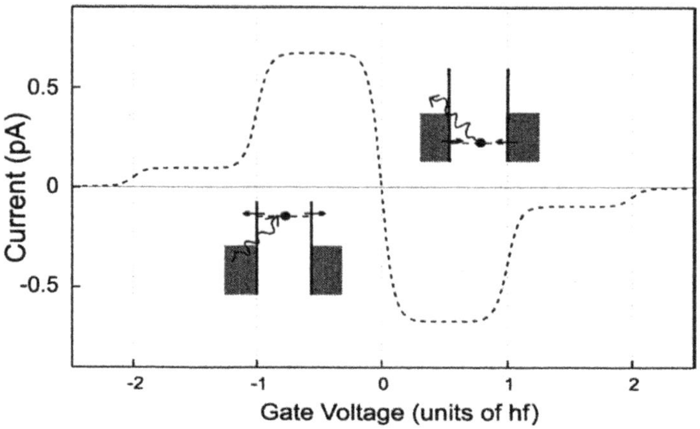

Figure 5. Calculation of the current for zero-bias voltage ($V = 0$) as a function of gate voltage in the case where the ac voltage drop over one barrier is 5% smaller than over the other barrier ($k_B T = 0.05hf, \Gamma = 510^8$ s^{-1}). The insets depict which tunneling events are responsible for the pumped current when the ground state of the quantum dot is below or above the Fermi levels of the leads.

however, an electron can only *leave* the dot to the left lead (upper right inset to Fig. 5). The dot can be filled from either lead once it is emptied. This time only the electron tunneling *in* from the right lead contributes to the net current. Therefore, there is a net current to the left. The difference between these two situations is the shift in the ground-state energy with respect to the Fermi levels of the leads. So, when the gate voltage is swept such that the ground state moves through these Fermi levels, the pumped current changes sign. The pumped current occurs over a width corresponding to the photon energy. The extra shoulders at the far left and far right are due to two-photon processes.

Asymmetric heating may induce a difference in the temperatures T_R and T_L of the two leads. This can also result in a finite transport current. The effects of asymmetric heating and asymmetric coupling of the microwave signal when a finite V_{SD} is applied across the sample, is illustrated in Fig. 6. The solid line is a reproduction of Fig. 3. The dashed-dotted line shows the same trace for $T_R/T_L =$ 0.95. The dashed line shows the influence of $\alpha_R/\alpha_L = 0.95$. It can be concluded that both effects can severely distort the data. At small V_{SD} the resonant peaks scale linearly with V_{SD}, while the pumped current does not change. Increasing the bias, while remaining in the linear regime, therefore improves the visibility of the resonant peaks.

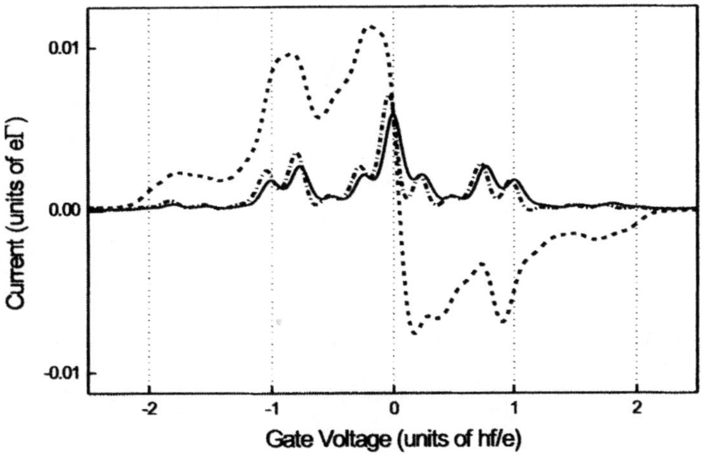

Figure 6. Simulation of the effects of both asymmetric heating and quantum rectification on the curve with $\alpha = 1$ from Fig. 3. The solid line represents the unperturbed curve. The dashed-dotted line shows the same curve for $T_R/T_L = 0.95$. The dashed line shows the influence of $\alpha_R/\alpha_L = 0.95$ on the unperturbed curve.

3. Experiments

3.1. SAMPLE GEOMETRY

Our measurements are performed on a quantum dot defined by metallic gates (see Fig. 7) in a GaAs/AlGaAs heterostructure containing a 2-dimensional electron gas (2DEG) 100 nm below the surface. The 2DEG has mobility 2.3 10^6 cm^2/Vs and electron density 1.9 10^{15} m^{-2} at 4.2 K. By applying negative voltages to the two outer pairs of gates, we form two quantum point contacts (QPCs). An additional pair of center gates between the QPCs confines the electron gas to a small dot. No electron transport is possible through the narrow channels between the center gates and the gates forming the QPCs. The center gate voltage, V_g, can shift the states in the dot with respect to the Fermi levels of the leads and thereby controls the number of electrons in the dot. The energy shift is given by $\Delta E = \kappa \Delta V_g$, with κ defined as the ratio between the dot-gate capacitance and the total capacitance of the dot [1]. A small dc voltage bias is applied between source and drain and the resulting dc source-drain current is measured. From standard dc measurements we find that the effective electron temperature is approximately $T = 200$ mK and the charging energy $E_c = 1.2 \pm 0.1$ meV. We independently determine the level splitting, $\Delta\varepsilon$, for different magnetic fields from current-voltage characteristics. In addition to the dc gate voltages, we couple a microwave signal (10-75 GHz) capacitively into one of the center gates. The microwave does not equally couple to the dot as to the leads, which results in an ac voltage drop over both barriers.

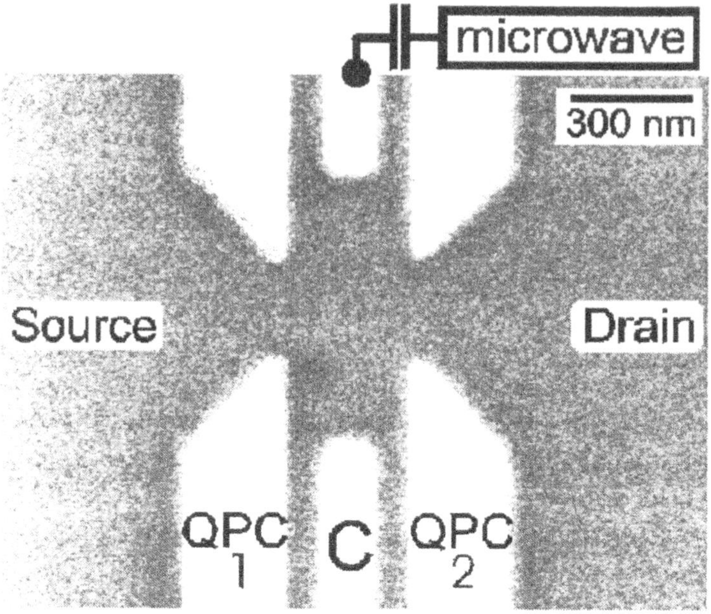

Figure 7. SEM photo of the sample. The lithographic size of the dot is (600×300) nm^2. Current can flow when we apply a voltage between source and drain. The microwave signal is capacitively coupled to one of the center gates.

3.2. FREQUENCY SENSITIVITY OF PUMPING

We first present experimental results with a strongly pumped current taken at B = 1.96 T for three frequencies around 47.4 GHz (the arrow denotes hf). The dashed line in Fig. 8 is the current without microwaves. For the lowest frequency the current is pumped in one direction, whereas for the highest frequency it is pumped in the opposite direction. At 47.33 GHz the left barrier apparently has the *smaller* ac voltage drop, while at 47.43 GHz the left barrier has the *larger* ac voltage drop. This illustrates that the asymmetry of the voltage drops over the two barriers sensitively depends on frequency. This sensitivity is ascribed to standing waves in the sample holder. The dotted line shows the current measured at an intermediate frequency, where we expect the ac voltage drop to be equal over both barriers. In contrast to the two solid curves, the dotted line is lower than the dashed line without microwaves over the whole gate voltage range. This cannot be explained by the pumping mechanism in our model. Our model only includes the oscillation of the potential of the leads relative to the dot, which always results in a pumped current which changes sign at the resonance. A negative pumped current over the whole gate voltage range, is attributed to the effect of the microwaves on

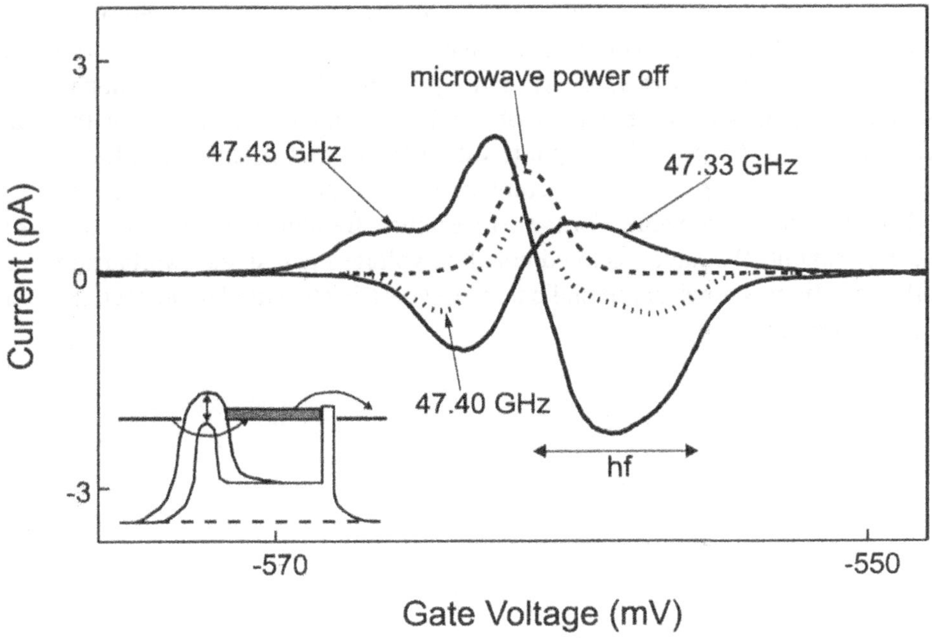

Figure 8. Measurements of the pumped current at B = 1.96 T, V_{SD} = 13 μV and frequencies are around 47.4 GHz. Dashed line is without microwaves. The dotted line shows the smallest asymmetry, but shows evidence for a pumping mechanism which is not included in our model.

the barrier height. The inset shows how a quantum dot can act as a pump when one tunnel barrier is periodically modulated in height. During one part of the cycle, when the left barrier is low, electrons enter the dot ($\Gamma_L^{low} > \Gamma_R$) while they escape the dot through the right barrier in the second half of the cycle when the left barrier is high ($\Gamma_L^{high} < \Gamma_R$). This is essentially a classical mechanism that has been verified experimentally in the MHz regime [3]. For observing clearly separated PAT sidebands, we first minimize pumping. For this we measure traces of current at zero bias voltage across a single Coulomb peak for slightly different frequencies. We finally choose those frequency values for which the pumped current is very small.

3.3. PAT: LOW FREQUENCY REGIME

First, we study the photon sidebands of the ground state at $B = 0.84$ T [34]. The main part of Fig. 9 shows measured curves of the current as a function of the gate voltage at different microwave powers for the case $hf < \Delta\varepsilon$. Here, current flows primarily via the ground state and its photon sidebands (i.e. upper diagrams in Fig. 4). On increasing the microwave power, we see that the height of the main resonance decreases to zero while additional resonances develop with increasing amplitude. When we convert gate voltage to energy, we find that the additional resonances are located at $\varepsilon_0 \pm hf$ and $\varepsilon_0 \pm 2hf$ [35]. The power dependence is in agreement with the behavior of the Bessel functions: $J_0^2(\alpha)$ for the main resonance ε_0, $J_1^2(\alpha)$ for the one-photon sidebands $\varepsilon_0 \pm hf$, and $J_2^2(\alpha)$ for the two-photon sidebands $\varepsilon_0 \pm 2hf$. For comparison, we show a calculation in the inset to Fig. 9 for the same values for the temperature, frequency and bias voltage as in the experiment. We have assumed equal ac voltages across the two barriers. The difference between measured and calculated data is attributed to an asymmetry in the ac coupling.

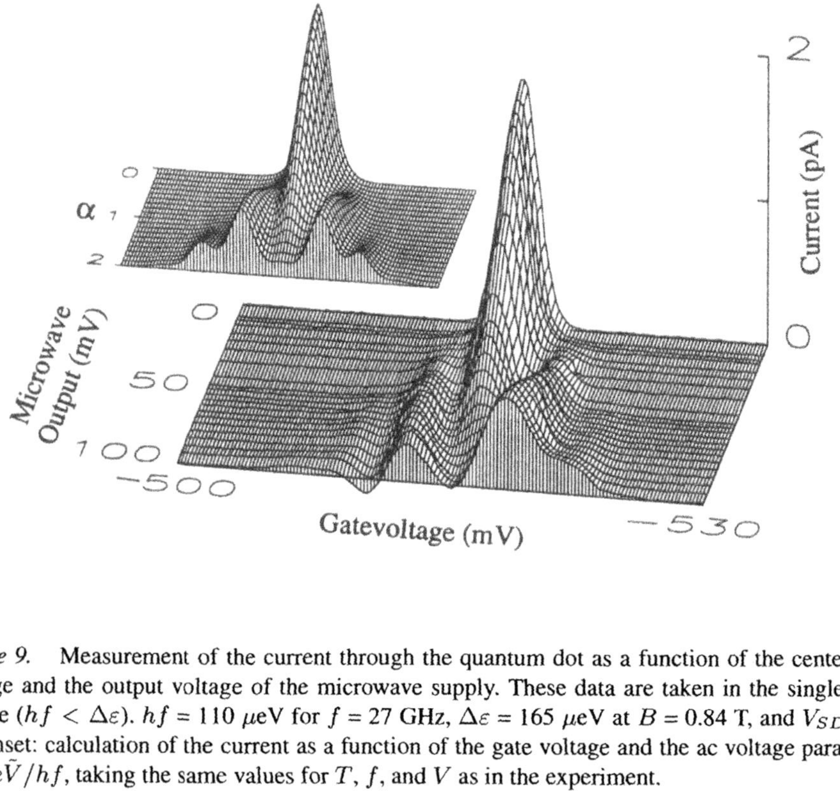

Figure 9. Measurement of the current through the quantum dot as a function of the center gate voltage and the output voltage of the microwave supply. These data are taken in the single-level regime ($hf < \Delta\varepsilon$). hf = 110 μeV for f = 27 GHz, $\Delta\varepsilon$ = 165 μeV at B = 0.84 T, and V_{SD} = 13 μV. Inset: calculation of the current as a function of the gate voltage and the ac voltage parameter $\alpha = e\tilde{V}/hf$, taking the same values for T, f, and V as in the experiment.

3.4. PAT: HIGH FREQUENCY REGIME

We now discuss the higher frequency regime where $hf > \Delta\varepsilon$, such that PAT can induce current through excited states. Figure 10 shows the current at B = 0.91 T (here $\Delta\varepsilon$ = 130 μeV). In the top section f = 61.5 GHz (hf = 250 μeV) and in the bottom section f = 42 GHz (hf = 170 μeV). As we increase the power, we see extra peaks coming up. We label the peaks as in Fig. 4. On the right side

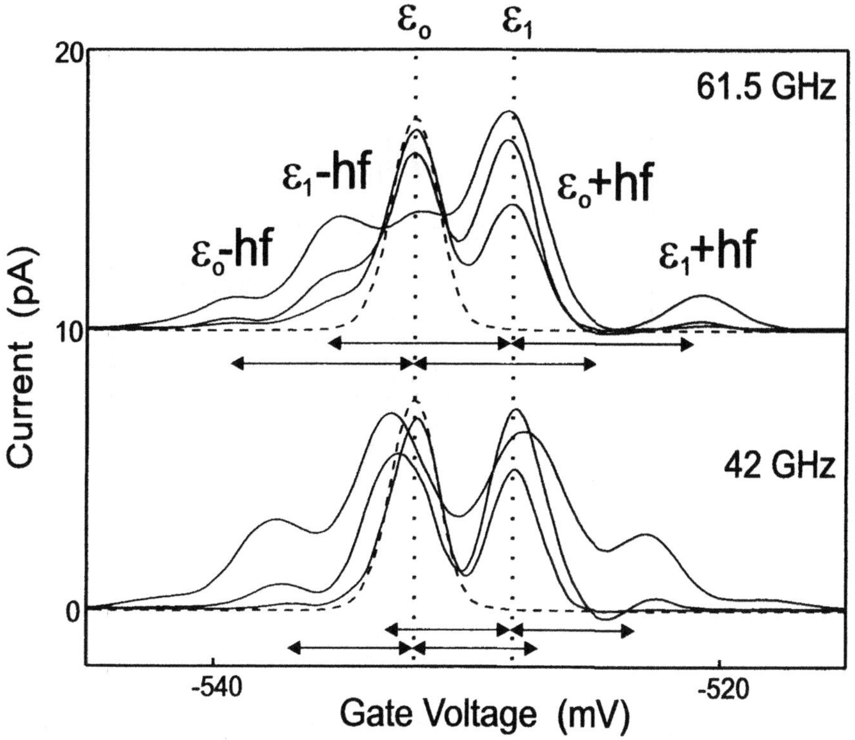

Figure 10. Measured current as a function of center gate voltage for different microwave powers. The dashed curve is without microwaves. $B = 0.91$ T, $V_{SD} = 13$ μV. $f = 61.5$ GHz in the top section, $f = 42$ GHz in the bottom section. As the frequency is reduced between top and bottom sections, the ground-state resonance ε_0 and the resonance attributed to the excited state ε_1 remain at the same gate voltage position. The other peaks, $\varepsilon_0 - hf$ and $\varepsilon_1 \pm hf$, shift inward by an amount which corresponds to the change in photon energy as indicated by the arrows. We do not observe $\varepsilon_0 + hf$ in this measurement.

of the main resonance a new peak appears, which we assign to photo-ionization, followed by tunneling through the first excited state. At higher powers the one-photon sidebands of the main resonance as well as those of the excited state resonance appear. We do not observe the peak for $\varepsilon_0 + hf$, in this measurement. This can be explained, at least in part, by the fact that here an electron can also tunnel into ε_1, which blocks the photon current through $\varepsilon_0 + hf$. Simulations confirm that the peak for $\varepsilon_0 + hf$ can be several times weaker than the peak for $\varepsilon_0 - hf$ [32]. Also, it is masked by the high peak for ε_1 right next to it. The arrows underneath the curves mark the photon energy. The peaks ε_0 and ε_1 remain in place when we change the frequency, since the photon energy evidently does not alter the energy splitting. The other peaks, $\varepsilon_0 - hf$ and $\varepsilon_1 \pm hf$, shift by

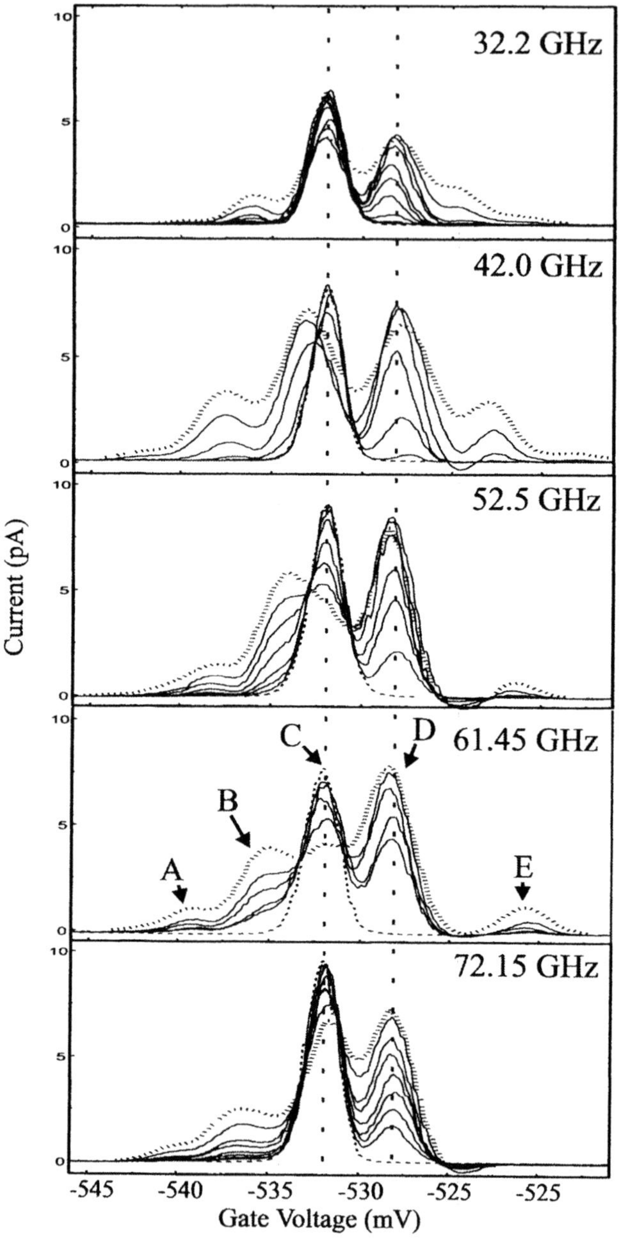

Figure 11. $I_{DC} - V_g$ curves for increasing frequency; $B = 0.91$ T, $V_{SD} = 13\ \mu$V and powers range between -30 and 0 dBm. The dashed curve is without microwave power. The dotted curve is with maximum power. Five different peaks can be distinguished in the data. These peaks, labelled A through E for the 61.45 GHz traces, correspond to the $\varepsilon_0 - hf$, $\varepsilon_1 - hf$, ε_0, ε_1 and $\varepsilon_1 + hf$ peaks, respectively.

an amount that corresponds to the change in photon energy as indicated by the arrows. This reflects that the sidebands originate from matching the states ε_0 and ε_1 to the Fermi levels of the leads by a photon energy hf. Figure 11 shows a large data set. In each panel, different traces are taken at different microwave power. The panels differ in frequency. We further substantiate the peak assignment below by studying detailed frequency, magnetic field and power dependence.

3.5. FREQUENCY DEPENDENCE

Figure 12 shows the spacing between a resonance and its photon sidebands as a function of the photon energy. Different markers correspond to different photon sidebands. The factor $\kappa = 35\ \mu eV/mV$, to convert the peak spacings in mV gate voltage into energy, is determined from dc measurements. The full width at half maximum (FWHM) of the resonance without microwaves, indicated by the arrow, is proportional to the effective electron temperature in the leads. Structure due to photon energies below this value is washed out by the thermal energy k_BT. The frequency scaling firmly establishes PAT as the transport mechanism [9,11-14]. The observation that the sidebands move linearly with frequency, while the ground and excited state resonances stay fixed, supports our identification of the different peaks.

3.6. MAGNETIC FIELD DEPENDENCE

We now use a magnetic field to change the energy separation between the ground state and the first excited state [1, 36], while keeping the distance to the sidebands fixed. Figure 13a shows the positions in gate voltage of all observed peaks for 52.5 GHz as a function of magnetic field. The filled circles reflect the evolution of ε_0 with magnetic field. This ground state weakly oscillates with a periodicity of ~80 mT which roughly corresponds to the addition of an extra flux quantum to the dot. The filled diamonds reflect the evolution of ε_1. The open circles (diamonds) show the sidebands $\varepsilon_0 \pm hf$ ($\varepsilon_1 \pm hf$). Figure 13b shows the magnetic field evolution of the excited state and the photon sideband peaks relative to the ground state (i.e. we have subtracted $\varepsilon_0(B)$ from the other curves). We see that the energy splitting decreases on increasing the magnetic field and for 0.54 T $< B <$ 0.58 T a degeneracy of the ground state is temporarily lifted and actually two excited states are observed [37]. The dashed lines denote the photon energy hf = 217 μeV for 52.5 GHz. The open circles close to these lines are the photon processes $\varepsilon_0 \pm hf$, demonstrating that they indeed move together with the ground state. The open diamonds are the $\varepsilon_1 - hf$ and $\varepsilon_1 - 2hf$ processes. Their motion follows the motion of ε_1. We have thus shown that we can vary the states ε_0 and ε_1 with the magnetic field and, independently, vary the separation to the sidebands with the microwave frequency.

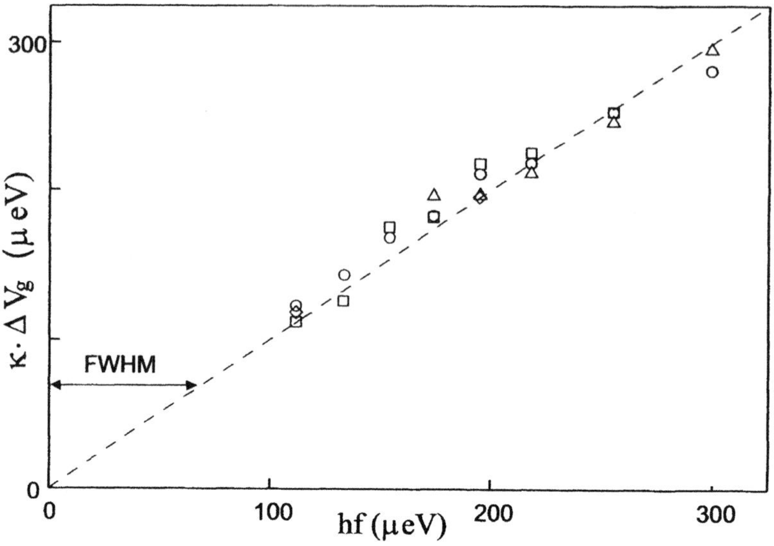

Figure 12. Peak spacings versus photon energy. □: spacing between ε_0 and $\varepsilon_0 - hf$. ◇: spacing between ε_0 and $\varepsilon_0 + hf$. △: spacing between ε_1 and $\varepsilon_1 - hf$. o: spacing between ε_1 and $\varepsilon_1 + hf$. The dashed line is based on the gate voltage to energy conversion factor κ determined independently from dc measurements, and has the theoretically expected slope equal to 1. The arrow indicates the FWHM of the main resonance.

3.7. POWER DEPENDENCE

Figure 14a shows a calculation of the peak heights as a function of the ac voltage drop across the barriers $\alpha = \frac{e\tilde{V}}{hf}$. Temperature, source-drain voltage, and frequency are taken from the experiment described below: $T = 200$ mK, $V_{SD} = 13$ μV, and $f = 52.5$ GHz. The tunnel rates from the leads to the ground state and the excited state are set to $\Gamma_{\varepsilon_0} = 5\ 10^8\ \mathrm{s}^{-1}$ and $\Gamma_{\varepsilon_1} = 14\ 10^8\ \mathrm{s}^{-1}$, respectively. The relaxation rate from the excited state to the ground state is assumed to be zero in the calculation. The effect of a finite relaxation rate is to reduce the height of ε_1 with respect to the other peaks. The calculated peak heights roughly follow the Bessel functions in Eq. (9). The ground-state resonance ε_0 follows $J_0^2(\alpha)$, since it involves only elastic tunnel events (see Fig. 4, diagram ε_0). The photon sidebands follow $J_1^2(\alpha)$, since they solely depend on the probability of photon absorption. For example, the process $\varepsilon_0 - hf$ is due to a photon assisted tunnel event which fills the dot. Once the dot is filled, however, it does not matter whether the dot is emptied via an elastic or an inelastic event. The process ε_1 follows the product of the Bessel functions $J_0^2(\alpha)J_1^2(\alpha)$ since it requires that the ground state is emptied via a PAT process, but also that the following tunneling processes through the excited state ε_1 are elastic.

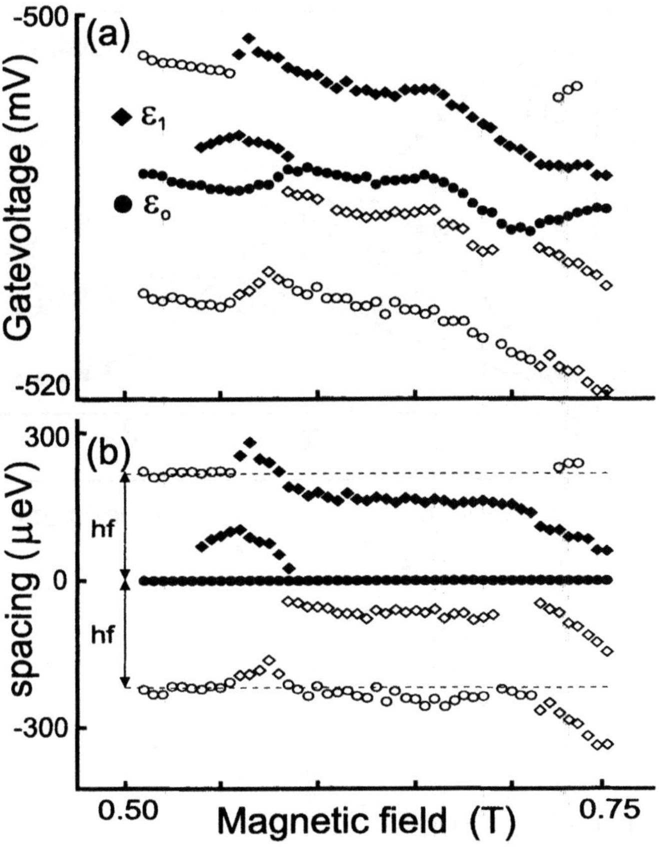

Figure 13. (a) Peak positions in gate voltage versus magnetic field at 52.5 GHz. Solid symbols denote peaks which are independent of frequency. Open symbols denote peaks that scale with frequency. (b) Peak spacings relative to the main resonance converted to energy. Closed circles: ε_0; open circles: $\varepsilon_0 \pm hf$; closed diamonds: ε_1; open diamonds: $\varepsilon_1 - hf$ and $\varepsilon_1 - 2hf$.

Figure 14b shows the experimental results for the peak heights at $B = 0.91$ T and $f = 52.5$ GHz as a function of the ac voltage amplitude at the output of the source. The measurements are in good qualitative agreement below an ac source voltage of 100 mV. At higher ac voltages the pumped current starts to become important. The values for the tunnel rates to ε_0 and to ε_1 derived from the dc current-voltage characteristic are $\Gamma_{\varepsilon_0} = 5\ 10^8\ \mathrm{s}^{-1}$ and $\Gamma_{\varepsilon_1} = 6\ 10^8\ \mathrm{s}^{-1}$. The value for Γ_{ε_1} in the calculation is larger than the experimentally determined value, but still the calculated value for the height of the ε_1 resonance is smaller than the experimental value. It is a general trend in most of our data that the peak ε_1 is higher than predicted by our model and that $\varepsilon_0 + hf$ is lower than expected from simulations. A possible explanation has been put forward in Ref. [38].

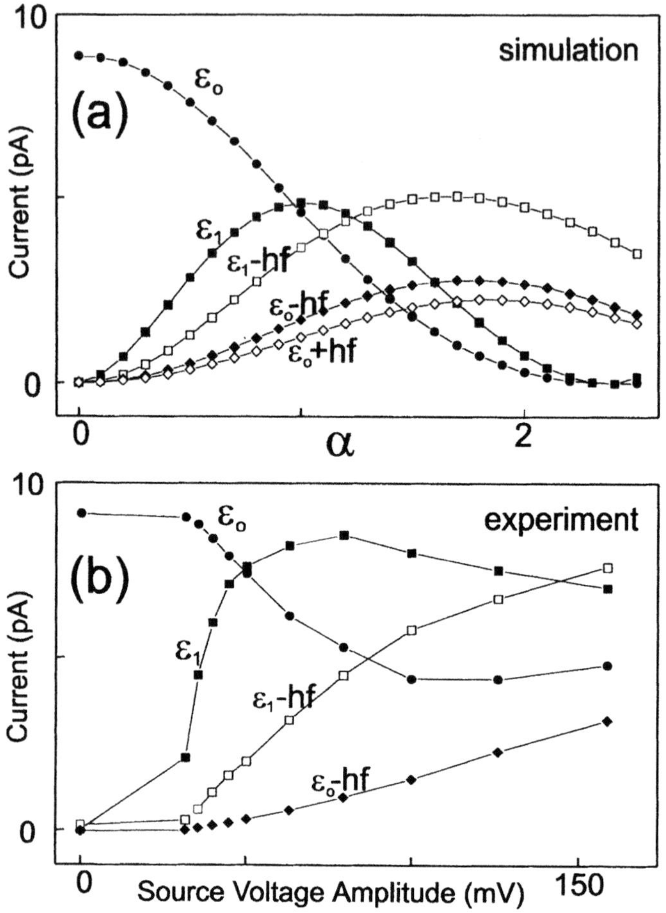

Figure 14. (a) Calculation of the peak heights as a function of the ac voltage drop across the barriers $\alpha = \frac{e\widetilde{V}}{hf}$ (T = 200 mK, V = 13 μV, f = 52.5 GHz). The tunnel rates from the leads to the ground state and the excited state are set to $\Gamma_{\varepsilon_0} = 5\ 10^8\ \mathrm{s}^{-1}$ and $\Gamma_{\varepsilon_1} = 14\ 10^8\ \mathrm{s}^{-1}$, respectively. The relaxation rate from the excited state to the ground state is assumed to be zero in the calculation. (b) Experimentally obtained peak heights as a function of the ac voltage amplitude (measured at the microwave source) for V = 13 μV and f = 52.5 GHz.

3.8. DISCUSSION AND CONCLUSIONS

The simulations described earlier, show that the pumped current is quite independent of the bias voltage when $eV_{SD} \ll hf$, while current due to the photon resonances increases linearly with the bias voltage when $eV_{SD} < k_BT$. Therefore, it is possible to improve the quality of our data by separating the pumped current from the photon resonances. This could be done by repeating a measurement at a particular microwave power for different bias voltages. In addition, the effective electron temperature of 200 mK reported here has now been improved to 50 mK. These two improvements would allow for a better comparison of future experiments with calculations over a wider range of microwave powers.

The conclusion of this work is that photon assisted tunneling is clearly observed in single-electron transport through small quantum dots. In addition, microwave irradiation can be used to perform spectroscopy on the discrete level spectrum. The parameter dependence of PAT is in reasonable agreement with calculations based on a master equation. Both, linear-frequency dependence and Bessel function power dependence are clearly observed. Recently, a non-equilibrium Green's function method applied to our PAT studies, has provided very good agreement with our results, including an explanation for the absence of the sideband at $\varepsilon_0 + hf$ [39]. Brune *et al.* [38] have analyzed the influence of intra-dot transitions. Including intra-dot transitions, they find good agreement with the height of the measured ε_1 peak in Figs. 10 and 11.

Qualitatively similar results have been obtained on superlattice structures irradiated by a free-electron laser. The dot and superlattice experiments have stimulated new theoretical interest in ac transport through non-superconducting structures. For instance, new results have been obtained for PAT across a single tunnel barrier for different sorts of oscillating potentials [40]; for PAT through superlattices in a self-consistent treatment [41], for the ac effects on transport through a QPC and double barrier structures in a numerical study [42], for the ac effects on correlated transport through Luttinger liquids [43], and for the Kondo effect in irradiated quantum dots [44].

These models and theories are very useful in explaining and predicting new transport mechanisms. We feel, however, that a word of caution is appropriate here. Although our PAT experiments give clean results, we have very little control over the oscillating potentials inside the sample. We simply couple in a microwave signal to one of the gates and measure the effect in the dc current. The dc response is very sensitive to the applied frequency. Sometimes the microwaves couple in more across the left barrier and sometimes more across the right barrier. The sensitivity of the asymmetry in coupling might be due to a complicated electric field pattern around the metallic gate structure. We think that to some degree our gate structure acts as a co-planar waveguide where the two QPC barrier gates serve as ground planes for the central microwave gate. Even under such conditions, it is not clear how this oscillating gate potential is carried over to the electron gas 100 nm below the surface. Pedersen and Büttiker [45] have recently analyzed this problem for the case of quantum dots. They stress that oscillating potentials not only wiggle energy levels but also generate alternating currents in the sample. The alternating currents affect the self-consistent potential of the tunnel barriers and the potential on the dot. In effect, all the voltages and capacitances become renormalized and depend on the occupation of the dot. Sideband positions and heights are not expected to precisely follow a linear frequency dependence nor a Bessel function behavior. An experimental, quantitative study of the parameter dependence of the sidebands might give valuable information about these screening effects. Even more interesting would be to study the frequency dependence of the current, including measurements of higher harmonics at multiples of f. An analysis for such a set-up in terms of a frequency dependent environment is given in Ref. [46].

PAT is intrinsically a coherent phenomenon. The PAT measurements described above, however, are insensitive to the phase of the transmitted electrons. Coherence in the presence of a time-dependent field is therefore not directly demonstrated. Jauho and Wingreen [47] have proposed a PAT measurement through a quantum dot situated in one of two branches of an Aharonov-Bohm ring. They find that coherent absorption and reemission of photons can be detected via a phase measurement at the sidebands. The proposed mesoscopic double-slit geometry has been successfully used before to demonstrate coherent transmission through a quantum dot for the time-*independent* case [48, 49].

By connecting two quantum dots in series, a double quantum dot system can be obtained. An overview of charging effects in double quantum dots is given in Ref. [1]. Theoretical studies on PAT in a double quantum dot are done by Stoof and Nazarov [50], Stafford and Wingreen [51] and Brune *et al.* [52]. PAT studies of a double quantum dot enable the characterization of the coupling between two discrete energy levels. Depending on the strength of the inter-dot coupling, the two dots can form 'ionic' or 'covalent' bonds. By varying the inter-dot coupling, Oosterkamp *et al.* [53] experimentally demonstrated the transition from ionic bonding to covalent bonding in a quantum dot 'artificial molecule' that is probed by microwave excitations. In the same frequency regime experiments have been performed by Blick *et al.* [54] and by Fujisawa and Tarucha [15].

The study of photon assisted tunneling in quantum dots described here, forms a valuable extension of the understanding of the dc properties. Microwave measurements have been proved to be a useful spectroscopy tool for quantum dot systems. Microwaves are also expected to play a role of importance in future experiments on quantum dots and the possible application of quantum dots as solid state quantum bits. An example would be a time-resolved measurement of Rabi oscillations in a double quantum dot. In this case, the microwave signal is used to tune the Rabi oscillation frequency [50, 51].

We thank R. Aguado, S.F. Godijn, A.E.A. Koolen, J.E. Mooij, Yu.V. Nazarov, R.M. Schouten, T. Fujisawa, S. Tarucha, T.H. Stoof, P. McEuen and N.C. van der Vaart for experimental help and useful discussions. This work was supported by the Dutch Organization for Research on Matter (FOM), by the EU via the TMR network (ERBFMRX CT98-0180), and by the NEDO joint research program (NTDP-98).

References

1. L.P. Kouwenhoven, C.M. Marcus, P.L. McEuen, S. Tarucha, R.M. Westervelt and N.S. Wingreen, in *Mesoscopic Electron Transport*, edited by L.L. Sohn, L.P. Kouwenhoven and G. Schön (Kluwer Dordrecht, Netherlands 1997), series E, vol. **345**, pp. 105-214; also available at http://vortex.tn.tudelft.nl/~leok/papers.
2. R. Ashoori, Nature **379**, 413-419 (1996).
3. L.P. Kouwenhoven, A.T. Johnson, N.C. van der Vaart, C.J.P.M. Harmans and C.T. Foxon, Phys. Rev. Lett. 67, 1626 (1991); and Zeitschrift für Physik **85**, 381 (1991).
4. U. Merkt, Physica B **189**, 165 (1993) and references therein.
5. R. Strenz, U. Bockelmann, F. Hirler, G. Abstreiter, G. Böhm and G. Weimann, Phys. Rev. Lett. **73**, 3022 (1994) and references therein.
6. K. Brunner, G. Abstreiter, G. Böhm, G. Tränkle and G. Weimann, Phys. Rev. Lett. **73**, 1138 (1994).
7. L. Jacak, P. Hawrylak and A. Wojs, *Quantum Dots* (Springer-Verlag Berlin Heidelberg 1998).
8. J.R. Tucker and M.J. Feldman, Rev. Mod. Phys. **57**, 1055 (1985).
9. P.K. Tien and J.R. Gordon, Phys. Rev. **129,** 647 (1963).
10. High frequency studies on quantum wells an quantum point contacts have not shown discrete photon features. See for microwave studies on quantum wells: V.A. Chitta, C. Kutter, R.E.M. de Bekker, J.C. Maan, S.J. Hawksworth, J.M. Chamberlain, M.Henini and G. Hill, J.Phys.: Condens. Matter **6**, 3945 (1994), and references therein. See for microwave studies on quantum point contacts: R.A. Wyss, C.C. eugster, J.A. del Alamo and Q. Hu, Appl. Phys. Lett. **63**, 1522 (1993); T.J.B.M. Janssen, J.C. Maan, J. Singleton, N.K. Patel, M. Pepper, J.E.F. Frost, D.A. Ritchie and G.A.C. Jones, J.Phys.: Condens. Matter **6**, L163 (1994).
11. P.S.S. Guimarães, B.J. Keay, J.P. Kaminski, S.J. Allen, Jr., P.F. Hopkins, A.C. Gossard, L.T. Florez and J.P. Harbison, Phys. Rev. Lett. **70**, 3792 (1993); B.J. Keay, S.J. Allen, Jr., J. Galán, J.P. Kaminski, K.L. Campman, A.C. Gossard, U. Bhattacharya and M.J.W. Rodwell, Phys. Rev. Lett. **75**, 4098 (1995); B.J. Keay, S. Zeuner, S.J. Allen, Jr., K.D. Maranowski, A.C. Gossard, U. Bhattacharya and M.J.W. Rodwell, Phys. Rev. Lett. **75**, 4102, (1995).
12. L.P. Kouwenhoven, S. Jauhar, K. McCormick, D. Dixon, P.L. McEuen, Yu. V. Nazarov, N.C. van der Vaart, and C.T.Foxon, Phys. Rev. B **50**, 2019 (1994).
13. L.P. Kouwenhoven, S. Jauhar, J. Orenstein, P.L. McEuen, Y. Nagamune, J. Motohisa and H. Sakaki, Phys. Rev. Lett. **73**, 3433 (1994).
14. R.H. Blick, R.J. Haug, D.W. van der Weide, K. von Klitzing, and K. Eberl, Appl. Phys. Lett. **67**, 3924 (1995).
15. T. Fujisawa and S. Tarucha, Superlattices Microstructures **21**, 247 (1997).
16. T.H. Oosterkamp, L.P. Kouwenhoven, A.E.A. Koolen, N.C. van der Vaart and C.J.P.M. Harmans, Phys. Rev. Lett. **78**, 1536-1539 (1997).
17. M. Büttiker and R. Landauer, Phys. Rev. Let. **49**, 1739 (1982).
18. M. Switkes, C.M. Marcus, K. Campman, and A.C. Gossard, to be published in Science, (March 1999).
19. Mixing is described by J. Iñarrea, G. Platero and C. Tejedor, Phys. Rev. B **50**, 4581 (1994).
20. D. Sokolovski, Phys. Rev. B **37**, 4201 (1988).
21. P. Johansson, Phys. Rev. B **41**, 9892 (1990).
22. *Single Charge Tunneling*, edited by H. Grabert and M. Devoret (Plenum Press London 1992), series B, vol. **294**.
23. D.V. Averin, A.N. Korotkov and K.K. Likharev, Phys. Rev. B **43**, 6199 (1991).
24. C.W.J. Beenakker, Phys. Rev. B **44**, 1646 (1991).
25. K.K. Likharev and I.A. Devyatov, Physica B **194-196** , 1341 (1994).
26. A. Hadicke and W. Krech, Physica B **193**, 256 (1994)

27. C. Bruder and H. Schoeller, Phys. Rev. Lett. **72**, 1076 (1994).
28. For a review see: G.L. Ingold and Yu. V. Nazarov, p. 21 in Ref. [22].
29. G.Y. Hu and R.F. O'Connell, Phys. Rev. B **49**, 16505 (1994).
30. J.M. Hergenrother, M.T. Tuominen, J.G. Lu, D.C. Ralph and M. Tinkham, Physica B **203**, 327 (1994).
31. T. Fujisawa, T.H. Oosterkamp, W.G. van der Wiel, B.W. Broer, R. Aguado, S. Tarucha and L.P. Kouwenhoven, Science **282**, 932-935(1998).
32. T.H. Oosterkamp, L.P. Kouwenhoven, A.E.A. Koolen, N.C. van der Vaart and C.J.P.M. Harmans, Semiconductor Science and Technology **11**, 1512 (1996); T.H. Oosterkamp, L.P. Kouwenhoven, A.E.A. Koolen, N.C. van der Vaart, and C.J.P.M. Harmans, Physica Scripta **T69,** 98 (1997).
33. J.-C. Wan, K.A. McGreer, L.I. Glazman, A.M. Goldman and R.I. Shekhter, Phys. Rev. B **43,** 9381 (1991).
34. PAT is also observed at $B = 0$ but with less resolution due to a higher effective electron temperature at $B = 0$.
35. Because $2hf > \Delta\varepsilon$ the excited state ε_1 is weakly visible at $V_g \sim -508$ mV in Fig. 9 where it overlaps with $\varepsilon_0 + 2hf$.
36. A.T. Johnson, L.P. Kouwenhoven, W. de Jong, N.C. van der Vaart, C.J.P.M. Harmans and C.T. Foxon, Phys. Rev. Lett. **69**, 1592 (1992).
37. Our measurement of the evolution of the excited state ε_1 agrees with the data in Fig. 4 of Ref. [36] and in Fig. 11 of P.L. Mc Euen *et al.*, Physica B **189**, 70 (1993). However, we do not know of a theory that properly describes the observed magnetic field dependence of excited states.
38. Ph. Brune, C. Bruder and H. Schoeller, Phys. Rev. B **56**, 4730 (1997).
39. Q. Sun, J. Wang and T. Lin, Phys. Rev. B **58**, 13007 (1998).
40. M. Wagner and W. Zwerger, Phys. Rev. B **55**, 10217 (1997), and references therein.
41. R. Aguado and G. Platero, Phys. Rev. Lett. **81**, 4971 (1998).
42. K. Yakubo, S. Feng and Q. Hu, Phys. Rev. B **54**, 7987 (1996).
43. G. Cuniberti, M. Sassetti and B. Kramer, Phys. Rev. B **57**, 1515 (1998).
44. R. López, R. Aguado, G. Platero and C. Tejedor, Phys. Rev. Lett. **81**, 4688 (1998), and references therein.
45. M.H. Pedersen and M. Büttiker, Phys. Rev. B **58** , 12993 (1998).
46. R. Aguado and L.P Kouwenhoven, Phys. Rev. Lett. **84**, 1986-1989 (2000).
47. A.P. Jauho and N.S. Wingreen, Phys. Rev. B **58,** 9619 (1998).
48. A. Yacoby, M. Heiblum, D. Mahalu and H. Strikhman, Phys. Rev. Lett. **74**, 4047 (1995).
49. R. Schuster, E. Buks, M. Heiblum, D. Mahalu, V. Umansky and H. Strikman, Nature **385**, 417 (1997).
50. T.H. Stoof and Yu.V. Nazarov, Phys. Rev. B **53**, 1050-1053 (1996).
51. C.A. Stafford and N.S. Wingreen, Phys. Rev. Lett. **76**, 1916-1919 (1996).
52. Ph. Brune, C. Bruder and H. Schoeller, Physica E **1**, 216-218 (1997).
53. T.H. Oosterkamp, T. Fujisawa, W.G. van der Wiel, K. Ishibashi, R.V. Hijman, S. Tarucha and L.P. Kouwenhoven, Nature **395**, 873 (1998).
54. R.H. Blick, R. J. Haug, J. Weis, D. Pfannkuche, K. v. Klitzing and K. Eberl, Phys. Rev. B **53**, 7899-7902 (1996); R.H. Blick, D.W. van der Weide, R.J. Haug and K. Eberl, Phys. Rev. Lett. 81, 689 (1998).

BOSONISATION AS THE HUBBARD–STRATONOVICH TRANSFORMATION

I. V. YURKEVICH
School of Physics & Astronomy, University of Birmingham, UK

1. Introduction

The problem of strongly correlated electrons in one dimension attracted attention of condensed matter physicists since early 50's. After the seminal paper of Tomonaga [1] who suggested the first soluble model in 1950, there were essential achievements reflected in papers by Luttinger [2] (1963) and Mattis and Lieb [3] (1963). A considerable contribution to the understanding of generic properties of the 1D electron liquid has been made by Dzyaloshinskii and Larkin [4] (1973) and Efetov and Larkin [5] (1976). Despite the fact that the main features of the 1D electron liquid were captured and described by the end of 70's, the investigators felt dissatisfied with the rigour of the theoretical description. The most famous example is the paper by Haldane [6] (1981) where the author developed the fundamentals of a modern bosonisation technique, known as the operator approach. This paper became famous because the author has rigourously shown how to construct the Fermi creation/anihilation operators out of the Bose ones. The most recent example of such a dissatisfaction is the review by von Delft and Schoeller [7] (1998) who revised the approach to the bosonisation and came up with what they called constructive bosonisation.

There exists another way to bosonise the problem. It was suggested 25 years ago by [8] and found its further elaboration in the paper by Lee and Chen [9] in 1988. This approach (sometimes called "functional bosonisation"), or more precisely its pieces, can be found in variety of papers. The main objective of this presentation is to assemble these pieces in a clear picture. The aim is purely methodological which is the reason why the number of complications (like backscattering, presence of lattice and spin degrees of freedom) are kept at the lowest possible level. Any generalisations can be easily incorporated into the scheme. The final formulation of the bosonised problem will look distinct from what can be found elsewhere. We believe that the approach present in this presentation is the

I. V. Lerner et al. (eds.),
Strongly Correlated Fermions and Bosons in Low-Dimensional Disordered Systems, 69–80.

"shortest shortcut" from the original fermionic problem of interacting electrons to the problem of non-interacting bosons.

2. Functional Integral Formulation

We start from a general Hamiltonian describing interacting electrons in one dimension:

$$\begin{aligned}\hat{\mathcal{H}} &= \frac{1}{2m}\int dx \partial_x \hat{\Psi}^\dagger(x)\partial_x\hat{\Psi}(x) \\ &+ \frac{1}{2}\int dx dx' \hat{\Psi}^\dagger(x)\hat{\Psi}^\dagger(x')\, V_0(x-x')\, \hat{\Psi}(x')\hat{\Psi}(x).\end{aligned} \tag{1}$$

Here $V_0(x-x')$ is the bare inter-electron interaction which is usually taken to be the Coulomb interaction screened by the carriers of the media surrounding the one-dimensional system.

The thermal Green function is the time-ordered product of the Fermion anihilation and creation operators

$$\mathcal{G}(x,\tau;x',\tau') = \left\langle T_\tau \hat{\Psi}(x,\tau)\hat{\bar{\Psi}}(x',\tau')\right\rangle \tag{2}$$

defined in the imaginary time Heisenberg representation

$$\hat{\Psi}(x,\tau) = e^{\hat{\mathcal{H}}'\tau}\hat{\Psi}(x)e^{-\hat{\mathcal{H}}'\tau}, \quad \hat{\bar{\Psi}}(x,\tau) = e^{\hat{\mathcal{H}}'\tau}\hat{\Psi}^\dagger(x)e^{-\hat{\mathcal{H}}'\tau}, \tag{3}$$

where $\hat{\mathcal{H}}' = \hat{\mathcal{H}} - \mu\hat{N}$. Here μ is the chemical potential and $\hat{N}$ is the particle number operator. This Green function can be written in the form of the coherent state functional integral over the Grassmann fields that reflects the fermionic nature of electrons:

$$\mathcal{G}(x,\tau;x',\tau') = \frac{\int \mathcal{D}\psi\mathcal{D}\psi^* \psi(x,\tau)\psi^*(x',\tau')\exp\{-\mathcal{S}[\psi,\psi^*]\}}{\int \mathcal{D}\psi\mathcal{D}\exp\{-\mathcal{S}[\psi,\psi^*]\}}, \tag{4}$$

where the action $\mathcal{S}[\psi,\psi^*]$ is defined by

$$\begin{aligned}\mathcal{S}[\psi,\psi^*] &= \int dx d\tau \psi^*(x,\tau)\left[\partial_\tau + \hat{\xi}\right]\psi(x,\tau) \\ &+ \frac{1}{2}\int dx dx' d\tau \psi^*(x,\tau)\psi^*(x',\tau)\, V_0(x-x')\, \psi(x',\tau)\psi(x,\tau).\end{aligned} \tag{5}$$

Here $\hat{\xi} = -(2m)^{-1}\partial_x^2 - \mu$ is the kinetic energy operator counted from the chemical potential μ, and $\psi(x,\tau)$ is the Grassmann function (or, equivalently, fermionic field) which is an anti-periodic function of the imaginary time τ, $\psi(\tau+\beta) = -\psi(\tau)$, with the inverse temperature $\beta = T^{-1}$ being the period.

Being aware that individual electrons are not good as a starting point for description of excitations in one dimension we choose to introduce an auxiliary field conjugated to the electron density. This can be achieved via the Hubbard-Stratonovich transformation which brings the fermionic problem to a Gaussian form at the price of complication caused by appearance of some new field and related to it extra integration. We start from the following representation of the exponential containing the interaction

$$\exp\left[-\mathcal{S}_{int}\right] = \frac{\int \mathcal{D}\phi \exp\left[-\frac{1}{2}\phi V_0^{-1}\phi + i\phi\psi^*\psi\right]}{\int \mathcal{D}\phi \exp\left[-\frac{1}{2}\phi V_0^{-1}\phi\right]}, \tag{6}$$

where we used symbolic expressions

$$\begin{aligned} \phi V_0^{-1}\phi &\equiv \int \mathrm{d}x\mathrm{d}x'\mathrm{d}\tau\phi(x,\tau)V_0^{-1}(x-x')\phi(x',\tau), \\ \phi\psi^*\psi &\equiv \int \mathrm{d}x\mathrm{d}\tau\phi(x,\tau)\psi^*(x,\tau)\psi(x,\tau) \end{aligned}$$

The notation $V_0^{-1}(x-x')$ stands for the inverse function understood as an integral kernel inverse to that corresponding to the bare interaction:

$$\int \mathrm{d}x_1 V_0^{-1}(x-x_1)V_0(x_1-x') = \delta(x-x'). \tag{7}$$

One can see that the field $\phi(x,\tau)$ should posess the same properties as a product of two Grassmann fields $\psi^*(x,\tau)\psi(x,\tau)$, i.e. it is a commuting field and periodic under the shift $\tau \to \tau+\beta$. Therefore, this field is of bosonic nature. Now we substitute the Hubbard-Stratonovich transform, Eq.(6), into the original representation Eq.(4) of the thermal fermionic Green function:

$$\mathcal{G}(x,\tau;x',\tau') = \frac{\int \mathcal{D}\psi\mathcal{D}\psi^*\mathcal{D}\phi\, \psi(x,\tau)\psi^*(x',\tau')\exp\left\{-\mathcal{S}[\psi,\psi^*;\phi]\right\}}{\int \mathcal{D}\psi\mathcal{D}\psi^*\mathcal{D}\phi \exp\left\{-\mathcal{S}[\psi,\psi^*;\phi]\right\}}. \tag{8}$$

Now the action is a functional of not only the fermionic field but also of the auxiliary field ϕ:

$$\begin{aligned} \mathcal{S}[\psi,\psi^*;\phi] &= \frac{1}{2}\int \mathrm{d}x\mathrm{d}x'\mathrm{d}\tau\phi(x,\tau)V_0^{-1}(x-x')\phi(x',\tau) \\ &+ \int \mathrm{d}x\mathrm{d}\tau\psi^*(x,\tau)\left[\partial_\tau + \hat{\xi} - i\phi(x,\tau)\right]\psi(x,\tau) \end{aligned} \tag{9}$$

Now we divide and multiply both numerator and denominator by

$$\int \mathcal{D}\psi\mathcal{D}\psi^* \exp\left\{-\int \mathrm{d}x\mathrm{d}\tau\psi^*\left[\partial_\tau + \hat{\xi} - i\phi\right]\psi\right\} = \mathrm{Det}\left[\partial_\tau + \hat{\xi} - i\phi\right] \tag{10}$$
$$= \exp\left\{\mathrm{Tr}\ln\left[\partial_\tau + \hat{\xi} - i\phi\right]\right\}.$$

Then one can see that the thermal Green function can be written as some auxiliary Green function $\tilde{\mathcal{G}}[\phi]$

$$\tilde{\mathcal{G}}(x,\tau;x',\tau';[\phi]) = \frac{\int \mathcal{D}\psi\mathcal{D}\psi^*\psi(x,\tau)\psi^*(x',\tau')\exp\left\{-\int \mathrm{d}x\mathrm{d}\tau\psi^*\left[\partial_\tau + \hat{\xi} - i\phi\right]\psi\right\}}{\int \mathcal{D}\psi\mathcal{D}\psi^*\exp\left\{-\int \mathrm{d}x\mathrm{d}\tau\psi^*\left[\partial_\tau + \hat{\xi} - i\phi\right]\psi\right\}} \tag{11}$$

averaged over the field ϕ with the new bosonic action $\mathcal{S}[\phi]$:

$$\mathcal{G}(x,\tau;x',\tau') = \mathcal{Z}^{-1}\int \mathcal{D}\phi\,\tilde{\mathcal{G}}(x,\tau;x',\tau';[\phi])\,e^{-\mathcal{S}[\phi]} \tag{12}$$

where

$$\mathcal{Z} = \int \mathcal{D}\phi\, e^{-\mathcal{S}[\phi]}, \qquad \mathcal{S}[\phi] = \frac{1}{2}\phi V_0^{-1}\phi - \mathrm{Tr}\ln\left[\partial_\tau + \hat{\xi} - i\phi\right]. \tag{13}$$

The auxiliary Green function is the one of the "free" electrons in the spatial and temporal dependent field $\phi(x,\tau)$ and it must be found from the equation

$$\left(\partial_\tau + \hat{\xi} - i\phi(x,\tau)\right)\tilde{\mathcal{G}}(x,\tau;x',\tau';[\phi]) = \delta(x-x')\,\delta(\tau-\tau'). \tag{14}$$

If we were able to solve this equation for arbitrary field $\phi(x,\tau)$ we would then have some expression for $\tilde{\mathcal{G}}$ which depends only on the bosonic field $\phi(x,\tau)$. Substituting this solution of Eq.(14) into Eq.(12) we would finally reduce our problem to another one where we must do some integral over the bosonic field governed by the action Eq.(13). And it should be pointed out that this procedure can be implemented for any observables constructed from any number of Fermi operators, not only for a bilinear combination defining the one-particle Green function Eq.(4). The only extra effort we would need is to use the Wick theorem in Eq.(11) which is valid because the action is quadratic in ψ-fields after the Hubbard-Stratonovich transformation is implemented.

It is evident enough, that this scheme is absolutely general and therefore absolutely useless. We should refer to some specifics of one-dimensional problems to develop it to the stage where it becomes a working tool. And that is what we will do next.

3. One Dimension

The main feature of one dimensional problems with Fermions is that their Fermi surface consists of just two points $-p_F$ and $+p_F$. Accordingly, electrons close to this Fermi surface can be divided into two groups, left- (with the momenta in the vicinity of $-p_F$) and right-movers (with the momenta in the vicinity of $-p_F$). That means that the Fermi fields over which we integrate in Eq.(4) might be represented in the form

$$\psi(x,\tau) = \psi_+(x,\tau)\, e^{ip_F x} + \psi_-(x,\tau)\, e^{-ip_F x}, \tag{15}$$

We are interested in large scale behaviour of our system which makes this distinction well defined because only small vicinity of the Fermi points should be taken into account. In other words, the Fermi fields $\psi_\pm(x,\tau)$ may be considered as smooth functions on the scale p_F^{-1}. It allows as when writing action in Eq.(4) to act by the kinetic energy operator $\hat{\xi}$ on the fast oscillating exponential mainly, neglecting second order derivatives of $\psi_\pm(x,\tau)$. Also, the integral over the whole space suppresses contribution from the "uncompensated" oscillatory functions proportional to $e^{\pm i2p_F x}$. Therefore, we may introduce two-vector

$$\Psi^T(x,\tau) = (\psi_+(x,\tau), \psi_-(x,\tau)) \tag{16}$$

and in these notations the action Eq.(13) becomes

$$\mathcal{S}[\phi] = \frac{1}{2}\phi\, V_0^{-1}\, \phi + \mathrm{Tr}\ln \begin{pmatrix} \partial_\tau - iv_F\partial_x - i\phi & 0 \\ 0 & \partial_\tau + iv_F\partial_x - i\phi \end{pmatrix}, \tag{17}$$

Writing down this action we neglected the backscattering of electrons due to Coulomb interaction. Generally speaking in the Coulomb field $\phi(x,\tau)$ there are two essential components related to two different scattering events. One of them contains the Fourier harmonics with $|q| \ll p_F$ and, therefore, is a smooth function of x, another one consists of the Fourier harmonics with $|q| \approx 2p_F$. The former is responsible for the forward scattering, i.e. when two colliding electrons keep their direction of motion which they had before collision. The latter causes backscattering when two colliding electrons make the U-turn after the collision. If we assume that for the Fourier transforms of the bare interaction we have the inequality $V_0(q \ll 2p_F) \ll V_0(q \approx 2p_F)$, we may neglect the backscattering which means that the Coulomb field $\phi(x,\tau)$ in the action Eq.(17) is a smooth function on the scale of the electron wavelength p_F^{-1}.

The Green function now becomes a 2×2 matrix and neglecting backscattering we find that it is a diagonal matrix with the diagonal elements $\tilde{\mathcal{G}}_+$ and $\tilde{\mathcal{G}}_-$ defined by the equations:

$$(\partial_\tau \mp iv_F\partial_x - i\phi(x,\tau))\, \tilde{\mathcal{G}}_\pm(x,\tau;x',\tau';[\phi]) = \delta(x-x')\,\delta(\tau-\tau'). \tag{18}$$

The formal solution of these equations can be easily written

$$\begin{aligned} \tilde{\mathcal{G}}_+(x,\tau;x',\tau';[\phi]) &= g_+(x-x',\tau-\tau')\, e^{i\theta(x,\tau)-i\theta(x',\tau')} \\ \tilde{\mathcal{G}}_-(x,\tau;x',\tau';[\phi]) &= g_-(x-x',\tau-\tau')\, e^{i\theta^*(x,\tau)-i\theta^*(x',\tau')}, \end{aligned} \tag{19}$$

where $\theta(x,\tau)$ should be determined from another equation

$$(\partial_\tau - iv_F\partial_x)\,\theta(x,\tau) = \phi(x,\tau) \tag{20}$$

and $g_\pm = (\partial_\tau \pm iv_F\partial_x)^{-1}$ are the Green functions of the chiral particles:

$$g_\pm(x,\tau) = \frac{T}{2v_F}\sin^{-1}\pi T\left(\tau \mp i\frac{x}{v_F}\right). \tag{21}$$

Now, when we have conveniently expressed the auxiliary Green function as an exponential of the linear functional of the ϕ-field, one should take a closer look at the action itself. The term written as Tr ln in Eq.(17) is a sum of two terms which are the contributions of the right- and left-movers. Let us analyze the right-movers contribution first. We may expand Tr ln in powers of ϕ:

$$\mathrm{Tr}\ln\left(\partial_\tau - iv_F\partial_x - i\phi(x,\tau)\right) = \mathrm{Tr}\ln\left(\partial_\tau - iv_F\partial_x\right) - \sum_{n=1}^{\infty}\frac{1}{n}\mathrm{Tr}\left(g_+\, i\phi\right)^n. \tag{22}$$

The term containing the n-th power of ϕ is the vertex containing the loop of n-th order Γ_n^+ and n external lines corresponding to ϕ's:

$$\mathrm{Tr}\left(g_+\,\phi\right)^n = \int\prod_{k=1}^{n} \mathrm{d}x_k\mathrm{d}\tau_k\;\Gamma_n^+(x_1,\tau_1;...;x_n,\tau_n)\prod_{i=1}^{n}\phi(x_i,\tau_i). \tag{23}$$

The loop is built from chiral Green functions g_+ in the following way:

$$\Gamma_n^+(x_1,\tau_1;...;x_n,\tau_n) = \prod_{i=1}^{n} g_+(x_i - x_{i+1},\tau_i - \tau_{i+1}) \tag{24}$$

and $(x_{n+1},\tau_{n+1}) = (x_1,\tau_1)$. It is convenient to calculate this loop in the frequency/momentum representation. We use notations ϵ and Ω for the fermionic and bosonic Matsubara frequencies correspondingly. Due to translational invariance of the system the loop Γ_n^+ will depend on $n-1$ momenta and frequencies

$$\begin{aligned}\Gamma_n^+(q_1,\Omega_1;...;q_n,\Omega_n) &= T\sum_\epsilon\int\frac{\mathrm{d}p}{2\pi}\,g_+(p,\epsilon)g_+(p+q_1,\epsilon+\Omega_1) \\ &\cdot ...\cdot\; g_+(p+q_1+...+q_{n-1},\epsilon+\Omega_1+...+\Omega_{n-1}).\end{aligned} \tag{25}$$

Using explicit expression of the Green function of right-moving free electrons

$$g_+(p,\epsilon) = (-i\epsilon + v_F q)^{-1} \tag{26}$$

and introducing temporarily the notations

$$z = -i\epsilon + v_F q, \qquad z_i = -i\Omega_i + v_F q_i \tag{27}$$

we can write the expression for the n-th loop in the following form

$$\Gamma_n^+ = \int \mathrm{d}z \, \frac{1}{z\,(z+z_1)\,(z+z_1+z_2)...(z+z_1+z_2+...+z_{n-1})}, \tag{28}$$

where we have used a symbolic notation for $\int \mathrm{d}z = T\sum_\epsilon \int \frac{\mathrm{d}p}{2\pi}$. There are very widely used in QED the Feynman identities which help calculating higher order diagrams. One of them,

$$\frac{1}{a_1...a_2} = \int\limits_{0\le x_{n-1}\le...\le x_1\le 1} \frac{(n-1)!\,\mathrm{d}x_1...\mathrm{d}x_{n-1}}{[a_1 x_{n-1} + a_2(x_{n-2}-x_{n-1}) + ... + a_n(1-x_1)]^n}$$

after some modification, may be brought to the form which is of help in our case, namely

$$\Gamma_n^+(z_1, ..., z_{n-1}) = (n-1)! \int \mathrm{d}z \int\limits_{Y_{n-1}} \mathrm{d}^{n-1}y \left[z + \sum_{i=1}^{n-1} y_i\, z_i\right]^{-n}. \tag{29}$$

where the inner integral runs over the area $Y_{n-1} = (0 \le y_{n-1} \le ... \le y_1 \le 1)$. All loops Γ_n^+ appear only in integrals with the products of ϕ's, Eq.(23), which are explicitely symmetric with respect to the permutation of all arguments (x_i, τ_i) (or, equivalently, with respect to the permutations of (q_i, Ω_i) in frequency/momentum representation). Therefore, only symmetrized form $\Gamma_n^{+\,(\mathrm{sym})}$ contributes to the integral in Eq.(23). Symmetrization of of the expression (29) leads to discarding the factorial and relaxing the ordering of y's:

$$\Gamma_n^{+\,(\mathrm{sym})}(z_1, ..., z_{n-1}) = \int \mathrm{d}z \int_0^1 \mathrm{d}^{n-1}y \left[z + \sum_{i=1}^{n-1} y_i\, z_i\right]^{-n}. \tag{30}$$

For $n > 2$ these integrals vanish. To see that we may take one integral over some y_i, say y_{n-1}:

$$\Gamma_n^{+\,(\mathrm{sym})} \propto \int \mathrm{d}z \int_0^1 \mathrm{d}^{n-2}y \left[\left(z + z_{n-1} + \sum_{i=1}^{n-2} y_i\, z_i\right)^{-n+1} - \left(z + \sum_{i=1}^{n-2} y_i\, z_i\right)^{-n+1}\right].$$

For $n > 2$ the integral of each term in the square brackets is finite. It means that we can make a shift of the integration variables in each term independently. Making the shift $z + z_{n-1} \to z$ (meaning $p + q_{n-1} \to p$, $\epsilon + \Omega_{n-1} \to \epsilon$) in the first term (integral) we find that it becomes identical to the second one and, thus, cancel it. Therefore, all loops containing more than two external lines are zero. That was the observation first made by Dzyaloshinskii and Larkin [4]. Above I gave the simplest proof to that fact. Nowdays this cancellation when it happens in many different one-dimensional problems is called the Loop Cancellation theorem.

Therefore, we are left with only two contributions from the loops with one and two legs. Calculating them one must excercise precaution because the corresponding integrals are ill-defined. These troubles have their roots in the linearization of the spectrum. However, we may predict the correct result even without healing these problems. It is clear that the loop with one leg is proportional to the zero mode of the Coulomb interaction and unperturbed particle density. This term is always cancelled against positive background. The loop with two legs is the standard polarization operator which is know to be (for the right-movers) in (q, Ω)-representation

$$\pi_+(q, \Omega) = \frac{1}{2\pi}\frac{q}{-i\Omega + v_F q} \tag{31}$$

Finally, back to the action Eq.(17), we collect two contributions coming from both chiral species and notice that the part independent of ϕ drops out of the expression Eq.(14). The action then becomes

$$\mathcal{S}[\phi] = \frac{1}{2}\phi V_0^{-1}\phi + \frac{1}{2}\phi\pi\phi, \tag{32}$$

where we introduced the overall polarization operator which is the sum of two chiral polarization operators π_+ and π_- and has the following familiar form in (q, Ω)-representation

$$\pi(q, \Omega) = \pi_+(q, \Omega) + \pi_-(q, \Omega) = \nu \frac{v_F^2 q^2}{\Omega^2 + v_F^2 q^2}, \tag{33}$$

with $\nu = (\pi v_F)^{-1}$ being the one-dimensional density of states.

The Gaussian action describing Coulomb bosons ϕ can be cast in a familiar form if we introduce effective time/frequency-dependent interaction

$$\mathcal{S}[\phi] = \frac{1}{2}\phi V^{-1}\phi, \tag{34}$$

where

$$V^{-1} = V_0^{-1} + \pi, \qquad V^{-1}(q, \Omega) = V_0^{-1}(q) + \pi(q, \Omega) \tag{35}$$

One can easily recognize the well-known result for the effective Coulomb potential with the screening calculated in the random phase approximation. It means that the random phase approximation becomes exact as we linearize the spectrum.

4. Calculation of Observables

After the Hubbard-Stratonovich transformation is done the action Eq.(9) becomes quadratic in the fermionic fields and Wick's theorem becomes applicable for any product of the ψ-fields. Consequently, any observable, originally a product of ψ-operators, decomposes into the product of the auxiliary Green functions $\tilde{\mathcal{G}}_{\pm}[\phi]$ which should be averaged over the auxiliary field ϕ. This field describes non-interacting bosons whose propagator is the screened Coulomb interaction Eq.(34). The auxilary Green functions $\tilde{\mathcal{G}}_{\pm}[\phi]$ are products of the free electron Green function $g_{\pm}$ and the exponentials with the exponent being a linear functional of the Coulomb bosons ϕ (see Eqs.(19)and (20)). Therefore, the problem of interacting electrons is reduced to the calculation of the simple Gaussian integrals.

In this way we, as an example, may calculate the Green function of the right movers

$$\left\langle \tilde{\mathcal{G}}_{+}(x,\tau;[\phi]) \right\rangle_{\phi} = g_{+}(x,\tau), \exp\left[-\mathcal{K}(x,\tau)\right] \tag{36}$$

where the exponent is given by

$$\mathcal{K}(x,\tau) = \frac{1}{2}\langle [\theta(x,\tau) - \theta(0,0)]^2 \rangle \tag{37}$$

$$= T\sum_{\Omega}\int \frac{\mathrm{d}q}{2\pi}\frac{V(q,\Omega)}{(-i\Omega + v_F q)^2}\left[e^{iqx - i\Omega\tau} - 1\right]. \tag{38}$$

The propagator of the "screened Coulomb field" (ϕ-field) is

$$V(q,\Omega) = V_0(q)\frac{\Omega^2 + v_F^2 q^2}{\Omega^2 + v^2 q^2}, \tag{39}$$

where the renormalized velocity v is defined as

$$v(q) = v_F\sqrt{1 + \frac{V_0(q)}{\pi v_F}}. \tag{40}$$

The simple arithmetics gives us the result which can be found in any review on one-dimensional bosonisation:

$$\mathcal{K}(x,\tau) = \int \frac{\mathrm{d}q}{q}\left\{\left[\frac{e^{-v_F q\tau}}{e^{-\beta v_F q} + 1} - \frac{e^{-vq\tau}}{e^{-\beta vq} + 1}\right] e^{iqx}\right. \tag{41}$$

$$\left. + \frac{(v - v_F)^2}{2vv_F}\frac{1 - \cos qx e^{-vq\tau}}{e^{-\beta vq} + 1}\right\}. \tag{42}$$

It is not our goal to describe the behaviour of different response functions. The results obtained with the use of the formalism present in this presentation are exactly the same as those obtained within operator formalism due to Haldane. The main distinction of the approach given in this presentation is its simplicity.

5. Electrons with Spin

We accept "g-ology" as our starting point, neglecting Umklapp and backscattering:

$$S[\psi] = \int \mathrm{d}x \mathrm{d}\tau \psi^*_{\sigma\alpha}(x,\tau)\left[\partial_\tau - ir_\alpha v_F \partial_x\right]\psi_{\sigma\alpha}(x,\tau) \tag{43}$$

$$+\frac{1}{2}\int \mathrm{d}x \mathrm{d}x' \mathrm{d}\tau g^{\alpha\alpha'}_{\sigma\sigma'}(x-x') n_{\sigma\alpha}(x,\tau) n_{\sigma'\alpha'}(x',\tau), \tag{44}$$

where $\sigma = \uparrow, \downarrow$ describes spin dependence and $\alpha = \pm$ refers to right- (+) or left-moving ($-$) particles ($r_\pm = \pm 1$). The coupling matrix has the structure

$$g^{\alpha\alpha'}_{\sigma\sigma'} = \delta_{\alpha,\alpha'}\left[\delta_{\sigma,\sigma'} g_4^{\parallel} + \delta_{\sigma,-\sigma'} g_4^{\perp}\right] + \delta_{\alpha,-\alpha'}\left[\delta_{\sigma,\sigma'} g_2^{\parallel} + \delta_{\sigma,-\sigma'} g_2^{\perp}\right]. \tag{45}$$

After Hubbard-Stratonovich transformation we arrive at

$$S[\phi] = \frac{1}{2}\phi_s g^{-1}_{ss'}\phi_{s'} - \sum_s \mathrm{Tr}\ln\left[\partial_\tau - ir_\alpha v_F \partial_x - i\phi_s\right] \tag{46}$$

Please note that we combine spin and chirality in one composite index $s = (\sigma, \alpha)$, where possible. Then expanding $'\mathrm{Tr}\ln'$ as in the previous section we are again left with the RPA result:

$$S[\phi] = S_{\mathbf{RPA}}[\phi] = \frac{1}{2}\phi_s V^{-1}_{ss'}\phi_{s'}, \quad V^{-1}_{ss'} = g^{-1}_{ss'} + \delta_{s,s'}\pi_\alpha \tag{47}$$

The matrix Green function is diagonal in the spin indices and is the sum of two chiral functions:

$$\begin{aligned} \mathcal{G}_{\sigma\sigma'}(x-x',\tau-\tau') &= \sum_{\alpha,\alpha'}\langle\psi_{\sigma\alpha}(x,\tau)\psi^*_{\sigma'\alpha'}(x',\tau')\rangle e^{ip_F(r_\alpha x - r_{\alpha'} x')} \\ &= \delta_{\sigma\sigma'}\sum_\alpha \mathcal{G}_{\sigma\alpha}(x-x',\tau-\tau') e^{ip_F r_\alpha (x-x')}, \end{aligned} \tag{48}$$

where

$$\mathcal{G}_{\sigma\alpha}(x-x',\tau-\tau') = g_\alpha(x-x',\tau-\tau')\left\langle e^{i[\theta_s(x,\tau)-\theta_s(x',\tau')]}\right\rangle. \tag{49}$$

The brackets stand for the averaging over θ_s with the action Eq.(47). The field θ_s itself is a linear functional of the "Coulomb field" ϕ_s, and must be found as the solution to the following equation

$$\left(\partial_\tau - ir_\alpha v_F \partial_x\right)\theta_s(\mathbf{x}) = \phi_s(\mathbf{x}). \tag{50}$$

Taking the Gaussian integral over ϕ we obtain

$$\left\langle e^{i[\theta_s(x,\tau)-\theta_s(x',\tau')]}\right\rangle = e^{-\mathcal{K}_s(x-x',\tau-\tau')}, \tag{51}$$

with the exponent related to the diagonal matrix element of the screened 'Coulomb interaction' $V_{ss}(q,\Omega)$ (see Eq.(47)):

$$\mathcal{K}_s(x,\tau) = T\sum_{\Omega}\int\frac{dq}{2\pi}\frac{V_{ss}(q,\Omega)}{(-i\Omega+v_F q)^2}\left[e^{iqx-i\Omega\tau}-1\right] \tag{52}$$

Due to the fact that the free Hamiltonian (and thus, density-density correlators $\pi_\pm(\mathbf{q})$) do not depend on spin, we can make a rotation in the spin sub-space only to diagonalize the coupling matrix $g_{ss'}$. The rotation can be performed by the matrix which has the form

$$L = \frac{1}{\sqrt{2}}\begin{pmatrix} 1 & 1 \\ 1 & -1 \end{pmatrix} \tag{53}$$

in the representation where the four-vector composed of components $\phi_{\sigma\alpha}$ is taken to be $\phi^{\mathrm{T}} = (\phi_{\uparrow+},\phi_{\uparrow-},\phi_{\downarrow+},\phi_{\downarrow-})$. This rotation, in fact, is equivalent to saying that we are writing the quadratic form in the new fields $\phi_{\rho\alpha}$ and $\phi_{\nu\alpha}$ which are the linear combinations of the old ones:

$$\phi_{\rho\alpha} = \frac{1}{\sqrt{2}}(\phi_{\uparrow\alpha}+\phi_{\downarrow\alpha}),\quad \phi_{\sigma\alpha} = \frac{1}{\sqrt{2}}(\phi_{\uparrow\alpha}-\phi_{\downarrow\alpha}), \tag{54}$$

corresponding to charge- (index ρ) and spin- (index σ) fields. After such a rotation the action becomes

$$S[\phi] = \frac{1}{2}\sum_{\nu,\alpha,\alpha'}\phi_{\nu\alpha}V^{-1}_{\nu,\alpha\alpha'}\phi_{\nu\alpha'}, \tag{55}$$

where $\nu = \rho,\sigma$. The new interaction matrix is block-diagonal in its spin/charge sectors and each block can be represented as a 2-by-2 matrix in the chiral indices:

$$\hat{V}_\nu = \frac{1}{(a_{4\nu}+\pi_+)(a_{4\nu}+\pi_-)-a_{2\nu}^2}\begin{pmatrix} a_{4\nu}+\pi_- & a_{2\nu} \\ a_{2\nu} & a_{4\nu}+\pi_+ \end{pmatrix} \tag{56}$$

where we introduced spin and charge coupling constants

$$g_{i\rho} = \frac{1}{2}(g_i^{\parallel}+g_i^{\perp}),\quad g_{i\sigma} = \frac{1}{2}(g_i^{\parallel}-g_i^{\perp}),\quad a_{i\nu} = \frac{g_{i\nu}}{2(g_{4\nu}^2-g_{2\nu}^2)}. \tag{57}$$

Substituting explicit expressions for the chiral polarization operators $\pi_\pm(q,\Omega)$ one can write it in the following form:

$$\hat{V}_\nu(q,\Omega) = 2\pi\frac{\Omega^2+v_F^2q^2}{\Omega^2+v_\nu^2q^2}\hat{A}(q,\Omega) \tag{58}$$

where $\hat{A}(q,\Omega)$ is 2-by-2 matrix

$$\hat{A}(q,\Omega) = \begin{pmatrix} (v-v_\nu)-2\gamma_\nu v_\nu\dfrac{-i\Omega+v_Fq}{i\Omega+v_Fq} & -\dfrac{v_\nu}{2}(K_\nu-K_\nu^{-1}) \\ -\dfrac{v_\nu}{2}(K_\nu-K_\nu^{-1}) & (v-v_\nu)-2\gamma_\nu v_\nu\dfrac{i\Omega+v_Fq}{-i\Omega+v_Fq} \end{pmatrix}$$

written in the notations standard for the literature on the Luttinger Liquid:

$$v_\nu^\pm = v_F + \frac{1}{\pi}(g_{4\nu} \pm g_{2\nu}), \quad K_\nu = \sqrt{\frac{v_\nu^-}{v_\nu^+}}, \tag{59}$$

$$v_\nu = \sqrt{v_\nu^+ v_\nu^-}, \quad \gamma_\nu = \frac{1}{4}\left(K_\nu + K_\nu^{-1} - 2\right)$$

The diagonal part of the interaction matrix V_{ss} needed for the average Green function does not actually depend on spin index σ:

$$V_{ss} = \frac{1}{2}\sum_\nu V_{\nu,\alpha\alpha} \tag{60}$$

Therefore, for the average Green function we have the representation:

$$\mathcal{G}(x,\tau) = \sum_\alpha g_\alpha(x,\tau)\, e^{-\mathcal{K}_\alpha(x,\tau)} e^{ip_F r_\alpha x}, \qquad \mathcal{K}_+(x,\tau) = \mathcal{K}_-(-x,-\tau) \tag{61}$$

where

$$\mathcal{K}_+(x,\tau) = \sum_\nu \int \frac{dq}{q}\left\{\frac{1}{2}\left[\frac{1 - e^{-v_\nu q\tau + iqx}}{e^{-\beta v_\nu q} + 1} - \frac{1 - e^{-v_F q\tau + iqx}}{e^{-\beta v_F q} + 1}\right] - \gamma_\nu \frac{e^{-v_\nu q\tau}}{e^{-\beta v_\nu q} + 1}\cos qx\right\} \tag{62}$$

Once again, the expression above is well known and can be found in any review on the Luttinger Liquid. The task of this presentation was to show that starting from the field-theoretical description of one-dimensional interacting electrons (by this I mean representation of the observables as functional integral over the Grassmann fields) one can derive the free-bosons action in an extremely simple, almost trivial, and elegant way.

References

1. S. Tomonaga, Prog. Theor. Phys. **5**, 544 (1950).
2. J. M. Luttinger, J. Math. Phys. **4**, 1154 (1963).
3. D. C. Mattis and E. H. Lieb, J. Math. Phys. **6**, 304 (1965).
4. I. E. Dzyaloshinskii and K. B. Larkin, Sov. Phys. JETP **38**, 202 (1973).
5. K. B. Efetov and A. I. Larkin, Sov. Phys. JETP **42**, 390 (1976).
6. F. D. M. Haldane, J. Phys. C **14**, 2585 (1981).
7. J. von Delft and H. Schoeller, AF **7**, 225 (1998).
8. H. C. Fogedby, J. Phys. –Condens. Matter **9**, 3757 (1976).
9. D. K. Lee and Y. Chen, J. Phys. A **21**, 4155 (1988).

QUASI ONE-DIMENSIONAL ORGANIC CONDUCTORS: DIMENSIONAL CROSSOVER AND SOME PUZZLES

S. BIERMANN[1,2], A. GEORGES [2,1], T. GIAMARCHI[1,2]
AND A. LICHTENSTEIN[3]
[1] *Laboratoire de Physique des Solides, CNRS-UMR 8502, UPS Bât. 510, 91405 Orsay France*
[2] *LPTENS CNRS UMR 8549 24, Rue Lhomond, 75231 Paris Cedex 05, France*
[3] *University of Nijmegen, NL-6525 ED Nijmegen, The Netherlands*

1. Introduction and scope of the paper

The nature of the metallic phase of interacting electron systems depends strongly on dimensionality. The interplay between interactions and dimensionality is an important issue for a large number of materials, ranging from cuprate superconductors to low-dimensional organic conductors [19]. For quasi-one dimensional conductors (such as the Bechgaard salts, which will be the main subject of this paper), these issues become crucial. Indeed in three dimensions, Fermi liquid (FL) theory applies, whereas in one dimension a different kind of low-energy fixed point known as a Luttinger liquid (LL) is found, with physical properties quite different from that of a FL. By varying the anisotropy of the system, or the energy scale at which it is probed, one can thus expect drastic changes in the physical properties.

This is even more true when the filling of the system is commensurate. In that case, interactions can lead to an insulating behavior via the Mott transition. This phenomenon occurs in all dimensions but the one-dimensional case is particularly favorable [28]. In quasi one-dimensional (Q1D) systems, interchain hopping can induce a (deconfinement) transition from the Mott insulating (MI) state to a metallic state, and crossovers between different metallic behaviors. Understanding how such a deconfinement transition can take place and what are the properties of the metallic phases is a particularly challenging problem, for reasons explained below.

This paper is organized as follows:

- In Sec. 2, *some* physical properties of the TMTSF and TMTTF organic compounds will be briefly reviewed [19, 18, 6], with an emphasis on those re-

I.V. Lerner et al. (eds.),
Strongly Correlated Fermions and Bosons in Low-Dimensional Disordered Systems, 81–102.

lated to the above issues. As described there, these compounds are three dimensional stacks of quarter- filled chains, which makes them wonderful laboratories, in which all the questions above can be addressed. Along the way, we shall point out some open questions and puzzles associated with the physics of these materials.

– In Sec. 3, we turn to theoretical models, and explain why *non-perturbative* methods are required to deal with interchain hopping. We shall review an extension of the dynamical mean-field approach [1, 2, 11] designed to deal with this problem (chain-DMFT), and describe its recent application to coupled Hubbard chains [3] and its bearing on the above issues.

2. Quasi one-dimensional organic conductors: some physical properties

The Bechgaard salts $TMTSF_2X$ were the first organic compounds to exhibit superconductivity, and have thus been the focus of intense experimental and theoretical studies. In addition to the superconducting phase, these materials have a remarkably rich phase diagram (cf. Fig. 1) and exhibit a host of remarkable properties (e.g non-FL metallic behaviour, quantized Hall conductance, Fröhlich conductivity), many of which are still poorly understood. Reviewing all these properties goes far beyond the scope of this lecture: we shall restrict ourselves to a discussion of the crossovers observed in the high-temperature regime, above the ordered phases. For more extensive reviews of the physical properties of quasi one-dimensional organics, see e.g [18, 6].

2.1. STRUCTURE

The basic building block of the Bechgaard salts (Fig. 2) is the flat molecule TMTSF (tetramethyltetraselenafulvalene), which contains four selenium atoms, surrounded by four methyl groups in a fulvalene type double ring structure. Also shown in Fig. 2 are the orbitals giving rise to the double bonds between the carbon atoms and the π orbitals of the selenium atoms. In the TMTTF compounds (Fabre salts), the selenium atoms are replaced by sulphur ones.

All these compounds crystallize in stacks of TM (short for TMTSF or TMTTF) molecules separated by the counter-ions X (e.g., PF_6 or ClO_4). The counter-ions are here to provide the charge neutrality, as in a standard salt. The transfer of charge between the ion and the TM stacks is total. Since there is one ion for two TM molecules, the chains are quarter filled. In addition the chains are slightly dimerized. This raises the important question whether these systems should be considered as half-filled, rather than quarter-filled, to which we shall come back in section 2.2. It is important to note that the commensurate filling is fixed by the chemistry of the compound, and so far it has not been possible to move away from

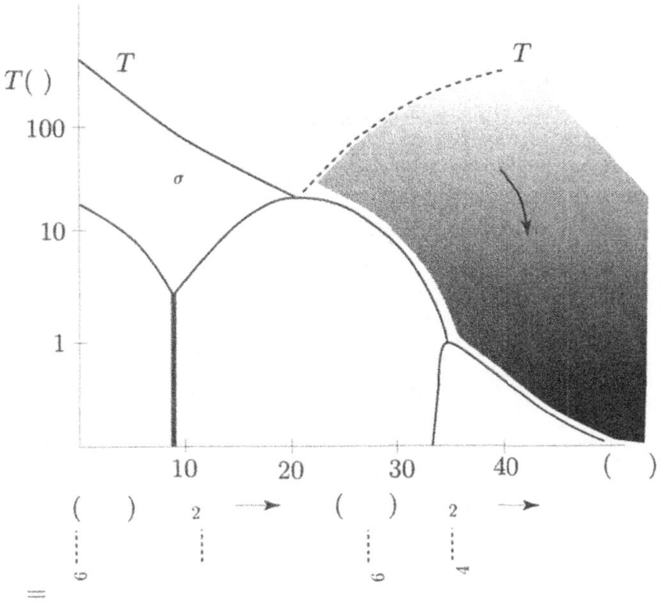

Figure 1. Unified experimental phase diagram for the TM compounds (from [6]). Either pressure or chemical changes (increasing pressure corresponds to going from the TMTTF to the TMTSF family and changing the anions) yields the same phases [MI: Mott insulator, LL: Luttinger liquid metal, FL: Fermi liquid metal, SP: spin-Peierls, AF: antiferromagnetic spin-density wave, SC: superconducting]. The TMTTF family is insulating at ambient pressure whereas the TMTSF family shows good metallic behavior at room temperature.

such a commensurate filling. No doubt that if this could be done (through e.g. a field effect transistor geometry [21]), this would prove very interesting.

The overlap of the π- orbitals of the selenium or sulfur atoms leads to a high mobility of electrons along the stacking direction; the hopping integrals in the perpendicular directions are indeed smaller by more than one order of magnitude. Estimated values of the hopping integrals along the stack direction (a-axis) and the two perpendicular axes pointing towards neighboring stacks (b-axis) and towards the anions (c-axis) respectively are: $t_a : t_b : t_c = 1000K : 100K : 30K$. Therefore one can think of these materials as one-dimensional chains coupled by small inter-chain hoppings. Given the hierarchy of transverse coupling the system is first expected to become two dimensional and then three dimensional at low temperatures. At very low temperatures the system has various ordered phases (spin-Peierls (SP), antiferromagnetic (AF), spin-density wave(SDW)) and superconducting (SC)). The nature of the molecule (TMTTF vs TMTSF) or of the ions slightly changes the interchain hopping and the dimerization. Such changes can also be induced by applying pressure to the system. This modifies the relative importance of the kinetic energy and Coulomb interation and leads to a very rich

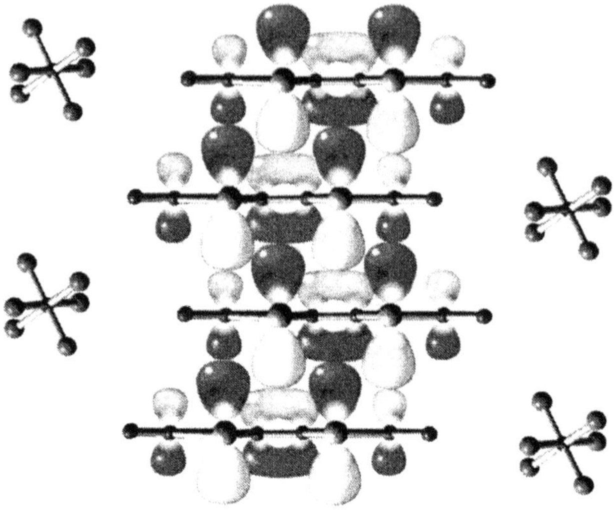

Figure 2. Structure of the Bechgaard salts $(TMTSF)_2X$ (after [6]). Electronic transport takes place preferentially along the stacks of TMTSF molecules (vertical, a-axis). The horizontal axis (c-axis) for which the stacks are separated by the counter-ions is the least conducting one.

phase diagram. The chemical and pressure changes have similar effects, which can be summarised by the unified phase diagram of Fig. 1.

2.2. MOTT INSULATORS AND LUTTINGER LIQUIDS

At ambient pressure, the $(TMTTF)_2PF_6$ compound displays insulating behavior (MI) . Upon increasing pressure, a transition to a metallic phase is found, and the properties of the TMTTF compounds evolve toward those of the compounds of the TMTSF family, which are good conductors. This evolution is clear from the a-axis resistivity measurements in Fig. 3. The minimum of the resistivity (followed by an activated law as temperature is lowered) defines the onset of the MI regime in Fig. 1. Such an insulating behavior in a quarter (or half filled) system suggests that it is due to the interactions and that the TMTTF family is a Mott-Hubbard insulator. It is thus clear that the interactions play a crucial role in the TMTTF family even at relatively high energies. For the TMTSF the question is more subtle in view of the metallic behavior at ambiant pressure and it was even suggested that such compounds could be described by a FL behavior with weak interactions [15]. Another important question is of course the reason for such a difference between the very close families TMTTF and TMTSF, for which the various characteristics (bandwidth, dimerization, interactions) vary relatively little.

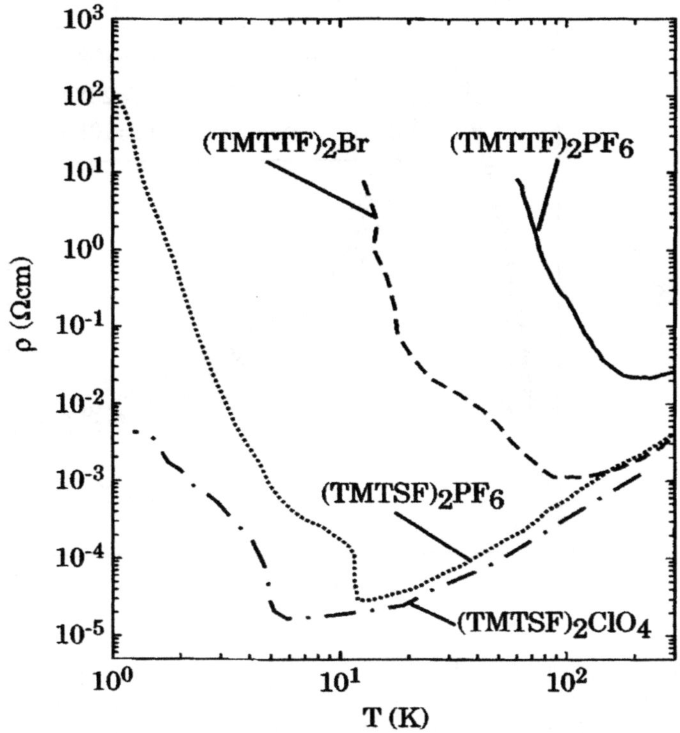

Figure 3. Resistivity along a-axis for a series of compounds showing the gradual evolution to metallic behavior as pressure is increased (from [27]).

A blatant proof of the importance of interactions for *both* the TF and SF compounds is provided by the optical conductivity [8, 24], as shown in Fig. 4. The optical conductivity clearly shows that the high energy structure is the one of a Mott insulator, with a decreasing gap (of the order of 2000 cm^{-1} for the $TMTTF_2(PF_6)$ to 200 cm^{-1} for $TMTSF_2(PF_6)$. Nearly all (99%) of the spectral weight is in this high energy structure. In the metallic compounds there is in addition a very narrow Drude peak. The optical conductivity shows thus clearly that these compounds are very far from simple Fermi liquids. In addition one can compare the optical data with the theoretical predictions [13, 14] for a LL. The data above the gap fits very well the power law LL behavior and thus shows quite convincingly that these compounds are indeed well described by a LL theory down to a scale of a few hundred Kelvin (temperature or frequency). This is also consistent [11] with the optical data along the c-axis [16], depicted in Fig. 6. These measurements directly probe the density of excited states in the $a - b$ plane. We note however that, although clearly revealing that electrons are confined in the chains above $\sim 100K$, the measurements of dc transport along the c-axis [20] (see Fig. 5) are not yet fully understood theoretically from a LL picture (see [11] for a discussion).

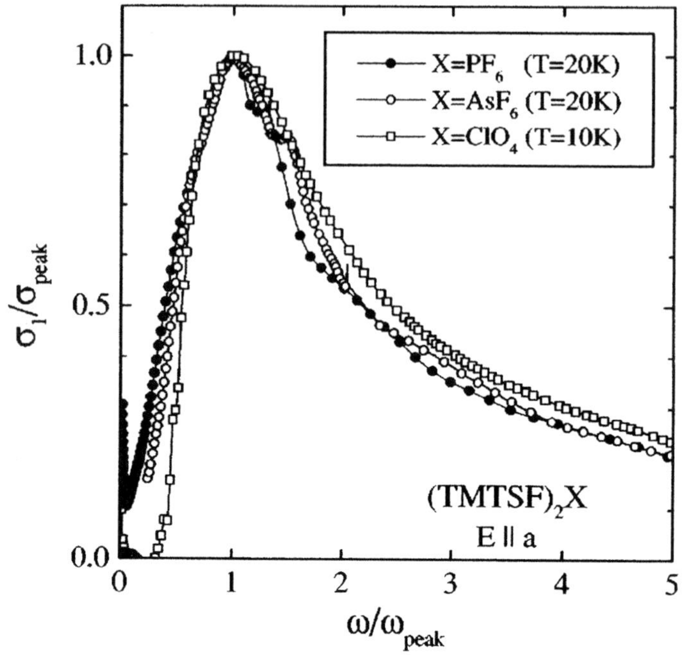

Figure 4. Optical conductivity in the TMTSF family. Although these compounds seem to have a rather "standard" d.c. conductivity, all the d.c. transport is in fact due to a very narrow Drude peak containing only 1% of the spectral weight, whereas 99% of the spectral weight is above an energy gap (of the order of 200 cm^{-1}), and is reminiscent of a Mott insulating structure. A fit of the ω dependence of the conductivity above the gap is well consistent with Luttinger liquid behavior (from [24]).

The a-axis optical measurements described above even allow for a quantitative determination [24] of the LL parameter K_ρ, yielding $K_\rho \simeq 0.23$, indicating very strong electron interactions [1]. This estimate of the LL parameter agrees reasonably well with measurements of the longitudinal resistivity in the range $100 - 300K$ [18]. Photoemission data [7, 27] are also consistent with this value.

In addition to providing strong evidence for the Luttinger liquid behavior, the optical data and its comparison to the theoretical predictions force one to reinvestigate the standard interpretation for the difference between TMTTF and TMTSF compounds. Indeed the commonly accepted point of view since the pioneering work of Ref. [9] was that the insulating behaviour of the TMTTF family is due to the stronger dimerization of these compounds, which effectively changes the band filling. As mentioned above, on average two TMTSF or TMTTF molecules donate one electron to the (monovalent) anion, so that the conduction band is nominally a quarter filled hole band. However, the dimerization between neigh-

[1] We use the conventions of Ref. [23], see also section 3 below. $K_\rho = 1$ corresponds to non-interacting electrons.

boring molecules opens a dimerization gap in the middle of the band. When this dimerization is large, the system might best be thought of as half-filled (TMTTF) rather than quarter-filled (TMTSF). From this point of view the Mott insulating behavior comes from the half filled nature of the system.

However the optical data are inconsistent with a half filled description in the SF family [24]. A more satisfactory explanation is thus to consider both series of compounds as quarter-filled, and the Mott insulator to be due to the quarter filled commensurability [14]. Increasing pressure or changing the chemistry reduces the relative strength of Coulomb interactions with respect to hopping, hence suppressing Mott localization. For this interpretation to be tenable, these compounds must be very close to the Mott transition point, which is indeed supported by the measured values of the LL parameter (indeed, $K_\rho^c = 0.25$ is the critical value for the opening of a Mott gap in a quarter-filled chain).

2.3. DIMENSIONAL CROSSOVER AND DECONFINEMENT

Due to the interchain hopping a dimensional crossover will take place at low energy between decoupled chains and a higher- dimensional behavior. Since the isolated chains would be insulators (because K_ρ is so small), the interchain hopping can induce a deconfinement transition provided that it becomes larger than the Mott gap. The system will thus crossover from a regime where one has essentially uncoupled (insulating) chains to that of metallic planes. Understanding the characteristics of such a transition (energy scale, critical values of the hopping, physical nature of the various phases) is one of the most challenging questions on these systems, on which we shall focus in the following.

A dimensional crossover is indeed observed in the transport along the c axis shown in Fig. 5. From this experiment, we see that the dimensional crossover takes place around $100K$ in $(TMTSF)_2PF_6$. This is in agreement with the change of behavior from T (LL behavior) to T^2 observed in dc transport along the a axis [18], and with the change of behavior in the transverse optical conductivity (see Fig. 6.)

The interpretation that the change of behavior between the insulating and metallic regimes is due to such deconfinement transition [14] can be strengthened by the optical data (see Fig. 7). A measure of the gap extracted from the optical conductivity shows that the change of nature occurs when the observed gap is roughly of the order of magnitude of the interchain hopping [26]

2.4. SUMMARY OF SOME OPEN ISSUES AND QUESTIONS

To summarize the experimental situation, here are some key questions raised by the physical properties of the TM family:

1. Are TMTSF strongly correlated systems ? What is the strength of interactions ?

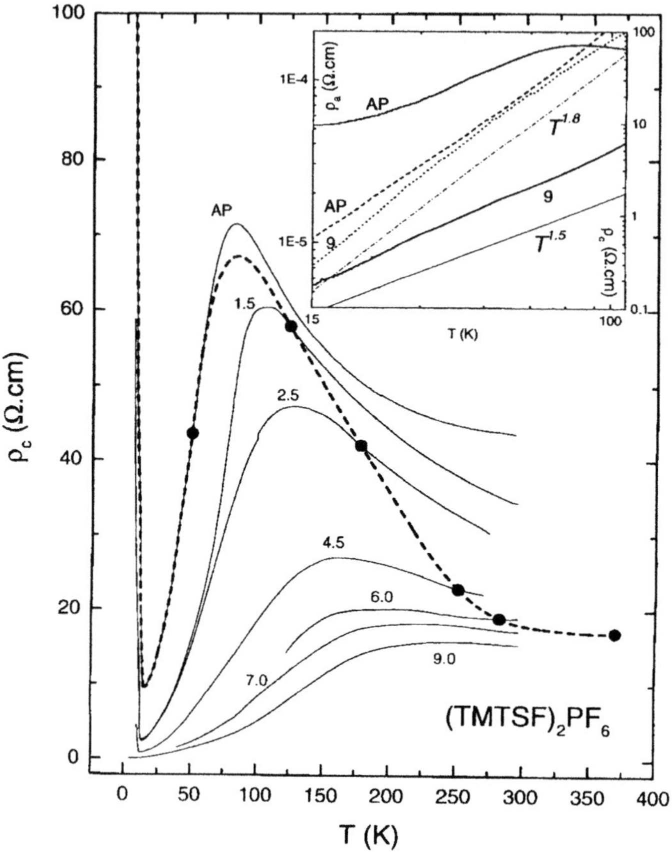

Figure 5. Transport along the least conducting axis. This is in effect a tunnelling experiment between the *ab* planes and thus a measure of the density of states in such planes. In the 1d regime there are no single particle excitations and the resistivity increases with decreasing temperature, whereas when the chains couple coherently one recovers a more conventional (metallic) behavior. The maximum in resistivity thus measures the scale for the dimensional crossover (from [20]).

2. What causes the difference between the various members of the organic families (TMTSF and TMTTF) ?
3. Is the high temperature metallic regime a Luttinger liquid ?
4. At what energy scale does the dimensional crossover take place ?
5. What is the nature of the metallic state of the TMTSF series in the temperature range from 10 to $100K$?
6. What are the physical properties of the low-T (Fermi-liquid?) metallic regime?

As we have seen, one is now in a position to have satisfactory answers to questions 1-4. In doing so one had to reexamine most of the commonly accepted point of views. The optical data shows that the dimerization plays little role and that the Mott insulating behavior is due to the *quarter filled* nature of the compounds [14, 24]. The estimate of the crossover scale of $\sim 100K$ questions early interpre-

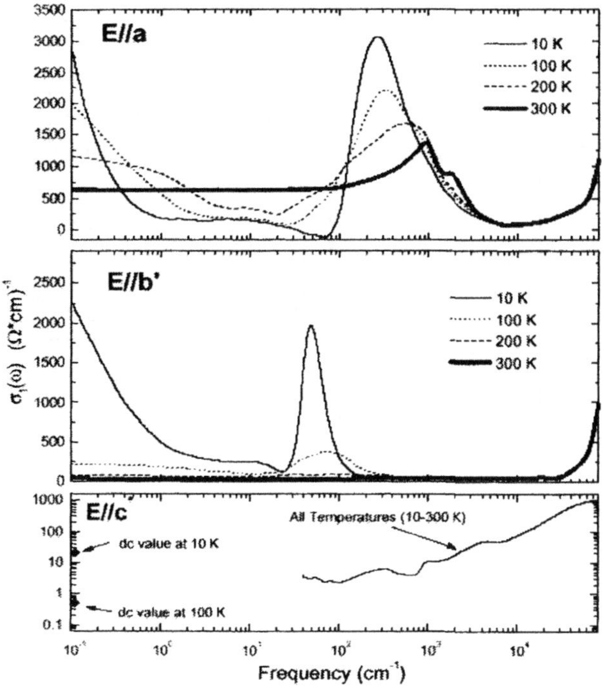

Figure 6. Optical conductivity along the three axis, at various temperatures. Depending on the temperature a different behavior is observed along the b direction which signals a dimensional crossover from essentially uncoupled chains to coherent planes. (from [16]))

tations of measurements of the NMR relaxation time [5] T_1 on $(TMTSF)_2ClO_4$. These showed the Korringa law typical of FL behavior $1/(T_1T)$ = const at very low temperatures. However, strong deviations from the Korringa law are observed already around $\sim 10K$, which is considerably smaller than the onset of FL behavior estimated from optics and transport. This makes question 5) particularly puzzling, since the NMR is anomalous in a temperature range for which one now knows that the compounds are *not* in a one-dimensional regime.

3. Modeling quasi-one-dimensional systems

3.1. NON-PERTURBATIVE EFFECTS OF INTER-CHAIN HOPPING

We now turn to simplified models in which one-dimensional chains of interacting electrons are coupled by an inter-chain hopping $t_\perp$. Later on we shall concentrate on the simplest case of the Hubbard model, but at this stage we would like to make some remarks with a broader degree of validity.

Let us consider first the case of a commensurate filling. For a single chain, strong enough interactions will then open a Mott gap, leading to an insulating state

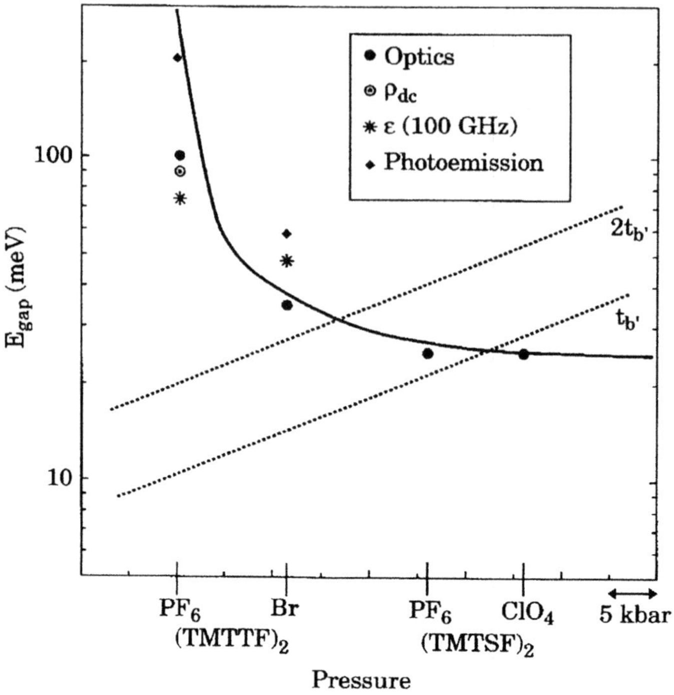

Figure 7. A comparison of the measured gap in the optical conductivity with the interchain hopping. The change of behavior from insulating to metallic occurs when the two quantities are of the same order of magnitude showing that the difference between the various members of the TM families is indeed linked to a deconfinement transition (from [27]).

in which the electrons are confined on the chains. Technically, this is due to umklapp scattering processes being relevant. This will actually happen for arbitrarily small repulsive interactions in the half-filled Hubbard model, while at quarter-filling a strong enough nearest-neighbour interaction is needed, in addition to a local Hubbard term. (This is one of the reasons why a simple Hubbard model is insufficient to describe the organic compounds, another one being the small values of K_ρ needed to fit the data). The interchain hopping will tend to delocalise the electrons, resulting in an insulator to metal transition at zero-temperature for a critical value of the inter-chain hopping. The generic situation is depicted on the schematic phase diagram of Fig. 8.

In addition to the $T = 0$ insulator to metal ("deconfinement") transition, finite-temperature crossovers have been depicted on this diagram. For temperatures smaller than the one-dimensional bandwidth (i.e the high-energy cutoff of the problem), but higher than the crossover scales depicted in the figure, the system is essentially insensitive to the interchain hopping. In this regime, the system is expected to behave as a Luttinger liquid (LL), which is the generic state for interacting one-dimensional electrons. However, one should bear in mind

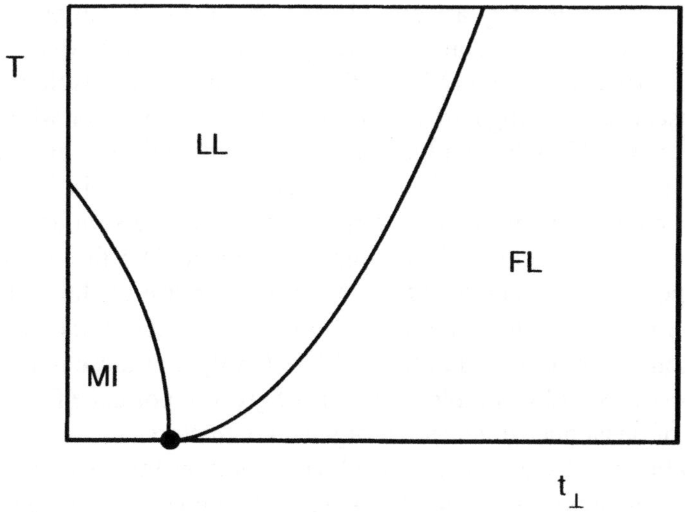

Figure 8. Schematic representation of the crossovers in the commensurate case. An insulator to metal transition occurs for a critical value of $t_\perp$ at zero-temperature. Crossovers are expected when reducing temperature, from a high-temperature Luttinger liquid behavior to either an insulating state or a Fermi liquid metal. At incommensurate fillings, there is no MI phase and only the LL to FL dimensional crossover line is present. Phases with long-range order have been omitted on this figure.

that, at least for small $t_\perp$, the Luttinger liquid parameter K_ρ will be gradually renormalized downwards by the (relevant) umklapp scattering as temperature is reduced. The temperature scale at which the effective K_ρ approaches zero signals the low-T Mott insulating regime: this is the first crossover line depicted in the figure. To the right of the $T = 0$ deconfinement transition, we expect a crossover from a high-T Luttinger liquid regime into a low-T Fermi liquid regime (FL). The crossover separates a high-T regime in which thermal fluctuations destroy coherent interchain transport from a low-T regime in which interchain coherence sets in. At incommensurate fillings, the situation is somewhat simpler: no Mott insulating state exists, and the decoupled chain is always in the metallic LL regime. Only the LL to FL crossover survives. Since for reasonable values of K_ρ the interchain hopping is a relevant perturbation [4, 29, 30], this crossover line starts at a $T = 0, t_\perp = 0$ and the ground-state is a FL for arbitrarily small but finite $t_\perp$.

Of course, this discussion and Fig. 8 will be complicated at very low temperatures by the intervening of phases with long-range order. This will depend however on the details of the specific model, in particular on the degree of deviation from perfect nesting. We do not address these issues here, being interested in compounds for which ordering temperatures are significantly smaller than the crossovers described above.

It is quite tempting to compare the schematic crossover diagram in Fig. 8 to the physical properties of quasi one-dimensional organic compounds described in the previous section and summarized by the unified phase diagram of Fig. 1. Indeed, it has been advocated [14, 26] that the change from insulating to metallic behavior observed in Bechgaard salts by increasing pressure or when going from the TMTTF to the TMTSF family, is associated with a deconfinement transition. Obviously, increasing pressure does not only affect interchain hopping in these compounds, but the deconfinement transition may be of a similar nature than the one induced by increasing $t_\perp$ in a simplified model. The first crossover would then correspond to the opening of the Mott gap (upturn of the resistivity in the TMTTF compounds), while the second one would correspond to the onset of interchain coherence observed in the TMTSF family. For these reasons, the study of these crossovers within simple models is of great importance for understanding the physics of quasi one-dimensional organic conductors.

The qualitative discussion above already hints at the fact that perturbation theory in the interchain hopping $t_\perp$ is only of limited use for the study of the crossovers. Let us make this statement more precise, and assess how far perturbative treatments can take us, starting with the simpler case of an incommensurate filling. In the non-interacting case, the dimensional crossover will occur when the energy (e.g the thermal energy kT, or the energy $\hbar\omega$ associated with probing the system at a given frequency) is of the order of the warping of the Fermi surface due to interchain hopping. Hence the crossover scale will simply be $kT^*, \hbar\omega^* \sim t_\perp$. In the interacting case, the situation is more complicated: the decoupled chains are Luttinger liquids, and $t_\perp$ is a relevant perturbation. A perturbative RG treatment will indicate that this effective inter-chain hopping grows as the energy scale is reduced: the crossover will occur when this running coupling reaches kT. This leads to the following estimate [5, 4] of the crossover scale $E^* \equiv kT^*, \hbar\omega^*$:

$$E^* \sim t_\perp (t_\perp/t)^{\alpha/(1-\alpha)} \tag{1}$$

where $\alpha = \frac{1}{4}(K_\rho + 1/K_\rho) - \frac{1}{2}$ is the exponent associated with the one-electron Green's function in the LL state. The important physical content of this expression is that interactions can significantly reduce the crossover scale [5, 29, 30], as compared to the non-interacting estimate $\sim t_\perp$. We note that (see Sec. 2) for the TMTSF Bechgaard salts, α appears to be close to $1/2$. Since $t_b \simeq 300$K and $t_a/t_b \simeq 10$, this would place E^* in the 10-30 K range according to the above RG estimate. Experimentally however, interchain coherence appears to set in at a significantly higher temperature, in the 100 K range: this is one of the puzzles in the field.

While this perturbative RG analysis allows to estimate a scale for the dimensional crossover, it breaks down for $T < E^*$ since the effective $t_\perp$ flows to large values. In particular, it does not provide informations on the detailed nature of the low-T Fermi liquid regime. Thus, a proper handling of the dimensional crossover in quasi one-dimensional systems has to resort to techniques which are not perturbative in $t_\perp$. This is even clearer in the case of a commensurate filling. If one starts from the 1D Mott insulator fixed point, the deconfinement transition is clearly a non-perturbative phenomenon since the inter-chain hopping is an irrelevant perturbation at this fixed point. If, on the other hand, one starts with the LL fixed point associated with the high-T regime, then one has to deal simultaneously with *two* relevant perturbations: the umklapp scattering (responsible for Mott physics) and the interchain hopping. The competition between these two relevant perturbations will determine the crossovers described above.

Non-perturbative studies are thus needed to investigate both the deconfinement transition and the dimensional crossover. Several authors have developed such methods in the case of *a finite number* of coupled chains. Below, we present a recently developed approach designed to handle *an infinite array of coupled chains* in a non-perturbative manner[1, 11, 3].

3.2. THE CHAIN- DYNAMICAL MEAN FIELD THEORY APPROACH

This non-perturbative method is inspired by the success of dynamical mean field theory (DMFT) in the description of lattice models of correlated fermions [12]. Ordinary DMFT maps a lattice model of interacting fermions onto a single site model in an effective time-dependent mean field, which has to be determined self-consistently. Thus the problem becomes equivalent to an Anderson impurity model in a self-consistent bath. It can be shown that in the limit of an infinite coordination number of the lattice, all quantum fluctuations become local and the dynamical mean field description is exact.

Very anisotropic systems of coupled one-dimensional chains lend themselves to a very natural extension of this approach (dubbed "chain-DMFT"), in which the array of chains is replaced by an effective chain in a self-consistent bath. It is crucial to retain a one-dimensional geometry of the effective problem, if the limit of decoupled chains (and hence 1D physics) is to be treated properly. The self-consistent bath describes the influence of all neighboring chains on the "impurity chain", thus freezing all quantum fluctuations that are non-local in the transverse direction, but retaining spatial and dynamical fluctuations along the chains. By the same arguments as in ordinary DMFT, the chain-DMFT approach becomes exact in the limit of an infinite transverse coordination number of the array. In the limit of decoupled chains chain-DMFT reduces to the single chain problem, which is solved numerically exactly.

Consider a system of coupled chains described by the Hamiltonian

$$H = \sum_m H_{1D}^{(m)} - \sum_{\langle m,m' \rangle} t_\perp^{mm'} \sum_{i\sigma} (c_{im\sigma}^+ c_{im'\sigma} + \text{h.c}) \qquad (2)$$

where $H_{1D}^{(m)}$ is the Hamiltonian for a single isolated chain and the sum in the second term runs over neighboring chains m, m', which are coupled by the transverse hopping $t_\perp^{mm'}$. Chain-DMFT maps this system onto a single-chain system described by the effective action

$$S_{\text{eff}} = -\int\int_0^\beta d\tau\, d\tau' \sum_{ij,\sigma} c_{i\sigma}^+(\tau)\mathcal{G}_0^{-1}(i-j,\tau-\tau')c_{j\sigma}(\tau') + \int_0^\beta d\tau H_{1D}^{int}[\{c_{i\sigma}, c_{i\sigma}^+\}] \tag{3}$$

where H_{1D}^{int} is the interacting part of the on-chain Hamiltonian. $\mathcal{G}_0$ is the *effective propagator* describing hopping processes, that destroy a fermion on site j at time τ' and create one at site i at time τ. In analogy to usual DMFT $\mathcal{G}_0$ has to be determined from a self-consistency condition that imposes that the Green's function $G(i-j,\tau-\tau') \equiv -\langle c(i,\tau)c^+(j,\tau')\rangle_{\text{eff}}$ calculated from S_{eff} coincides with the on-chain Green's function of the original problem. Since the self-energy is $\Sigma = \mathcal{G}_0^{-1} - G^{-1}$, this condition reads:

$$G(k, i\omega_n) = \int d\epsilon_\perp \frac{D(\epsilon_\perp)}{i\omega_n + \mu - \epsilon_k - \Sigma(k, i\omega_n) - \epsilon_\perp} \tag{4}$$

where $\epsilon_\perp(k_\perp)$ denotes the Fourier transform of the inter-chain hopping $t_\perp^{mm'}$, $D(\epsilon_\perp) = \sum_{k\perp} \delta[\epsilon_\perp - \epsilon_\perp(k_\perp)]$ the corresponding density of states, k the momentum in chain direction and ω_n the Matsubara frequencies. The chain-DMFT equations (3,4) fully determine the self-energy and Green's function of the coupled chains. In particular, once Σ has been calculated, the full Green's function for the coupled chains is obtained from: $G(k, k_\perp, i\omega_n)^{-1} = i\omega_n + \mu - \epsilon_k - \epsilon_\perp(k_\perp) - \Sigma(k, i\omega_n)$. We note that a key approximation of this approach is that the self-energy is considered to be independent of the transverse component of the momentum. In the following we specialize the discussion to a model of Hubbard chains coupled by a perpendicular hopping $t_\perp$. Then the Hamiltonian is of the form (2) with $t_\perp^{mm'} = t_\perp$ and $H_{1D}^{(m)}$ being the Hamiltonian of the one-dimensional Hubbard model.

Different choices of the lattice are possible. Choosing a semi-circular density of states

$$D(\epsilon) = \frac{1}{2\pi t^2}\sqrt{\epsilon^2 - 4t^2} \tag{5}$$

in Eq. (4) corresponds to a Bethe lattice with infinite coordination number, in which each lattice site is replaced by a one-dimensional chain and hopping between the chains follows the topology of the Bethe lattice. This choice is particularly appealing, since the chain-DMFT formalism is exact in this case.

Another possibility is to consider a two dimensional array of chains with the square lattice geometry. In this case, each chain has exactly two nearest neighbors and chain-DMFT must be viewed as an approximation. The transverse dispersion becomes:

$$\epsilon_\perp(k_\perp) = -2t_\perp \cos k_\perp \tag{6}$$

and the corresponding density of states is the one of a one-dimensional lattice:

$$D(\epsilon) = \frac{1}{\pi} \frac{1}{\sqrt{\epsilon^2 - 4t^2}}. \tag{7}$$

As for the Bethe lattice, the self-consistency condition simplifies in the 2D geometry in such a way that the Weiss field can directly be written in terms of the Green's function:

$$\mathcal{G}_0^{-1}(k, i\omega_n) = i\omega_n + \mu + G^{-1}(k, i\omega_n) - \sqrt{G^{-2}(k, i\omega_n) + 4t_\perp^2}. \tag{8}$$

This model does not provide a controlled limit, in which chain-DMFT becomes exact. Still, from a conceptual point of view, a two-dimensional picture seems more appealing if one aims at a comparison with the experimental situation in the Bechgaard salts. Therefore the calculations presented below were performed for this choice of the model. We stress, however that in practice, the specific choice of the transverse dispersion does not have a significant qualitative influence (as long as one does not address long-range ordering).

A practical implementation of the chain-DMFT approach requires to solve the effective one-dimensional interacting problem described above. This is a rather formidable task, and numerical methods are required. Even though other techniques are conceivable, we have chosen to use a Quantum Monte Carlo algorithm, which is a straightforward generalization of the Hirsch-Fye algorithm used in single-site DMFT [17, 12]. It relies on a Trotter discretization of the effective action in imaginary time, and on a discrete Hirsch transformation of the interaction term:

$$e^{-\Delta\tau U n_{i\uparrow} n_{i\downarrow}} = \frac{1}{2} \sum_{s=\pm 1} e^{\lambda s \left(n_{i\uparrow} - n_{i\downarrow}\right) - \frac{\Delta\tau U}{2} \left(n_{i\uparrow} + n_{i\downarrow}\right)} \tag{9}$$

with $\lambda = \operatorname{arccosh}\left(e^{\frac{\Delta\tau U}{2}}\right)$. An Ising field s is introduced at each time slice and each site of the chain. Monte Carlo sampling of the Ising fields then allows for the direct calculation of the on-chain Green's function and on-chain correlation functions. In practice, chains of 16 to 32 sites with periodic boundary conditions are sufficient to access the 1d Luttinger Liquid regime.

In the following we will present numerical evidence for the deconfinement transition at half filling and the dimensional crossover as a function of temperature in the doped case.

3.3. LUTTINGER-LIQUID TO FERMI-LIQUID CROSSOVER

We summarize some of our QMC results for an array of coupled Hubbard chains in the chain-DMFT approach, starting with the doped (incommensurate) case. At high temperature, the model is expected to display LL behaviour. In order to

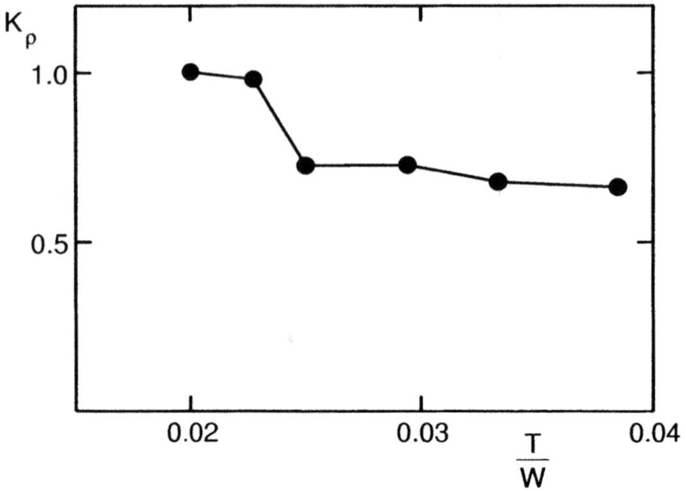

Figure 9. Effective K_ρ *vs.* temperature in the doped case (filling $n \simeq 0.8$) for $U/W = 1.0$, $t_\perp/W = 0.14$.

measure numerically the LL parameter K_ρ, we have computed the local spin-spin correlation:

$$\chi_s(\tau) = \langle S^z(j,0)S^z(j,\tau)\rangle = \sum_{k,k\perp} \chi_s(k,k_\perp,\tau) \tag{10}$$

In a LL, the asymptotic behavior of this quantity reads, in imaginary time:

$$\chi_s(\tau) = \chi_s(\beta/2)\left(\frac{1}{\sin \pi\tau/\beta}\right)^{1+K_\rho}, \tag{11}$$

valid for β, τ and $\beta - \tau$ larger than the inverse of the high-energy cutoff (i.e in some range around $\tau = \beta/2$). We emphasize that this is a much better manner of accessing K_ρ in a QMC simulation than by looking at the single-electron Green's function, whose asymptotics involves the exponent $\alpha = (K_\rho + 1/K_\rho)/4 - 1/2$ which is never very large for the Hubbard model. This makes it very hard numerically to distinguish LL from FL behavior at the level of one-electron Green's functions.

In Fig. 9, the effective value of K_ρ obtained by fitting the local spin correlation function to the above asymptotic form is plotted as a function of temperature [2], for $U/W = 1$, a filling $n \simeq 0.8$ and $t_\perp/W = 0.14$. This plot clearly reveals the expected dimensional crossover from a high-T LL regime with $K_\rho < 1$ to a low-T FL regime with $K_\rho = 1$. The numerically measured value $K_\rho \simeq 0.7$ in the LL regime is in very good agreement with the exact value known from analytical (Bethe ansatz) calculations on the one-dimensional Hubbard model, for that value of U/W [22].

[2] We give all energies in units of the longitudinal band width $W = 4t$.

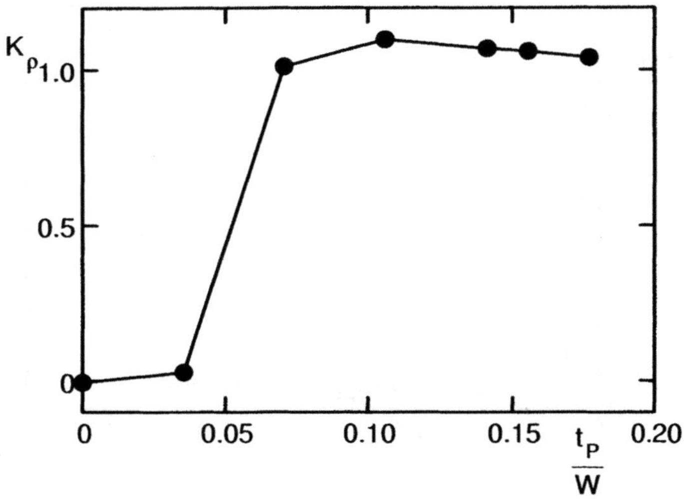

Figure 10. K_ρ as a function of the perpendicular hopping $t_\perp$ for a half filled system.

Our results for the crossover scale are consistent with $T^* \simeq Ct_\perp/\pi$ with $C \simeq 0.5$. We cannot meaningfully test the RG prediction $T^* = \frac{t_\perp}{\pi}C(t_\perp/t)^{\alpha/(1-\alpha)}$ for the reduction of this scale due to interactions [5, 4], because of the very small value of $\alpha \simeq 0.03$ for the present model. As discussed in the last section, extensions of our calculations to models with smaller values of K_ρ are required both for their theoretical interest and for a meaningful comparison to organics.

3.4. DECONFINEMENT TRANSITION

We now turn to the half-filled case. In Fig.10, we display the effective K_ρ (determined as above) as a function of interchain hopping, for $U/W = 0.65$ and at a rather low temperature $T/W = 0.025$. For small $t_\perp/W$, the value $K_\rho = 0$ indicating a Mott insulating behaviour (with a decay of the spin-spin correlation similar to that of a Heisenberg spin chain). In that regime, the calculated charge correlation function (not shown) clearly displays the exponential decay associated with a finite charge gap. Beyond a critical value of $t_\perp/W$, we find $K_\rho \simeq 1$, signalling a FL regime (and a corresponding behavior for the charge correlation function). Hence, the expected deconfinement transition (Fig.8) is clearly revealed by our calculations. In principle, it should be possible to identify first a Luttinger Liquid phase and then, with increasing interchain coupling, a Fermi Liquid phase. However, in this parameter range the Luttinger Liquid phase is too narrow to become visible. The location of the deconfinement transition is in reasonable agreement with the naive criterion $\Delta_{1D} \sim t_\perp^{eff}$, with $t_\perp^{eff}$ the renormalised inter-chain hopping. [14, 26, 25].

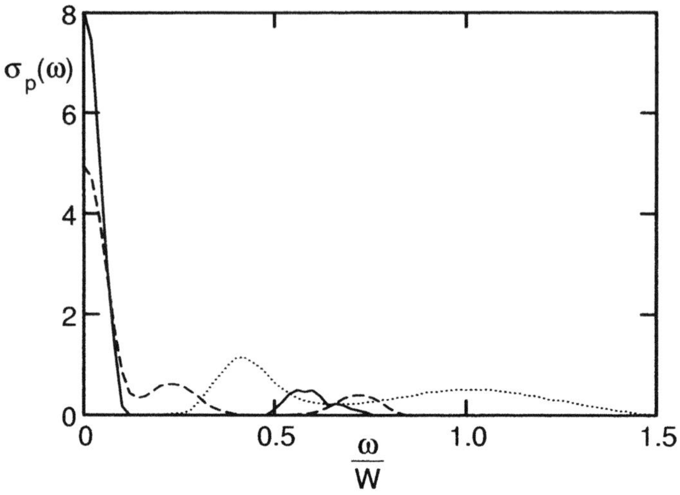

Figure 11. Inter-chain optical conductivity at half-filling for $U = 0.65W$, $\beta = 40/W$ and $t_\perp = 0.14$ and 0.07 (solid and dashed lines) and for $U = 1.0W$, $\beta = 40/W$ and $t_\perp = 0.$ (dotted line).

3.5. INTERCHAIN OPTICAL CONDUCTIVITY

Inter-chain optical conductivities within chain-DMFT can be obtained from the one-particle Green's functions. Vertex corrections drop out for analogous reasons as in single-site DMFT, and therefore:

$$\begin{aligned} \mathrm{Re}\,\sigma_\perp(\omega, T) \propto \;& t_\perp^2 \int \frac{dk_\perp}{2\pi} \sin^2 k_\perp \int \frac{dk}{2\pi} \int d\omega' A(\epsilon_\perp, k, \omega') \\ & \times A(\epsilon_\perp, k, \omega + \omega') \frac{f(\omega') - f(\omega' + \omega)}{\omega} \end{aligned} \tag{12}$$

where $A(\epsilon_\perp, k, \omega) = -\frac{1}{\pi}\mathrm{Im}G(\epsilon_\perp, k, \omega)$ is the single-particle spectral function of the coupled chains system. Note that we have taken into account the $k_\perp$- dependence of the current vertex in this formula. Fig. 11 shows the inter-chain optical conductivity at half-filling for $U = 0.65W$, $\beta = 40/W$ and $t_\perp = 0.14$ and 0.07 (solid and dashed lines) and for $U = 1.0W$, $\beta = 40/W$ and $t_\perp = 0.$ (dotted line). As expected, in the case of uncoupled chains ($t_\perp = 0.$) the system exhibits a Mott gap, followed by an onset of absorption starting at approximately the gap and extending up to a scale of order U, where a broad second peak is apparent. As the insulator to metal transition is crossed, a low-frequency Drude peak develops. Close to the transition the weight of the Drude peak is small, while the Hubbard band feature is still visible and carries a significant part of the spectral weight.

To make comparison to the Bechgaard salts, where the Drude peak was shown to carry less than 2% of the spectral weight with 98% of the weight being contained in the Hubbard features, more realistic models should be considered. In

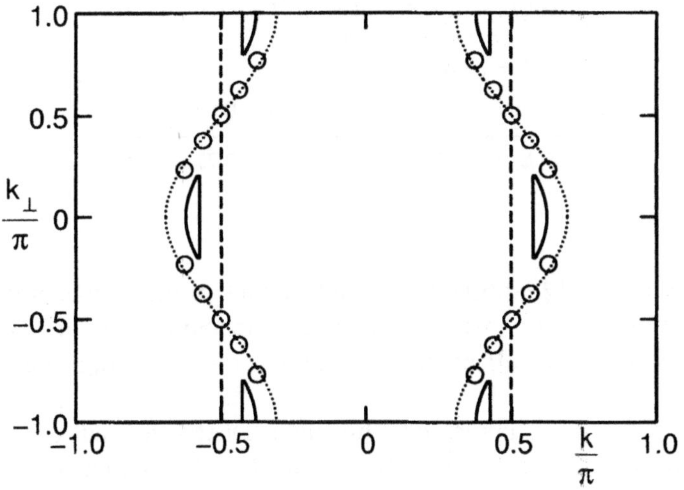

Figure 12. FS in the half-filled case with $t_\perp/W = 0.14$, $U/W = 0.65$ (circles), compared to the FS of the non-interacting case (dotted line) and of the purely 1d case ($t_\perp = 0$ -dashed-.). The solid line depicts schematically the FS obtained within the RPA ($\Sigma = \Sigma_{1D}$) [10].

particular, those should allow for stronger effects of the interactions, that is smaller values of K_ρ. We note, however, that the general tendency of coexistence of a small Drude peak with strong Hubbard bands is already visible in the Hubbard model.

3.6. THE LOW-TEMPERATURE FERMI LIQUID REGIME

At strong enough transverse coupling the system becomes a FL. In our numerical simulations, the onset of the FL regime is identified from the behavior of K_ρ (see Fig. 10) and from a linear behavior of the imaginary part of the self-energy in Matsubara space: $\Sigma(k, i\omega) \sim i\omega$. The equation defining the Fermi surface $\mu - \epsilon_k - \Sigma(k, 0) - \epsilon_\perp = 0$ then yields a relation $k_\perp(k)$ for the points $(k, k_\perp)$ that lie on the Fermi surface. These are visualized in Fig. 12 for the half-filled case. For the uncoupled (1d) system the Fermi surface consists of straight lines (dashed lines in the figure); the transverse hopping induces some cosine-like modulation but does not change the topology drastically. Indeed, the Fermi surface of the interacting coupled system (circles in Fig. 12) is very close to the one of non-interacting ($U = 0.$) coupled chains (dotted line in Fig. 12).

This is in striking contrast with predictions of an RPA approach, as pointed out in [10]. The RPA consists of replacing the self-energy of an interacting coupled system by the self-energy of a 1d chain. Since the 1d chain is a Mott insulator at half-filling, Σ diverges at low frequency for $k = \pi/2$ and the above equation has no solution for this k-value. Thus, within the RPA the Fermi surface cannot cross the points $(\pm\pi/2, \pm\pi/2)$ but consists of disconnected pockets as depicted

TABLE I. QP weights $Z(k_\perp)$ for different points on the FS (half-filled case, $t_\perp = 0.14W$, $U/W = 0.65$).

$k_\perp/\pi$	0.23	0.38	0.50	0.62	0.77
$Z(k_\perp)$	0.79	0.77	0.76	0.77	0.79

schematically in Fig. 12 (solid lines). The feedback of the effects of the interchain hopping on the self-energy, which is taken into account by chain-DMFT, regularizes the behavior of the self-energy near these points, leading to an open Fermi surface.

The QP residue $Z_{k_\perp}$ calculated within chain-DMFT depends only very weakly on the Fermi surface point (Table I). Again, this shows that the regularization by interchain hopping feedback effects is very efficient: a nearly divergent self-energy as could be imagined to result from an only slightly regularized 1d self-energy would lead to very small $Z(k_\perp)$ close to $\pm\pi/2$. Therefore approximations based on the 1D self-energy are likely to predict "hot spots" at those FS points corresponding to a vanishing inter-chain kinetic energy. Our results, in agreement with those of Arrigoni [2], do not support such a picture. Rather, the $k_\perp$-dependence of $Z(k_\perp)$ is very weak, with very shallow minima at $k_\perp \sim \pm\pi/2$. This small variation is however on the scale of our error bars.

4. Outlook

In these lecture notes, we have reviewed a few of the fascinating physical properties of quasi one-dimensional organic conductors. Our point of view has been that a unified description of both the TMTTF and TMTSF compounds is possible, as strongly interacting *quarter-filled* chains with weak interchain couplings. In that picture, the physical changes under pressure (or from TMTTF to TMTSF) are viewed as a deconfinement transition from a Mott insulating regime to a metallic regime [14, 26]. Still, we have emphasized that it is hardly possible to understand these compounds using a purely one-dimensional picture. Inter-chain coherence sets in at low energy in the metallic compounds, and this leaves us with a quite difficult theoretical problem, which cannot be handled by perturbative techniques in the inter-chain hopping.

We have reviewed the recently developed chain-DMFT approach which is in our opinion a promising route to handle these issues. QMC calculations on weakly coupled Hubbard chains have clearly demonstrated that this method is able to reproduce both the deconfinement transition and the dimensional crossover from a LL to a FL as the energy scale is reduced. We have also emphasized that a purely local Hubbard interaction is not appropriate for modelling quasi one-dimensional organics. Indeed, the phenomenology requires i) the possibility of a Mott insulating state at quarter filling and ii) rather small values of K_ρ ($\simeq 0.23$) which cannot be reached in the Hubbard model. A minimal theoretical model obeying these requirements is that of weakly coupled quarter-filled chains with both an on-site and a nearest-neighbour interaction. We hope to be able to deal with this model in the chain-DMFT framework in the near future.

Many outstanding questions on the physics of organics remain unanswered at this stage. We have tried to emphasize some of these puzzles in this paper.

Acknowledgements. We acknowledge many discussions with D. Jerome, as well as P. Auban-Senzier, J. Moser, P. Wzietek and C. Pasquier, and with L. Degiorgi and G. Grüner on the physics of quasi one-dimensional organic conductors. This research is partially supported by a Marie Curie Fellowship of the European Community Programm "Improving Human Potential" under contract number MCFI-2000804 and a grant of supercomputing time at NIC Jülich.

References

1. Arrigoni, E.: 1999. *Phys. Rev. Lett.* **83**, 128.
2. Arrigoni, E.: 2000. *Phys. Rev. B* **61**, 7909.
3. Biermann, S., A. Georges, A. Lichtenstein, and T. Giamarchi: 2001. *Phys. Rev. Lett.* **87**, 276405.
4. Bourbonnais, C. and L. G. Caron: 1991. *Int. J. Mod. Phys. B* **5**, 1033.
5. Bourbonnais, C., F. Creuzet, D. Jérome, K. Bechgaard, and A. Moradpour: 1984. *J. de Phys. (Paris) Lett.* **45**, L755.
6. Bourbonnais, C. and D. Jerome: 1999, 'The normal phase of quasi-one-dimensional organic superconductors'. In: P. Bernier, S. Lefrant, and G. Bidan (eds.): *Advances in Synthetic Metals, Twenty years of Progress in Science and Technology*. New York, p. 206. preprint cond-mat/9903101.
7. Dardel, B., D. Malterre, M. Grioni, P. Weibel, Y. Baer, J. Voit, and D. Jérôme: 1993. *Europhys. Lett.* **24**, 687.
8. Dressel, M., A. Schwartz, G. Grüner, and L. Degiorgi: 1996. *Phys. Rev. Lett.* **77**, 398.
9. Emery, V. J., R. Bruinsma, and S. Barisic: 1982. *Phys. Rev. Lett.* **48**, 1039.
10. Essler, F. H. L. and A. M. Tsvelik: 2001. preprint cond-mat/0108382.
11. Georges, A., T. Giamarchi, and N. Sandler: 2000. *Phys. Rev. B* **61**, 16393.
12. Georges, A., G. Kotliar, W. Krauth, and M. J. Rozenberg: 1996. *Rev. Mod. Phys.* **68**, 13.
13. Giamarchi, T.: 1991. *Phys. Rev. B* **44**, 2905.
14. Giamarchi, T.: 1997. *Physica B* **230-232**, 975.
15. Gor'kov, L.: 1997. *Physica B* **230-232**, 970.
16. Henderson, W., V. Vescoli, P. Tran, L. Degiorgi, and G. Grüner: 1999. *Eur. Phys. J. B* **11**, 365.
17. Hirsch, J. E. and R. M. Fye: 1986. *Phys. Rev. Lett.* **56**, 2521.
18. Jerome, D.: 1994, *Organic Superconductors: From $(TMTSF)_2PF_6$ to Fullerenes*, Chapt. 10, p. 405. Marcel Dekker, Inc.
19. Jérome, D. and H. J. Schulz: 1982. *Adv. Phys.* **31**, 299.
20. Moser, J., M. Gabay, P. Aubin-Senzier, D. Jerome, K. Bechgaard, and J. Fabre: 1998. *Eur. Phys. J. B* **1**, 39.
21. Schön, J. H., A. Dodabalapur, Z. Bao, C. Kloc, O. Schenker, and B. Batlogg: 2001. *Nature* **410**, 189.

22. Schulz, H. J.: 1990. *Phys. Rev. Lett.* **64**, 2831.
23. Schulz, H. J.: 1995, 'Fermi liquids and non–Fermi liquids'. In: E. Akkermans, G. Montambaux, J. L. Pichard, and J. Zinn-Justin (eds.): *Mesoscopic quantum physics, Les Houches LXI*. Amsterdam, p. 533.
24. Schwartz, A., M. Dressel, G. Grüner, V. Vescoli, L. Degiorgi, and T. Giamarchi: 1998. *Phys. Rev. B* **58**, 1261.
25. Tsuchiizu, M., P. Donohue, Y. Suzumura, and T. Giamarchi: 2001. *Eur. Phys. J. B* **19**, 185. and references therein.
26. Vescoli, V., L. Degiorgi, W. Henderson, G. Grüner, K. P. Starkey, and L. K. Montgomery: 1998. *Science* **281**, 1191.
27. Vescoli, V., F. Zwick, W. Henderson, L. DeGiorgi, M. Grioni, G. Gruner, and L. K. Montgomery: 2000. *Eur. Phys. J. B* **13**, 503.
28. Voit, J.: 1995. *Rep. Prog. Phys.* **58**, 977.
29. Wen, X. G.: 1990. *Phys. Rev. B* **42**, 6623.
30. Yakovenko, V. M.: 1992. *JETP Lett.* **56**, 510.

PROXIMITY INDUCED AND INTRINSIC SUPERCONDUCTIVITY IN CARBON NANOTUBES

M. KOCIAK[1], A. KASUMOV[1,2], R. DEBLOCK[1], S. GURON[1], B. REULET[1] AND H. BOUCHIAT[1]
[1]*Laboratoire de Physique des Solides, Associé au CNRS, Bâtiment 510, Université Paris-Sud, 91405, Orsay, France.*
[2]*Institute of Microelectronic Technology & High Purity Materials, Russian Academy of Sci., Chernogolovka, 142432 Moscow Region*

We have developed a technique for measuring transport measurements on carbon nanotubes at very low temperature. Carbon nanotubes are connected through low contact resistance to metallic pads. When the contact pads are superconducting, nanotubes become superconducting due to proximity effect with surprising large values of supercurrent. It has also been recently possible to observe observe intrinsic superconductivity in ropes of single wall carbon nanotubes mounted on normal contacts provided that the distance between the normal electrodes is large enough, other wise superconductivity is destroyed by proximity effect. These experiments indicate the presence of attractive interactions in carbon nanotubes which overcome coulomb repulsive interactions at low temperature.

1. Introduction

Metallic carbon nanotubes are known to be model systems for the study of 1D electronic transport. Depending on diameter and helicity, single wall carbon nanotubes (SWNT) have at most 2 conducting channels with minimum resistance in the absence of disorder predicted to be $h/4e^2 = 6.5k\Omega$. [10, 17, 40] Electronic correlations are expected to lead to a breakdown of the Fermi liquid state. Nanotubes should then be described by Luttinger Liquids (LL) theories [11, 21], with collective low energy excitations and no long range order. Proof of the validity of LL description with repulsive interactions in SWNT was given by the measurement of a resistance diverging as a power law with temperature down to 10 K [4]. The extrapolation of this behavior down to very low temperature would indicate an insulating state. However, this measurement was done on nanotubes separated from measuring leads by tunnel junctions. Because of Coulomb blockade [15], the low temperature and voltage regime were not explored. In contrast, we have

I.V. Lerner et al. (eds.),
Strongly Correlated Fermions and Bosons in Low-Dimensional Disordered Systems, 103–116.

developed a technique in which measuring pads are connected through low contact resistance to suspended nanotubes [22]. We first show that when the contact pads are superconducting, a large supercurrent can flow through nanotubes [23]. These experiments indicate that quantum coherent transport can take place on carbon nanotubes on micron scales. However the surprisingly high values found for critical currents in individual nanotubes rise the question of possible intrinsic superconductivity. Furthermore our experiments on ropes of SWNT connected to normal contacts show indeed evidence of such superconductivity below 0.5 K provided that the distance between normal electrodes is long enough.

2. Sample description and preparation

The SWNT are prepared by an electrical arc method with a mixture of nickel and yttrium as a catalyst [20, 37]. SWNT with diameters of the order of 1.4 nm are obtained. They are purified by the cross-flow filtration method [37]. The tubes are usually assembled in ropes of a few hundred parallel tubes, but individual tubes can also be obtained after chemical treatment with a surfactant [6]. Isolation of an individual rope and connection to measuring pads are performed according to the following procedure. A focused laser beam releases a nanotube from a target onto the sample with a Si_3N_4 membrane coated with a metal film. A submicron width slit about $100\mu m$ in length has previously been cut in the membrane using a focused ion beam; the nanotube connects the edges of the slit and shorts the electric circuit whose resistance was over $1G\Omega$ before the nanotube was "shaken off". The resulting junctions have low resistivity ohmic contacts between the tube and the metal electrodes molten during the process, and can be characterized using transmission microscopy. The use of superconducting bilayer electrodes (Re/Au or Ta/Au) gives rise to the proximity effect. Re or Ta ensures good adhesion to the substrate and superconductivity of contacts; the imiscibility of the Re/Au and Ta/Au systems (both solid and liquid) and the very high melting temperature of Re and Ta prevents electrode damage during laser welding of the tubes to the top gold layer. Au is an ideal solder for nanotubes because it neither reacts with carbon, nor oxidizes. The samples were measured in a dilution refrigerator, at temperatures ranging from 300 K to 0.05 K, through filtered lines [33]. Magnetic fields up to 5 T could be applied perpendicularly to the contacts and the tubes. The resistance was measured by applying a small (1 nA to 10 nA, 30 Hz) a.c. current though the sample and measuring the a.c. voltage using lock-in detection.

The room temperature value R_0 of such junctions varies considerably from sample to sample, typically between 100Ω and $100k\Omega$. For ropes, when R_0 is of the order or larger than $10k\Omega$, a sharp divergence of the resistance is observed below $T = 50K$ corroborating experimental results obtained by other groups (12). This low-temperature insulating behavior has been attributed to the helicity of the tubes or to the existence of defects such as heptagon-pentagon pairs. It has

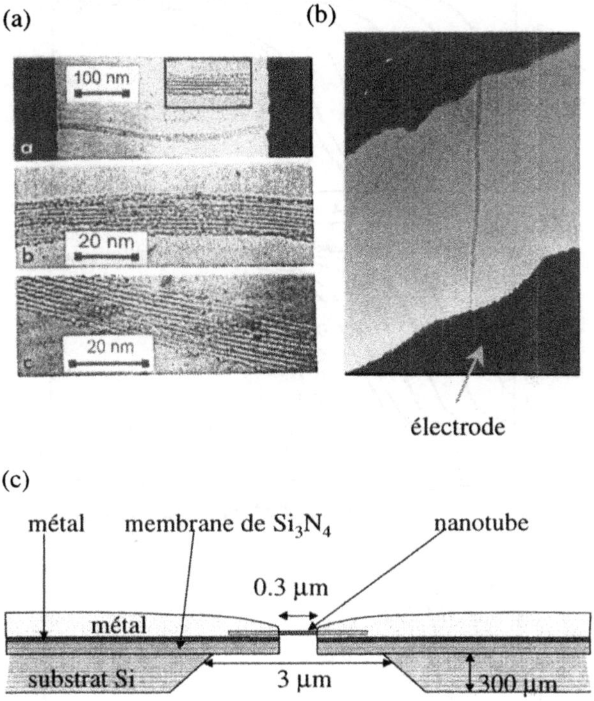

Figure 1. Transmission electronic micrograph of the nanotubes, suspended across a slit between 2 metallic pads. From these pictures it is possible to confirm that the metal of the contacts neither covers nor penetrates the tubes. A) Ropes. B) Sample Au_1, consists of a single SWNT $0.3\mu m$ long which is completely isolated at one end, on a length of 50 nm, but merges into a narrow rope of 7 tubes (0.2 μm long) on the other end. C)Schematic picture of a typical sample suspended across a slit and soldered to metallic pads.

also been theoretically shown (13) that Van der Waals interaction between individual nanotubes in a rope can induce a band gap of about 0.1 eV, making a rope consisting in individually metallic tubes semiconducting. We have found that it is also possible to obtain samples with resistance much lower than $6.5k\Omega$, indicating that a large fraction of the tubes participate in transport. Such low values of R_0 and metallic behavior down to low temperature indicate good metallic contacts. Moreover, intrinsic electronic transport properties of SWNT are more likely to be observed in suspended nanotubes than in deposited tubes where they could be altered by the substrate. Metallic behavior is observed for single SWNT when $R_0 < 30k\Omega$ and for ropes when $R_0 < 10k\Omega$.

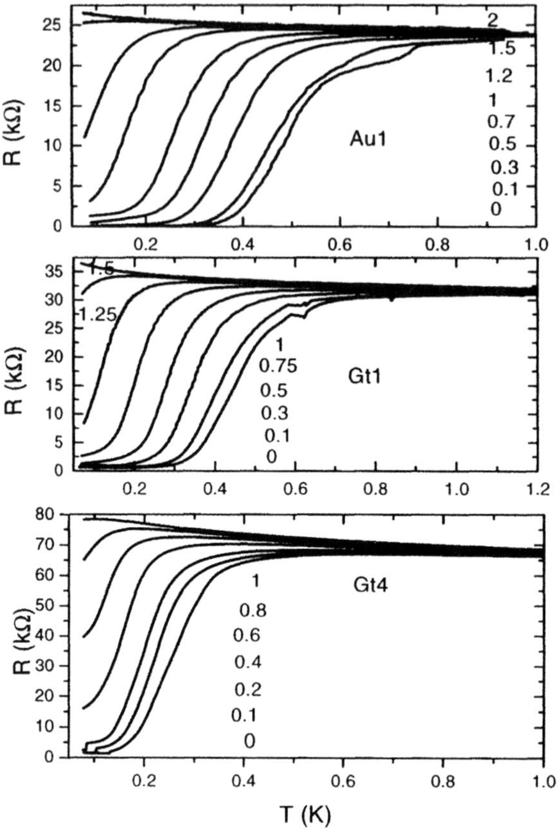

Figure 2. Temperature dependence of the resistance of individual single wall nanotubes mounted on Au/Ta measured for different values of the magnetic field perpendicular to the tube axis. (The labels on the curves correspond to the value of magnetic field in Tesla).

3. Proximity induced superconductivity

A normal metal (N) in contact with a macroscopic superconductor (S) exhibits superconductivity on a characteristic length which is smaller than both the phase coherence length in the normal metal L_ϕ and the thermal length L_T, (in a clean metal $L_T = \hbar v_f / k_B T$ and in a dirty metal $L_T = \sqrt{\hbar D / k_B T}$ where D is the electron diffusion coefficient). Both

lengths, of the order of a few μm, can be much longer than the superconducting coherence length. The superconducting transition temperature T_c and the gap Δ in the normal metal can be of the order of their values in S if the resistance of the SN interface is sufficiently small. This proximity effect has been extensively studied in multilayered planar SN junctions [9] and more recently in lithographically fabricated micron scale metallic wires made of normal noble metals between two macroscopic superconducting electrodes [38]. The maximum low temperature value of the supercurrent (critical current) in such SNS junctions is $\pi\Delta/eR_N$, where R_N is the normal resistance of the junction [26]. We show that the supercurrent sustained by a single nanotube exceeds this prediction by a large factor.

We have observed proximity induced superconductivity in isolated SWNT samples like $Au1$ depicted in Fig.1 with $R_0 = 27k\Omega$. This value is of the order of the inverse quantum conductance $h/e^2 = 25.8k\Omega$, the maximum value for which one expects low temperature metallic behavior in a coherent 1D system. The tube is mounted on Ta/Au electrodes (5nm Ta, 100nm Au) and has a superconducting transition at $T_c = 0.4K$, close to the transition temperature of the contacts. The transition is shifted to lower temperature when a magnetic field is applied in the plane of the contacts (perpendicular to the tube axis). The critical field is surprisingly high (of the order of 1 Tesla) and is ten times larger than the measured critical field of the contact. Above 2 Tesla the resistance becomes field independent and a small increase is apparent when the temperature is lowered below $0.2K$ (Fig.2).

The most striking signature of induced superconductivity is the existence of Josephson supercurrents through the samples. The results of non-linear transport measurements performed on these superconducting samples are exemplified by the voltage vs. current characteristics on sample $Au1$ depicted in Fig.3. The transition between the superconducting state (zero voltage drop through the sample) and the dissipative state is quite abrupt and displays hysteresis at low temperature. It is characterized by a critical current i_c whose value at $T \sim 0$ is $i_0 = 0.104\mu A$ for $Au1$. We note that the junction does not recover its normal resistance above the critical current and the curve V(I) exhibits subsequent hysteretic jumps at higher currents. This behaviour is similar to that observed in long superconducting filaments (13), where above i_c, small normal regions (of size comparable to the inelastic length) appear located around defects in the sample (phase slip centers). However, the precise understanding of the physical mechanism involved in a SWNT with similar aspect ratio, but with dimensions 100 times smaller, requires further experimental and theoretical study.

It is very difficult to understand these results in the frame-work of conventional proximity induced superconductivity. The value of the product $R_N i_c$ at $T = 0$ is equal to 2.5 mV, 40 times larger than the expected value (0.06mV) assuming the gap is identical to its value in the superconducting electrodes (as estimated from T_c). More over signs of superconductivity persist for voltage drop through the sample much higher than the gap of pure tantalum. Similar results are found in similar individual SWNT samples with values of normal state resistance up to $70k\Omega$. Proximity induced superconductivity was also observed in ropes of SWNT with also suprisgly hifgh values of supercurrents.

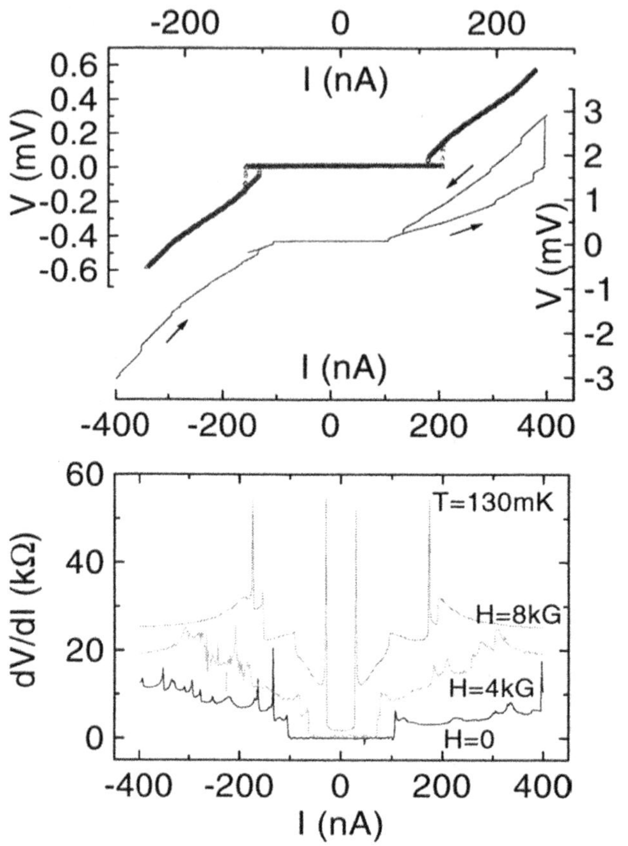

Figure 3. A) Upper curve, V(I) characteristics for the individual SWNT Au1 , from which the critical current is deduced. Lower curve, the same data on a wider current scale showing the existence of voltage steps for currents higher than i_c. B) Differential resistance measured with a small ac modulation of the current for various values of the magnetic field. It is clear from this data that the normal resistance $R_N = 25k\Omega$ is only recovered at currents much higher than i_c.

An explanation of these phenomena is not straightforward. The problem of an SNS junction constituted of a Luttinger liquid between two superconducting reservoirs has been considered theoretically by Fazio et al. [12] for weakly transmitting contacts and Maslov et al. [29] for perfectly transmitting contacts. In both cases, these authors show that it is indeed possible to induce superconductivity by a proximity effect. In the case of repulsive interactions and perfectly transmitting interfaces the zero temperature value of the critical current is not changed by the interactions. More recently it has been shown that the presence of attractive interactions in the Luttinger liquid can result in a significant increase of the critical current [1], [14].

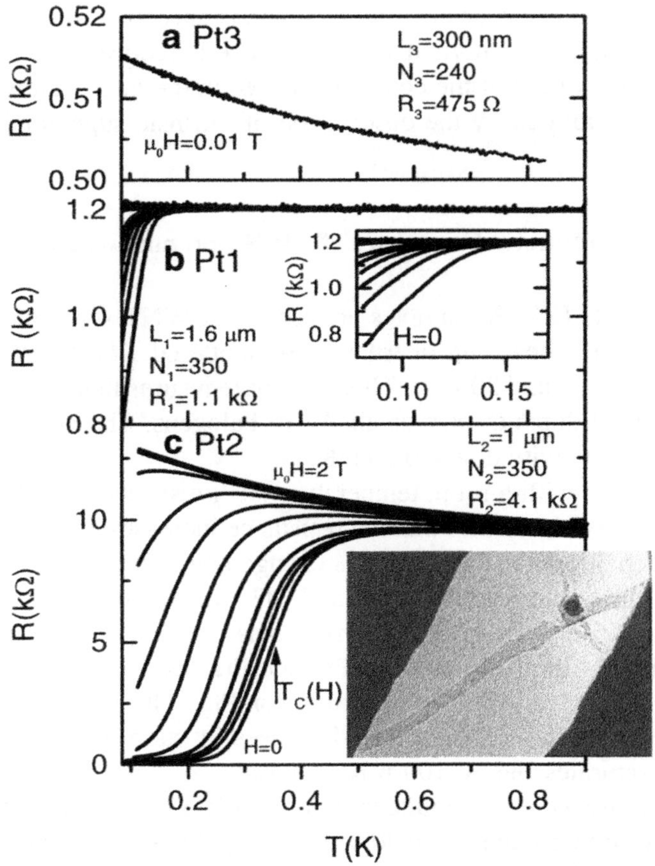

Figure 4. Resistance as a function of temperature for the three samples. The length L, number of tubes N and room temperature resistance R of each sample are given in the corresponding panel. a: Sample Pt3. b: Resistance of Pt1 in applied magnetic fields of $\mu_0 H=$ 0, 0.02, 0.04, 0.06, 0.08, 0.1, 0.2, 0.4, 0.6, 0.8 and 1 T from bottom to top. Inset is a zoom of the low temperature region. c: Resistance of Pt2 at $\mu_0 H$=0, 0.05, 0.1, 0.2, 0.4, 0.6, 0.8, 1, 1.25, 1.5, 1.75, 2, 2.5 T from bottom to top. Inset: TEM micrograph of sample Pt2, from which we deduce L_2 and N_2. N_2 is estimated from the measured diameter D_2, through $N_2 = (D_2/(d+e))^2$, where d is the diameter of a single tube (d=1.4 nm), and e is the typical distance between tubes in a rope (e=0.2 nm). The dark spot is a Ni/Y catalyst particle.

Our data could indeed be explained by the existence of superconducting fluctuations intrinsic to SWNT. For an infinite nanotube, because of its 1D character, these fluctuations are not expected to give rise to a superconducting state at finite temperature. However, the superconducting state might be stabilized by the macroscopic superconductivity of the contacts. In such a situation, it is conceivable to expect that i_c could be enhanced compared to its value in a conventional SNS junction. The existence of superconducting fluctuations intrinsic to nanotubes may also explain the positive magneto-resistance (not shown) observed in all our samples where the normal state resistance is recovered at fields much higher than the critical field of the contacts. We see in the next section that these statements are corroborated by the observation of intrinsic superconductivity in ropes of SWNT.

4. Intrinsic superconductivity in ropes of SWNT on normal contacts

For these experiments [24] the samples are ropes of SWNT suspended between normal metal contacts consisting of trilayers of sputtered $Al_2O_3/Pt/Au$ of respective thicknesses 5, 3 and 200 nm. This procedure insures that the tubes do not contain any chemical dopants such as alkalis or halogens. The contacts showed no sign of superconductivity down to 50 mK.

We select samples with a room temperature (RT) resistance less than 10 kΩ. As is generally observed, we find that the resistance increases as the temperature is lowered between 300 K and 1K . Things change however below 1K, as shown in Fig.4 for the three samples Pt1, Pt2, and Pt3, measured in magnetic fields ranging from 0 to 2.5 T. At zero field, the zero-bias resistance of Pt3 increases as T is reduced, whereas the resistances of Pt1 and Pt2 decrease drastically below $T_1^* = 140$ mK for Pt1 and $T_2^* = 550$ mK for Pt2. The resistance of Pt1 is reduced by 30% at 70 mK. That of Pt2 decreases by more than two orders of magnitude, and saturates below 100 mK at a value $R_r = 74\ \Omega$. We define a transition temperature T_{C_2} by the inflexion point of R(T). T_{C_2} is 370 mK at zero field, decreases at higher magnetic fields, and extrapolates to zero at 1.35 T (Fig 7c). At fields above 1.25 T, the resistance increases with decreasing temperature, similarly to Pt3, and becomes independent of magnetic field. The resistance of Pt1 follows qualitatively the same trend, but the full transition did not occur down to 70 mK. Figures 4 and 5 show that in the temperature and field range where the zero-bias resistance drops, the differential resistance is strongly bias-dependent, with lower resistance at low bias. These data suggest that the rope Pt2 (and, to a lesser extent, Pt1) is intrinsically superconducting. Although the experimental data of Pt2 seem similar to those of SWNT connected to superconducting contacts [23], there are major differences. In particular the $V(I), dV/dI(I)$ do not show any supercurrent because of the existence of a finite residual resistance.

We now analyse the superconductivity in these systems, taking into account several features: the large normal contacts, the coupling between tubes within the rope, the 1D character of each tube, and their finite length compared to relevant mesoscopic and superconducting scales. The resistance of any superconducting wire measured through normal contacts (an NSN junction) cannot be zero because the number of channels in the wire is much smaller than in the contacts [25]: a metallic SWNT, with 2 conducting channels, has a contact resistance of half the resistance quantum, $R_Q/2$ (where $R_Q = h/(2e^2)$=12.9 kΩ), even if it is superconducting. A rope of N_m parallel metallic SWNT will have a minimum resistance of $R_Q/(2N)$. Therefore we use the residual resistance $R_r = 74\ \Omega$ of Pt2 to deduce that Pt2 has at least $N_m = R_Q/2R_r \approx 90$ metallic tubes. This is approximately one quarter of the number of tubes in the rope, measured by TEM (Fig 4c). Similarly, R_Q is also the maximum resistance of any phase coherent metallic wire [35]. As a consequence, the high value (9.2 kΩ) of the resistance at 1K (which corresponds to an average resistance per metallic tube of $9.2\ \mathrm{k}\Omega * N_m = 830\ \mathrm{k}\Omega = 130\ R_Q$) cannot be understood if the nanotubes are independent, unless considering a very short (unphysical) phase coherence length $L_\varphi(1K) = L/130 = 8$ nm. On the other hand if the electrons are free to move from tube to tube[28], the resistance is simply explained by the presence of disorder. The mean free path is deduced from the RT resistance $R_2 = 4.1\ \mathrm{k}\Omega$ through [19] $l_{e2} \approx \frac{L}{R_2}\frac{R_Q}{N_m} \approx 18$ nm. We conclude that Pt2 is a diffusive conductor with a few hundred conduction channels.

With such a small number of channels, we expect the superconductivity to differ from 3D superconductivity. In particular, we expect to observe a broad resistance drop starting at the 3D transition temperature [13] T^* and going eventually to R_r at zero temperature. This is what is observed in Pt2 (see figure 4.c). We estimate the gap through the BCS relation $\Delta = 1.76\ k_B T^* : \Delta \approx 85\ \mu eV$ for Pt2. We can then deduce the superconducting coherence length along the tube in the diffusive limit $\xi_2 = \sqrt{\hbar v_f l_e/\Delta} \approx 0.3\ \mu$m where v_f is the longitudinal Fermi velocity 8×10^5 m/s [5]. Consistent with 1D superconductivity, ξ_2 is ten times larger than the diameter of the rope.

Finally, reminiscent of measurements of narrow superconducting metal wires [13], we find jumps in the differential resistance as the current is increased (Figures 5 and 6). For Pt2 the differential resistance at low currents remains equal to R_r up to 50 nA, where it strongly rises but does not recover its normal state value until 2.5 μA (fig 6a). The jump in resistance at the first step corresponds approximately to the normal state resistance of a length ξ_2 of Pt2. Each peak corresponds to a hysteretic feature in the V-I curve (fig 6b). Above 1 T the differential resistance is peaked at zero current. This is also the case for Pt3 (data not shown). The variations of the differential resistance of Pt1 are similar to those of Pt2 close to its transition temperature. These jumps are identified as phase slips [13, 31, 36], which are the occurrence of normal regions located around defects in the sample. Such phase slips can be thermally activated (TAPS), leading to an exponential decrease of the resistance instead of a sharp transition, in qualitative agreement with our experimental observation (fig 7a). At sufficiently low temperature, TAPS are replaced by quantum phase slips (QPS), which, when tunneling through the sample, contribute an additional resistance to the zero temperature resistance. Moreover, QPS are predicted to supress the transition when the normal state resistance of the sample on the phase coherence scale is larger than $R_Q/2$ [41](as confirmed by recent experiments [3]). Our data on Pt2 show no evidence

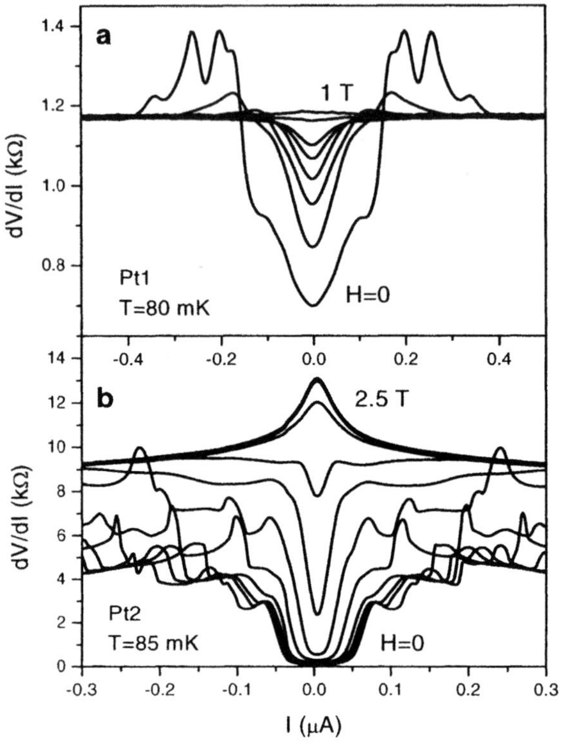

Figure 5. Differential resistance as a function of current for samples Pt1 and Pt2, in different applied fields. a: Sample Pt1. Fields are 0, 0.02, 0.04, 0.06, 0.08, 0.1, 0.2 and 1 T. b: Sample Pt2. Fields are 0, 0.2, 0.4, 0.6, 0.8, 1, 1.25, 1.5, 1.75, 2, and 2.5 T.

of such an effect, even though the normal state resistance, measured above T*, is 40% larger than $R_Q/2$.

The current above which the jumps disappear, 2.5 μA, is close to the critical current $I_C = \Delta/R_r e \approx 1$ μA of a superconducting wire without disorder and with the same number of conducting channels [36]. This large value of critical current would also be the maximum supercurrent in a structure with this same wire placed between superconducting contacts (with gap Δ_S), and is much larger than the Ambegaokar-Baratoff prediction $R_N I_C = \Delta_S/e$. This might explain the anomalously large supercurrent measured in a previous experiment [23], where nanotubes were connected to superconducting contacts. We now discuss the effect of the magnetic field. The field at which the resistance saturates to its normal value and at which the critical current vanishes, 1.25 T, coincides with the field obtained

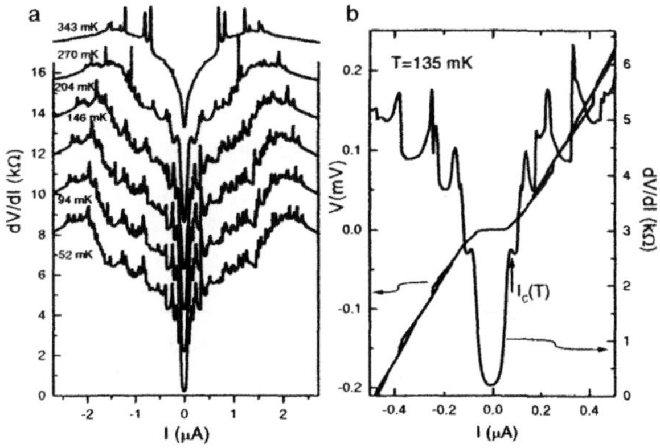

Figure 6. Left panel: Differential resistance of Pt2 vs. current for a larger current amplitude than in Fig.5, at different temperatures. Curves are offset vertically for clarity. Right panel: V(I) and $\frac{dV}{dI}(I)$ curves showing the hysteretic behavior in V(I) at each peak in the $\frac{dV}{dI}(I)$ curve.

by extrapolation of $T_C(H)$ to zero temperature (fig. 7b).

It is difficult to say what causes the disappearance of superconductivity. The value of Hc(0) should be compared to the depairing field in a confined geometry [30], and corresponds to a flux quantum Φ_0 through a length ξ of an individual SWNT of diameter d, $\mu_0 H_C = \Phi_0/(2\sqrt{\pi}d\xi) = 1.35$ T. But $H_C(0)$is also close to the field $\mu_0 H_p = \Delta/\mu_B = 1.43$ T at which a paramagnetic state becomes more favorable than the superconducting state [8, 7]. Note that this value is of the same order as the critical field that was measured on SWNT connected between superconducting contacts, i.e. much higher than the critical field of the contacts.

We now estimate the superconducting coherence length of the two other samples, to explain the extent or absence of observed transition. Indeed, investigation of the proximity effect at high-transparency NS interfaces has shown that superconductivity resists the presence of normal contacts only if the length of the superconductor is much greater than ξ [2], i.e. if the wire contains a superconducting reservoir. This condition is nearly fulfilled in Pt2 ($\xi_2 \approx L_2/3$). Using the high temperature resistance values of Pt1 and Pt3, and assuming a gap Δ and number of metallic tubes equal to those of Pt2 we find $\xi_1 \approx L_1/2$ and $\xi_3 \approx 2L_3$. These values explain qualitatively a reduced transition temperature for Pt1 and the absence of a transition for Pt3. Moreover we can argue that the superconducting transitions we see are not due to a hidden proximity effect : if the $Al_2O_3/Pt/Au$ contacts were made superconducting by the laser pulse, the shortest nanotube (Pt3) would become superconducting at temperatures higher than the longer tubes (Pt1 and Pt2). The main result, i.e. no visible transition with a short rope, and a visible transition in a long rope, are confirmed by measurements on two other samples which are not presented here.

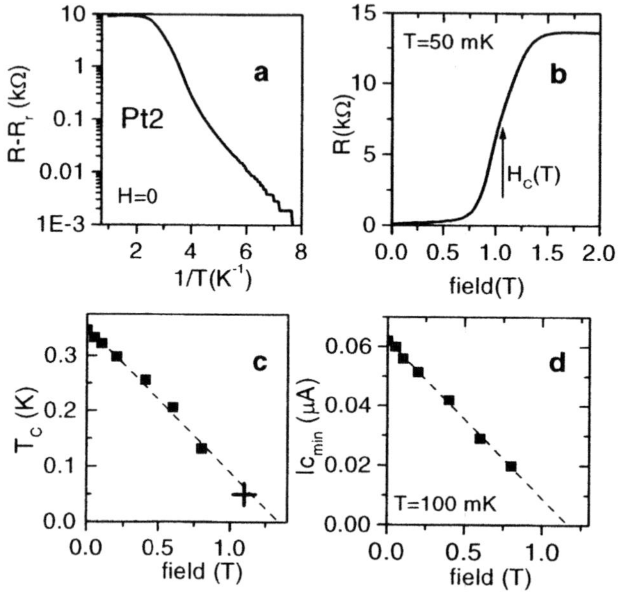

Figure 7. **a Resistance of Pt2 plotted on a log scale as a function of the inverse temperature at H=0. We have subtracted the low temperature residual resistance (contact resistance). The slope yields an approximate activation energy of 0.8 K. b Magnetoresistance of Pt2 at 50 mK. We define the critical field as the inflection point of R(H): $\mu_0 H_C$(T= 50 mK)=1.1 T. c Transition line of Pt2 defined in the H,T plane by the inflection point of R(T) or equivalently by the inflection point of R(H). d Field dependence of the critical current of Pt2 defined as the current at which the first resistance jump occurs in the dV/dI curves of Fig.5. $I_C(H)$ extrapolates to a critical field of 1.2 T, in agreement with the linear extrapolation 1.3 T of $T_C(H)$.**

We now consider the possible mechanism of superconductivity. It has been suggested that coupling with low energy phonons can turn repulsive interactions in a Luttinger liquid into attractive ones and drive the system towards a superconducting phase [27]. Such low energy phonons have been experimentally observed in the form of mechanical bending modes of a suspended SWNT rope [34]. It was also shown that superconducting fluctuations can dominate at low temperature in ladders such as tubes [11]. In this case the system must be away from half-filling, a condition probably fulfilled in our experimental situation, due to hole doping from the contacts [39, 32]. Finally, the superconductivity reported here recalls that of graphite intercalated with alkalis (Cs,K), which also occurs between 0.2 and 0.5 K [18]. Much higher temperatures were observed in alkali doped fullerenes [16] because of the coupling to higher energy phonons. This suggests the possibility of increasing the transition temperature by chemically doping the nanotubes.

We have shown that ropes of carbon nanotubes are intrinsically superconducting. This is the first observation of superconductivity in a system with such a small number of conduction channels. The understanding of this superconductivity calls for future experimental and theoretical work and motivates in particular a search of superconducting fluctuations in a single SWNT.

A.K. thanks the Russian foundation for basic research and solid state nanostructures for financial support, and thanks CNRS for a visitor's position. We thank M. Devoret, N. Dupuis, T. Martin, D. Maslov and C. Pasquier for stimulating discussions.

References

1. Affleck I., J.B. Caux and Zagoskin A. *Phys. Rev.*, **B 62,** 1433 (2000).
2. Belzig W., Bruder C. and Schön G., Phys. Rev. B **54**, 9443 (1996).
3. Bezryadin A., Lau C. N. and Tinkham M. Nature **404**, 971 (2000).
4. Bockrath M. *et al.*, Nature **397**, 598 (1999).
5. Note however that the transverse velocity is certainly much smaller, yielding a transverse coherence length $\xi_\perp$ smaller than the longitudinal one, but the 1D character of the transition depicted below indicates that $\xi_\perp$ is certainly larger than the diameter of the rope. Such anisotropic transport is observed in organic conductors where a very small coupling between 1D chains restores a 3D behavior of the electrons, and allows superconductivity in a macroscopic sample. (see e.g. Bourbonnais C., Jérôme D., in Advance in Synthetic Metals, eds P.Bernier, S. Lefrant and G. Bidan) 206 (1998).
6. Burghard M. et al, *Electronic Properties of Novel Materials* (Eds H. Kuzmany et al,) AIP, New York (1998).
7. Chandrasekhar B.S., Appl. Phys. Lett. **1**, 7 (1962).
8. Clogston A.M., Phys. Rev. Lett. **9**, 266 (1962).
9. Deutscher G. and de Gennes P.G. in *Superconductivity* (Ed. R.D. Parks) Marcel Dekker Inc. (1969).
10. Dresselhaus M.S., Dresselhaus G., and Eklund P.C., *Science of Fullerenes and Carbon nanotubes* (Academic, San Diego, 1996).
11. Egger R., Gogolin A., Phys. Rev. Lett. **79**, 5082 (1997). Egger R., Phys. Rev. Lett. **83**, 5547 (1999).
12. Fazio R., J Hekking F.W. and Odintsov A.A., *Phys. Rev.*, **B 53**, 6653 (1995).
13. Giordano N., Phys. Rev. B **50**, 160 (1991).
14. Gonzalez J., Phys. Rev. Lett. **87**, 136401 (2001).
15. Grabert H. and Devoret M. H. (eds), Single Charge Tunneling (Plenum, New-York, 1992); Tans J. T. *et al.*, Nature **386**, 474 (1997).
16. Gunnarsson O., Rev. Mod. Phys. **69**, 575 (1997).
17. Hamada N., Sawada S. I. and Oshiyama A., Phys. Rev. Lett. **68**, 1579 (1992).
18. Hannay N. B. *et al.*, Phys. Rev. Lett. **14**, 225 (1965).

19. Imry Y., Europhys. Lett. **1**, 249 (1986).
20. Journet C., *et al.*, Nature **388**, 756 (1997).
21. Kane C., Balents L., Fisher M. P., Phys. Rev. Lett **79**, 5086 (1997).
22. Kasumov A.Yu., Khodos I.I., Ajayan P.M., Colliex C., Europhys. Lett. **34**, 429 (1996); Kasumov A.Yu. *et al.*, Europhys. Lett. **43**, 89 (1998).
23. Kasumov A.Yu. *et al.*, Science **284**, 1508 (1999).
24. Kociak M. *et al.*, Phys. Rev. Lett. **86** 2416, (2001).
25. Landauer R., IBM Res. Dev. 1, 223 (1957).
26. For a review of suppercurrents through superconducting junctions see: Likharev K., *Rev. Mod. Phys.* **51**,101 (1979).
27. Loss D., and Martin T., Phys. Rev. B **50**, 12160 (1994).
28. Electronic transfer between tubes in a disordered rope is indeed suggested in: Maarouf A. A., Kane C. L., and Mele E. J., Phys. Rev. B**61**, 11156 (2000); H.R. Shea, R. Martel and Ph. Avouris, Phys. Rev. Lett. **84**, 4441 (2000).
29. Maslov D., Stone M., Goldbart P.M.,Loss D. *Phys. Rev.*, **B 53,** 1548 (1995).
30. Meservey R.,Tedrow P.M., Phys. Rep. **238**, no.4, 173 (1994).
31. Meyer J., Minnigerode G. V., Physics Letters, **38A**, 7, 529 (1972).
32. Odintsov A.A., Phys. Rev. Lett. **85**, 150 (2000).
33. Reulet B., Bouchiat H., and Mailly D., Europhys. Lett. **31**, 305 (1995).
34. Reulet B. *et al.*, Phys. Rev. Lett **85**, 2829 (2000).
35. Thouless D. J. Phys. Rev. Lett. **39**, 1967 (1977).
36. Tinkham, M., Introduction to superconductivity, McGraw-Hill, 2d Ed. (Singapore, 1996).
37. Vaccarini L., *et al.*, C.R.Acad.Sci.**327**, 925 (1999).
38. van Dover R.B., de Lozanne A. and Beasley M.R. *J.Appl. Phys.* **52**, 7327 (1981), Courtois H., Gandit Ph.,Panetier B., *Phys. Rev.* **B 52** 1162 (1995).
39. Venema L. C. *et al.*, Science **283**, 52 (1999).
40. Wildoer J. W. G. *et al.*, Nature **391**, 59 (1998).
41. Zaikin A.D. *et al.*, Phys. Rev. Lett. **78**, 1552 (1997).

QUANTUM IN-PLANE MAGNETORESISTANCE IN 2D ELECTRON SYSTEMS

J.S. MEYER[1], V.I. FAL'KO[2] AND B.L. ALTSHULER[3]
[1] *Institut für Theoretische Physik, Universität zu Köln, 50937 Köln, Germany*
[2] *Physics Department, Lancaster University, LA1 4YB, Lancaster, UK*
[3] *Physics Department, Princeton University, NJ 08544, USA and NECI, 4 Independence Way, Princeton, NJ 08540, USA*

1. Introduction

Studies of low-dimensional electron systems subject to an in-plane magnetic field have come to the focus of intense attention recently. In several experiments the influence of an in-plane magnetic field on two- and one-dimensional electrons in semiconductor heterostructures [1, 2] as well as in lateral quantum dot devices [3] has been investigated. The use of the fairly unconventional in-plane field geometry aimed at achieving a stronger magnetic field influence on the 2D electron spin, thus, compensating the dominance of orbital effects in most of the known semiconductor materials. Then, by manipulating the field-induced spin polarization, one may extract information about the ground state properties of interacting electrons. At low temperatures – i.e., in the regime where electron-electron interactions open the possibility of a metal-insulator transition or the ferromagnetic Stoner instability – interaction effects coexist with single-particle interference effects. The aim of this article is to provide a theoretical overview of possible influences of an in-plane magnetic field on single-particle quantum transport phenomena in semiconductor heterostructures, quantum wells, and lateral dots.

Quantum transport phenomena have been the subject of extensive experimental and theoretical investigations during recent decades [4–7]. There exists a broad variety of electronic systems, where spectacular effects produced by the interference of electron waves have been observed. These include weak localization (WL) [8], leading to the low-temperature magnetoresistance in low-dimensional electron structures, such as two-dimensional electron and hole gases in semi-

I.V. Lerner et al. (eds.),
Strongly Correlated Fermions and Bosons in Low-Dimensional Disordered Systems, 117–164.

conductor heterostructures and quantum wells, as well as universal conductance fluctuations (UCF) [9–12] in quantum wires and dots.

Weak localization of electron waves is the result of an enhancement of back-scattering from a scattering center in a disordered environment. This enhancement of back-scattering suppresses electrical conductivity of a quantum conductor, as compared to the value expected on the basis of a classical Drude formula. The Drude result is determined by the mean free path corresponding to a single-impurity scattering cross section whereas multiple scattering events as depicted in Fig. 1 are responsible for the enhanced back-scattering. It is generated by the constructive interference between pairs of electron waves that scatter from the same impurities, i.e., two electron waves arrive together at a particular impurity, scatter from it to follow a random-walk path (in the form of a closed loop) visiting the same set of surrounding impurities – either in clockwise or anti-clockwise direction – and finally scatter into the exactly backward direction. An illustration of electron paths involved in such a non-local scattering process is shown in Fig. 2. Obviously, the observation of such an interference effect requires phase-coherent propagation along the closed-loop part of the geometrical trajectory. Furthermore, the magnitude of the observable contribution to the conductivity is sensitive to the fundamental symmetries of the system, in particular, to the presence or absence of time-reversal ($\mathcal{T}$) symmetry.

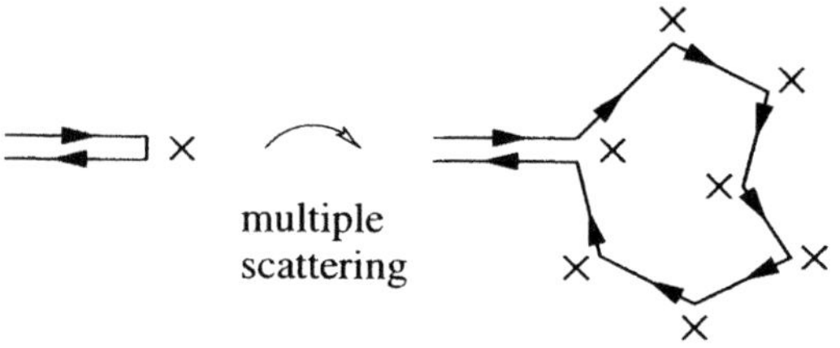

Figure 1. Back-scattering from an impurity: Contributions due to multiple scattering events.

The effect of enhanced back-scattering off an impurity inside a disordered metal can be discussed in the same terms as the enhancement of back-scattering of waves from a diffusive medium. The probability $w_{\mathbf{k},-\mathbf{k}}$ of a wave to be reflected back from a medium after multiple scattering inside is determined as the modulus of the sum of amplitudes A_j corresponding to particular paths j:

$$w_{\mathbf{k},-\mathbf{k}} = \left| \sum_j \left[A_j^{\circlearrowright}(\mathbf{k}, -\mathbf{k}) + A_j^{\circlearrowleft}(\mathbf{k}, -\mathbf{k}) \right] \right|^2 . \tag{1}$$

Here, $\circlearrowright$ ($\circlearrowleft$) stands for clockwise (anti-clockwise) propagation along the path.

In an ideally time-reversible system, the two amplitudes related to the electron propagation along the same geometrical loop 'j' in opposite directions, $A_j^{\circlearrowright}(\mathbf{k}, -\mathbf{k})$

and $A_j^{\circlearrowright}(\mathbf{k},-\mathbf{k})$, are identical. Thus, $A_j^{\circlearrowright}A_j^{\circlearrowleft *} = |A_j^{\circlearrowright}|^2$, though for each particular loop A_j contains a possibly large random semiclassical phase of propagation. The cancellation of phase factors happens because for an electron described by a $\mathcal{T}$-symmetric Hamiltonian, (a) the phases acquired along the same 'free' segment of the path are equal for waves propagating with the wave number $\mathbf{k}$ and $-\mathbf{k}$, and (b) the amplitudes of intermediate scattering processes, $\mathbf{k}_a \to \mathbf{k}_b$ for $\circlearrowright$ and $-\mathbf{k}_b \to -\mathbf{k}_a$ for $\circlearrowleft$ also equal each other at each node of the loop j. Therefore, after averaging over different loops which is equivalent to the averaging over disorder, one can identify two phase-insensitive contributions to the back-scattering probability,

$$\begin{aligned} w_{\mathbf{k},-\mathbf{k}} &= \sum_j \left\langle |A_j^{\circlearrowright}|^2 + |A_j^{\circlearrowleft}|^2 \right\rangle + 2\,\mathrm{Re} \sum_j \left\langle A_j^{\circlearrowright} A_j^{\circlearrowleft *} \right\rangle \\ &= w^{\mathrm{class}} + \delta w^{\mathrm{WL}}, \end{aligned} \tag{2}$$

where w^{class} is the back-scattering probability of a classical particle whereas δw^{WL} is an additional quantum contribution due to constructive interference between time-reversed paths. This second term in Eq. (2) increases the return probability, i.e., the probability for an electron to revisit the same scatterer again – which explains the name 'weak localization' given to this effect.

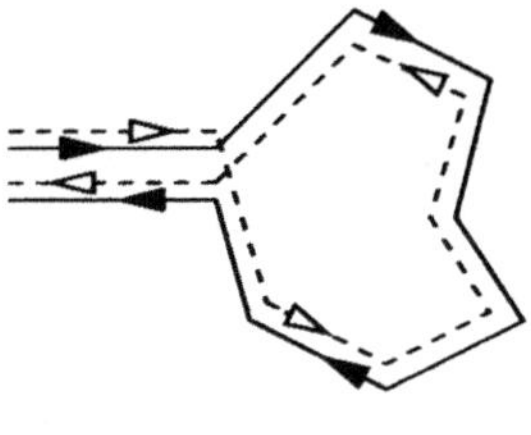

Figure 2. Enhanced back-scattering caused by the interference of electrons traveling along the same geometrical path in opposite directions.

The length of closed paths contributing to the interference between back-scattered waves is limited by the phase coherence time τ_ϕ, yielding a maximal length $\mathcal{L}_{\max} = v_{\mathrm{F}}\tau_\phi$. In practice, this time scale is determined either by phase relaxation within the system or the electron escape from a fully phase-coherent conductor to bulk reservoirs. Estimates for the weak localization corrections to the conductance read

$$\delta g^{\mathrm{WL}} \sim -\frac{e^2}{h} \times \begin{cases} \ln \dfrac{\tau_\phi}{\tau} & \text{2D}, \\ \dfrac{1}{\sqrt{D\tau_\phi}} & \text{Q1D wire}, \\ 1 & \text{0D dot}. \end{cases} \tag{3}$$

For the case of a dot, by g we understand its full conductance. Furthermore, τ is the elastic mean free path, and $D = v_{\rm F}^2\tau/d$ ($v_{\rm F}$ Fermi velocity, d dimensionality of the system) is the diffusion constant.

The violation of $\mathcal{T}$-invariance lifts the equivalence between the amplitudes $A_j^{\circlearrowright}$ and $A_j^{\circlearrowleft}$ and, thus, suppresses contributions to the enhanced back-scattering probability from longer paths. An obvious reason for $t \to -t$ symmetry breaking is an external magnetic field: thus, provided the orbital electron motion is coupled to the field, it affects the interference effects in the phase-coherent electron transport. For two-dimensional electrons in a heterostructure subject to a magnetic field perpendicular to the 2D plane, the difference between the amplitudes $A_j^{\circlearrowright}$ and $A_j^{\circlearrowleft}$ arises due to the Aharonov-Bohm phases accumulated along the path. For clockwise and anti-clockwise propagation around the oriented area S_j, the Aharonov-Bohm phases, $\delta\varphi_{\rm AB} = B_z S_j/\phi_0$ (ϕ_0 flux quantum), have opposite signs. As for random-walk trajectories the encircled area S_j is random, contributions from paths encircling a magnetic flux larger than typically the flux quantum ($|B_z S_j| \gtrsim \phi_0$) cancel out after disorder averaging: $\langle A_j^{\circlearrowright} A_j^{\circlearrowleft *}\rangle = \langle\, |A_j^{\circlearrowright}|^2 \times \exp[2i\delta\varphi_{\rm AB}^{\circlearrowright}]\,\rangle \to 0$.

When the orientation of the applied magnetic field lies within the plane of the heterostructure where an effectively two-dimensional electron moves (i.e., $B_z = 0$), an analysis of its influence on the weak localization properties requires taking into account more subtle arguments than a straightforward consideration of the Aharonov-Bohm effect. In fact, a strictly 2D system is insensitive to the orbital $\mathcal{T}$-breaking effect of an in-plane magnetic field. A possibility to couple a 2D orbital motion of the electrons to the in-plane magnetic field and, therefore, to break the time-reversal symmetry in a 2D system appears only after taking into account [13–16] the finite extent of electronic wave functions in the confinement direction of a quantum well as well as subband mixing by the magnetic field [17–20] and possibly by disorder.

In Chapter 3, we discuss this purely orbital effect of an in-plane magnetic field on quantum interference in 2D semiconductor structures. As for the confinement of the transverse motion, we also describe the crossover from the quantum limit (where the width of the system is of order of the Fermi wavelength) to the semiclassical regime corresponding to thin films of disordered and pure metals [21–24, 13–16]. The calculation aims at identifying the time scale τ_B after which the time-reversal symmetry breaking sufficiently affects the phase of an electron to destroy constructive interference. Correspondingly, only paths with lengths shorter than $\mathcal{L}(B) \sim v_{\rm F}\tau_B$ contribute to weak localization. Thus, τ_B is the time scale that determines the estimates for the weak localization correction to the conductivity, i.e., it replaces the phase coherence time in Eq. (3). Therefore the conductance is field-dependent, and the so-called magnetoconductance

$\Delta g_B = \delta g^{\mathrm{WL}}(B) - \delta g^{\mathrm{WL}}(0)$ obtains as

$$\Delta g_B \sim \frac{e^2}{h} \times \begin{cases} \ln \dfrac{\tau_\phi}{\tau_B} & \text{2D,} \\ \dfrac{1}{\sqrt{D\tau_\phi}} - \dfrac{1}{\sqrt{D\tau_B}} & \text{Q1D wire.} \end{cases} \tag{4}$$

More details about the multi-subband case can be found in chapter 6.

Another possibility for the in-plane magnetic field to affect the interference of 2D electron waves arises due to the coupling between orbital motion and electron spin, which is a pronounced feature of the electronic band structure of most III-V semiconductors with the zinc-blend type crystalline lattice and a unit cell without inversion symmetry. In a heterostructure or quantum well, spin-orbit coupling for a spin-$\frac{1}{2}$ carrier is linear, both in the planar momentum $\mathbf{k}$ and the spin operator $\underline{\sigma}$. For a given initial spin-polarized state, it produces an electron spin precession, where the precession axis as well as the precession frequency depend on the direction and value of the Fermi momentum of the propagating electron. For a diffusing electron, such a precession randomly changes direction, thus, providing spin relaxation for an initially spin-polarized particle – the mechanism known as Dyakonov-Perel relaxation [25]. However, spin-orbit coupling alone does not violate time-reversal symmetry, though it does modify the above-described interference effects. In particular, strong spin-orbit coupling changes the constructive interference between electrons encircling the same closed geometrical path in opposite directions into destructive interference [26–28]. To trace the origin of this change, let us consider the extreme limit of dominant spin-orbit coupling, where an electron always remain in the same chiral state along its orbital motion (for instance, $\xi = \frac{\mathbf{k}}{|\mathbf{k}|} \cdot \underline{\sigma} = +1$). Thus, an electron approaching the scattering medium in the initial state $|\mathbf{k}, \xi\rangle$ (which can be viewed as $|\mathbf{k}, \rightarrow\rangle$) ends up in the final backward moving state $|-\mathbf{k}, \xi\rangle$ (which can be viewed as $|-\mathbf{k}, \leftarrow\rangle$). This implies a rotation of its spin within the $x-y$ plane by the angle $|\theta_j| = (2n+1)\pi$ while moving along a diffusive path connecting these two states. For two spin-$\frac{1}{2}$ electrons encircling the same closed loop in opposite directions, $\theta_j^{\circlearrowleft} = -\theta_j^{\circlearrowright} = (2n+1)\pi$. As a result, their interference produces a negative contribution to the back-scattering,

$$A_j^{\circlearrowleft} A_j^{\circlearrowright *} = |A_j^{\circlearrowleft}|^2 \exp\left[\frac{i}{2}(\theta_j^{\circlearrowleft} - \theta_j^{\circlearrowright})\right] = -|A_j^{\circlearrowleft}|^2, \tag{5}$$

which increases the quantum conductance as compared to its classical Drude value – the effect known as weak anti-localization.

Now, the application of an in-plane magnetic field causes a Zeeman splitting of spin-$\uparrow$ and spin-$\downarrow$ electron states which – in combination with spin-orbit coupling – affects the interference of back-scattered waves. Chapter 4 is devoted to the description of the interplay between Zeeman splitting and spin-orbit coupling in determining quantum transport characteristics (in particular, of lateral dots).

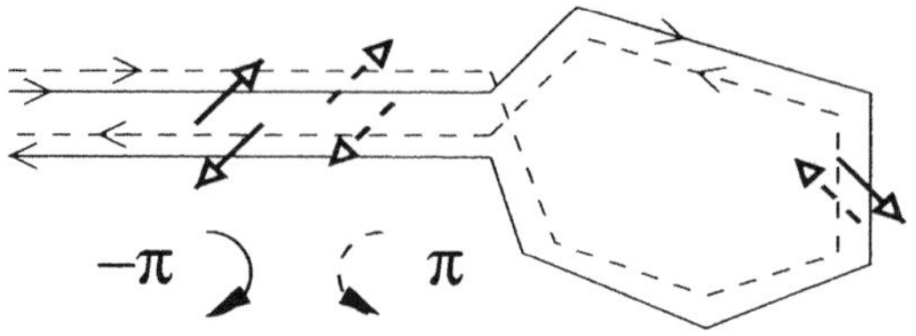

Figure 3. Suppressed back-scattering in the presence of strong spin-orbit coupling.

Interference between different paths not only affects the mean value of the conductance (i.e., $\langle g\rangle = g_{\text{class}} + \delta g^{\text{WL}}$) but also characteristic deviations from that mean value, $g = \langle g\rangle + \delta g$. Universal conductance fluctuations (UCF) represent another interference phenomenon observed in mesoscopic conductors and quantum dots at low temperatures. To discuss this effect, let us consider the transmission probability through a disordered piece of metal. For a diffusing electron wave, there are many scenarios leading it from the same initial state $|\mathbf{k}\rangle_{\text{left}}$ on the left to the same final state $|\mathbf{k}'\rangle_{\text{right}}$ on the right, thus, creating the possibility for interference. Depending on the individual distribution of scatterers in the sample, different paths j have different weights – and, thus, the waves following them acquire different phases – which makes the transmission probability,

$$P_{\mathbf{k},\mathbf{k}'} = \left|\sum_j A_j(\mathbf{k},\mathbf{k}')\right|^2 = \sum_j |A_j|^2 + \sum_{ji} A_j A_i^* \,, \tag{6}$$

sample-specific as well as dependent on specific interference conditions, namely the energy (wavelength) of the incident electron or the presence of an external magnetic field. In measurements, this manifests itself in a random but reproducible dependence of the sample conductance on the Fermi energy E_{F} and the applied field B_z. Furthermore, gate voltages (varying the shape of a semiconductor dot) represent an additional control parameter.

In a phase-coherent system, the variance of conductance fluctuations has a value,

$$\langle \delta g^2\rangle = c_{\text{geom}}\, \alpha \times \left(\frac{e^2}{h}\right)^2 , \tag{7}$$

which is universal [9–11, 29, 30] up to a geometry-dependent numerical factor $c_{\text{geom}} \sim 1$. Furthermore, α reflects the fundamental symmetries of the system, expressed by three coefficients,

$$\alpha = \frac{s}{\beta\Sigma}. \tag{8}$$

The symmetry plays such an important role here as it determines the number of independently fluctuating components of the generically random amplitudes

A_j related to each particular geometrical path. The higher the symmetry, the more correlations are implicit between real and imaginary parts as well as spin components of A_j. In particular, $s = 1$ or 2 depending on whether Kramers' degeneracy is lifted or not. When the orbital electron motion is independent of the electron spin state, i.e., when the system possesses spin-rotation invariance, the $\mathcal{T}$-symmetric system is described by $\beta = 1$ whereas $\beta = 2$ if time-reversal symmetry is broken. The value $\beta = 4$ (with $s = 1$) represents the case of efficient spin-relaxation due to spin-orbit scattering. Finally, an additional symmetry class number Σ [30] characterizes whether a violation of time-reversal symmetry in the spin-sector affects ($\Sigma = 2$ – or not, $\Sigma = 1$) the interference in the electron orbital motion via spin-orbit coupling. The crossover between the latter two symmetry classes may be achieved by varying the Zeeman splitting. Its features will be described in Chapter 4, together with weak localization effects.

As phase coherence is crucial for the occurrence of interference effects, the influence of symmetry breaking on conductance measurements is not observable when the corresponding time scale is larger than the decoherence time, $\tau_B > \tau_\varphi$. As for finite systems, when decoherence of electrons in a dot or a short wire happens faster than the escape into bulk reservoirs, $\tau_\varphi < \tau_{\text{esc}}$, both the enhanced back-scattering effect and UCFs get suppressed [31, 32] as compared to the conductance quantum, e^2/h. In these situations, no influence of an in-plane magnetic field on quantum transport characteristics of the system would be expected – unless the field affects the decoherence time.

This is the case when the latter is caused by spin-flip scattering at paramagnetic impurities [33]. In all foreseeable regimes [33, 34], degenerate paramagnetic impurities destroy weak localization via spin-flip scattering, the relevant time scale being the spin-flip scattering time τ_{s}. Furthermore, they suppress conductance fluctuations [35–39] as the flipping of impurity spins effectively changes the realization of disorder in the sample and, thus, leads to a self-averaging of the conductance. However, a Zeeman splitting of the magnetic impurity levels by a magnetic field causes an energy threshold for electron spin-flip processes which slows down the electron spin relaxation. At a high field, the electron spin relaxation is possible only due to very few thermally activated impurity spins, so its rate is reduced by the exponential factor, $\tau_s^{-1}(B) \propto \tau_s^{-1} \times \exp[-g^{(\text{imp})}\mu_{\text{B}}B/T]$, where μ_{B} is the Bohr magneton and $g^{(\text{imp})}$ the g-factor of the impurities. Thus, a sufficiently strong magnetic field restores UCFs [36–39] as well as weak localization (if the orbital effect of the magnetic field is excluded). The in-plane magnetoresistance of magnetically contaminated 2D semiconductors and other features related to the dynamics of paramagnetic impurities are discussed in Chapter 5.

For all practical purposes, the study of symmetry breaking rates, determining the parallel field effect on the weak localization corrections to the conductivity of a 2D electron gas or a semiconductor wire, can be completed using the diagrammatic perturbation theory for disordered systems. Although one can find more details about this technique in several textbooks and review articles (see, e.g., [40]), for the sake of completeness, we describe its necessary elements in Chapter 2. As an alternative, field theoretic methods may be used. In chapter 6, we illustrate the supersymmetric non-linear sigma-model in application to one of the problems listed above, i.e., the orbital effect of an in-plane magnetic field (via subband mixing) on the transport properties of 2D electrons.

2. The diagrammatic technique: diffusons and Cooperons

Diagrammatics is a perturbative method which allows one to a) classify different contributions to the perturbation series and b) sum up the relevant terms. To be specific, here – for simplicity – we choose the random disorder potential V to be drawn from a Gaussian white noise distribution:

$$\langle V(\mathbf{r})\rangle = 0, \qquad \langle V(\mathbf{r})V(\mathbf{r}')\rangle = \frac{1}{2\pi\nu\tau}\delta(\mathbf{r}-\mathbf{r}'), \tag{9}$$

where ν is the density of states (DoS) at the Fermi level and τ is the mean free scattering time. Note that throughout this review we will use units where $\hbar = c = e = 1$.

The starting point of the perturbative approach is the representation of the Green function as a series in powers of V, i.e., $G = G_0 + G_0\sum_n (VG_0)^n$. Under averaging, diagrams with single impurity lines vanish (due to $\langle V\rangle = 0$). Thus, the diagrammatic expansion involves only diagrams with paired impurity lines. The average Green function is then given by a Dyson equation, depicted schematically in Fig. 4:

$$\langle G\rangle = G_0 + G_0\sum_{n=1}^{\infty}(\Sigma G_0)^n = G_0 + G_0\Sigma\langle G\rangle \quad \Longleftrightarrow \quad \langle G\rangle = \frac{G_0}{1-\Sigma G_0},$$

where the self-energy Σ contains all irreducible diagrams, i.e., diagrams that cannot be split into two by cutting one G_0-line.

Figure 4. Diagrammatic representation of the Dyson equation.

The dominant contribution to the self-energy reads

$$\Sigma_0^{\pm}(\mathbf{p}) = \left\langle \int d\mathbf{q}\, V(\mathbf{q})G_0^{\pm}(\mathbf{p}-\mathbf{q};\epsilon)V(-\mathbf{q})\right\rangle, \tag{10}$$

where the shorthand notation $d\mathbf{q} = d^dq/(2\pi)^d$ has been introduced. Since the potential is δ-correlated in real space, its Fourier transform does not depend on momentum. Hence, $\Sigma_0^{\pm} = \int d\mathbf{p}\, G_0^{\pm}(\mathbf{p};\epsilon)/(2\pi\nu\tau)$.

The real part of Σ, leading to a shift in energy, can be absorbed in the ground state energy. Using that the unperturbed Green function reads $G_0^{\pm}(\mathbf{p};\epsilon) = (\epsilon \pm i0 + \xi_{\mathbf{p}})^{-1}$, where $\xi_{\mathbf{p}} = \epsilon_F - p^2/(2m)$, and the definition of the density of states, $\nu(\epsilon) = L^{-d}\sum_{\mathbf{p}}\delta(\epsilon + \xi_{\mathbf{p}})$, the imaginary part of Σ obtains

$$\Im\,\Sigma_0^{\pm} = \mp\frac{1}{2\nu\tau}\int d\mathbf{p}\,\delta(\epsilon + \xi_{\mathbf{p}}) = \mp\frac{1}{2\tau}. \tag{11}$$

The associated length scale ℓ is the decay length of the average Green function as we will see shortly. In the case of weak disorder, $\tau^{-1} \ll \epsilon_F$, all other contributions are small in $1/(k_F\ell)$, and Eq. (11) determines the self-energy Σ in the self-consistent Born approximation (SCBA).

Inserting the above result into the Dyson equation yields

$$\langle G^{\pm}\rangle(\mathbf{p};\epsilon) = \frac{1}{\epsilon + \xi_{\mathbf{p}} \pm \frac{i}{2\tau}}. \tag{12}$$

In real space representation, this leads to a decay of the average Green function on the scale of the mean free path,

$$\langle G\rangle(\mathbf{r},\mathbf{r}') = G_0(\mathbf{r},\mathbf{r}')\,e^{-\frac{|\mathbf{r}-\mathbf{r}'|}{2\ell}}. \tag{13}$$

Having found the average Green function, the next step is to calculate correlation functions of the form

$$F(\mathbf{p}_1,\mathbf{p}_1',\mathbf{p}_2,\mathbf{p}_2';\omega) = \langle G^-(\mathbf{p}_1,\mathbf{p}_1';\epsilon)G^+(\mathbf{p}_2,\mathbf{p}_2';\epsilon+\omega)\rangle. \tag{14}$$

Again – as for the averaged Green function – the diagrammatic perturbation series involves summing up diagrams with paired impurity lines. The dominant contributions are series of ladder diagrams and maximally crossed diagrams, see Fig. 5. With the notation $\langle AB\rangle_c = \langle AB\rangle - \langle A\rangle\langle B\rangle$, i.e., subtracting the disconnected part of the correlator, the contribution of connected diagrams can be written as

$$\begin{aligned} &\langle G^-(\mathbf{p}_1,\mathbf{p}_1';\epsilon)G^+(\mathbf{p}_2,\mathbf{p}_2';\epsilon+\omega)\rangle_c \\ = &\ \langle G^-\rangle(\mathbf{p}_1,\epsilon)\,\langle G^-\rangle(\mathbf{p}_1',\epsilon)\,\langle G^+\rangle(\mathbf{p}_2,\epsilon+\omega)\,\langle G^+\rangle(\mathbf{p}_2',\epsilon+\omega)\rangle\,\times \\ &\times\,\Gamma(\mathbf{p}_1,\mathbf{p}_2,\mathbf{p}_1',\mathbf{p}_2';\omega)\,\delta(\mathbf{p}_1-\mathbf{p}_2-\mathbf{p}_1'+\mathbf{p}_2'), \end{aligned} \tag{15}$$

thus defining the reducible[1] vertex function $\Gamma(\mathbf{p}_1,\mathbf{p}_2,\mathbf{p}_1',\mathbf{p}_2';\omega)$.

Consider first the sum of ladder diagrams depicted in the upper part of Fig. 5. Due to momentum conservation at each vertex, the momentum difference $\mathbf{q} \equiv \mathbf{p}_1 - \mathbf{p}_2 (= \mathbf{p}_1' - \mathbf{p}_2')$ is constant, and Γ depends on this difference only. Then

[1] In the case of two-particle functions, a diagram is called reducible if it can be split into two separate diagrams by cutting two $\langle G\rangle$-lines.

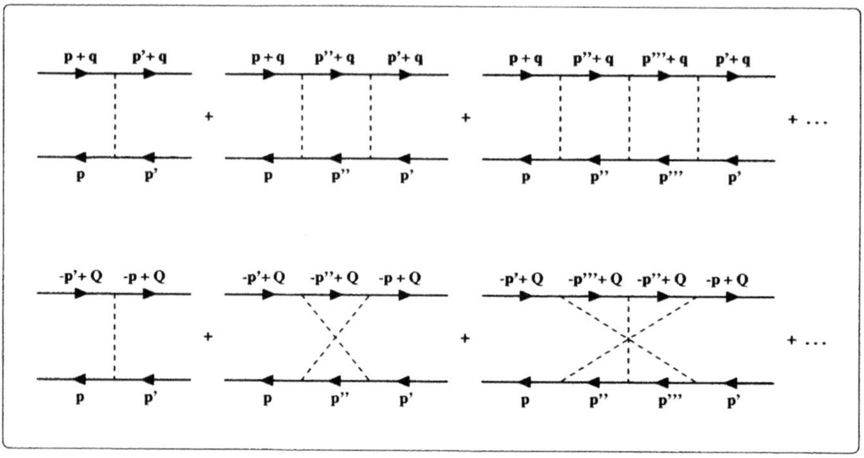

Figure 5. Ladder and maximally crossed diagrams.

$\Gamma(\mathbf{q},\omega)$ is given by a Bethe-Salpeter equation, the two-particle analogue of the Dyson equation:

$$\Gamma(\mathbf{q},\omega) = \gamma + \gamma \underbrace{\Big[\int d\mathbf{p}''\,\langle G^-\rangle(\mathbf{p}''+\mathbf{q},\epsilon+\omega)\langle G^+\rangle(\mathbf{p}'',\epsilon+\omega)\Big]}_{\equiv\,\Pi(\mathbf{q},\omega)}\Gamma(\mathbf{q},\omega),$$

where $\gamma = 1/(2\pi\nu\tau)$. Anticipating the result that Γ diverges for $\mathbf{q},\omega \to 0$, we can approximate the irreducible vertex function Π for small $|\mathbf{q}|,\omega$:

$$\Pi(\mathbf{q},\omega) \simeq 2\pi\nu\tau(1 + i\omega\tau - D\mathbf{q}^2\tau), \tag{16}$$

where $D = v_\mathrm{F}^2\tau/d$ is the diffusion constant (d dimensionality of the system). Thus, the sum of ladder diagrams finally yields

$$\Gamma(\mathbf{q},\omega) = \frac{\gamma}{1-\gamma\,\Pi(\mathbf{q},\omega)} = \frac{1}{2\pi\nu\tau^2}\frac{1}{D\mathbf{q}^2 - i\omega} \equiv \mathcal{D}(\mathbf{q},\omega). \tag{17}$$

This is a diffusion pole or so-called diffuson. It can easily be seen that the diffuson is not affected by a weak magnetic field: Minimal coupling implies that the magnetic field shifts all momenta by the corresponding vector potential, $\mathbf{p} \to \mathbf{p} - \mathbf{A}$; however, $(\mathbf{p}_1 - \mathbf{A}) - (\mathbf{p}_2 - \mathbf{A}) = \mathbf{p}_1 - \mathbf{p}_2 = \mathbf{q}$.

A maximally crossed diagram [41] (for a detailed discussion see e.g. [42]) can be converted to a ladder by reversing one G-line. Since now all the arrows point in the same direction, momentum conservation at the vertices requires the momentum sum $\mathbf{Q} = \mathbf{p}_1 + \mathbf{p}_2'$ to be constant. Apart from that, the structure of all equations is the same as for the ladder diagrams. Thus,

$$\Gamma(\mathbf{Q},\omega) = \frac{1}{2\pi\nu\tau^2}\frac{1}{D\mathbf{Q}^2 - i\omega} \equiv \mathcal{C}(\mathbf{Q},\omega). \tag{18}$$

This is called a Cooperon – in analogy to superconductivity, as it corresponds to a correlator in the particle-particle channel (as opposed to the particle-hole channel for the diffuson). Now the presence of a magnetic field requires replacing $\mathbf{Q} \to \mathbf{Q} - 2\mathbf{A}$ which leads to a decaying of the Cooperon.

The connected part of averages of the form $\langle G^+G^+\rangle$ or $\langle G^-G^-\rangle$ vanishes, i.e., no long-ranged correlations exist.

Taking into account the spin of the electrons, diffusons and Cooperons split into singlet and triplet modes. This will become important when considering spin-orbit coupling or spin scattering. Details will be given in the corresponding chapters 4 and 5.

In coordinate representation, the expression obtained in Eq. (18) has the form of a diffusion equation,

$$[D(\nabla + i2\mathbf{A})^2 - i\omega]\mathcal{C} = \delta(\mathbf{x} - \mathbf{x}'). \tag{19}$$

(A similar equation holds for the diffuson, Eq. (17).)

In a finite size geometry, this equation should be complemented by boundary conditions. For an 'open' surface, as, e.g., the interface with a bulk reservoir, the boundary condition reads

$$\mathcal{C} = 0. \tag{20}$$

For an insulating surface, for example, the edge of a metallic dot, the corresponding boundary condition expresses the lack of a phase-coherent current density across the sample surface,

$$\mathbf{n}_{\|} \cdot (\nabla + i2\mathbf{A})\mathcal{C} = 0, \tag{21}$$

where $\mathbf{n}_{\|}$ is a unit vector normal to the sample surface. In the case of zero magnetic field, this boundary condition indicates that the lowest Cooperon and diffuson modes correspond to the constant ($Q = 0$) solution and are gapless. This choice of the lowest mode, being constant in space, is called the zero-dimensional (0D) approximation. In the presence of a magnetic field, one has to give the analysis of the lowest modes in closed systems more attention, applying a gauge transformation to the gauge where $\mathbf{n}_{\|} \cdot \mathbf{A} = 0$, and only then making the 0D approximation. Subsequently, the discrete spectrum of relaxational modes describing the Cooperon and diffuson can be found using the standard Hamiltonian perturbation theory methods. This is a useful trick which enables us find the Cooperon and diffuson functions and, in particular, their lowest eigenmodes in finite size geometries.

3. In-plane magnetoresistance effect due to intersubband mixing

As pointed out in the introduction, the influence of an in-plane magnetic field on the quantum transport characteristics of a 2D or quasi-2D system can be divided into orbital and spin-related effects. In this section, we consider the orbital effect – treating electrons as spinless.

There are two features that make the issue of the influence of an in-plane magnetic field on quantum interference effects in inversion layers and thin metallic films non-trivial. The first issue is related to the fact that a finite spatial extent of electron wave functions along the confinement direction, i.e., the z-direction perpendicular to the 2D plane, is needed in order that the electron can accumulate an additional magnetic field-induced phase if the field is exactly parallel to the 2D plane. Furthermore, the possible accumulation of such an Aharonov-Bohm phase is sensitive to the inversion symmetry properties of the system. In particular, for a quantum well with a potential profile that is symmetric with respect to the transformation $\mathcal{P}_z : z \to -z$, an in-plane magnetic field cannot induce any dephazing, and the field-dependent rate τ_B^{-1} is equal to zero. This property reflects the Berry-Robnik symmetry effect [43]: Consider a system described by a Hamiltonian $\hat{\mathcal{H}}$ that at $B = 0$ is invariant under a discrete symmetry transformation, such as $\mathcal{P}_z$, and at finite fields remains invariant under the combined time and z-coordinate inversion transformation, $\mathcal{P}_z\mathcal{T}$, i.e., $\mathcal{P}_z\mathcal{T}\hat{\mathcal{H}}(B) = \hat{\mathcal{H}}(B)$. Then there always exist pairs of paths having exactly the same value of the semiclassical action which, therefore, interfere constructively even in the presence of a magnetic field. As a result, the suppression of weak localization by an in-plane magnetic field is possible only if either the scattering potential in the quantum well is z-dependent [13] (since a generic z-dependence does not respect $\mathcal{P}_z$-symmetry), or if the form of the confinement potential in z-direction has no inversion symmetry [15, 44, 16] (as usually is the case for a inversion layer in a heterostructure). In the first subsection, we provide a quantitative analysis of this effect. In the second subsection, this analysis is extended to multi-subband systems [15, 44], and the results are compared to the magnetoresistance in thin films with diffusive transverse motion.

The second non-trivial issue concerns the so-called geometrical flux cancellation characteristic for ballistic metallic films (wires), where electrons scatter only from surface (edge) defects at two parallel surfaces. In such a 'quasi-ballistic' film, the finite resistance is provided by the diffusive scattering of carriers from the surface. Quantum corrections to the conductivity are related to the interference between waves propagating along paths consisting of a sequence of ballistic flights between the film edges. Among those paths, the trajectories containing ballistic segments that are almost parallel to the film surface and, therefore, much longer than the film width d play an important role. These so-called Lévy flights (in the terminology of anomalous diffusion theory [45]) produce a logarithmically large contribution to the sample conductance [46]. The oriented area encircled by a closed loop made of quasibalistic paths is exactly equal to zero [22], thus suppressing the magnetic field effect on the interference of electron waves propagating along them. An electron can accumulate a magnetic flux only due to a curving of its trajectory by the magnetic field itself. This makes the role of Lévy flights important for the interference-related magnetoresistance of ballistic films and wires: curving of the longest ballistic flights – which is a purely classical effect – switches their scattering at the surface roughness from one film surface (or wire edge) to the other [24]. This peculiar situation and the resulting unexpected influence of a magnetic field on the localization properties in quasi-ballistic wires are discussed at the end of this section.

3.1. PHASE-BREAKING BY AN IN-PLANE MAGNETIC FIELD IN 2D INVERSION LAYERS

The influence of an in-plane magnetic field on the orbital motion of carriers in a heterostructure or quantum well is a result of the finite width d of any 2D layer. This effect has been discussed previously in various contexts [17–20]. The Lorentz force on the electrons generated by the in-plane field $\mathbf{B}_{\|}$ mixes up the electron motion along and across the confinement direction, thus resulting in a modification of the 2D dispersion, $E(p) \rightarrow E(\mathbf{B}_{\|}, \mathbf{p})$. In particular, a) the 2D electron mass increases in the direction perpendicular to $\mathbf{B}_{\|}$. Furthermore, in heterostructures with a confining potential that does not possess inversion symmetry, b) the in-plane field also lifts the $\mathbf{p} \rightarrow -\mathbf{p}$ symmetry in the dispersion law [17, 18]:

$$E(\mathbf{B}_{\|}, \mathbf{p}) - E(\mathbf{B}_{\|}, -\mathbf{p}) \propto \left(\mathbf{p} \cdot [\mathbf{B}_{\|} \times \mathbf{e}_z]\right)^3 \neq 0,$$

where $\mathbf{e}_z$ is a unit vector perpendicular to the plane. This leads to a field-dependent suppression of the WL signal as will be discussed below. By contrast, for the z-inversion symmetric problem, due to the Berry-Robnik phenomenon [43], the Cooperon mode remains massless, and there is no magnetoresistance.

In the presence of an in-plane magnetic field, the effective 2D Hamiltonian for electrons in a heterostructure with the confining potential profile $W(z)$ can be obtained from the 3D Hamiltonian,

$$\hat{\mathcal{H}}_{3D} = \frac{(-i\nabla - \mathbf{A})^2}{2m} - \frac{\partial_z^2}{2m} + W(z) + u(\mathbf{r}, z), \tag{22}$$

using a plane wave representation, $\Psi_{\mathbf{p}} = e^{i\mathbf{pr}}\phi_0^{(\mathbf{p})}(z)$ for the electrons in the lowest subband. Here, $\phi_k^{(\mathbf{p})}(z)$ are the eigenfunctions of the z-dependent part of the Hamiltonian. In particular, at $B_{\|} = 0$, the in-plane and perpendicular motion are separable, and the corresponding eigenfunctions are given by $|k\rangle \equiv \phi_k^{(0)}$. Due to the mixing between subbands $|0\rangle$ and $|k > 0\rangle$ by an in-plane magnetic field, the eigenfunction $\phi_0^{(\mathbf{p})}(z)$ in the presence of the field depend on the in-plane momentum $\mathbf{p}$. In order to find $\phi_0^{(\mathbf{p})}(z)$ and the corresponding energy $E(\mathbf{B}_{\|}, \mathbf{p})$ for each plane wave state a perturbation theory analysis – as described in the following – or a numerical self-consistent-field technique may be used.

Furthermore, in Eq. (22), the vector potential reads $\mathbf{A} = (z - z_0)\mathbf{B}_{\|} \times \mathbf{e}_z$, where $z_0 = \langle 0|z|0\rangle$ is the center of mass position of the electron wave function in the lowest subband (at $B_{\|} = 0$), and the potential $u(\mathbf{r}, z)$ is a combination of the Coulomb potential of impurities and the lateral potential forming the quantum dot.

For weak to intermediate magnetic fields $\mathbf{B}_{\|}$, the effective 2D Hamiltonian takes the form

$$\hat{\mathcal{H}}_{2D} = \frac{\mathbf{p}^2}{2m} - p_\perp^2\gamma(B_{\|}) + p_\perp^3\beta(B_{\|}) + u(\mathbf{r}). \tag{23}$$

Here, $\mathbf{p} = -i\nabla - \mathbf{a}(\mathbf{r})$ is a purely 2D momentum operator and $p_\perp = \mathbf{p}\cdot[\frac{\mathbf{B}_{\|}}{B_{\|}} \times \mathbf{e}_z]$ its component perpendicular to $\mathbf{B}_{\|}$.

Thus, the application of an in-plane magnetic fields results in a) an effective vector potential $\mathbf{a}(\mathbf{r})$, and b) two additional terms in the free electron dispersion due to $p_\perp$-dependent inter-subband mixing. The first term, $\delta\hat{\mathcal{H}}^{(2)} = -p_\perp^2\gamma$, lifts the rotational symmetry by causing an anisotropic mass enhancement [17, 18]. It increases the 2D density of states and, for a 2D gas with fixed sheet density, it reduces the Fermi energy calculated from the bottom of the 2D conduction band, i.e., $E_{\mathrm{F}}(B_{\|}) = E_{\mathrm{F}}^0 - \gamma(B_{\|})p_{\mathrm{F}}^2/2$. The second term, $\delta\hat{\mathcal{H}}^{(3)} = p_\perp^3\beta$, is related to the time-reversal symmetry breaking by $B_{\|}$. A perturbative analysis of the problem results in the field dependences $\gamma \sim B_{\|}^2$ and $\beta \sim B_{\|}^3$. Indeed, an expansion of the plane-wave energy up to the third order in $p_\perp$ yields

$$\gamma(B_{\|}) \approx \frac{B_{\|}^2}{m^2}\sum_{k\geq 1}\frac{\langle k|z|0\rangle^2}{\epsilon_k - \epsilon_0} \sim \frac{1}{m}\left(\frac{d}{\lambda_B}\right)^4,$$

$$\beta(B_{\|}) \approx \frac{B_{\|}^3}{m^3}\sum_{k,k'\geq 1}\frac{\langle 0|z|k\rangle\langle k|z|k'\rangle\langle k'|z|0\rangle}{(\epsilon_k - \epsilon_0)(\epsilon_{k'} - \epsilon_0)} \sim \frac{d}{m}\left(\frac{d}{\lambda_B}\right)^6,$$

where ϵ_k are the subband energies and $\lambda_B = 1/\sqrt{B_{\|}}$ is the magnetic length.

In $\hat{\mathcal{H}}_{2D}$ given by Eq. (23), the "in-plane" disorder is incorporated in the form of a scattering potential $u(\mathbf{r}) \approx \langle 0|u(\mathbf{r}, z)|0\rangle$. It can be characterized by the mean free path, $\ell \gg 1/p_{\mathrm{F}}$ or, equivalently, the momentum relaxation time τ, related to the diffusion coefficient $D = v_{\mathrm{F}}^2\tau/2$. The modification of the electron density of states by $B_{\|}$ only slightly affects the value of the electron mean free path. However, the presence of a parallel field also changes the symmetry of the Born scattering amplitudes between plane waves, $f_{\mathbf{pp}'} = \langle\Psi_{\mathbf{p}}|u(\mathbf{r}, z)|\Psi_{\mathbf{p}'}\rangle$. Due to the momentum-dependent subband mixing, $f_{\mathbf{pp}'}$ acquires an additional contribution, $f_{\mathbf{pp}'} = f_{\mathbf{pp}'}^{(0)}\left\{1 + (p_\perp + p'_\perp)B_{\|}\zeta\right\}$, where

$$\zeta = \frac{1}{m\langle 0|u(\mathbf{p}-\mathbf{p}', z)|0\rangle}\sum_{k>1}\frac{\langle 0|u(\mathbf{p}-\mathbf{p}', z)|k\rangle\langle k|z|0\rangle}{\epsilon_k - \epsilon_0}.$$

This corresponds to the presence of a random gauge field in the effective 2D Hamiltonian [13, 14, 47],

$$\mathbf{a}(\mathbf{r}) = 2\sum_{k\geq 1} \frac{\langle 0|u(\mathbf{r},z)|k\rangle\langle k|z|0\rangle}{\epsilon_k - \epsilon_0}[\mathbf{B}_{\|} \times \mathbf{e}_z],$$

which can be interpreted as a result of an effective 'curving' of the 2D plane by impurities with a z-dependent scattering potential in the system. In the presence of an in-plane magnetic field this curving generates a random effective perpendicular field component, $b_\perp = [\nabla \times \mathbf{a}]_z$. In systems, where scattering is dominated by Coulomb centers separated from the 2D plane by a spacer and, thus, is almost independent of z, a smaller effect may be taken into account, namely $\delta\mathbf{a} \sim [\mathbf{B}_{\|} \times \mathbf{e}_z]([\mathbf{B}_{\|} \times \mathbf{e}_z] \cdot \nabla)^2 u(\mathbf{r})$. However, $\delta\mathbf{a}$ has a negligible influence on the quantum transport characteristics of 2D electrons as compared to the effect of the field-dependent electron dispersion.

The perturbative calculation of two-particle correlation functions, i.e., diffusons $\mathcal{D}(\mathbf{x},\mathbf{x}';\omega)$ and Cooperons $\mathcal{C}(\mathbf{x},\mathbf{x}';\omega)$ [9–11], requires taking into account all field-dependent terms in the effective 2D Hamiltonian Eq. (23). Admitting for different magnetic fields $\mathbf{B}_{\|1}$ and $\mathbf{B}_{\|2}$, the result for the diffuson has the form

$$\left[-D\nabla^2 - i\tilde{\omega} + \tau_{\rm d}^{-1}\right]\mathcal{D} = \delta(\mathbf{x}-\mathbf{x}'), \tag{24}$$

where $\tilde{\omega} = \omega + \delta$ with $\delta = p_{\rm F}^2[\gamma(B_{\|1}) - \gamma(B_{\|2})]/2$. The field-dependent rate $\tau_{\rm d}^{-1}(B_{\|1}, B_{\|2})$ reads

$$\tau_{\rm d}^{-1} = \frac{\tau p_{\rm F}^4}{8}[\gamma(B_{\|1}) - \gamma(B_{\|2})]^2 + \frac{\zeta^2}{2\tau}p_{\rm F}^2[B_{\|1} - B_{\|2}]^2 \tag{25}$$

which vanishes for $B_{\|1} = B_{\|2}$.

The first term in Eq. (25) comes from a deformation of the Fermi circle by the magnetic field $B_{\|}$. The second term takes into account the field effect on the scattering of plane waves. Furthermore, Eq. (24) contains the difference between the electron kinetic energies in two measurements of conductance, $\tilde{\omega} = E_{\rm F}(B_{\|1}) - E_{\rm F}(B_{\|2})$, each of them shifted by the magnetic field with respect to the Fermi energy $E_{\rm F}^0$ at $B_{\|} = 0$ of the electron gas with the same sheet density. The latter fact is important, since, for lateral dots where the electron density is fixed, one should substitute $\tilde{\omega} = 0$.

The Cooperon equation can be represented in the form

$$\left[D(-i\nabla - \mathbf{A}_0)^2 - i\tilde{\omega} + \tau_{\rm d}^{-1} + \tau_{\rm c}^{-1}\right]\mathcal{C} = \delta(\mathbf{x}-\mathbf{x}'). \tag{26}$$

It contains an additional decay rate $\tau_{\rm c}^{-1}(B_{\|1}, B_{\|2})$ given as

$$\tau_{\rm c}^{-1} = \frac{\tau p_{\rm F}^6}{8}\left[\frac{\beta(B_{\|1}) + \beta(B_{\|2})}{2}\right]^2 + \frac{\zeta^2 p_{\rm F}^2}{2\tau}[B_{\|1} + B_{\|2}]^2$$

which accounts for the dephazing of electrons encircling the same chaotic trajectory in opposite directions. Thus, the result of lifting time-reversal symmetry by an in-plane magnetic field can be described by the Cooperon phase-breaking rate

$$\tau_{B\|}^{-1} \equiv \tau_{\rm c}^{-1}(B_\|, B_\|) = \frac{\tau p_{\rm F}^6}{8}\beta^2(B_\|) + \frac{2\zeta^2 p_{\rm F}^2}{\tau}B_\|^2 \sim aB_\|^6 + bB_\|^2. \qquad (27)$$

Note the gauge shift in the Cooperon equation (26). The 'vector potential' $\mathbf{A}_0 = \frac{3}{2}p_{\rm F}^2 m\beta(B_\|)[\frac{\mathbf{B}_\|}{B_\|} \times \mathbf{e}_z]$ results from the following artifact: the cubic term in the effective electron dispersion not only lifts the $\mathbf{p} \to -\mathbf{p}$ inversion symmetry of the line $E(B_\|, \mathbf{p}) = E_{\rm F}(B_\|)$, but also shifts its geometrical center with respect to the bottom of the 2D conduction band. However, since for the conductance only electrons with energy $\epsilon = E_{\rm F}$ matter, such a shift can be eliminated by choosing a slightly modified initial gauge. This can be corrected for now by applying a gauge transformation $\mathcal{C} \to e^{i\mathbf{q}\mathbf{x}}\mathcal{C}$ directly to the Cooperon. In other words, the phase-coherent transport is only affected by the $B_\|$-induced $\mathbf{p} \to -\mathbf{p}$ asymmetric distortion of the Fermi circle into an oval, but not by a shift of its center in the momentum space. That is, only the $B_\|$-dependent anisotropy of the electron wavelength along the Fermi line affects the interference pattern of current carriers. Note that both terms in Eq. (27) are non-vanishing only in the absence of $z \to -z$ inversion symmetry.[2]

The phase-breaking rate in Eq. (27) can also be used to describe magnetoresistance and universal conductance fluctuations (UCF) in lateral semiconductor dots. For quantum dots, weak localization corrections and the variance of UCF can be represented as

$$g_{\rm WL}(B_\|) \propto \int (dx)\; W(\mathbf{x})\,\mathcal{C}(\mathbf{x},\mathbf{x};0) \equiv \langle g(B_\|)\rangle - \langle g\rangle_u,$$

where the subscript 'u' stands for unitary, and

$$\langle \delta g(B_{\|1})\delta g(B_{\|2})\rangle \propto \int (dx)(dx')\; W(\mathbf{x})W(\mathbf{x}') \sum_{\mathcal{P}=\mathcal{D},\mathcal{C}} |\mathcal{P}(\mathbf{x},\mathbf{x}';\omega)|^2,$$

where $\omega = E_{\rm F}(B_{\|1}) - E_{\rm F}(B_{\|2})$, and $E_{\rm F}(B_\|)$ is the Fermi energy of the 2D electron gas calculated from the bottom of the conduction band. The dispersionless weight factors $W(\mathbf{x})$ take care of the particle number conservation upon diffusion inside the dot [48] and incorporate the coupling parameters to the leads. In the zero-dimensional limit, both, $g_{\rm WL}$ and $\langle \delta g(B_{\|1})\delta g(B_{\|2})\rangle$, are dominated by the lowest (spatially homogeneous) Cooperon (diffuson) relaxation mode λ_0, which is determined by the rate of escape to the reservoirs, $\tau_{\rm esc}^{-1} \ll D/L^2$ (Thouless

[2] In fact, the first term requires an asymmetric confining potential whereas for the second term a generic z-dependence of the impurity potential is sufficient.

energy), and the Cooperon suppression by time-reversal symmetry breaking described by $\tau_{B\|}^{-1}$. The latter expresses the efficient reduction of the fundamental symmetry of the system from orthogonal (o) to unitary (u). In the absence of spin-orbit effects, the parameters $\tau_{\rm esc}^{-1}$ and $\tau_{B\|}^{-1}$ describe the value of WL corrections as well as the variance of UCF, $\langle\delta g^2(B_\|)\rangle$, as compared to their nominal values, $g_{\rm WL}(0) \equiv \langle g\rangle_o - \langle g\rangle_u$ and $\langle\delta g^2\rangle_u$:

$$g_{\rm WL}(B_\|) = g_{\rm WL}(0)\frac{1}{1+\tau_{\rm esc}/\tau_B}, \qquad (28)$$

$$\langle\delta g^2(B_\|)\rangle = \langle\delta g^2\rangle_u\left(1+\frac{1}{(1+\tau_{\rm esc}/\tau_B)^2}\right).$$

Thus, these two parameters can be studied from the UCF fingerprints measured by changing the shape of a dot in multi-gate devices or by varying the Fermi energy in back-gated dots.

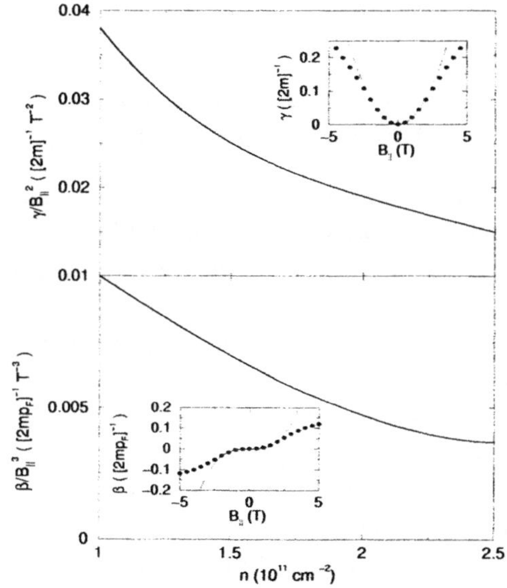

Figure 6. Calculated dependence of the parameters γ and β on the sheet density of 2D electrons. The insets show the effect of $B_\|$ on the symmetric, $[E(\mathbf{p}) + E(-\mathbf{p})]/2$, and anti-symmetric, $[E(\mathbf{p}) - E(-\mathbf{p})]/2$, parts of the 2D electron dispersion in a broader range of $B_\|$, where a perturbative expansion is not applicable.

In ballistic billiards or heterostructures with a z-independent impurity scattering potential, the in-plane magnetic field effects are governed by the unusual $B_\|^6$-dependence of $\tau_{B\|}^{-1}$ [15, 44, 16] that we attribute to the effect of a cubic term in the 2D electron dispersion generated by the field. This should give rise to a relatively sharp crossover between 'flat' regions related to regimes with orthogonal and unitary symmetry, respectively. To get an idea about the relevance of this effect for the large-area ($\sim 8\mu\text{m}^2$) quantum dots studied by Folk *et al* [3], quantitative estimates for the parameters $\gamma(B_{\parallel})$ and $\beta(B_{\parallel})$ can be obtained by evaluating the electron dispersion within a fully self-consistent numerical method [16]. Using the nominal growth parameters of the $Al_{.34}Ga_{.66}As/GaAs$ sample studied in [3], the dependences of γ and β on the in-plane magnetic field for an electron sheet density of 2×10^{11} cm^{-2} shown in the insets of Fig. 6 a) and b) were found. At low fields, $\gamma \sim B_{\parallel}^2$ and $\beta \sim B_{\parallel}^3$, as anticipated in the perturbation theory treatment. The proportionality coefficients are plotted in Fig. 6 a) and b) versus the electron sheet density. Both, the effective mass renormalization in the quadratic term of the energy dispersion and the time-reversal symmetry breaking cubic term are larger at lower 2D electron gas densities, due to the weaker confining electric field, i.e., the increased width d of the potential well. From this analysis, one estimates the field necessary to suppress the weak localization effect completely as $B_{\parallel} = 0.6 \div 0.8$T, which is in agreement with the observed complete suppression of WL corrections beyond $B_{\parallel} \approx 0.6$T in experiment [C. Marcus, private communication].

3.2. IN-PLANE MAGNETORESISTANCE IN INVERSION LAYERS WITH SEVERAL FILLED SUBBANDS AND IN DIFFUSIVE METALLIC FILMS.

If the width of the quantum well exceeds the Fermi wavelength of the electrons, several subbands become occupied – as shown in Fig. 7 – and, thus, contribute to the lateral transport. As for the one-subband case discussed above, the effect of an in-plane magnetic field depends sensitively on the microscopic structure of the wavefunctions in the z-direction perpendicular to the plane, that determines possible couplings between the subbands. In particular, in the absence of z-dependent impurity scattering, the influence of $B_{\parallel}$ is different for quantum wells with a symmetric, $W(z) = W(-z)$, and asymmetric, $W(z) \neq W(-z)$, confinement potential due to the Berry-Robnik symmetry effect.

In the absence of an external magnetic field ($B_{\parallel} = 0$), the subbands are uncoupled and, thus, each subband contributes separately to the conductivity. The weak localization correction for a film with M occupied subbands is, therefore, given as

$$g_{\text{WL}} = M\frac{e^2}{\pi h}\ln(\tau/\tau_\phi),$$

where phase coherent propagation in each subband is limited by the inelastic decoherence time $\tau_\phi^{-1} \propto T^p$.

An external magnetic field now plays two complementary roles: a) it breaks time-reversal ($\mathcal{T}$) symmetry and b) it couples the different subbands (i.e., if the random disorder potential is z-independent, it represents the only possible coupling between subbands). In fact, these two roles are linked: if the subbands

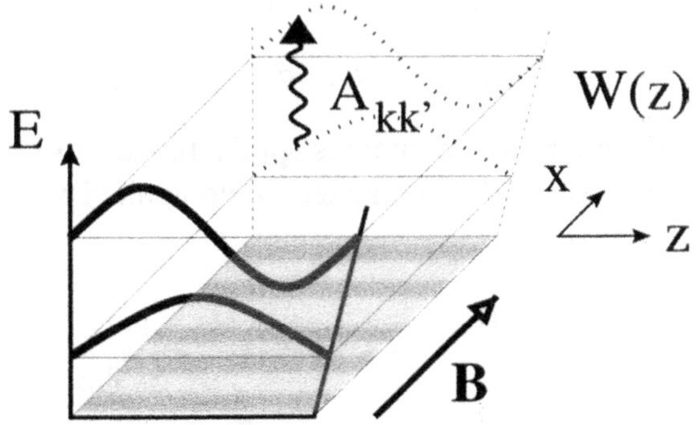

Figure 7. Schematic picture of the quantum well. Two exemplary subband wavefunctions are shown. The spatial profile of the impurity potential is sketched on the bottom of the well.

remained completely decoupled, $\mathcal{T}$-invariance of the 2D motion in each of the M subbands would be preserved because the vector potential of the parallel field can be gauged out in each particular subband. The field-induced subband mixing results in a spectrum of effective dephazing rates $1/\tau_B^{(k)}$ (where $k = 0, \ldots, M-1$) with the following properties: at least $M - 1$ dephazing rates are quadratic in the magnetic field, $1/\tau_B^{(k>0)} \propto B_\parallel^2$. The remaining $1/\tau_B^{(0)}$ is exactly equal to zero for a $\mathcal{P}_z$-symmetric system whereas it acquires a quadratic field dependence as well if the symmetry is broken (due to an asymmetric confining potential or z-dependent impurity scattering). As a result, the field and temperature dependence of the magnetoconductivity are different for a symmetric vs asymmetric confinement potential,

$$g_{\mathrm{WL}}(B_\parallel, T) = \frac{e^2}{\pi h} \begin{cases} p \ln T + 2(M-1) \ln B_\parallel & \text{for } W(z) = W(-z), \\ 2M \ln B_\parallel & \text{for } W(z) \neq W(-z). \end{cases} \tag{29}$$

The derivation of the above weak localization magnetoconductivity formula is most conveniently performed within the non-linear sigma-model formulation that will be presented in chapter 6.

For strong inter-subband impurity scattering, i.e., when the inter-subband scattering rate exceeds the subband splitting at the Fermi level, the weak localization suppression by the in-plane magnetic field happens in the same way as in thin metallic films with diffusive transverse motion, $d \gg \ell$. In a film with diffusive transverse motion, the electron encircles a random flux of order $\Phi \sim B_\parallel d^2$ during its diffusive transverse flight, where the time of flight is given as $\tau_z \sim d^2/D$. While diffusing along the wire for a longer time, $t \gg \tau_z$, the electron encircles a random flux with the variance

$$\langle \Phi^2 \rangle \sim \frac{t}{\tau_z} \times (B_\parallel d^2)^2 = tD(B_\parallel d)^2,$$

which results in a field-induced phase breaking at the rate [21]

$$\tau_B^{-1} \sim D(B_{\|}d)^2. \tag{30}$$

Note that the same geometrical argument is applicable to a thin diffusive wire with cross sectional dimensions $\sim d$ in a magnetic field applied along the wire axis

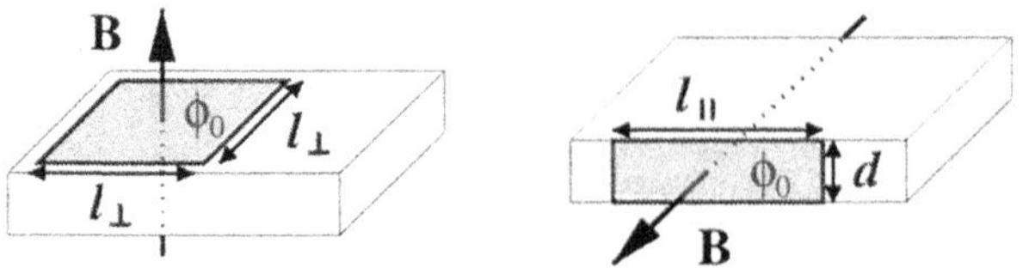

Figure 8. 'Relevant areas for perpendicular and parallel magnetic fields.

Quantitatively, τ_B can be found from an analysis of the lowest eigenvalue of the Cooperon propagator. The Cooperon propagator is defined by the (diffusion) equation

$$\left\{D(i\nabla_{\mathbf{r}} - 2\mathbf{A}(\mathbf{r}))^2 + \frac{1}{\tau_\phi}\right\} \mathcal{C}(\mathbf{r}, \mathbf{r}') = \delta(\mathbf{r} - \mathbf{r}')$$

with the boundary condition

$$(\nabla_{\mathbf{n}} + 2i\mathbf{A} \cdot \mathbf{n})\,\mathcal{C} = 0.$$

Choosing the gauge $\mathbf{A} = B_{\|} z \mathbf{e}_y$, one can separate variables in the eigenfunction equation for the Cooperon modes,

$$-D\left[\frac{\partial^2}{\partial z^2} - Q_x^2 - \left(Q_y - 2B_{\|}z\right)^2\right] \phi_{k,\mathbf{Q}}(z) = \Omega_{k,\mathbf{Q}} \phi_{k,\mathbf{Q}}(z),$$

and find the lowest eigenvalue $\Omega_{0,\mathbf{Q}} = DQ^2 + \tau_B^{-1}$ within perturbation theory. This yields the magnetic decoherence rate [21]

$$\tau_B^{-1} = \frac{1}{3} D(B_{\|}d)^2$$

which can be substituted into the weak localization formula for a diffusive film. A similar calculation gives

$$\tau_B^{-1} = \frac{1}{8} D(B_{\|}d)^2$$

for a circular diffusive wire with diameter d.

3.3. IN-PLANE MAGNETORESISTANCE IN PURE METALLIC FILMS AND QUANTUM MAGNETORESTANCE IN QUASI-BALLISTIC WIRES WITH ROUGH EDGES.

For a clean metallic film or wire, where the dominant scattering processes involve surface or edge roughness, the estimate of τ_B given in Eq. (30) is incorrect due to the exact geometrical phase cancellation described in [22]. Thus, a magnetic field $B_{\|}$ cannot affect the interference pattern, i.e., neither WL magnetoresistance nor magneto-fingerprints are to be expected in these systems at any orientation of the magnetic field. As a result, magnetic field effects acquire several unusual features.

Here, we consider a quasi-ballistic wires with width d and length L. In such a wire, a diffusive path may contain ballistic segments with length $\eta = d/\sin\alpha$ much longer than the typical length scale d, crossing the wire at the small angle α. These segments appear with the probability

$$\rho(\eta) \sim \pi^{-1} \left| \frac{d\eta(\alpha)}{d\alpha} \right|^{-1} \sim \eta^{-2},$$

for $L > \eta \gg d \gg \lambda_F$. Thus, the mean free path ℓ in such a wire exceeds the width of the wire by a large logarithmic factor, $\ell \sim \langle\eta\rangle \sim d\,\ln(L/d)$.

Quantum in-plane magnetoresistance in such a wire may appear only due to a curving of the longest ballistic paths by the field itself. The effect of this curving leads to a switching of the longest segments between the edges of the wire (or surfaces of a film) and, thus, to a sudden suppression of the interference contribution from paths including these longest ballistic flights. However, the change in the weak localization part of the conductivity in such a situation is overcome by the change in the classical part of the conductivity: due to the cyclotron curving of free electron trajectories in combination with the finite width of the wire ,the longest ballistic flights are suppressed, i.e., $\eta < \sqrt{r_c d}$ – yielding a mean free path $\ell(B_{\|}) \sim \frac{1}{2} d \ln(r_c/d)$, where $r_c = p_F/B_{\|}$. Thus, the shortening of the longest ballistic flights happens simultaneously with their involvement into phase accumulation, immediately bringing in a large time-irreversible phase of order of the total flux penetrating through the entire sample. On the other hand, paths that encircle a large flux can participate in the formation of a random interference pattern. Therefore, starting from the field value $B_{\text{defl}} = \phi_0 d/(L^2\lambda_F)$ (implying $\sqrt{r_c d} < L$) – that coincides with the set in of a classical (positive) magnetoresistance – the quantum conductance should acquire a universal random dependence on the field with a characteristic scale related to a flux change of order of the flux quantum. By the same reason, the magnetic field effect on the localization properties of quasi-ballistic wires is opposite to the commonly expected crossover between orthogonal and unitary symmetry classes, i.e., the magnetic field tends to shorten the localization length in a quasi-ballistic wire.

These qualitative expectations have been verified numerically in Ref. [24]. The results of Ref. [24] are based upon the numerical solution of a two-dimensional

Anderson Hamiltonian on a square lattice, $H = \sum_i |i\rangle\epsilon_i\langle i| - V\sum_{\langle ij\rangle}|i\rangle\langle j|$, where $\langle ij\rangle$ denotes nearest neighbor sites i and j. The structure considered consists of two ideal leads attached to a scattering region that is W sites wide and L sites long (all lengths are in units of the lattice constant a). For simulating bulk disorder, the energies ϵ_i in the scattering region are taken uniformly from the interval $-U/2 < \epsilon_i - \epsilon_0 < U/2$, where U is the disorder strength. For sites on the boundary, $\epsilon_i = \epsilon_0 + \epsilon_B$ with $\epsilon_B = 10^4$. The rough structure of the boundary was generated by having an equal probability of either 0, 1 or 2 sites at each edge with on-site potential $\epsilon_0 + \epsilon_B$. To simulate the effect of a magnetic field, a Peierls phase factor has been incorporated into V in the scattering region.

The effect of Lévy flights has been identified in the numerical results in several ways. One was to study the distribution of transmission coefficients through the wire and the distribution $P(\xi^{-1})$ of the corresponding Lyapunov exponents ξ^{-1}. This is shown in Fig. 9(a) (taken from Ref. [24]) for 4 series of quasi-ballistic wires (A-D) as well as a series of samples with bulk 'defects' (U). As pointed out by Tesanovich *et al* [49] and verified numerically in Ref. [50], the length of Lévy flights in quantum systems is limited. The limitation is due to the fact that the uncertainty in the transverse momentum, $\delta k_\perp \sim 1/d$, in a wire with a finite width sets a quantum limit to the angles $\alpha \sim d/\eta$ which can be assigned to a classically defined ballistic segment, namely $\alpha > \delta\alpha \sim \delta k_\perp/k_{\rm F} \sim \lambda_{\rm F}/d$. This sets the cut-off $\eta_{\max} = d^2/\lambda_{\rm F}$ and entails a finite localization length $L_c \sim \frac{\pi d^2}{\lambda_{\rm F}}$. Samples from the series A and B meet the criterion $L < L_c$ which manifests itself in the distribution of Lyapunov exponents by a pronounced peak at small ξ^{-1} corresponding to eigenvalues $T_n \approx 1$. This should be compared to the plateau-like distribution [51] obtained in a sample from the series U using the same numerical procedure. Note the finite width of the ballistic peak in $P(\xi^{-1})$ and that $P(0) = 0$. The enhanced density of small ξ^{-1} can be identified even in samples from the series C and D with $L \approx L_c$. In these samples, $P(\xi^{-1})$ starts to show a periodic modulation specific to the localized regime, where the spectrum of Lyapunov exponents tends to crystallize [52, 53].

The distribution of eigenvalues of the transmission matrix results in a finite-width distribution of conductances and, hence, conductance fluctuations. The statistics of conductance fluctuations, $P(g)$, for wires with edge disorder is shown in Fig. 9(b): the results of numerical simulations [24] are plotted for series A, where $\langle g\rangle \approx 2.0$ and the distribution is almost Gaussian, and series C, where $\langle g\rangle \approx 0.71$. (The conductance is expressed in quantum unitse^2/h.) The distribution function $P(g)$ is the result of the analysis of various realizations of disorder. When calculating correlation functions, there is an additional averaging over energy, since the energy dependence of conductance fluctuations is random on the scale of the Thouless energy $E_{\rm Th}$. The latter can be determined from the wire conductance as $g \sim E_{\rm Th}(L)/\Delta(L)$, where Δ is the mean level spacing in the wire. Taking into account the logarithmic multiplier in the dependence of the classical conductance

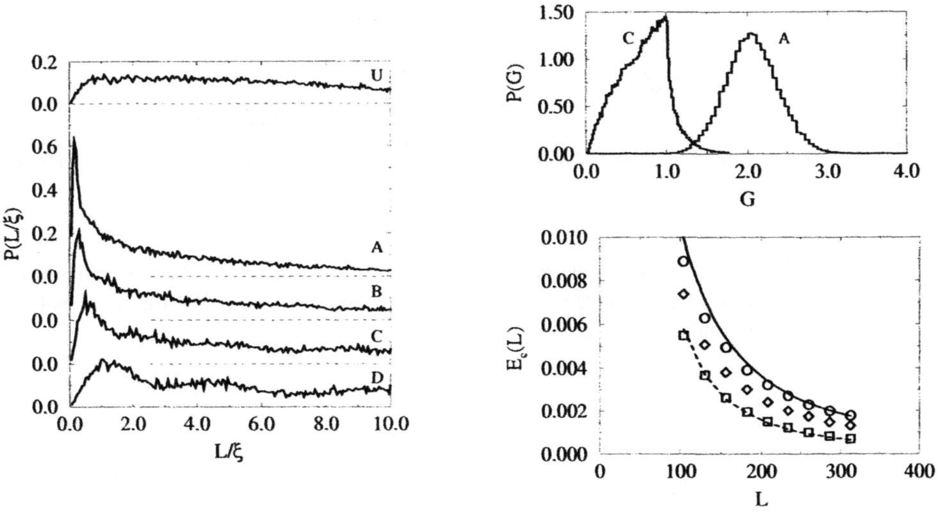

Figure 9. (a) Distribution function $P(L/\xi)$ for 4 quasi-ballistic samples with W =15 and L =52(A), 104(B), 208(C) and 416(D) and for a wire (U) with on-site bulk disorder $U = 2.0$ (nominal mean free path $\ell \approx 8.5$, $W = 15$ and $L = 52$). (b) Distribution function $P(g)$ for the two quasi-ballistic structures A and C. (c) $E_{\mathrm{Th}}(L)$ for quasi-ballistic wires with $\epsilon_0 \in [1.5, 1.7]$ (circles) and $\epsilon_0 \in [1.0, 1.2]$ (diamonds) and for a wire from the U series (squares). Also shown is the analytical result (cf. Eq. (31) with $hv_{\mathrm{F}} = 12.3$ and $W = 9.3$ (solid curve). The dashed curve corresponds the ergodic law $E_{\mathrm{Th}} \sim hD/L^2$ with $hD = 62.5$.

on sample length due to the Lévy flights, one obtains

$$g \sim \frac{e^2}{h} \frac{d^2}{L\lambda_{\mathrm{F}}} \ln(\frac{L}{1.7d}). \tag{31}$$

This result for the wire conductance implies an anomalous scaling of the Thouless energy, $E_{\mathrm{Th}} = \pi d v_{\mathrm{F}} L^{-2} \ln(\frac{L}{1.7d})$. In Fig. 9(c), it is shown that, indeed, the Levy flights manifest themselves in the correlation energy of UCF in the quasi-ballistic regime. For comparison, it is demonstrated that the results for bulk-disordered samples (U) coincide with the standard scaling law $E_{\mathrm{Th}}(L) \sim L^{-2}$.

Coming now to the magnetic field effects. The expected deflection of Lévy flights at $B > B_{\mathrm{defl}}$ entailing a break of geometrical flux cancellation is observable in Fig. 10 as pronounced magnetoconductance fluctuations $\delta g(B) = g(B) - \langle g \rangle$ beginning together with a gradual decrease of the average conductance value – which is a purely classical effect. Their variance corresponds to the usual UCF value, and the auto-correlation function as a function of magnetic field gives the correlation field B_c corresponding to a change of magnetic flux through the sample area of order $2.5\phi_0$ as compared to $1.5\phi_0$ in the bulk-disordered case. This seems to explain an earlier experimental observation of UCF in quasi-ballistic semiconductor wires [54].

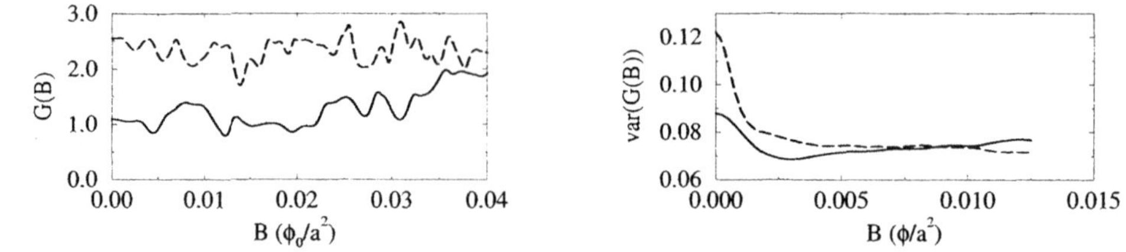

Figure 10. (a) Conductance for a quasi-ballistic wire (solid curve) from the B series and a disordered wire (dashed curve) with mean free path $\ell \approx 16$). (b) Variance $\mathrm{var}(g(B))$ for quasi-ballistic wires from the B series and disordered wires.

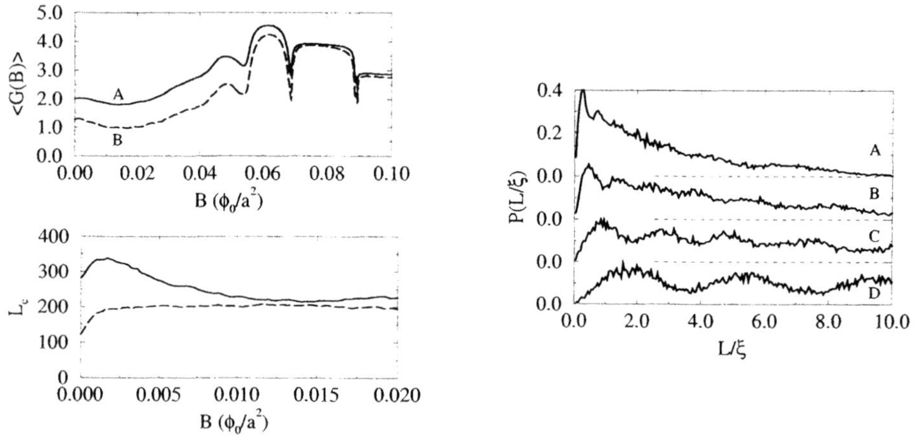

Figure 11. (a) The ensemble averaged conductance as a function of magnetic field for wires from the A and B series. (b) The same curves (A-D) as in Fig. 9(b), but at finite magnetic field $B = 0.02\phi_0/a^2$. (c) The localization length for quasi-ballistic wires (solid curve) and disordered wires (dashed curve) with mean free path $\ell \approx 8.5$.

Furthermore, Fig. 11 shows the tendency of the localization length in a quasi-ballistic wire to increase with the magnetic field – as opposed to the behavior of $L_c(B)$ in systems with diffusive transverse motion [40, 52]. The relevant field scale at which the the localization length in quasi-ballistic wires starts to shorten, $B_c \sim \phi_0 \lambda_F / W^3$, can be obtained from B_{defl} by replacing the sample length L with L_c. Formally, this yields a similar field scale as the one leading to a crossover in L_c between two symmetry classes due Aharonov-Bohm type phase effects. Thus, the latter is hindered by the deflection effect and does not appear as an intermediate crossover regime.

4. In-plane magnetoresistance due to spin-orbit coupling

4.1. SPIN-ORBIT COUPLING IN 2D HETEROSTRUCTURES AND QUANTUM DOTS

Coupling between the electron spin and its orbital motion is a relativistic effect inherent to many metallic and semiconductor systems. For example, in zinc-blend type III-V semiconductors, such as GaAs, which have no inversion symmetry in the unit cell, the 3D bulk dispersion of conduction band electrons contains terms that are cubic in the electron momentum and linear in its spin operator [55]. This generates a linear spin-orbit coupling for electrons confined to the 2D plane in a heterostructure or a quantum well [56, 57],

$$\begin{aligned} H_0 &= \frac{\mathbf{p}^2}{2m} + \frac{\alpha_{\rm br}}{2m}[\mathbf{p}\times\mathbf{e}_z]\cdot\underline{\sigma} + \frac{\alpha_{\rm cr}}{2m}(p_x\sigma_x - p_y\sigma_y) \\ &= \frac{\mathbf{p}^2}{2m} + \frac{v_{\rm F}}{2\lambda_{\rm SO}}\mathbf{n_p}\cdot\underline{\sigma}, \end{aligned} \tag{32}$$

where $\mathbf{n_p}$ is a unit vector in $\mathbf{p}$-direction.

The form of the spin-orbit (SO) coupling in Eq. (32) is specified for a GaAs heterostructure grown along the (001) plane which has the symmetry of a square lattice without inversion symmetry within the unit cell. It includes two possible combinations of spin and momentum operators that are invariant under the symmetry transformation corresponding to such a lattice symmetry, parameterized by two constant, $\alpha_{\rm br}$ and $\alpha_{\rm cr}$.

For a particle moving diffusively in an infinite heterostructure, spin-orbit coupling provides an efficient mechanism for spin relaxation. In the frame of the moving electron, its spin undergoes a precession with angular frequency $\underline{\Omega} \sim \mathbf{n_p} v_{\rm F}/\lambda_{\rm SO}$ which randomly changes direction each time the particle changes its direction of propagation due to scattering at impurities. Using the parameterization of the SO coupling by a spin-orbit length, $\lambda_{\rm SO} \sim \alpha^{-1}$, as in Eq. (32), the precession angle acquired between successive scattering events can be estimated as $\delta\varphi_{\rm so} \sim \tau v_{\rm F}/\lambda_{\rm SO} = \ell/\lambda_{\rm SO}$. For weak SO coupling, $\ell/\lambda_{\rm SO} \ll 1$, the electron spin precession can be viewed as a random walk of the electron spin polarization vector around the unit sphere with a typical step size $\delta\varphi_{\rm so}$. This causes a loss of spin memory at the (Dyakonov-Perel) rate [25]

$$\tau_{\rm DP}^{-1} \sim D/\lambda_{\rm SO}^2. \tag{33}$$

In a 2D electron gas, this parameter limits the diffusion time in the triplet Cooperon channel (whereas the singelt Cooperon channel is unaffected). Therefore, it suppresses the triplet contribution to the weak localization correction formula and controls the crossover between the weak localization (at weak SO coupling) and weak anti-localization (at strong SO coupling) regimes in the quantum correction to the conductivity [26, 27].

When the electron motion within the 2D plane is also confined laterally, the use of the so called Dyakonov-Perel formula, Eq. (33), dramatically overestimates the rate of spin relaxation. In particular, if the electron motion is confined to a single 1D channel (i.e., a quantum wire), no real spin relaxation takes place – despite the electron spin precession. In this case, the spin precession is a reversible process that rotates the initial electron spin to the same definite state at each point along the wire, no matter how many times the electron moves forwards and backwards. Mathematically, the above-mentioned property of the SO coupling becomes transparent in the following way: Consider a one-dimensional electron moving, say, along the x-direction. Applying the unitary transformation,

$$U = \exp\left(\frac{ix}{2\lambda_{\rm SO}}\mathbf{n}(\mathbf{l}_x)\cdot\underline{\sigma}\right),$$

where $\mathbf{l}_x$ is a unit vector along the wire axis, the SO coupling is eliminated completely from the electron dispersion represented in a rotated spin-coordinate frame [58, 59]. Therefore, it does not affect any charge transport properties. By contrast, the Dyakonov-Perel relaxation of the spin of a 2D electron (with the rate described in Eq. (33)) is the result of the diversity of paths the electron may use to move between two points in the sample.

A small quantum dot with dimensions $L \ll \lambda_{\rm SO}$ represents an intermediate situation between the 1D and the 2D system. Here, we shall consider a dot coupled to metallic leads via two contacts, left (l) and right (r), each with $N_{l,r} \gtrsim 1$ open orbital channels. Then the escape rate [29] from the dot into the leads is given as

$$\gamma \equiv \tau_{\rm esc}^{-1} = (N_l + N_r)\Delta/(2\pi),$$

where $\Delta = 2\pi/(m\mathcal{A})$ is the mean level spacing ($\mathcal{A}$ area of the dot).

Although there is no unitary transformation that eliminates the SO coupling completely by changing to a locally rotated spin coordinate system, the linear term in the small SO coupling constant can be cancelled by a proper spin-dependent gauge transformation [30], thus, generating only contributions to the free electron Hamiltonian of order $\lambda_{\rm SO}^{-2}$ (or higher). As a result, the influence of spin-orbit coupling on the transport characteristics of a small dot is suppressed as compared to the infinite 2D system with similar material parameters.

However, the application of an in-plane magnetic field changes the situation. The transformation to the locally rotated spin-coordinate frame turns an initially constant external magnetic field, $\mathbf{B} = B\,\mathbf{l}$, into an inhomogeneous Zeeman field that accelerates the loss of spin memory of an electron passing through the dot. Below, this effect will be described quantitatively.

For the sake of convenience, we choose the coordinate system (x_1, x_2) with axes along the crystallographic directions $\mathbf{e}_1 = [110]$ and $\mathbf{e}_2 = [1\bar{1}0]$ such that we can exploit the C_{2v} lattice symmetry of the (001) plane of GaAs. Then, the single-particle Hamiltonian in the presence of an in-plane magnetic field $\mathbf{B} = B\,\mathbf{l}_{\parallel}$ with

$\mathbf{l}_{\|} = (l_1, l_2, 0)$ can be rewritten as

$$H_0 = \frac{1}{2m}\left[\left(p_1 - \frac{\sigma_2}{2\lambda_1}\right)^2 + \left(p_2 + \frac{\sigma_1}{2\lambda_2}\right)^2\right] + \frac{\epsilon_Z}{2}\mathbf{l}_{\|}\cdot\underline{\sigma}, \tag{34}$$

where $\lambda_{1,2} = 1/(\alpha \pm \varrho)$ characterize the length scales associated with the strength of the SO coupling for electrons moving along the principal crystallographic directions. Here, $\sigma_{1,2,3}$ are Pauli matrices with $\sigma_2 = -\sigma_2^T$ and $\sigma_{1,3} = \sigma_{1,3}^T$. $\epsilon_Z = g\mu_B B_{\|}$ is the Zeeman energy associated with the parallel magnetic field. Furthermore, $\mathbf{p} = \mathbf{P} - \mathbf{A}$ is the kinetic momentum with $\mathbf{P} = -i\nabla$ and the vector potential $\mathbf{A} = eB_z[\mathbf{r} \times \mathbf{n}_z]/2$ corresponding to an additional perpendicular magnetic field.

In the following, we assume the dot to be sufficiently small to fulfill the conditions $L_{1,2}, \ll \lambda_{1,2}$ and $\gamma, \epsilon_Z \ll E_{\text{Th}}$, where $E_{\text{Th}} = hD/L^2$ is the Thouless energy.

4.2. EFFECT OF SPIN-ORBIT COUPLING ON WEAK LOCALIZATION AND CONDUCTANCE FLUCTUATIONS

For a spin-$\frac{1}{2}$ particle in a quantum dot connected to adiabatic ballistic contacts, the WL corrections to the conductance can be related to the lowest-lying modes of Cooperons in the singlet and triplet channels. The expressions for the different Cooperon modes can summarized in the equation

$$\mathcal{C}^{LM} = \frac{1}{2}\left\langle \text{tr}\left[\sigma_L\sigma_2\hat{G}_R^T(\varepsilon)\sigma_2\sigma_M\hat{G}_A(\varepsilon-\omega)\right]\right\rangle, \tag{35}$$

where $L, M = 0, \ldots, 3$. Note that the use of the (perturbative) diagrammatic technique in the description of quantum dots is justified if the number of channels in the leads is large, $N_{l,r} \gg 1$.

In the absence of SO coupling and Zeeman splitting, the Cooperon modes $\mathcal{C}^{LM}$ can be separated into four completely independent channels, i.e., one singlet channel $\mathcal{C}_0 = \mathcal{C}^{00}$ and three triplet channels $\mathcal{C}_1^M = \mathcal{C}^{MM}$ $(M = 1, 2, 3)$. Or, in a matrix representation ,

$$\hat{\mathcal{C}} = \mathcal{C}\hat{\delta}, \qquad \text{where } \hat{\delta} \equiv \delta_{LM},$$

and $\mathcal{C}$ obeys the conventional diffusion equation. This leads to the familiar result for the WL corrections [26],

$$g^{\text{WL}} \propto \frac{e^2}{h}\left(\mathcal{C}_0 - \sum_{M=1,2,3}\mathcal{C}_1^M\right).$$

However, the SO coupling and Zeeman splitting mix the various components [27] and, thus, split their spectra. This changes the conventional diffusion equation into the matrix equation

$$\hat{\Pi}\,\hat{C}(\mathbf{x},\mathbf{x}') = \delta(\mathbf{x}-\mathbf{x}')\,\hat{\delta}, \tag{36}$$

where

$$\hat{\Pi} = -D\left((\partial_{x_1}+2iA_1)\hat{\delta} - \frac{i}{\lambda_1}\hat{S}_2\right)^2 - D\left((\partial_{x_2}+2iA_2)\hat{\delta} + \frac{i}{\lambda_2}\hat{S}_1\right)^2 + \\ + \gamma\,\hat{\delta} + i\epsilon_Z\,\hat{\eta}. \tag{37}$$

Here, $\hat{S}_K^{LM} = -i\varepsilon^{KLM}$ are spin-1 operators ($K, L, M = 1, 2, 3$), and ε^{KLM} is the antisymmetric tensor. As a 4×4 matrix, $\hat{S}$ also has zero elements when $L = 0$ or $M = 0$. The other relevant matrix $\hat{\eta}$ is defined as

$$\eta^{LM} = l_L\,\delta_{0M} + l_M\,\delta_{0L}$$

indicating that coherence between electrons with opposite polarization is lost on the time scale ϵ_Z^{-1} which affects the singlet channel.

Eq. (37) is supplemented with a boundary condition at the edge of the dot characterized by the normal direction $\mathbf{n}_{\|} = (n_1, n_2)$,

$$\left[\mathbf{n}_{\|}\cdot(\nabla + 2i\mathbf{A})\hat{\delta} - in_1\lambda_1^{-1}\hat{S}_2 + in_2\lambda_2^{-1}\hat{S}_1\right]\hat{C} = 0. \tag{38}$$

Because of the boundary condition, the lowest mode of the Cooperon cannot be a mere constant solution, such as $\mathbf{q} = 0$. To find the lowest Cooperon mode in this problem, one should make such a gauge transformation (that is, a unitary rotation of the Cooperon spin components)

$$\hat{C} = \hat{U}\hat{C}'\hat{U}^{-1},$$

that would transform the original boundary condition into $\mathbf{n}_{\|}\cdot\nabla\hat{C}' = 0$. In the new spin co-ordinate system, one can now approximate the lowest Cooperon mode for $\hat{C}'$ by a $\mathbf{q} = 0$ solution, i.e., $\hat{C}' = \mathrm{const.}$, and evaluate its eigenvalue using the standard methods of the Hamiltonian perturbation theory with respect to the terms generated by such a rotation in the initial differential operator $\hat{\Pi}$. This program can be realised by applying the transformation

$$\hat{U} = \exp\left[i\left(\frac{x_2}{\lambda_2}\hat{S}_1 - \frac{x_1}{\lambda_1}\hat{S}_2\right)\right]\exp[-i\varphi_s(\mathbf{x})\hat{S}_3]\exp[-i\varphi_a(\mathbf{x})],$$

where function $\varphi = \varphi_s + \varphi_a$ transforms from the symmetric gauge to a gauge, where the normal component of the vector potential at the boundary of the dot vanishes. This eliminates the lowest orders SO coupling terms from the boundary condition, and, thus, (in a small dot, $L_{1,2} \ll \lambda_{1,2}$) can be followed by a perturbative analysis of extra terms in Eq. (37) generated by the rotation $\hat{U}$. This results in the 0D matrix equation for the Cooperon,

$$\hat{C} = \left[\gamma\hat{\delta} + i\epsilon_Z\hat{\eta} + \left(\frac{1}{\sqrt{\tau_B}}\hat{\delta} - \sqrt{\epsilon_\perp^{so}}\hat{S}_3 \right)^2 + \epsilon_\perp^Z(\hat{\delta} - \hat{S}_3^2) + \epsilon_\parallel^{so}(\hat{\mathbf{S}}^2 - \hat{S}_3^2) \right]^{-1}, \quad (39)$$

which depends on six different energy scales to be discussed in the following. (γ and ϵ_Z have been introduced earlier.)

In Eq. (39), the two parameters

$$\tau_B^{-1} = \frac{4\pi B_z^2}{\Delta} \left\langle |M_{\alpha\beta}|^2 \right\rangle = \kappa E_{\mathrm{Th}}(2eB_z\mathcal{A})^2,$$
$$\epsilon_\perp^{so} = \kappa E_{\mathrm{Th}} \left(\mathcal{A}/\lambda_1\lambda_2\right)^2,$$

possess the same dependence on the shape of the dot and the disorder in the sample. Here, κ is a geometry-dependent coefficient. Furthermore, the random quantities $M_{\alpha\beta}$ are the non-diagonal matrix elements of the magnetic moment of the electron in the dot. Note that the difference in the third term in brackets of Eq. (39), containing τ_B^{-1} and $\epsilon_\perp^{so}$, reflects the addition or subtraction of the Berry and Aharonov-Bohm phases, as was pointed out in Ref. [58, 59].

The parameter $\epsilon_\perp^Z$ in Eq. (39) is the result of a parallel field induced additional Zeeman splitting,

$$\epsilon_\perp^Z = \frac{\epsilon_Z^2}{2\Delta} \sum_{i,j=1,2} \frac{l_i l_j}{\lambda_i\lambda_j} L_i L_j\, \Xi_{ij}, \qquad \text{where } \Xi_{ij} = \frac{\pi}{L_i L_j} \langle x_i^{\alpha\beta} x_j^{\beta\alpha} \rangle,$$

and $x_{1,2}^{\alpha\beta}$ are the non-diagonal matrix elements of the dipole moment of the electron in the dot. The quantities Ξ_{ij} depend on geometry and the disorder and may be estimated as $\Xi \simeq \Delta/E_{\mathrm{Th}} = g^{-1}$. This yields $\epsilon_\perp^Z \ll \epsilon_Z$. A similar energy scale has been found in other recent publications [60].

Finally, the parameter

$$\epsilon_\parallel^{so} \sim \left[(L_1/\lambda_1)^2 + (L_2/\lambda_2)^2 \right] \epsilon_\perp^{so} \ll \epsilon_\perp^{so}$$

introduces the smallest energy scale through which the SO coupling affects the Cooperon propagator.

In principle, the form of Eq. (39) is applicable beyond the diffusive approximation as it follows purely form symmetry considerations. Now, the weak localization corrections can be found from Eq. (39) as

$$g^{\mathrm{WL}} \propto \mathrm{tr}\left[\hat{C}\left(\hat{\delta} - \hat{\mathbf{S}}^2 \right) \right]. \quad (40)$$

In the absence of a perpendicular magnetic field ($\tau_B^{-1} = 0$) and using the assumptions made above, namely $\gamma, \epsilon_Z, D/\lambda_{1,2}^2 \ll E_{\mathrm{Th}}$, this yields

$$g^{\mathrm{WL}} \approx \frac{e^2}{h}\frac{a_{lr}}{4} \left(-\frac{\gamma}{\gamma + \epsilon_\perp^{so}} - \frac{\gamma}{\gamma + \epsilon_\perp^Z + 2\epsilon_\parallel^{so}} + \frac{\epsilon_\perp^{so}}{\gamma + \epsilon_\perp^{so} + \frac{\epsilon_Z^2}{\gamma}} \right), \quad (41)$$

where the fact that $\epsilon_\perp^Z \ll \epsilon_Z$ has been used. Furthermore, we introduced the notation

$$a_{lr} = \frac{4N_l N_r}{(N_l + N_r)^2}.$$

The formula in Eq. (41) describes the average tendency of the in-plane magnetoresistance of a dot due to the interplay between SO coupling and Zeeman splitting.

It is interesting to note that the application of a Zeeman field (i.e., an in-plane magnetic field [3]) alone does not suppress the weak localization corrections completely as long as $\epsilon_Z \ll E_{\rm Th}$ – whereas in the opposite case, $\epsilon_Z \gtrsim E_{\rm Th}$, it does. However, the orbital effect of such a strong in-plane field is already sufficient to suppress weak localization as discussed in the previous section.

In the regime $\epsilon_\parallel^{\rm so}/\gamma, \epsilon_\perp^{\rm so}/\gamma \to 0$ relevant for the experiments on small dots [3], the form of Eq. (41) is dominated by the first two terms,

$$g^{\rm WL}(B) \approx -\frac{e^2}{h}\frac{a_{lr}}{4}\left(1 + \frac{1}{1 + \epsilon_\perp^Z(B)/\gamma}\right). \tag{42}$$

This suggests a possible procedure for measuring the ratio λ_1/λ_2. By fitting the experimental magnetoresistance data to $g^{\rm WL}(B)$, one can determine the characteristic in-plane magnetic field $\mathcal{B}$ at which the weak localisation part of the dot conductance gets suppressed by the factor of two. For a dot with a strongly anisotropic shape, this parameter would depend on the orientation of the in-plane magnetic field. In particular, $\mathcal{B}$ should be measured for two orientations of the in-plane field: namely $\mathcal{B}_{[110]}$ for $\mathbf{l} = [110]$ and $\mathcal{B}_{[1\bar{1}0]}$ for $\mathbf{l} = [1\bar{1}0]$. Furthermore, one should perform a simultaneous measurement of the two characteristic fields $\mathcal{B}'_{[110]}$ and $\mathcal{B}'_{[1\bar{1}0]}$ in a dot produced on the same chip by rotating the same lithographic mask by 90°. The anisotropy of the SO coupling is then obtained directly from the ratio

$$\left(\mathcal{B}_{[110]}\mathcal{B}'_{[110]}/\mathcal{B}_{[1\bar{1}0]}\mathcal{B}'_{[1\bar{1}0]}\right) = (\lambda_1/\lambda_2)^4,$$

that is, independently of details of the sample geometry.

The results for weak localization part of a two-terminal conductance of a dot and the variance of its universal fluctuations are summarized for the limiting regimes in the following equation,

$$g^{\rm WL} = -\frac{2-\beta}{2\beta\Sigma}a_{lr}\frac{e^2}{h}, \qquad \langle\delta g^2\rangle = \frac{s}{16\beta\Sigma}a_{lr}^2\left(\frac{e^2}{h}\right)^2. \tag{43}$$

Here, the conventional parameter β describes time-reversal symmetry of the orbital motion, s is the Kramers' degeneracy parameter, and Σ is an additional parameter characterizing the mixing of states with different spins for strong Zeeman splitting. In a small dot [3], where $\epsilon_\parallel^{\rm so}/\gamma \to 0$ as well as $\epsilon_\perp^{\rm so}/\gamma \to 0$, one obtains the following values for the different parameters: $\beta = 1$ indicates that for weak SO coupling the electron spin splitting cannot violate the time-reversal symmetry of the orbital motion; Kramers' degeneracy is preserved ($s = 2$) for $\epsilon_Z < \gamma$, but lifted ($s = 1$) for $\epsilon_Z > \gamma$; finally, $\Sigma = 1$ for $\epsilon_\perp^Z < \gamma$ and $\Sigma = 2$ for $(\epsilon_Z \gg)\, \epsilon_\perp^Z > \gamma$ at very strong Zeeman splitting.

5. In-plane magnetoresistance in systems with magnetic impurities

In addition to a purely potential disorder, metals and semiconductors may contain magnetic impurities [39, 38]. In fact, most of real materials certainly do have them to some extent [61, 62]. In this section, we discuss the influence of a dilute magnetic contamination on the quantum transport characteristics of disordered conductors. In particular, we describe the suppression of weak localization, and its restoration by an in-plane magnetic field due to a polarization of the localized magnetic moments – which slows down the decoherence of conducting electrons and produces an observable magnetoresistance, $g(B_\parallel) = g_{\text{class}} + \delta g^{\text{WL}}(B_\parallel)$.

5.1. ELECTRON SPIN RELAXATION AND KORRINGA TIME FOR MAGNETIC IMPURITIES

The coupling between a spin-$\frac{1}{2}$ electron and the magnetic impurity spin originates from the exchange interaction and can be described by the Hamiltonian

$$\hat{\mathcal{H}}_{\text{s}} = J\ \mathbf{S}(\mathbf{r}) \cdot \underline{\sigma},$$

where $\mathbf{S}$ is the impurity spin and J is the exchange coupling. Furthermore, $\underline{\sigma}$ is a vector of Pauli matrices. When scattering on a magnetic impurity, the electron can change its spin state. Thus, electrons diffusing in a disordered conductor containing magnetic impurities loose their spin memory after the time scale τ_{s} determined by the spin-flip relaxation rate

$$\tau_{\text{s}}^{-1} \sim \nu\, J^2\, S(S+1) \sim \nu\, n_{\text{s}}\, j^2\, S(S+1).$$

Here, j the exchange coupling due to a single impurity and n_{s} is the density of magnetic impurities. For dilute magnetic impurities, $\tau_{\text{s}} \gg \tau$.

To incorporate spin-flip scattering into the quantitative Cooperon/diffuson analysis of weak localization and UCFs, one has to make assumptions concerning correlation properties of this additional source of disorder. Similarly to the non-magnetic impurities, we assume the magnetic disorder to be Gaussian δ-correlated with zero mean and variance

$$\langle JS_\alpha(\mathbf{r},t) JS_\beta(\mathbf{r}',t)\rangle = \frac{1}{6\pi\nu\tau_{\text{s}}} \delta_{\alpha\beta}\, \delta(\mathbf{r} - \mathbf{r}')$$

at coinciding moments of time.

There is, however, an important difference between potential disorder, which is static for a fixed configuration of impurities in the sample, and magnetic scatterers, which change their spin state after each electron-impurity spin-flip event. As a result, magnetic disorder has its own dynamics, and a magnetic scatterer forgets about its initial spin state with the so-called Korringa relaxation rate [63]

$$\tau_{\rm K}^{-1} \sim \nu\, n_{\rm e}\, j^2 \frac{T}{\epsilon_{\rm F}}, \tag{44}$$

where $n_{\rm e}$ is the electron density.

Korringa relaxation, as it is known, takes place due to spin-flip scattering at the impurity of any electron with energy $|\epsilon - \epsilon_{\rm F}| \sim T$ close to the Fermi level. Therefore, thecorresponding relaxation rate is temperature-dependent, whereas the single-electron spin relaxation rate $1/\tau_{\rm s}$ is temperature independent. As a result, for temperatures $T > T_{\rm s}$ higher than a certain temperature

$$T_{\rm s} \sim \frac{n_{\rm s}}{n_{\rm e}} \epsilon_{\rm F},$$

an impurity changes its spin state faster than any free electron whose individual propagation we may follow, i.e., $1/\tau_{\rm K} > 1/\tau_{\rm s}$. By contrast, for $T < T_{\rm s}$, the single-electron spin relaxation is faster, $1/\tau_{\rm s} > 1/\tau_{\rm K}$, and the exchange field of the impurities can be treated as static when analysing weak localization effects.

The relaxation of the impurity spin sub-system should be taken into account in the correlation properties of this source of randomness. This can be achieved by saying that, for the same magnetic scatterer,

$$\langle S_\alpha(\mathbf{r},t) S_\beta(\mathbf{r},0)\rangle = \delta_{\alpha\beta}\, S(S+1)\, e^{-|t|/\tau_{\rm K}}.$$

5.2. SPIN-FLIP SCATTERING, IMPURITY SPIN DYNAMICS AND POLARIZATION OF IMPURITIES IN WEAK LOCALIZATION

After an electron flips its spin a few times at encountered magnetic impurities, the coherence between waves propagating in clockwise and anti-clockwise directions along the same geometrical path is destroyed. In the regime of fast Korringa relaxation, $1/\tau_{\rm K} > 1/\tau_{\rm s}$, this happens since electron waves traveling towards each other along a loop encounter the same magnetic scatterer at different moments of time and, therefore, in uncorrelated initial states. In the regime of slow Korringa relaxation, $1/\tau_{\rm K} < 1/\tau_{\rm s}$, decoherence between these two waves accumulates due to a non-commutativity of electron spin operators, whose product would appear due to the coupling to to spin-flip transitions between the same initial and final states of the same set of magnetic impurities visited in the opposite order.

As a result, in a magnetically contaminated conductor, the time of flight along trajectories forming the enhancement of backscattering is limited by the spin-relaxation time, $\tau_{\rm s}$. As in the previous section, a quantitative description of weak localization corrections to the conductivity,

$$\delta\sigma = \frac{e^2}{2\pi h}\int_{\tau}^{\infty} dt\ \left[\mathcal{C}_0(t;0) - 3\,\mathcal{C}_1(t;0)\right], \tag{45}$$

requires a diagrammatic evaluation of the singlet and triplet Cooperon correlation functions. In Eq. (45), the variable of integration t is the difference between moments of time when clockwise and anti-clockwise propagating waves pass through the same point in space. As shown in Fig. 12, when deriving the equation of motion for Cooperons, one should include magnetic impurity scattering into both, (a) the self-energy of the impurity-averaged electron Green functions, i.e., as a correction to the mean free time, and (b) the two-particle correlation functions as additional impurity lines in the ladder diagram. Taking into account the Clebsh-Gordan coefficients that appear upon splitting the Cooperon into singlet and triplet channels, the equation for the Cooperon takes the form

$$\left\{\partial_t - D\nabla^2 + \frac{1}{\tau_{\mathrm{S}}}\left(1 + c_J e^{-|t|/\tau_{\mathrm{K}}}\right)\right\}\mathcal{C}_J(t,t';\mathbf{r},\mathbf{r}') = \delta(t-t')\delta(\mathbf{r}-\mathbf{r}'), \tag{46}$$

where $J = 0$ (spin-singlet) or 1 (spin-triplet), and $c_0 = 1$, $c_1 = -1/3$. The time-dependence accounts for the fact that, for clockwise and anti-clockwise trajectories, electron waves test the same scatterer at different moments of time. If the relevant time scale determining the value of the integral in Eq. (45) is large such that $|t| > \tau_{\mathrm{K}}$, the states of the same impurity seen by waves propagating in opposite directions are uncorrelated. In this regime, the solution of Eq. (46) is

$$\mathcal{C}(\mathbf{q},\omega) = \frac{1}{D\mathbf{q}^2 - i\omega + \tau_{\mathrm{S}}^{-1}}, \tag{47}$$

with the same relaxational pole τ_{S}^{-1}, i.e., the single-particle spin relaxation rate, in all spin channels. This regime is realized when $\tau_{\mathrm{K}} < \tau_{\mathrm{S}}$, that is, at high temperatures $T > T_{\mathrm{S}}$.

By contrast, for $\tau_{\mathrm{K}} > \tau_{\mathrm{S}}$, the integral in in Eq. (45) is determined by the behavior of Cooperons at $|t| < \tau_{\mathrm{K}}$, where the decoherence in the triplet channel takes about thrice longer than in the singlet channel,

$$\mathcal{C}_0(\mathbf{q},\omega) = \frac{1}{D\mathbf{q}^2 - i\omega + 2\tau_{\mathrm{S}}^{-1}},\quad \mathcal{C}_1^{\alpha\beta}(\mathbf{q},\omega) = \frac{\delta_{\alpha\beta}}{D\mathbf{q}^2 - i\omega + \frac{2}{3}\tau_{\mathrm{S}}^{-1}}.$$

The latter feature indicates that a single spin-flip does not completely destroy coherence between clock and anti-clockwise paths, given that in both propagation scenarios the magnetic impurity state undergoes exactly the same transition.

Eq. (46) can be solved using the following substitution,

$$\mathcal{C}_J(t,t';\mathbf{r},\mathbf{r}') = e^{-f(t)}\,\tilde{\mathcal{C}}_J(t,t';\mathbf{r},\mathbf{r}')\,e^{f(t')},$$

Figure 12. Cooperon for $|t| < \tau_K$ (both contributions) and $|t| > \tau_K$ (only left diagram). (The thin dashed line represent scattering from the non-magnetic impurities while the thick dashed lines stand for (spin-flip) scattering from the magnetic impurities.)

with

$$f(t) = \frac{1}{\tau_s}\left(t + c_J\tau_K \,\mathrm{sign}\, t\,(1 - e^{-|t|/\tau_K})\right).$$

Then the 'modified' Cooperon $\tilde{C}_J(t,t';\mathbf{r},\mathbf{r}')$ obeys the conventional diffusion equation $(\partial_t - D\nabla^2)\tilde{C}_J(t,t';\mathbf{r},\mathbf{r}') = \delta(t-t')\delta(\mathbf{r}-\mathbf{r}')$. After substituting the resulting expression for the Cooperon C_J back into Eq. (45), we arrive at

$$g^{\mathrm{WL}} = \frac{e^2}{2\pi h}(D\tau_s)^{1-\frac{d}{2}} \int_{\tau/\tau_s}^{\infty} \frac{d\theta}{\theta^{d/2}} \left\{ \exp\left[-2\theta - \frac{2\tau_K}{\tau_s}(1 - e^{-\theta\tau_s/\tau_K})\right] \right.$$
$$\left. -3\exp\left[-2\theta + \frac{2\tau_K}{3\tau_s}(1 - e^{-\theta\tau_s/\tau_K})\right]\right\}. \quad (48)$$

In the 'low-temperature' limit, where $\tau_K^{-1} \to 0$,

$$g^{\mathrm{WL}}(\tau_K^{-1}) \to 0 = -\frac{e^2}{\pi h}(D\tau_s)^{1-\frac{d}{2}} \begin{cases} \frac{\pi}{\sqrt{2}}(3^{3/2} - 1) & d = 1, \\ \ln \frac{3^{3/2}\tau_s}{4\tau} & d = 2. \end{cases}$$

For comparison, in the limit $\tau_K^{-1} \to \infty$, the above equation changes into

$$g^{\mathrm{WL}}(\tau_K^{-1} \to \infty) = -\frac{e^2}{\pi h}(D\tau_s)^{1-\frac{d}{2}} \begin{cases} \sqrt{\frac{\pi}{2}} & d = 1, \\ \ln \frac{\tau_s}{\tau} & d = 2. \end{cases}$$

The difference between the dynamic and static results in Eqs. (49) and (49) manifests itself in a weak temperature dependence of the conductivity,

$$\sigma(T > T_s) - \sigma(T \ll T_s) \approx \frac{e^2}{2\pi h}(D\tau_s)^{1-\frac{d}{2}} \begin{cases} 1.2 & d = 1, \\ 1.9 & d = 2. \end{cases} \quad (49)$$

So far in this section, we addressed only the suppression of weak localization by spin-flip scattering, whereas our main goal was to identify possible effects of an external magnetic field leading to an observable in-plane magnetoresistance. An in-plane magnetic fields tends to polarize the paramagnetic impurities, thus, opening a gap $\epsilon_Z^{(\mathrm{imp})} = g^{(\mathrm{imp})}\mu_B B_\parallel$ for the electron spin-flip relaxation process. At high magnetic fields, $\epsilon_Z^{(\mathrm{imp})} \gtrsim T$, the allowed electron energies are insufficient to flip the impurity spin, i.e., spin-flip scattering is suppressed. Thus, at $B_{\|} > B_{\rm s}$, where

$$B_{\rm s}(T) = T/(g^{(\rm imp)}\mu_{\rm B}), \tag{50}$$

magnetic impurities act as non-magnetic 'potential' scatterers, with the only difference that their potential profiles are different for spin-↑ and spin-↓ electrons. Treating the spin-↑ and spin-↓ electron subsystems separately, one may conclude that, for $T \ll \epsilon_{\rm Z}^{(\rm imp)}$, each of them contributes a 'restored' value towards the weak localization correction to the conductivity (conductance), which results in its overall drop across a broad range of in-plane magnetic fields,

$$g(B_{\|} \gg B_{\rm s}) - g(B_{\|} = 0) \sim -\frac{e^2}{\pi h} \times \begin{cases} \ln \frac{\tau_\varphi}{\tau_s}, & \text{2D;} \\ \sqrt{\frac{1}{D\tau_s}} & \text{Q1D wire;} \\ 1 & \text{0D dot.} \end{cases} \tag{51}$$

In Eq. (51), $\tau_\varphi > \tau_{\rm s}$ is a 'true' inelastic decoherence rate due to the electron-electron interaction or external electromagnetic noise.

The quantitative analysis of a system with fully polarized static magnetic impurities implies the evaluation of all Cooperon channels, $\mathcal{C}_0$, $\mathcal{C}_{10}$, and $\mathcal{C}_{1\pm1}$. The first two happen to have a gap in the spectrum, both due to i) the splitting between up/down-spin bands caused by the combination of $B_{\|}$ and the mean exchange splitting due to impurities, and ii) the difference between the scattering conditions for up/down-spin electron. By contrast, $\mathcal{C}_{11}$ and $\mathcal{C}_{1-1}$ remain massless. When $\tau_\varphi \gg \tau_{\rm s}$, the crossover between the two limits in Eq. (51) can be roughly described by substituting the electron spin-flip rate at intermediate fields,

$$\tau_{\rm s}^{-1}(B_{\|}) = \frac{u}{e^u - e^{-u}} \times \tau_{\rm s}^{-1}, \quad \text{where } u = B_{\|} g^{(\rm imp)}\mu_{\rm B}/T = B_{\|}/B_{\rm s}, \tag{52}$$

into Eq. (49). The field-dependent electron spin-flip rate in the latter expression manifests the activational character of the impurity spin flip process and was calculated after takinbg into account thermal occupancy of spin-split impurity states. For a2D electron gas or a thin metallic film, this yields a magnetoconductivity with the characteristic form,

$$g(B_{\|}) - g(0) = \frac{e^2}{h} \ln \frac{u}{e^u - e^{-u}} \approx -\frac{e^2}{h} \times \begin{cases} \frac{1}{6}\left(\frac{B_{\|}}{B_{\rm s}}\right)^2 & B_{\|} < B_{\rm s}, \\ \frac{B_{\|}}{B_{\rm s}} & B_{\|} > B_{\rm s}. \end{cases} \tag{53}$$

5.3. EFFECT OF IMPURITY SPIN DYNAMICS ON MESOSCOPIC CONDUCTANCE FLUCTUATIONS

At $B_{\|} = 0$, contamination by magnetic impurities also suppresses mesoscopic conductance fluctuations – as a result of the combination of electron spin-flip scattering with the impurity spin dynamics. In a small sample, the mesoscopic part of the interference correction to the conductance, $\delta g(t)$, is finite for any instant configuration of magnetic scatterers, but it is specific for each particular snapshot of the magnetic subsystem. As the localized magnetic moments undergo a random temporal evolution, this interference correction to the conductance also fluctuates in time, thus, leading to a self-averaging of the UCFs to much smaller values. By contrast, in the regime of strong magnetic fields, $B_{\|} \gg B_{\mathrm{s}}$, the impurity spins stay polarized, such that the random potentials for spin-$\uparrow$ and spin-$\downarrow$ electrons are static (though different), and UCFs are fully restored.

To describe the reproducible part of conductance fluctuations in a DC-current measurement over the entire range of fields, one has to analyse the time-averaged conductance,

$$\bar{g} = \lim_{\Upsilon\to\infty} \frac{1}{\Upsilon}\int_0^{\Upsilon} dt\; g(t) = \langle g\rangle + \delta\bar{g},$$

and its variance,

$$\langle \delta\bar{g}^2\rangle = \lim_{\Upsilon\to\infty} \frac{1}{\Upsilon}\int_0^{\Upsilon} dt\; \langle \delta g(t)\delta g(0)\rangle = \lim_{t\to\infty} \langle \delta g(t)\delta g(0)\rangle .$$

The correlation function $\langle \delta g(t)\delta g(0)\rangle$ determining $\langle \delta\bar{g}^2\rangle$ can be expressed in terms of diffusons $\mathcal{D}_J(\eta,\eta';t;\mathbf{r},\mathbf{r}')$ with an additional time variable t and Cooperons as in Eq. (46) in the usual way – via the standard set of perturbation theory diagrams.

The diffusons $\mathcal{D}_J(\eta,\eta';t;\mathbf{r},\mathbf{r}')$ describe the quantum diffusion of electrons across two distinctive time intervals separated by t. They obey the following equation,

$$\left\{\partial_\eta - D\nabla^2 + \frac{1}{\tau_{\mathrm{s}}}\left(1 - c_J e^{-|t|/\tau_{\mathrm{K}}}\right)\right\}\mathcal{D}_J(\eta,\eta';t;\mathbf{r},\mathbf{r}') = \delta(\eta-\eta')\delta(\mathbf{r}-\mathbf{r}'),$$

where the loss of spin memory of the magnetic impurities due to the Korringa relaxation is taken into account. In the relevant limit of $|t|/\tau_{\mathrm{K}} \to \infty$, where $e^{-|t|/\tau_{\mathrm{K}}} \to 0$, the same simplification occur as the ones leading to Eq. (47) for the Cooperon: the poles of all (singlet and triplet) diffuson components acquire the same relaxational gap equal to the single-electron spin-relaxation rate, i.e.,

$$\mathcal{D}(\mathbf{q},\omega) = \frac{1}{D\mathbf{q}^2 - i\omega + \tau_{\mathrm{s}}^{-1}}.$$

As a result, one arrives at strongly suppressed fluctuations. For a wire or dot geometry, their variance for a given value of the magnetic field follows the field-dependence of the spin-flip scattering rate for an electron in Eq. (52), as

$$\langle \delta \bar{g}^2 \rangle \sim \left(\frac{e^2}{h} \right)^2 \times \begin{cases} (\tau_s/\tau_{fl})^{3/2} & \text{wire geometry,} \\ (\tau_s/\tau_{esc})^2 & \text{dot geometry,} \end{cases}$$

where $\tau_s = \tau_s(B_\parallel)$, τ_{esc} is the electron escape rate from a dot defined in the previous chapter, and $\tau_{fl} = L^2/D$.

The suppression of mesoscopic fluctuation in magnetically contaminated conductors as well as their restoration by a magnetic field have been observed, both in metallic [38] and semiconductor [39] microstructures. It is worth mentioning that this characteristic behavior of weak localization and mesoscopic fluctuations may be used for testing the nature of decoherence [64, 65] even in purified materials – in particular, when its suspected origin is spin-flip scattering at residual magnetic impurities [61, 62].

6. 2D layer symmetry, in-plane magnetoresistance and the non-linear sigma-model

In the previous chapters, we have shown how to analyze "simple" problems using the diagrammatic perturbation theory technique. In this section, we give an example of a problem that is more convenient to tackle with the supersymmetric (SUSY) sigma-model method: the analysis of the interplay between spatial symmetry of a quasi-2D electron gas with few occupied subbands and an in-plane magnetic field.[3] The main results have been discussed in chapter 3. Here, we first give a brief overview of the SUSY technique and, then, present the derivation of the SUSY sigma-model action for a multi-subband 2D system subject to an in-plane magnetic field. Finally, we discuss the properties of the Cooperon matrix and the consequent implications for the magnetoresistance.

6.1. THE FIELD THEORETIC TECHNIQUE

In Sec. 2 the diagrammatic approach to calculating correlation functions in the perturbative regime has been introduced. An alternative approach is provided by the coherent state path integral. The (retarded) Green function can be represented as a field integral,

$$\begin{aligned} G^+(\mathbf{r}, \mathbf{r}'; \epsilon) &= \langle \mathbf{r} | (\epsilon + i0 - \hat{\mathcal{H}})^{-1} | \mathbf{r}' \rangle \\ &= \frac{i}{\mathcal{Z}} \int Ds\, Ds^*\; s^*(\mathbf{r}) s(\mathbf{r}')\; e^{i \int d\mathbf{r}\, s^* (\epsilon^+ - \hat{\mathcal{H}}) s}, \end{aligned}$$

[3] In fact, the present, perturbative problem is accessible by diagrammatic methods as well. However, to get the interplay between inter-band correlations and disorder scattering reliably under control, the formalism of field integration has the advantage that the fully microscopic aspects of the problem are processed in the early stages of the derivation [66].

where $\mathcal{Z} = \int Ds\, Ds^* \exp[i \int d\mathbf{r}\, s^*(\epsilon^+ - \hat{\mathcal{H}})s]$, and $\hat{\mathcal{H}} = \hat{\mathbf{p}}^2/(2m) + V(\mathbf{r})$. Here V represents the impurity potential which is assumed to be drawn from a Gaussian white noise distribution, cf. (9).

Unfortunately, in this form it is not possible to carry out the disorder averaging: Due to the presence of the partition function as a normalization factor, $\mathcal{Z}^{-1}[V]$, the random potential appears in the numerator as well as in the denominator. One method to circumvent this problem is supersymmetry [66]. When one is considering only single-particle properties of a system, there are two equivalent formulations of the path integral, namely by using bosonic or fermionic fields. Supersymmetry now exploits the following property of commuting (s) versus anti-commuting or Grassmann (χ) variables:

$$\int ds^*\, ds\, e^{-s^* M s} = \det{}^{-1} M, \qquad \int d\bar{\chi}\, d\chi\, e^{-\bar{\chi} M \chi} = \det M.$$

Thus, combining both variables into a 'supervector', $\psi^T = (s, \chi)$, yields the result $\int d\psi^\dagger\, d\psi\, \exp[-\psi^\dagger\, M \otimes \mathbf{1}^{\rm bf}\, \psi] = 1$, where the superscript 'bf' stands for 'boson-fermion'. Applying this to the partition sum, it is automatically normalized to unity, $\mathcal{Z} = 1$, and the impurity averaging is straightforward. The evaluation of a two-particle correlation function requires the introduction of two sets of fields, covering the advanced and retarded sector. With $\psi^T = (s_1, \chi_1, s_2, \chi_2)$ the correlator can be written as

$$\langle G^+(\epsilon + \frac{\omega}{2}) G^-(\epsilon - \frac{\omega}{2}) \rangle = -\left\langle \int D[\psi, \bar{\psi}]\, s_1^* s_1 s_2^* s_2\, e^{-i \int d\mathbf{r}\, \bar{\psi}(\epsilon - \frac{\omega^+}{2} \sigma_3^{\rm ar} - \hat{\mathcal{H}})\psi} \right\rangle,$$

where $\omega^+ = \omega + i0$ and $\sigma_3^{\rm ar}$ is a Pauli matrix in advanced/retarded space. Furthermore, $\bar{\psi} = \psi^\dagger L$ with $L = \sigma_3^{\rm ar} \otimes E_{\rm bb} + \mathbf{1}^{\rm ar} \otimes E_{\rm ff}$, where $E_{\rm bb}$ and $E_{\rm ff}$ are projectors onto the boson-boson and fermion-fermion block, respectively. By introducing a source term, $S_J = -\int d\mathbf{r}\, (J^\dagger \psi + \bar{\psi} J)$, different correlators of Green functions can be obtained from the generating functional $\mathcal{Z}[J]$ by taking derivatives with respect to the source field J. In the following, we will suppress the sources and consider only $\mathcal{Z}[0]$.

Now the impurity averaging of the partition function leads to a quartic term in the fields ψ,

$$\langle e^{i \int d\mathbf{r}\, \bar{\psi} V \psi} \rangle = \exp[-\frac{1}{4\pi\nu\tau} \int d\mathbf{r}\, (\bar{\psi}\psi)^2].$$

By Fourier transformation to momentum representation, one can identify the slow modes,

$$\int d\mathbf{r}\, (\bar{\psi}\psi)^2 = \sum_{\sum \mathbf{p}_i = 0} (\bar{\psi}_{\mathbf{p}_1} \psi_{\mathbf{p}_2})(\bar{\psi}_{\mathbf{p}_3} \psi_{\mathbf{p}_4}) \approx \sum_{\mathbf{p}, \mathbf{p}'; \mathbf{q}} \Big((\bar{\psi}_{\mathbf{p}} \psi_{-\mathbf{p}+\mathbf{q}})(\bar{\psi}_{-\mathbf{p}'} \psi_{\mathbf{p}'-\mathbf{q}}) +$$
$$+ (\bar{\psi}_{\mathbf{p}} \psi_{-\mathbf{p}'})(\bar{\psi}_{\mathbf{p}'-\mathbf{q}} \psi_{-\mathbf{p}+\mathbf{q}}) + (\bar{\psi}_{\mathbf{p}} \psi_{\mathbf{p}'-\mathbf{q}})(\bar{\psi}_{-\mathbf{p}+\mathbf{q}} \psi_{-\mathbf{p}'}) \Big),$$

where $|\mathbf{q}| \ll \ell^{-1}$.

The first term corresponds to slow fluctuations of the energy which can be absorbed by a local redefinition of the chemical potential. Thus, we concentrate on the remaining terms: The second term generates the diffuson contribution while the third term yields the Cooperon contribution. Enlarging the field space[4] by defining $\Psi^T = (\psi^T, \psi^\dagger)/\sqrt{2}$, the last two terms can be rewritten into a single contribution, $\mathrm{Str}\,[\sum_{\mathbf{q}} \zeta(\mathbf{q})\zeta(-\mathbf{q})]$, where $\zeta(\mathbf{q}) = \sum_{\mathbf{p}} \Psi(\mathbf{p}-\mathbf{q})\bar{\Psi}(-\mathbf{p})$. The components of the newly defined vector Ψ fulfill the symmetry relation $\Psi^\dagger = (C\Psi)^T$, where $C = \sigma_1^{\mathrm{tr}} \otimes E_{\mathrm{bb}} + i\sigma_2^{\mathrm{tr}} \otimes E_{\mathrm{ff}}$. This symmetry corresponds to time-reversal ($\psi \to \psi^*$, $\hat{\mathcal{H}} \to \hat{\mathcal{H}}^T$). In the absence of the symmetry-breaking energy difference, $\omega = 0$, the action is invariant under rotations $\Psi \to U\Psi$, where $ULU^\dagger = L$ and $U^T = CU^\dagger C^T$. Thus, $U \in \mathrm{Osp}(4|4)$.

As a next step the quartic interaction is decoupled by a Hubbard-Stratonovich transformation, introducing the new (supermatrix-) fields Q:

$$\exp\left[-\frac{1}{4\pi\nu\tau}\int d\mathbf{r}\,(\bar{\Psi}\Psi)^2\right] = \int DQ\,\exp\left[\frac{\pi\nu}{8\tau}\int d\mathbf{r}\;\mathrm{Str}\,Q^2 - \frac{1}{2\tau}\int d\mathbf{r}\;\bar{\Psi}Q\Psi\right],$$

where $\mathrm{Str}\,M = \mathrm{tr}\,M_{\mathrm{bb}} - \mathrm{tr}\,M_{\mathrm{ff}}$. The symmetries of Q reflect the symmetries of the dyadic product $\Psi \otimes \bar{\Psi}$, namely $Q = CLQ^T(CL)^T$. Now the resulting exponent is only quadratic in the original Ψ-fields. Therefore, the Gaussian integral can be readily evaluated, yielding the action

$$S[Q] = -\frac{\pi\nu}{8\tau}\int d\mathbf{r}\;\mathrm{Str}\,Q^2 + \frac{1}{2}\int d\mathbf{r}\;\mathrm{Str}\,\ln\mathcal{G}^{-1}, \tag{54}$$

where $\langle\mathcal{Z}\rangle = \int DQ\,\exp(-S[Q])$ and $\mathcal{G}^{-1} = \frac{1}{2m}\hat{\mathbf{p}}^2 - \epsilon_F + \frac{\omega^+}{2}\sigma_3^{\mathrm{ar}} + \frac{i}{2\tau}Q$.

To extract an effective low-energy, long-wavelength field theory from this action, a saddle point analysis has to be performed. Variation of (54) with respect to Q yields $Q_{\mathrm{sp}}(\mathbf{r}) = i\mathcal{G}(\mathbf{r},\mathbf{r})/(\pi\nu)$. Neglecting the small energy ω, the Ansatz Q_{sp} constant and diagonal leads to

$$Q_{\mathrm{sp}} = -\frac{i}{\pi}\int d\xi\,\frac{1}{\xi - \frac{i}{2\tau}Q_{\mathrm{sp}}} = \mathrm{sgn}(Q_{\mathrm{sp}}). \tag{55}$$

Thus, the saddle point Q_{sp} has the meaning of a self-energy. Analytic properties of the Green function single out the solution $Q_{\mathrm{sp}} = \sigma_3^{\mathrm{ar}}$.

In fact, the action is invariant under transformations $Q \to TQT^{-1}$, where T constant: instead of one saddle point one obtains – at $\omega = 0$ and in the absence of symmetry breaking sources – a degenerate saddle point manifold $Q^2 = 1$. Fluctuations around the saddle point can be subdivided into longitudinal modes, $[\delta Q_l, Q_{\mathrm{sp}}] = 0$, and transverse modes, $\{\delta Q_t, Q_{\mathrm{sp}}\} = 0$. The longitudinal modes

[4] Generally, each discrete symmetry leads to a doubling of the low-lying modes and, thus, should be incorporated by doubling the field space [67].

δQ_l leave the saddle point manifold $Q^2 = 1$. Therefore, they are massive and do not contribute to the low-energy physics of the system. In the following, we concentrate on the transverse modes δQ_t. The parameter which stabilizes this distinction is $k_F\ell \sim \epsilon_F\tau$, i.e., the following considerations are valid in the quasi-classical limit. We proceed by expanding the action around the saddle point in the slowly varying fields $Q(x) = T(x)Q_{\rm sp}T^{-1}(x)$. Separating the fast and slow degrees of freedom with $\hat{\mathbf{p}} \to \mathbf{p} + \hat{\mathbf{q}}$, this expansion yields

$$S \simeq \frac{1}{2}\int d\mathbf{r}\int d\mathbf{p}\ \mathrm{Str}\left[\frac{\omega^+}{2}\mathcal{G}_0 T^{-1}\sigma_3^{\rm ar}T - \frac{1}{2m^2}\left(\mathcal{G}_0\, T^{-1}\,\mathbf{p}\hat{\mathbf{q}}\,T\right)^2\right].$$

The integral over fast momenta, $\mathbf{p}$, can be performed using the following representation for the Green function,

$$\mathcal{G}_0(\mathbf{p}) = \frac{1}{2}\sum_{s=\pm}\frac{1+s\sigma_3^{\rm ar}}{-\xi_p + s\frac{i}{2\tau}} \equiv \frac{1}{2}\sum_{s=\pm}(1+s\sigma_3^{\rm ar})G_0^s(\mathbf{p}).$$

Then, $\int d\mathbf{p}\ \mathcal{G}_0(\mathbf{p}) = -i\pi\nu\sigma_3^{\rm ar}$, and

$$\int d\mathbf{p}\ \mathrm{Str}\left[\mathcal{G}_0(\mathbf{p})\,T^{-1}\,\mathbf{p}\hat{\mathbf{q}}\,T\,\mathcal{G}_0(\mathbf{p})\,T^{-1}\,\mathbf{p}\hat{\mathbf{q}}\,T\right]$$
$$= \underbrace{\frac{\nu p_F^2}{4d}\int d\xi\ G_0^+G_0^-}_{=m^2\pi\nu D/2}\ \mathrm{Str}\left[(1+s\sigma_3^{\rm ar})T^{-1}\,\hat{\mathbf{q}}\,T(1-s\sigma_3^{\rm ar})T^{-1}\,\hat{\mathbf{q}}\,T\right].$$

Finally, summing over s and using the cyclic invariance under the trace, the effective action takes the form of a non-linear σ model,

$$S[Q] = -\frac{\pi\nu}{8}\int d\mathbf{r}\ \mathrm{Str}\left[D(\partial Q)^2 + 2i\omega^+\sigma_3^{\rm ar}Q\right]. \tag{56}$$

Note that the effect of a weak magnetic field is to generalize the derivatives to $\tilde{\partial} = \partial - i\mathbf{A}[\sigma_3^{\rm tr}, .]$. For the quasi-twodimensional system in an in-plane magnetic field this will be discussed in more detail in the following.

In the perturbative regime, an expansion of the effective action around the saddle point in the generators W, where $Q = e^{-W/2}\sigma_3^{\rm ar}e^{W/2}$ and $\{W, \sigma_3^{\rm ar}\} = 0$, reproduces the diagrammatic results. By contrast, in the non-perturbative regime, the action is dominated by zero-modes which require an integration over the whole saddle point manifold Osp(4|4)/(Osp(2|2)⊗Osp(2|2)). For our purposes, a perturbative expansion will be sufficient.

6.2. MAGNETORESISTANCE IN A MULTI-SUBBAND ELECTRON LAYER WITH A POSSIBLE BERRY-ROBNIK SYMMETRY

As pointed out earlier, the magnetoresistance in quasi-twodimensional electron systems subject to parallel fields is very sensitive to the presence or absence of $\mathcal{P}_z$-symmetry. While in chapter 3 we restricted ourselves to a qualitative discussion of the effect, here the quantitative results are presented. To explore such type of phenomena one needs to construct an approach which on the one hand is sensitive to microscopic details in z-direction while on the other hand should be capable of efficiently describing large scale in-plane properties. This task can efficiently be addressed within a field integral formalism. The starting point of the derivation is a supersymmetric field integral with action

$$S[\psi] = i\int d^3r\,\bar{\psi}\Big(\epsilon_{\rm F} - \omega^+\sigma_3^{\rm ar} + \frac{1}{2m}(\partial_x^2 + (\partial_y - iBz)^2 + \partial_z^2) - -W(z) - V(x,y)\Big)\psi, \tag{57}$$

where W is the confining potential of the 2DEG, and V a disorder potential. To simplify the analysis, it is assumed that the disorder potential does not depend on the z -coordinate. Given the typical architecture of 2DEGs, this is certainly a justified zeroth order assumption. In, e.g., high mobility 2DEGs (or 2DHGs) in GaAs/AlGaAs heterostructures, the mobility is limited by a long-range random potential, created by charged impurities located far from the plane. Later on, this condition will be relaxed by generalizing V according to $V(x,y) \to V(x,y) + U(x,y,z)$, where U is weak and can be treated perturbatively in a sense to be specified below.

The confinement in z-direction is responsible for the size quantization which entails a subband structure of the system. To make progress with the action, Eq. (57), an orthonormalized set of wavefunctions $\{\phi_k\}$, diagonalizing the z-dependent part of the problem, is introduced:

$$\left(-\frac{1}{2m}\partial_z^2 + W(z) - \epsilon_k\right)\phi_k = 0. \tag{58}$$

Expanding the original fields ψ in the complete set of eigenfunctions ϕ_k, that is $\psi(x,y,z) = \sum_k \psi_k(x,y)\phi_k(z)$, the action takes the form

$$S[\psi] = i\int d^2r\,\bar{\psi}_k\Big([\epsilon_{\rm F} - \omega^+\sigma_3^{\rm ar} - \epsilon_k + \frac{\partial_x^2}{2m} - V(x,y)]\delta_{kk'} + +\frac{1}{2m}\left((\partial_y - i\hat{A})^2\right)_{kk'}\Big)\psi_{k'},$$

where the integration extends over the x-y–plane, summation over k, k' is implied, and

$$A_{kk'} \equiv B\int dz\,\phi_k(z) z \phi_{k'}(z) \tag{59}$$

is the vector potential. I.e. the magnetic field couples to the dipole matrix elements $d_{kk'} = \int dz\, \phi_k(z) z \phi_{k'}(z)$ that contain detailed information about the microscopic symmetry properties of the system.

Using the same steps as explained in the previous section , the action in terms of the slow supermatrix fields $Q_{kk'}$ reads

$$\begin{aligned} S[Q] &= -\frac{\pi\nu}{8\tau}\int d^2r \,\mathrm{Str}\, Q^2 + \\ &+\frac{1}{2}\int d^2r \,\mathrm{Str}\,\ln\left(\epsilon_{\mathrm{F}}-\omega^+\sigma_3^{\mathrm{ar}}-\hat{\epsilon}+\frac{1}{2m}(\partial_x^2+(\partial_y-i\hat{A}\sigma_3^{\mathrm{tr}})^2)-\frac{i}{2\tau}Q\right), \end{aligned} \tag{60}$$

where a compact k-index free notation has been introduced. Here the energy matrix $\hat{\epsilon} \equiv \mathrm{diag}(\epsilon_0, \epsilon_1, \ldots)$ contains the subband energies. The next step in the construction of the effective theory is the saddle point analysis. Functional differentiation of the action with respect to Q obtains the equation

$$\Lambda_k = \frac{i}{\pi\nu}\int d^2p \frac{1}{i\delta\sigma_3^{\mathrm{ar}} - \epsilon_{\mathrm{F}} - \epsilon_k + \frac{p^2}{2m} + \frac{i}{2\tau}\Lambda_k}$$

for the diagonal elements Λ_k of the saddle point matrix $Q_{kk'}$. At this stage one has to specify the relative position of the Fermi energy E_{F} and the subband energies ϵ_k. Below, we will explore the case where M bands with energy $\epsilon_k < E_{\mathrm{F}}$ ($k = 0, \ldots, M-1$) exist. This leads to $\Lambda_k = \sigma_3^{\mathrm{ar}}$ for $k < M$, and $\Lambda_k = 0$ otherwise, where it has been assumed that the highest occupied subband ϵ_{M-1} lies well below (farther than τ^{-1}) the Fermi level. Based on this solution, the low-lying fields of the theory can be represented as $Q = T\Lambda T^{-1}$, where $\Lambda = \{\Lambda_k \delta_{kk'}\}$, and the final expression for the general slow action reads [15, 44]

$$\begin{aligned} S[Q] &= -\frac{\pi\nu}{8}\int dS \tilde{\sum_k} \mathrm{Str}\left(4i\omega\sigma_3^{\mathrm{ar}}Q_k + D_k(\tilde{\partial}_k Q_k)^2\right) + \\ &+\frac{\pi\nu}{4}\int dS \tilde{\sum_{k,k'}} \mathrm{Str}\left(\mathcal{X}_{kk'}\sigma_3^{\mathrm{tr}}Q_k\sigma_3^{\mathrm{tr}}Q_{k'}\right), \end{aligned} \tag{61}$$

where $\tilde{\partial}_k = \partial - i\mathbf{e}_y A_{kk}[\sigma_3^{\mathrm{tr}}, .]$. The sum $\tilde{\sum}_k$ involves only the occupied subbands $k = 0, \ldots, M-1$. Furthermore,

$$\mathcal{X}_{kk'} = \frac{1}{2}(D_k + D_{k'})\frac{1}{(\epsilon_{kk'}\tau)^2 + 1}A_{kk'}A_{k'k}(1 - \delta_{kk'}). \tag{62}$$

Here D_k is the diffusion constant of subband k, and $\epsilon_{kk'} = \epsilon_k - \epsilon_{k'}$. The first line of (61) gives the conventional result of a $2d$ system while the second line describes the coupling of the subbands induced by the magnetic field.

To prepare the one Cooperon approximation to the conductivity, one has to expand the fields Q to lowest non-trivial order in some generators. It is convenient to decompose the generators into 'diffuson' (d) and 'Cooperon' (c) blocks. It is clear from the structure of the second order action, that it does not couple between the 'd' and the 'c' sector, i.e., $S = S^d + S^c$. The diffuson action S^d is generated by those fields that commute with $\sigma_3^{\rm tr}$ and, thus, do not couple to the magnetic field, while the Cooperon action S^c is generated by those fields that anticommute with $\sigma_3^{\rm tr}$ and, thus, are field-sensitive.

Here we are only interested in the Cooperon action being responsible for WL corrections. The kernel appearing in S^c is the 'inverse of the Cooperon'. More explicitly, the Cooperon $\mathcal{C}$, which in our formulation is a matrix in the discrete space of k-indices and diagonal in $\mathbf{q}$-space, is obtained by inverting the matrix

$$(\mathcal{C}_{\mathbf{q}}^{-1})_{kk'} = \left(-\frac{2i\omega}{D_k} + (\mathbf{q}-2\mathbf{A}_{kk})^2 + \frac{2}{D_k}\sum_{k''}\mathcal{X}_{kk''}\right)\delta_{kk'} + \frac{2}{\sqrt{D_k D_{k'}}}\mathcal{X}_{kk'}.$$

The magnetoconductance is determined by the specific form of this matrix. The conductivity is given as $\sigma = \sigma^0 + \Delta\sigma$ with $\sigma^0 = \sum_{k=0}^{M-1}\sigma_k^0$, where $\sigma_k^0 = \nu D_k$ is the Drude conductivity of subband k, and

$$\Delta\sigma = -\frac{2}{\pi}\sum_{k=0}^{M-1}\sum_{\mathbf{q}}(\mathcal{C}_{\mathbf{q},\omega=0})_{kk}. \tag{63}$$

In general, the field-dependent terms will render $\mathcal{C}$ massive, i.e., the weak localization corrections will suffer from a field induced suppression. However, there is the situation mentioned above, where z-inversion, $\mathcal{P}_z : z \mapsto -z$, is an (approximate) symmetry of the Hamiltonian, $[\mathcal{H}, \mathcal{P}_z] \approx 0$.

For systems with an exact $\mathcal{P}_z$-symmetry, the eigenfunctions ϕ_k obey $\mathcal{P}_z\phi_k = (-)^k\phi_k$. The definition of the vector potential matrix $\hat{A}$ then implies

$$A_{kk'} = \begin{cases} A_{kk'} & k+k' \text{ odd}, \\ 0 & k+k' \text{ even}, \end{cases}$$

and, thus, the same holds true for $\mathcal{X}_{kk'}$.

This structure bears consequences on the Cooperon mass. To analyze this point, consider the spatial Cooperon zero-mode,

$$(\mathcal{C}_{\mathbf{0},\omega=0}^{-1})_{kk'} = 2\left(\frac{1}{D_k}\sum_{k''}\mathcal{X}_{kk''}\delta_{kk'} + \frac{\mathcal{X}_{kk'}}{\sqrt{D_k D_{k'}}}\right). \tag{64}$$

This matrix has determinant zero implying that there is a Cooperon mode which is not affected by the field. In fact, it is straightforward to verify that the M-dimensional vector

$$\mathbf{X} \equiv \mathcal{N}^{-1/2}\sum_k(-)^k\sqrt{D_k}\,\mathbf{e}_k, \tag{65}$$

where $\mathcal{N} = \sum_k D_k$, is annihilated by $C^{-1}_{\mathbf{0},\omega=0}$. As the zero-mode matrix is symmetric, one can, in principle, construct a complete set of orthonormal eigenvectors, $\{\mathbf{X}_0 \equiv \mathbf{X}, \mathbf{X}_1, \ldots \mathbf{X}_{M-1}\}$, with eigenvalues $\{0, \lambda_1, \ldots, \lambda_{M-1}\}$.

Furthermore, due to $A_{kk} = 0$, the full Cooperon kernel is separable (i.e., it is the sum of a spatial and an 'internal' operator). Inserting the result into Eq. (63) yields

$$\Delta\sigma(B) = -\frac{2}{\pi} \tilde{\sum_{k;\mathbf{q}}} \frac{1}{q^2 + \lambda_k}. \tag{66}$$

This is our final result for the conductivity. Notice that, due to $\lambda_0 = 0$, the weak localization corrections do survive the magnetic field; carrying out the $\mathbf{q}$-summation leads to the usual logarithmic correction to the Drude conductance. Thus, even at high magnetic fields, a logarithmic temperature dependence – see Chap. 3 – of the conductance should be observable. All other eigenvalues are proportional B^2 and, thus, display the usual field dependence.

In the following, let us concentrate on the behavior of the lowest eigenvalue if the system is not exactly inversion symmetric. An asymmetry can be caused either by the confining potential or by a z-dependence of the random impurity potential.

In the case of an asymmetric confining potential, Eq. (64) generalizes to $A_{kk'} \to A_{kk'} + \delta A_{kk'}$, where $\delta A_{kk'}$ is assumed to be much weaker than the symmetry allowed elements $A_{kk'}$, $k + k'$ odd. Similarly, there are non-vanishing but small matrix elements $\delta\mathcal{X}_{kk'}$ for both $k + k'$ even and odd. To lowest order in perturbation theory, the presence of these matrix elements shifts the zero-mode eigenvalue $\lambda_0(\mathbf{q})$ of the unperturbed Cooperon mode at momentum $\mathbf{q}$ by the amount $\delta\lambda_0^{(\mathrm{as})}(\mathbf{q}) = \mathbf{X}^T \delta C_{\mathbf{q}}^{-1} \mathbf{X}$, where $\delta C_{\mathbf{q}}^{-1}$ is the perturbation contribution to the Cooperon operator. Explicitly,

$$\left(\delta C_{\mathbf{q}}^{-1}\right)_{kk'} = \Big[-4q_y \delta A_{kk} + 4\delta A_{kk}^2 + 2\sum_{k''} \delta\mathcal{X}_{kk''}\Big]\delta_{kk'} + 2\delta\mathcal{X}_{kk'}. \tag{67}$$

Combining these equations and making use of the definition of the zero-mode eigenvectors (65) yields

$$\delta\lambda_0^{(\mathrm{as})}(q_{\min}) = \frac{2}{\mathcal{N}^2}\sum_{k,k'} D_k D_{k'}(\delta A_{kk} - \delta A_{k'k'})^2 + \frac{4}{\mathcal{N}} \sum_{k+k'\ \mathrm{even}} \delta\mathcal{X}_{kk'}.$$

Thus, the Cooperon acquires a mass term $\sim B^2$.

The influence of z-dependent impurities has a similar effect. A potential with a generic z-dependence will not be inversion symmetric, implying that, somehow, the Cooperon must pick up a mass. Assuming a Gaussian distributed potential $U(\mathbf{r}, z)$,

$$\langle U(\mathbf{r}, z)\rangle = 0, \qquad \langle U(\mathbf{r}, z)U(\mathbf{r}', z')\rangle = \gamma^2\delta(\mathbf{r} - \mathbf{r}')\delta(z - z'),$$

for small γ, it is sufficient to consider the lowest order non-vanishing contribution in U to the action. We arrive at

$$S[Q] = S_0[Q] + \left(\frac{\pi\gamma\nu}{2}\right)^2 \int d\tilde{S} \sum_{kk'} \Gamma_{kk'} \,\mathrm{Str}\,(Q_k(\mathbf{r})Q_{k'}(\mathbf{r})), \tag{68}$$

where the coefficient $\Gamma_{kk'k''k'''} = \int dz\, \phi_k(z)\phi_{k'}(z)\phi_{k''}(z)\phi_{k'''}(z)$, and $\Gamma_{kk'} \equiv \Gamma_{kkk'k'}$ is positive. This expression tells that the z-dependent scattering tends to lock the fields Q_k. For γ large, only field configurations $\{Q_k \equiv Q\}$ with no k-dependence survive. The physical mechanism is the following: Scattering in z-direction leads to a coupling between the different k -bands. Thus, the formerly independent diffusons and Cooperons are coupled, too. The formerly massless Cooperon channel does not survive this coupling as the k-space eigenvector $\mathbf{X}_0$ associated with the eigenvalue $\lambda_0 = 0$ is staggered in k, cf. Eq. (65), i.e., it stands orthogonal on the field configurations that are compatible with the locking.

If the coupling due to the impurity scattering is smaller than the field induced subband coupling, the shift of the lowest eigenvalue is again obtained by first order perturbation theory. Then,

$$\delta\lambda_0^{(\mathrm{imp})} = \frac{1}{\mathcal{N}}\pi\nu\gamma^2 \sum_{k+k'\ \mathrm{odd}} \Gamma_{kk'}. \tag{69}$$

Or, $\delta\lambda_0^{(\mathrm{imp})} \sim 1/(\mathcal{N}\tau')$, where τ' has the meaning of a scattering time perpendicular to the plane, i.e., between the subbands.

This result which does not depend on the magnetic field holds true only for sufficiently large fields. For smaller fields, the disorder induced mass term fixes the preferred eigenvector. To compute the mass of the completely locked Cooperon, consider $\lambda_l \equiv \mathbf{X}_l^T C_{\mathbf{0},\omega=0}^{-1}\mathbf{X}_l$, where the Cooperon operator is given by Eq. (64), and the 'locked' vector $\mathbf{X}_l$ reads $\mathbf{X}_l \equiv \mathcal{N}^{-1/2}\sum_k \sqrt{D_k}\,\mathbf{e}_k$. Explicitly computing the matrix element leads to

$$\lambda_l = \frac{2}{\mathcal{N}^2}\sum_{k,k'} D_k D_{k'}(A_{kk} - A_{k'k'})^2 + \frac{4}{\mathcal{N}}\sum_{k,k'} \mathcal{X}_{kk'}. \tag{70}$$

At low fields, the mass of the Cooperon increases quadratically with B according to Eq. (70), but then, due to Eq. (69), it levels off at large fields. The characteristic field B_c can be estimated by comparing Eqs. (69) and (70) which yields $B_c \sim E/v_{\mathrm{F}}\sqrt{\tau/\tau'}/d$, where E stands for the typical energy separation between subbands and d sets the scale for the width of the quantum well.

As the expressions obtained above are rather lengthy, it is helpful to consider some specific examples. We concentrate on the experimentally most relevant case $M = 2$ and, for simplicity, choose[5] $D_0 = D_1 \equiv D$. Diagonalization of the 2×2

[5] Admitting for different diffusion constants $D_0 \neq D_1$ does not change the results qualitatively.

Cooperon matrix,

$$\mathcal{C}^{-1} = \begin{pmatrix} (\mathbf{q}-\mathbf{A})^2 + \frac{2}{D}\mathcal{X}_{01} & \frac{2}{D}\mathcal{X}_{01} \\ \frac{2}{D}\mathcal{X}_{01} & (\mathbf{q}+\mathbf{A})^2 + \frac{2}{D}\mathcal{X}_{01} \end{pmatrix},$$

yields

$$\lambda = q^2 + A^2 + \frac{2}{D}\mathcal{X}_{01} \pm 2\sqrt{(\mathbf{A}\mathbf{q})^2 + \frac{1}{D^2}\mathcal{X}_{01}^2}, \tag{71}$$

where $\mathbf{A} = \mathbf{A}_{00} - \mathbf{A}_{11}$, and $\mathcal{X}_{01} = DA_{01}^2/(1+(\epsilon_{10}\tau)^2)$ obtains from (62). The corresponding magnetic decoherence times read $1/\tau_B = D\lambda$.

Note that at small magnetic fields, $2\mathcal{X}_{01} \ll 1/\tau_\phi$, the symmetry mechanism is ineffective. Irrespective of A, the magnetoconductance yields

$$\sigma(B) - \sigma(0) \simeq 2\frac{e^2}{\pi h}\mathcal{X}_{01}\tau_\phi,$$

which shows the usual low-field quadratic dependence on B. However, the coefficient is diminished by the factor $1/(1+(\epsilon_{10}\tau)^2)$.

At large magnetic fields, $2\mathcal{X}_{01} \gg 1/\tau_\phi$, if the confining potential is fully symmetric ($\mathbf{A} = 0$), the result reduces to $1/\tau_B = 2\mathcal{X}_{01}(1 \pm 1)$. While $1/\tau_B = 4\mathcal{X}_{01}$ leads to a logarithmic field dependence (see Eq. (29)), due to the field-insensitive $1/\tau_B = 0$, the conductance maintains its temperature dependence through τ_ϕ even at large fields. A slight asymmetry of the confining potential entails a finite $\mathbf{A}$, which leads to $1/\tau_\phi(B) \simeq DA^2 + 2\mathcal{X}_{01}(1 \pm 1)$. Thus, the temperature dependence remains as long as $DA^2 < 1/\tau_\phi$.

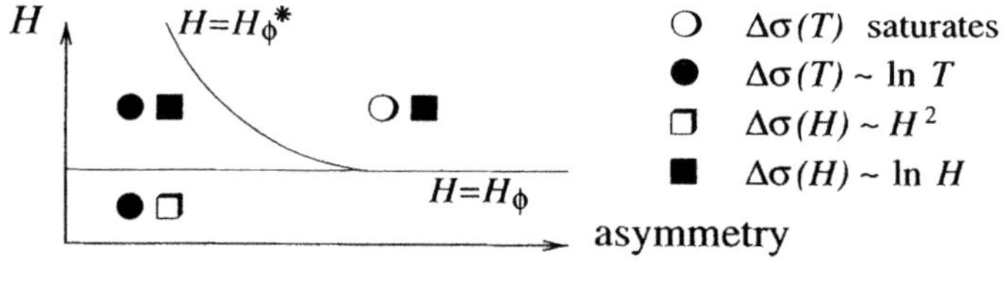

Figure 13. Different regimes of B- and T-dependence of the weak localization corrections.

The regimes with different field and temperature dependences of the conductance are schematically shown in Fig. 13 [15, 44]. The fields B_ϕ and B_ϕ^* are defined through $2\mathcal{X}_{01}\tau_\phi = 1$ and $DA^2\tau_\phi = 1$, respectively. I.e.

$$B_\phi \equiv \frac{1}{\sqrt{D\tau_\phi}\, d_{01}\sqrt{\frac{2}{1+(\epsilon_{10}\tau)^2}}}, \qquad B_\phi^* \equiv \frac{1}{\sqrt{D\tau_\phi}\,(d_{00}-d_{11})},$$

where $d_{kk'}$ are the dipole matrix elements defined above.

7. Summary

In summary, we presented three mechanisms of quantum in-plane magnetoresistance in 2D electron systems, such as quantum wells, heterostructures, and inversion layers in field transistors, in lateral quantum dots prepared of these materials, and in thin metallic films. These mechanisms include: the purely orbital effects possible due to subband mixing in a 2D structure by the magnetic field, the effect of an interplay between the spin-orbit coupling in the 2D electron dispersion and Zeeman splitting by a magnetic field, and the field effect on the efficiency of spin-flip processes in materials contaminated by paramagnetic impurities.

ACKNOWLEDGMENTS: We would like to thank I. Aleiner, A. Altland, A.K. Geim, C. Marcus, and B.D. Simons for useful discussions. Some of the recent work described in this article has been supported by EPSRC and NATO CLG Programme. We also thank the Max-Planck-Institut für Physik komplexer Systeme in Dresden for their hospitality at an early stage of preparation of this review during one of the MPI-PKS research workshops.

References

1. T. Ando, A. Fowler, and F. Stern, Rev. Mod. Phys. **54**, 437 (1982).
2. K. J. Thomas, J. T. Nicholls, M. Y. Simmons, M. Pepper, D. Mace, and D. Ritchie, Phys. Rev. Lett. **77**, 135 (1996).
3. J. A. Folk, S. R. Patel, K. M. Birnbaum, C. M. Marcus, C. I. Duruoz, and J. S. Harris, Phys. Rev. Lett. **86**, 2102 (2001).
4. B. L. Altshuler, P. A. Lee, and R. A. Webb, *Mesoscopic Phenomena in Solids*, North-Holland, Amsterdam, Oxford, New York, Tokio (1991).
5. *Mesoscopic Quantum Physics (Les Houches Session LXI)*, Elsevier, North-Holland, Amsterdam (1995).
6. *Mesoscopic Electron Transport*, Kluwer, Dordrecht (1997).
7. B. D. Simons and A. Altland, in *Theoretical Physics at the End of the XXth Century, Banff*, edited by, Springer (2001).
8. L. P. Gor'kov, A. I. Larkin, and D. E. Khmel'nitskii, JETP Lett. **30**, 228 (1979).
9. B. L. Altshuler and D. E. Khmel'nitskii, JETP Lett. **42**, 291 (1985).
10. P. A. Lee and A. D. Stone, Phys. Rev. Lett. **55**, 1622 (1985).
11. P. A. Lee, A. D. Stone, and H. Fukuyama, Phys. Rev. B **35**, 1039 (1987).
12. R. A. Webb, S. Washburn, C. P. Umbach, and R. B. Laibovitz, Phys. Rev. Lett. **54**, 2696 (1985).
13. V. I. Fal'ko, J. Phys. Cond. Matt. **2**, 3797 (1990).
14. V. I. Fal'ko, Phys. Rev. B **50**, 17406 (1994).
15. J. S. Meyer, A. Altland, and B. L. Altshuler. preprint cond-mat/0105623.
16. V. I. Fal'ko and T. Jungwirth, Phys. Rev. B **65**, 81306 (2002).
17. L. Smrcka, P. Vasek, J. Kolacek, T. Jungwirth, and M. Cukr, Phys. Rev. B **51**, 18011 (1995).
18. J. M. Heisz and E. Zaremba, Phys. Rev. B **53**, 13594 (1996).
19. I. Kukushkin, V. I. Fal'ko, K. von Klitzing, D. Heitmann, and K. Ploog, JETP Lett. **53**, 334 (1991).
20. V. Kirpichev, I. Kukushkin, V. Timofeev, and V. I. Fal'ko, JETP Lett. **51**, 436 (1990).
21. B. L. Altshuler and A. G. Aronov, JETP Lett. **33**, 499 (1981).

22. V. K. Dugaev and D. E. Khmel'nitskii, Sov. Phys. JETP **59**, 1038 (1984).
23. C. W. J. Beenakker and H. van Houten, Phys. Rev. B **37**, 6544 (1988).
24. M. Leadbeater, V. I. Fal'ko, and C. J. Lambert, Phys. Rev. Lett. **81**, 1274 (1998).
25. M. Dyakonov and V. Perel, Sov. Phys. JETP **33**, 1053 (1971).
26. S. Hikami, A. I. Larkin, and Y. Nagaoka, Prog. Theor. Phys. **63**, 707 (1980).
27. S. Iordanskii, Y. Lyanda-Geller, and G. E. Pikus, JETP Lett. **60**, 207 (1994).
28. Y. Lyanda-Geller and A. D. Mirlin, Phys. Rev. Lett. **72**, 1894 (1994).
29. C. W. J. Beenakker, Rev. Mod. Phys. **69**, 731 (1997).
30. I. L. Aleiner and V. I. Fal'ko, Phys. Rev. Lett. **87**, 256801 (2001).
31. A. G.Huibers, M. Switkes, C. M. Marcus, K. Capman, and A. C. Gossard, Phys. Rev. Lett. **81**, 200 (1998).
32. A. G. Huibers, J. A. Folk, S. R. Patel, C. M. Marcus, C. I. Duruoz, and J. S. Harris, Phys. Rev. Lett. **83**, 5090 (1999).
33. G. Bergmann, Phys. Rep. **101**, 1 (1982).
34. V. I. Fal'ko, JETP Lett. **53**, 342 (1991).
35. V. Chandrasekhar and *et al*, Phys. Rev. B **42**, 6823 (1990).
36. A. A. Bobkov, V. I. Fal'ko, and D. E. Khmel'nitskii, Sov. Phys. JETP **71**, 393 (1990).
37. V. I. Fal'ko, J. Phys. Cond. Matt. **4**, 3943 (1992).
38. A. Benoit and *et al*, Superlatt. Microstr. **11**, 313 (1992).
39. A. K. Geim, S. V. Dubonos, and I. Y. Antonova, JETP Lett. **52**, 247 (1990).
40. K. B. Efetov, Phys. Rev. Lett. **74**, 2299 (1995).
41. J. S. Langer and T. Neal, Phys. Rev. Lett. **16**, 984 (1966).
42. G. Bergmann, Phys. Rep. **107**, 2 (1984).
43. M. Robnik and M. V. Berry, J. Phys. A **19**, 669 (1986).
44. J. S. Meyer. Mesoscopic phenomena driven by parallel magnetic fields. PhD thesis, 2001.
45. J. P. Bouchaut and A. Georges, Phys. Rep. **195**, 127 (1990).
46. A. B. Pippard, *Magnetoresistance in Metals*, Cambridge University Press, New York (1989).
47. H. Mathur and H. U. Baranger, Phys. Rev. B **64**, 235325 (2001).
48. R. Serota and *et al*, Phys. Rev. B, 5031 (1987).
49. Z. Tesanovich and *et al*, Phys. Rev. Lett. **57**, 2760 (1986).
50. K. Nikolic and A. MacKinnon, Phys. Rev. B, 11008 (1994).
51. J. T. Chalker and A. Macedo, Phys. Rev. Lett. **71**, 3693 (1993).
52. J.-L. Pichard and *et al*, J. Phys. (Paris) **51**, 587 (1990).
53. K. Frahm, Phys. Rev. Lett. **74**, 4706 (1995).
54. H. van Houten and *et al*, Superlatt. Microstruct. **3**, 497 (1988).
55. G. Dresselhaus, Phys. Rev. **100**, 580 (1958).
56. Y. Bychkov and E. Rashba, JETP Lett. **39**, 78 (1984).
57. F. Malcher, G. Lommer, and U. Rossler, Phys. Rev. Lett. **60**, 729 (1988).
58. H. Mathur and A. D. Stone, Phys. Rev. Lett. **68**, 2964 (1992).
59. A. G. Aronov and Y. B. Lyanda-Geller, Phys. Rev. Lett. **70**, 343 (1993).
60. B. I. Halperin, A. Stern, Y. Oreg, J. Cremers, J. A. Folk, and C. M. Marcus, Phys. Rev. Lett. **86**, 2106 (2001).
61. A. B. Gougam and *et al*, J. Low Temp. Phys. **118**, 447 (2000).
62. A. Anthore and *et al*. preprint cond-mat/0109297.
63. J. Korringa, Physica **16**, 601 (1959).
64. P. Mohanty and R. A. Webb, Phys. Rev. Lett. **84**, 4481 (2000).
65. P. Mohanty, E. M. Q. Jariwala, and R. A. Webb, Phys. Rev. Lett. **78**, 3366 (1997).
66. K. B. Efetov, *Supersymmetry in Disorder and Chaos*, Cambridge University Press, New York (1997).
67. B. D. Simons, O. Agam, and A. V. Andreev, J. Math. Phys. **38**, 1982 (1997).

DISORDERED WIGNER CRYSTALS

T. GIAMARCHI[(1,2)]
[1] *Laboratoire de Physique des Solides, CNRS-UMR 8502, UPS Bât. 510, 91405 Orsay France*
[2] *LPTENS CNRS UMR 8549 24, Rue Lhomond 75231 Paris Cedex 05, France*

1. Introduction

Disorder effects in quantum electronic systems have led to a variety of novel phases. Fermionic systems have played a special role in our understanding of such effects. Indeed for fermions, the Pauli principle prevent the fermions to be trapped macroscopically in the minima of the random potential, making the non interacting case worthwhile to study. Disorder then leads to the rich physics of Anderson localization. Using both scaling theories and sophisticated field theoretical techniques, it is now known that electrons are localized by disorder in one and two dimensions, whereas a mobility edge exists in three dimensions. [7, 3, 1, 35, 12, 11]

The situation becomes much more complicated when one wants to take into account the electron-electron interaction. Such a question was crucial for the understanding of doped semiconductors [33]. In addition recent experiments in two dimensional electron gas systems have prompted the question of whether a metal-insulator transition could exist in interacting systems (see [2] and references therein), stimulating further interest in this problem.

On the theoretical side the question is extremely complicated. Most of the theoretical approaches used for free electrons either fail or become much more complicated when interactions are included which makes it more difficult to obtain unambiguous answers. Perturbative calculations or renormalization group calculations can be made for weak interactions. Unfortunately they scale to strong coupling, which leaves the question of the large scale/low energy physics still open [4, 14, 27].

In fact, even the pure system is interesting. Indeed for weak interactions it is reasonable to expect to have a Fermi liquid behavior, at least for three dimensional systems. Upon increasing the interaction is was predicted long ago by Wigner that the electrons would crystallize [36]. Such crystallization can be induced by other

I.V. Lerner et al. (eds.),
Strongly Correlated Fermions and Bosons in Low-Dimensional Disordered Systems, 165–183.

means, decreasing the density of particles or quenching their kinetic energy by applying a magnetic field.

This suggests another line of attack for the fermionic disordered problem: start from the Wigner crystal phase and study the effect of disorder on such a phase. In the crystalline phase one can expect the statistics to be less important and thus the problem to be more tractable. Such a problem falls in the more general category of disordered elastic systems, which exhibit competition between elastic forces that like some ordered structure (perfect crystal for a periodic structure, flat structures for manifolds) and disorder. Physical systems entering in this category range from manifold (magnetic domain walls, wetting interfaces etc.), periodic classical systems (vortex lattice, colloids, magnetic bubbles, charges spheres, etc.), quantum systems (Luttinger liquids, stripe phases, charge and spin density waves, Wigner crystal). There has been an immense activity in these various domains and many recent progress and it is out of question in these few pages to cover such a vast topic. I will thus avoid completely here the topic of classical disordered systems. The reader who want to know more on that subject and see the links between the quantum systems discussed here and the classical problems is referred to the review [16], where this topic is discussed. This review provides references to other relevant review papers on this vast subject.

For the quantum problems I will confine the discussion to the two dimensional Wigner crystal. Rather than to try to embark on a review on the subject, I will focuss on some specific points of this problem. I will mostly insist on the basic concepts and present some little additional complements on the question of the compressibility in such systems. These notes are thus *not* self contained and rely heavily on other published material, both for contents and references. These few pages can thus be viewed as an "appendix" of the following papers:

- The basic concepts of quantum disordered elastic systems are summarized in the review paper [22]. The reader is thus referred to this paper for the basic physical ideas in this field and for the technology needed to treat such problems. In addition this paper contains of course references to further material. This paper also treats in detail the interesting case of the one dimensional disordered electron gas. This is a situation relevant for systems such as quantum wire or nanotubes. In that case electrons are known to lead to a non fermi liquid state, the Luttinger liquid, and the disorder effects are particularly drastic.
- The Wigner crystal is examined in details in [9, 10]. These papers contain a detailed discussion of the physical issues, problems and comparison with the experiments. They describes all the technical details of the approach that we have used and that I will briefly discuss in the paper, and give references to other relevant papers published on the subject.

The plan of these notes is the following: In Sec. 2 I will recall the experimental questions as well as the minimal ingredients underlying the description of a disordered Wigner crystal. Sec. 3 discusses the standard approach used to treat such problems in the past and its comparison with the recent experiments. Sec. 4 will very briefly present the results of [9, 10] with an emphasis on the differences between the approach used in Sec. 3. Sec. 5 discusses in details the question of the compressibility and capacitance measurements. Some aspects of the dynamics are presented in Section 6. Conclusions and perspectives are presented in Section 7.

2. Basic Questions

2.1. EXPERIMENTS

Although the theoretical concept of a Wigner crystal is easy to grasp it is much more difficult to check experimentally that such a phase is realized in nature. Indeed for other crystal states such as the vortex lattice or the charge density waves, imaging techniques (decoration, neutrons, X-rays) allow to directly see the crystal structure. In the case of the Wigner crystal no direct imaging has been possible so far, although it might become feasible in a near future. One is thus forced to see whether indirect measurements (mostly transport, sound absorption) can be consistently interpreted by assuming a Wigner crystal. The first evidences for a Wigner crystal are shown in Fig. 1. The optical conductivity, an example of which is shown in Fig. 2, provides detailed information as we will discuss it in more details in Sec. 4.

In all these experiments there is no direct evidence of the crystal structure. In order to know whether the transport experiments can be considered as a proof or not of the existence of the Wigner crystal is it thus specially important to have a reliable theory that allows to compute the transport properties. Such a task is far from being trivial given the complexity of the problem.

2.2. ELASTIC DESCRIPTION

Starting from the full electronic Hamiltonian (fermions with interactions and disorder) is a near impossible task. In the crystal phase some simplifications can be made since the particles are now discernable by their position. This allows for a minimal phenomenological model to describe such a crystal: one assumes that the particle are characterized by an equilibrium position R_i^0 and a displacement u_i relative to this equilibrium position. In order to define uniquely the displacement one should not have topological defects such as dislocations in the crystal. I will come back to this point in Sec. 3. From the original quantum problem one has to define the "particles" of the crystal. If the wavefunction is localized enough then one can indeed ignore the exchange between the various sites and thus define "particles" that have a size given by the extension of the wavefunction as indicated in Fig. 3. Of course the density fixes the lattice spacing a. These two lengthscales

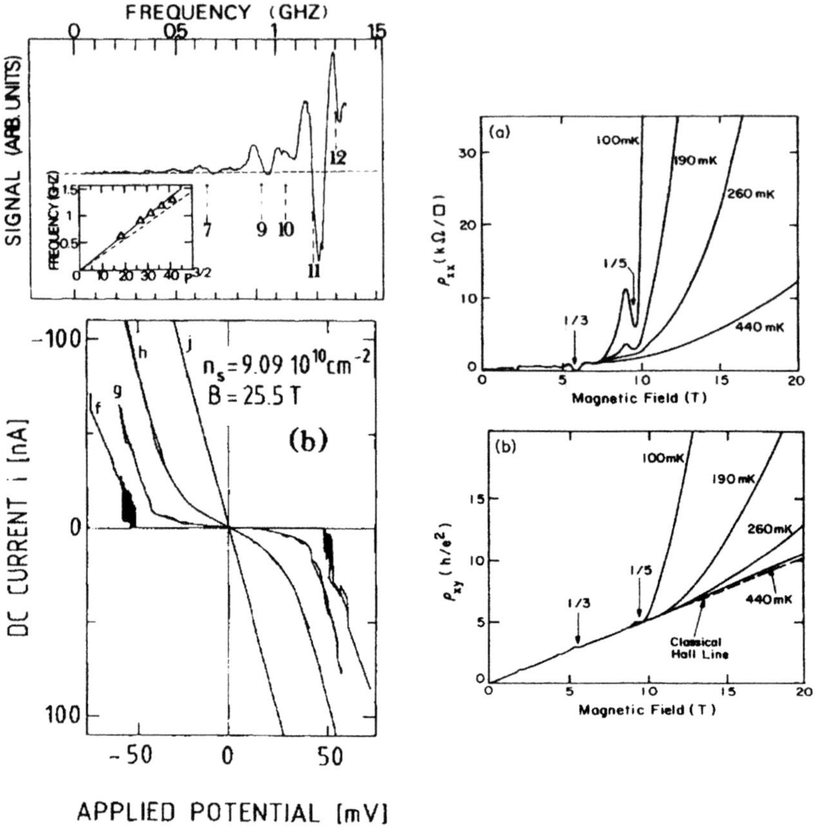

Figure 1. (Top Left) Soundwave absorption by a two dimensional electron gas (2DEG) under strong magnetic field. The frequencies at which the sound is absorbed correspond to the eigenmodes of the crystal (see Sec. 2.3), and are interpreted as evidence of a Wigner crystal in this system (from [5]); (Right) Transport properties of a 2DEG. At strong magnetic field an insulating phase appears, again suggestive of the formation of a pinned Wigner crystal (from [37]). (Bot. Right) Current vs. voltage characteristics. One clearly sees a threshold field needed to have conduction. This is again reminiscent of what one expects of a pinned crystal (From [38])

(size of particle, lattice spacing) are independent and should be kept. Then one retains for the energy the phonon modes of the crystal. This leads to the action (see [10] for more details):

$$
\begin{aligned}
S = {} & \frac{1}{2}\int_{\mathbf{q}} \frac{1}{\beta}\sum_{\omega_n} u_L(q,\omega_n)(\rho_m\omega_n^2 + c_L(q))u_L(-q,-\omega_n) \qquad (1) \\
& + u_T(q,\omega_n)(\rho_m\omega_n^2 + c_T(q))u_T(-q,-\omega_n) \\
& + \rho_m\omega_c\omega_n[u_L(q,\omega_n)u_T(-q,-\omega_n) - u_T(q,\omega_n)u_L(-q,-\omega_n)] \\
& + \int d^2r \int_0^\beta d\tau V(r)\rho(r,\tau)
\end{aligned}
$$

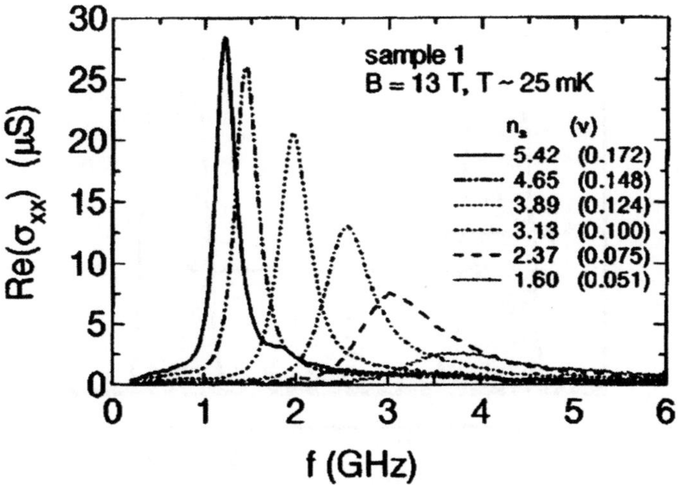

Figure 2. Optical conductivity for various densities for a 2DEG under strong magnetic field. The peak at a characteristic frequency (pinning frequency) is again an expected characteristics of a pinned crystalline structure (see Sec. 3.2) (from [29]).

where we have used the decomposition of the displacements in longitudinal and transverse modes $\vec{u} = \frac{\vec{q}}{q} u_L(q) + (\frac{\vec{q}}{q} \wedge \vec{z}) u_T(q)$. $\int_{\mathbf{q}}$ denote the integration over the Brillouin zone $\int_{BZ} \frac{d^2q}{(2\pi)^2}$, and the ω_n are the standard Matsubara frequencies. The third term in (1) comes from the Lorentz force. $\rho_m \simeq \frac{m}{\pi a^2}$ is the mass density and $\omega_c = eB/m$ the cyclotron frequency.

$C_{L,T}(q)$ are the elastic coefficients for the longitudinal and transverse modes respectively. These coefficients can be obtained from an expansion of the coulomb correlation energy of the WC in terms of the displacements [8, 30]. Since the longitudinal mode describes compressional modes, it is drastically affected by the coulomb repulsion thus $c_L(q) \propto q$, whereas the transverse mode describes shear and thus $c_T(q) \propto q^2$ as in elastic media with only short range interactions.

Finally, the last term describes the coupling to disorder, modelled here by a random potential V. The density of particles

$$\rho(r) = \sum_i \bar{\delta}(r - R_i - u_i) \tag{2}$$

where $\bar{\delta}$ is a δ-like function of range l_c (see Figure 3) and $u_i \equiv u(R_i)$. Since the disorder can vary at a lengthscale r_f *a priori* shorter or comparable to the lattice spacing a, the continuum limit $u_i \to u(r)$, valid in the elastic limit $|u_i - u_{i+1}| \ll a$ should be taken with care in the disorder term [17, 18]. This can be done using the decomposition of the density in terms of its Fourier components

$$\rho(r) \simeq \rho_0 - \rho_0 \nabla \cdot u + \rho_0 \sum_{K \neq 0} e^{iK\cdot(r - u(r))} \tag{3}$$

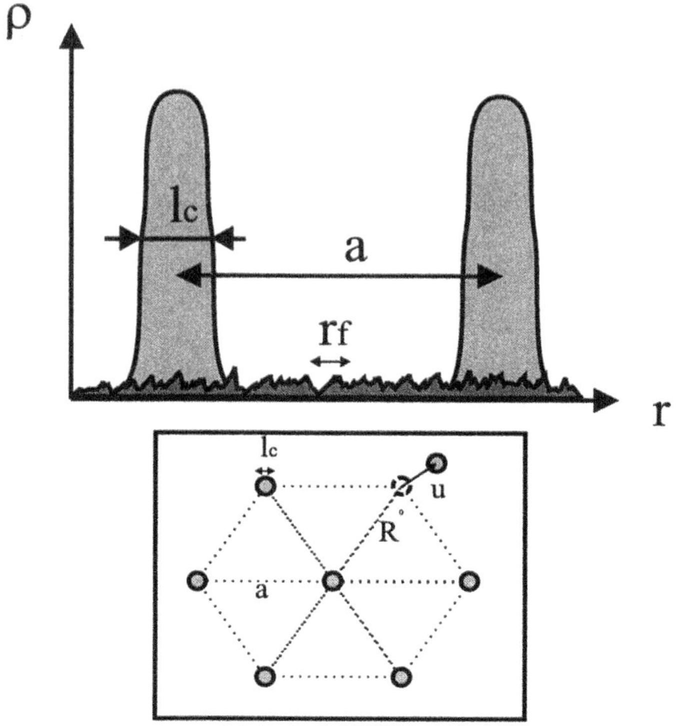

Figure 3. The three length characterizing the Wigner crystal. The size l_c of the "particles" in the crystal (at low temperature it is essentially given by the extension of the wavefunction around the equilibrium position, at large temperatures it is controlled by the thermal fluctuations and is the Lindemann length), a the lattice spacing is controlled by the density of particles, and the disorder is correlated over a length r_f. The inset shows the triangular structure of the Wigner crystal. Particles are labeled by an equilibrium position R_i and a displacement u.(From [10])

where ρ_0 is the average density and K are the reciprocal lattice vectors of the perfect crystal. The finite range of $\overline{\delta}$ is recovered [18] by restricting the sum over K to momentum of order $K_{\max} \sim \pi/l_c$ The disorder is often assumed gaussian, a limit valid when there are many weak pins

$$\overline{V(r)V(r')} = \Delta_{r_f}(r - r') \tag{4}$$

Δ_{r_f} is a delta-like function of range r_f which is the characteristic correlation length of the disorder potential (see Figure 3).

These characteristics lengthscales and Hamiltonian define the minimal model needed to describe a pinned Wigner crystal.

TABLE I. q dependence of the eigenmodes in a Wigner crystal in the absence of magnetic field or for a very strong field.

Mode	zero field	High magnetic field
$\omega_-(q)$	$\propto q^2$ (trans.)	$\propto \frac{q^{3/2}}{\omega_c}$
$\omega_+(q)$	$\propto q$ (long.)	$\sim \omega_c$

2.3. CONSEQUENCES FOR PURE SYSTEM

For a pure system the consequences of the quadratic Hamiltonian (1) are easy to carry out. The eigenmodes of the system are easy to compute. In the absence of field the longitudinal one is plasmon like, whereas the transverse one is phonon-like. In the presence of a large magnetic field the two modes are mixed, giving the eigenmodes of Table I. These are the modes that were probed in the sound absorption experiment shown in Fig. 1.

For the crystal, the current is simply given by $J = e\rho_0\partial_t u$, making thus the conductivity very simple to compute since it is essentially the correlator of the displacements (up to a factor ω). In the absence of magnetic field the optical conductivity is a simple δ function at zero frequency, which traduces the fact that the crystal slides when submitted to an external force. In the presence of a finite magnetic field the electrons describes cyclotron orbits and the peak in conductivity is pushed to the cyclotron frequency ω_c.

Of course these results are for the pure system only and the crucial question is to determine how the disorder changes the above results, in order to make contact with the experiments.

3. Conventional wisdom

3.1. BASIC IDEAS AND CONVENTIONAL WISDOM

In order to know how disorder can modify the above results and lead to pinning, it is necessary to solve the full problem (1), an herculean task. People have thus resorted to approximations. Various such approximation are presented in the review [22]. Based on the various approximate solutions a conventional wisdom on how a disordered elastic system should behave has emerged, as shown in Fig. 4. It was believed that because of disorder the crystal is "broken" into crystallites whose characteristic size is the pinning length. Topological defects (dislocations etc.) in the crystals would be generated at about the same lengthscale. All positional order in the crystal is thus lost beyond the "crystallites". Each crystallite can thus be seen as pinned practically individually.

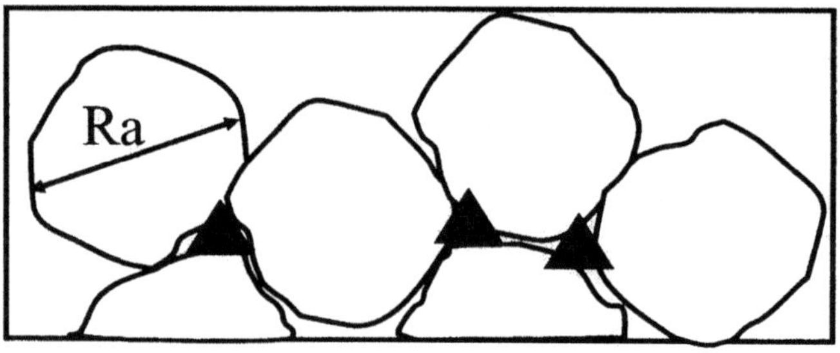

Figure 4. The traditional image of a disordered elastic system. The system is "broken" into crystallites of size R_a. The size corresponds to relative displacements of the order of the lattice spacing between edges of the "crystallite". Topological defects (e.g. dislocations) are argued to be induced by the disorder at the same characteristic lengthscale R_a. All positional order is thus lost beyond R_a. As we now know (see Sec. 4) this physical image is *incorrect*.

3.2. COMPARISON WITH EXPERIMENTS

This physical image inspired from pioneering theories used for charge density waves or for the pinning of vortex lattices (see [22] for more details and references) allows to compute, using some approximations, the optical conductivity. Essentially the crystallites will be held by a pinning potential and respond at a given frequency, the pinning frequency, related to the pinning length, as shown in Fig. 5. The coupling between the various crystallites lead to a broadening of the peak, very often assumed to be lorentzian.

When one compares the experiments, with these theoretical predictions the agreement is *qualitatively* wrong. Let us show for example the density dependence of the pinning peak (see Fig. 5). The predicted density dependence of the pinning frequency would be from the above mentioned approximations $\omega_p \propto n^{1/2}$. This result would be totally opposite to the data which shows a *decrease* of the pinning frequency with the density. Such important problems when one tries to compare with the data could cast serious doubts on the interpretation of the insulating phase in terms of a Wigner crystal and quite naturally other interpretations for this phase have been proposed [39].

Moreover note that in fact the conventional description shown in Fig. 4 would in fact invalidate the very use of an elastic approximation such as (1) to compute the peak in the optical conductivity. Indeed the peak in such approach would be controlled by the pinning length R_a of the crystallites. But at that lengthscale topological defects are argued to occur. Such defects are not taken into account in the elastic description and could in principle modify the results for the peak. Strictly speaking the elastic theory could thus only be applied for frequencies $\omega \gg \omega_p$ (see Fig. 6).

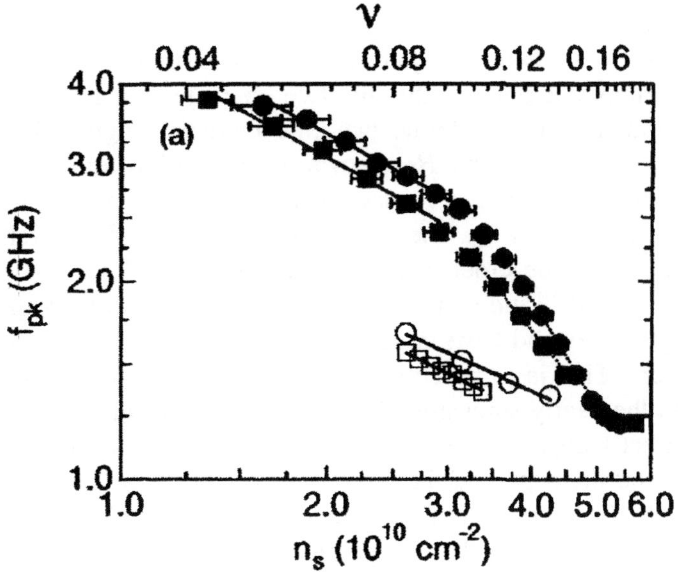

Figure 5. Variation of the pinning frequency with the density for a 2DEG under a strong magnetic field. The pinning frequency decreases with the density in all systems (different symbols), in contradiction with the naive calculations based on the physical image shown in Fig. 4. Such calculations would lead to $\omega_p \propto n^{1/2}$. (From [28])

4. Bragg glass and disordered Wigner crystal

The situation is in fact much better than one could think based on the naive approach exposed in Sec. 3. In fact the discrepancy lies in the fact that the theory used to connect the elastic Hamiltonian (1) to the transport properties and based on the physical ideas shown in Fig. 4, is in fact incorrect. Fortunately, it has been possible with recent "theoretical technology" to obtain a quite complete solution of (1). I will not review here the method or solution but refer the reader to [19, 22] for the general technology and to [9, 10] for the Wigner crystal solution. The agreement with experiments is now quite good (see [10] for a full discussion). In particular one finds a decrease of the pinning frequency with the density as $\omega_p \propto n^{-3/2}$, as well as a good magnetic field dependence of the pinning frequency. This very good agreement gives a good confirmation that the insulating phase in the 2DEG under strong magnetic field is indeed a Wigner crystal collectively pinned by impurities.

Of course despite this good agreement some points still remain open. Among them the question of the low frequency behavior and the magnetic field dependence of the width of the peak. I will not discuss these questions further and refer the reader to [15, 10] for further discussions of these issues.

I want to insist here on two important physical features which are apparent in the solution [9, 10] and whose physics deserves to be explained in detail.

The first important point is that the two characteristic lengthscales (size of particle and lattice spacing) define *two* characteristic lengthscales via the displacement field. The first one R_c, known as the Larkin length, correspond to the distance for which relative displacements are of the order of the size of the particle $u(R_c) - u(0) \sim l_c$. The second R_a is the one for which the relative displacements are of the order of the lattice spacing $u(R_a) - u(0) \sim a$. Since in the Wigner crystal l_c and a are quite different R_c and R_a corresponds to quite different lengthscales and have in general quite different dependence in the various parameters. This is to be contrasted with charge density waves for which $l_c \sim a$ due to the nearly sinusoidal density modulation and thus $R_c \sim R_a$. Thus borrowing directly approximate solutions that have been developed for this case is dangerous and gives part of the physics incorrectly. As can be checked from the solution [9, 10] the pinning frequency ω_p is controlled by the lengthscale R_c and *not* the lengthscale R_a. This is physically reassuring since one knows [24] for classical systems that R_c is the length that controls the threshold force, and thus is naturally associated with pinning. The distinction is important since R_c depends on the size of the particle l_c. This gives, for example for the case of strong magnetic field for which l_c is just the cyclotron orbit, additional magnetic field dependence to the pinning frequency.

The second important point concern the possibility to use the elastic theory. The elastic theory is in fact much more stable to the presence of topological defects than initially anticipated. In $d = 3$ it is now known that below a certain threshold of disorder *no* topological defects can be induced by the disorder. The disordered elastic system is in a Bragg glass state [18] with a quasi long range positional order, much more ordered than the image of Fig. 4 suggests (see e.g. [16] for a discussion and references on this point). In $d = 2$ the situation is marginal, and defects appear in the ground state, but at distance R_d much larger (for weak disorder) than the lengthscale R_a and *not* at that lengthscale [18, 26]. Dislocations will thus spoil the results of the elastic theory for the optical conductivity only well below the peak, as shown on Fig. 6. This implies that the theory is a *reliable tool* to compute the characteristics of the peak and above, and thus most of the a.c. transport.

5. Compressibility

5.1. COMPRESSIBILITY IN CHARGED SYSTEMS

Naively one relates the compressibility to the density-density correlation function by

$$\kappa = \lim_{q\to 0} \langle \rho(q, \omega_n = 0)\rho(-q, \omega_n = 0)\rangle \tag{5}$$

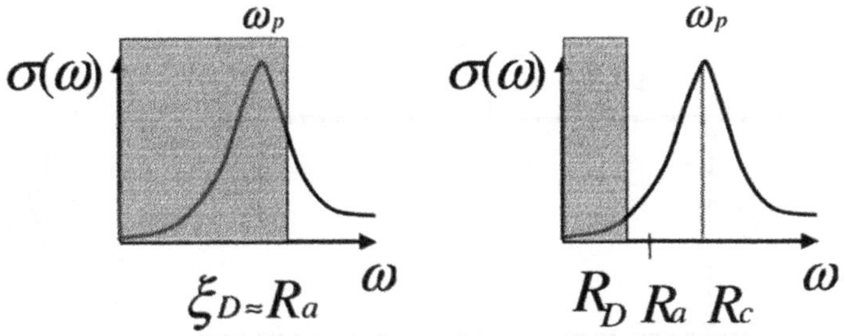

(a) (b)

Figure 6. (a) If dislocations occurred at scale R_a and the pinning frequency was controlled by R_a, as was naively believed, the elastic theory is incapable of giving any reliable information on the pinning peak. It would be necessary to include dislocations from the start. (b) As was discussed in the text, dislocations occur in fact at $R_D \gg R_a$ and the pinning peak depends on $R_c \ll R_a$. Thus the pinning peak is given quantitatively by a purely elastic theory. It is necessary to take into account topological defects such as dislocations only at much lower frequencies, and in particular if one wants to obtain reliable results for the d.c. transport.(from [10])

The case of the disordered system will be discussed in Sec. 5.4, but let us look first at the pure system. The compressibility is simply (only the longitudinal mode plays a role)

$$\kappa(q) = \lim_{q \to 0} \frac{q^2}{c_L(q)} \tag{6}$$

If only short range interactions are present in the system the longitudinal mode is a phonon-like mode $c_L(q) \propto q^2$ and the one recovers a finite compressibility. On the other hand if one has long range Coulomb interactions $c_L(q) \propto |q|$ and the compressibility becomes zero. This is simply due to the fact that (5) measures the density response to a change of chemical potential while *keeping the neutralizing background unchanged.* A charged systems thus does not remain neutral, hence the infinite compressibility.

One has thus to define the compressibility more precisely. Based on the correlations a standard substraction procedure consists in keeping only the "irreducible" part of the density-density correlation function, i.e. define the compressibility as

$$\kappa_{\text{irr}} = \frac{\kappa}{1 - V_q \kappa} \tag{7}$$

where V is the long range Coulomb potential. However it is unclear how this procedure is related to the standard way of measuring the compressibility, i.e. the capacitance measurements (see below). Many derivations of the compressibility use instead directly a derivative of the free energy with respect to the number of particles. The free energy can be computed for a neutral system for an arbitrary number of particles which solves the above-mentioned problem. Unfortunately very often the calculation is only possible in some sort of approximate way such as

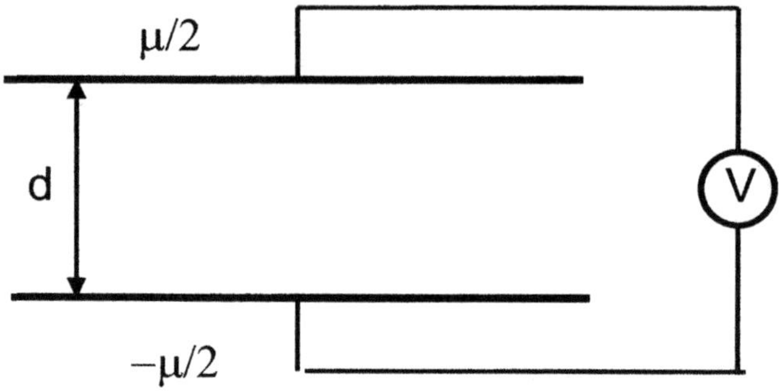

Figure 7. Capacitance measurement, which gives access to the compressibility of the system. A voltage difference μ is applied to a capacitor. Here for simplicity the capacitor is made of two plates of the 2DEG.

an Hartree-Fock approximation. Here again the link with the direct measurements of the compressibility is not clear. Using such procedures, so called "negative" compressibilities are found for some range of the interactions, for interacting electrons. Similarly, experiments measure such negative "compressibilities" [13, 23] (see [2] for further references and discussion on this question).

5.2. CAPACITANCE MEASUREMENTS

In order to make the physics of such negative compressibility more transparent, I will discuss now a very simple way to compute them. This way is hopefully more physically transparent than the standard derivations [1], and has the advantage to be easily extensible to the Wigner crystal in presence of disorder. It is in fact a direct calculation of the quantity that is actually measured to determine the "compressibility", i.e. the capacitance of a system made by with the 2DEG [13]. For simplicity I take here a capacitor formed of two identical systems, as shown in Fig. 7. Taking one system and one metallic plate would not change the results in an essential way. The Hamiltonian of the system is thus

$$
\begin{aligned}
H &= H_1^0 + H_2^0 + \\
& \sum_{(\alpha,\beta)=1,2} \int_{r,r'} \frac{1}{2} V(r-r')[\rho_\alpha(r) - \rho_0][\rho_\beta(r') - \rho_0] + \frac{\mu}{2} \int_r [\rho_1(r) - \rho_2(r)]
\end{aligned}
\tag{8}
$$

If one assumes that the system is neutral in the absence of μ, then the charge on one plate when a potential μ is applied is

$$
\langle \rho_1 \rangle = \frac{\mu}{2} [\langle \rho_1 \rho_1 \rangle - \langle \rho_1 \rho_2 \rangle] \tag{9}
$$

[1] This is of course a personal and probably biased opinion !

in linear response. (9) give directly the capacitance $\langle \rho_1 \rangle / \mu$.

As a warmup let us shown that in the RPA approximation (9) leads back to the standard results for the compressibility. With (8) it is easy to check that the susceptibilities $\chi_{\alpha\beta} = \langle \rho_\alpha \rho_\beta \rangle$ are given, in RPA, by

$$\begin{pmatrix} \chi_{11} \\ \chi_{12} \end{pmatrix} = - \begin{pmatrix} \chi^0 V_{11} & \chi^0 V_{12} \\ \chi^0 V_{12} & \chi^0 V_{11} \end{pmatrix} \begin{pmatrix} \chi_{11} \\ \chi_{12} \end{pmatrix} + \begin{pmatrix} \chi^0 \\ 0 \end{pmatrix} \tag{10}$$

where χ^0 is the bare (i.e. for H^0 only) density-density correlation function in one of the systems. It is easy to solve (10) to obtain for the (q dependent) capacitance

$$\chi_{11} - \chi_{12} = \frac{\chi^0}{1 + \chi^0 (V_{11} - V_{12})} \tag{11}$$

The Fourier transform of the Coulomb potentials are given by

$$V_{11} - V_{12} = \int d^2 r e^{iqr} [\frac{1}{r} - \frac{1}{\sqrt{r^2 + d^2}}] = \frac{(2\pi)(1 - e^{-qd})}{q} \tag{12}$$

since the two plates are at a distance d. The true capacitance is the limit $q \to 0$ which leads to

$$C = \frac{1}{(2\chi^0(q=0))^{-1} + 4\pi d} \tag{13}$$

On thus recovers that the capacitance is the sum of a geometrical one C_{geom} and one due to the electron gas inside the plates C_{el}

$$\frac{1}{C} = \frac{1}{C_{\text{geom}}} + \frac{1}{C_{\text{el}}} \tag{14}$$

The geometrical one is the standard $1/(4\pi d)$ result. For a simple electron gas $\chi^0(q = 0)^{-1}$ is simply the screening length λ of the electron gas. One thus recovers that the geometrical distance d between the plates is increased by the screening length λ on each side.

5.3. WIGNER CRYSTAL

One can use the general formula (9) to compute the capacitance for the Wigner crystal. One substitutes in (1) the density decomposition (3). The ∇u terms give directly the contribution of the long range part of the Coulomb interaction

$$H_{\text{long-range}} = \frac{1}{2}\rho_0^2 \sum_q \sum_{\alpha\beta=1,2} [V_{\alpha\beta}(q) u_L^\alpha(q) u_\beta(-q)] \tag{15}$$

Since $V_{11}(q) \sim 1/q$, (15) gives obviously the part proportional to q in the elastic coefficients for an isolated plane. The higher harmonics give the regular part (i.e.

the part proportional to q^2 in the elastic coefficients. Such a way to determine the coefficient is equivalent the calculation of the coefficients in [8].

Taking a pure system the Hamiltonian becomes (only the $\omega_n = 0$ term of the longitudinal part needs to be computed to have the compressibility)

$$H = \begin{pmatrix} u_L^1(q) \\ u_L^2(q) \end{pmatrix} = \begin{pmatrix} c_L^{\rm SR}(q) + \rho_0^2 q^2 V_{11}(q) & \rho_0^2 q^2 V_{12}(q) \\ \rho_0^2 q^2 V_{12}(q) & c_L^{\rm SR}(q) + \rho_0^2 q^2 V_{11}(q) \end{pmatrix} \begin{pmatrix} u_L^1(-q) \\ u_L^2(-q) \end{pmatrix} \tag{16}$$

where $c_L^{\rm SR}(q)$ is the "short range" part of the elastic coefficients. Using (9) and the expression of the density for small q from (3) $\rho_L(q) = -\rho_0 q u_L(q)$ one gets for the capacitance an equation like (14), where now

$$\frac{1}{C_{\rm el}} = \lim_{q\to 0} \frac{2c_L^{\rm SR}(q)}{\rho_0^2 q^2} \tag{17}$$

(again the factor of 2 comes from the fact that here I took two identical plates). The electronic one corresponds to the propagator where *only* the short range part of the elastic coefficients is kept. Using [8] $c_L^{\rm SR}(q) = -\omega_0^2(0.18..)(aq)^2$, where $\omega_0 = \frac{4\pi e^2}{\sqrt{3ma^3}}$, one finds for the Wigner crystal a "negative" compressibility. The fact that a system of discrete charges can lead to such effects has been noted before for classical Wigner crystals (see e.g. [31] and references therein). The method presented here allows to easily determine this "compressibility" and is trivially applicable to the disordered case.

One should note that if the distance d between the plates is large compared to the lattice spacing of the Wigner crystal is it safe to throw away the higher harmonics in the coupling term $\rho_1\rho_2$ since they behave as e^{-Kd} where K are the vectors of the reciprocal lattice. The capacitance is then indeed only given by the properties of a single system as in (17). This is not true if $d \sim a$ (which can be the case experimentally) in that case the higher harmonics of the coupling will also contribute to the capacitance, which then cannot be trivially related to the intrinsic properties of a single Wigner crystal. These questions will be addressed in more details elsewhere.

5.4. VARIATIONAL COMPRESSIBILITY

Finally let us consider the effects of disorder. The method discussed in Sec. 4 gives straightforwardly the density-density correlation function which is simply related to the displacement-displacement correlation. Since it is a Gaussian approximation one can easily use the capacitance method shown above. The capacitance is thus given by (17) but where one should use the propagator in the presence of disorder. It is given by (within the variational approximation used [9, 10]) by:

$$\langle uu \rangle = \frac{1}{\rho_m \omega_n^2 + c_L(q) + \Sigma(1-\delta_{n,0}) + I(\omega_n)} \tag{18}$$

where Σ and $I(\omega_n)$ are respectively a constant and a function related to the disorder verifying $I(0) = 0$.

At $\omega_n = 0$ (18) leads to a compressibility in identical to the one of the pure system, and thus *also* "negative". This is the thermodynamics results, but the disorder can in principle lead to an additional twist. The thermodynamic compressibility is obviously given by the correlation function in imaginary time at $\omega_n = 0$. For the Wigner crystal this leads back to the compressibility of the pure system. If however one makes the analytic continuation $i\omega_n \to \omega + i\epsilon$, one gets a quite different result. (18) becomes

$$\langle uu \rangle = \frac{1}{-\rho_m \omega^2 + c_L(q) + \Sigma + I(\omega)} \tag{19}$$

Taking then the limit $\omega \to 0$ first as should be done to get the compressibility (and not the transport which corresponds to the opposite limit $q \to 0$ first $\omega \to 0$ after) one has

$$\langle uu \rangle = \frac{1}{c_L(q) + \Sigma} \tag{20}$$

which leads to a zero compressibility *because* of the presence of the mass Σ due to the disorder. One thus may expect in an experiment for a disordered Wigner crystal a drastic change in the behavior of the "compressibility" (i.e. the measured capacitance) of the system as a function of the frequency. At extremely low frequency, one measures the thermodynamic compressibility (negative correction for the Wigner crystal as discussed above). When the frequency increases one measures (20), i.e. a zero compressibility. What is the characteristic frequency separating these two behavior is still unclear the moment, although it is obviously related to the coupling to the external environment. This phenomenon seems to be the equivalent for the Wigner crystal of the Coulomb gap, in usual interacting electronic systems, where a gap linked to the Fermi level appears.

6. Dynamics

We have seen in Sec. 4 that the a.c. transport is a very efficient way to probe the crystalline nature of the electronic system. Getting the d.c. transport is a much more complicated task. For the full quantum system this problem is still a tough cookie, although it is clear that the same methods than used for classical system (functional renormalization group for example) should give good results. One can however gain considerable intuition on what to expect by looking at the classical equivalent.

Let us thus consider the case of a two dimensional classical crystal, submitted to an external force (here due to the electric field). As was shown in [20], periodic moving periodic systems have a quite specific dynamics. Indeed due to the existence of the periodicity in the direction *transverse* to the direction of motion, the

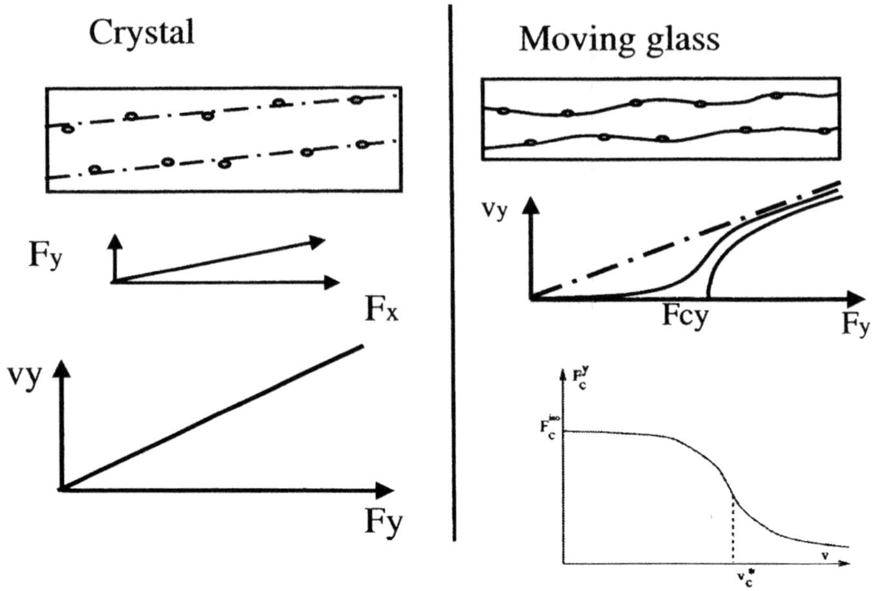

Figure 8. (Left) for a pure crystal the particles move in straight lines. When the particles move an applied transverse force tilts the trajectories leading to linear response. (Right) In a disordered system the moving system is still in a glassy state (see text). The particle move along rough channels that are the best compromise between the elasticity and the disorder. Although the particles themselves move the channels themselves are pinned, and a transverse pinning force F_{cy} exists. Thus if a transverse force is applied there is not response (at $T = 0$) in the transverse direction until $F_y > F_{cy}$. The transverse pinning force decreases with the longitudinal velocity as shown in the figure above.

motion cannot average completely over the disorder. The moving system is thus submitted to a random potential, which leads to a channel like motion as shown in the right hand side of Fig. 8.

The channels are the best compromise between the elastic energy and the remaining disorder. Their very existence have an important consequence if one tries to make the system move in the *transverse* direction. Indeed although the particles do move along the channels, the channels themselves are pinned. This means that even above the longitudinal threshold field, if one tries to apply a force in the transverse direction a transverse critical force still exists, as shown on Fig. 8. The value of this transverse critical force can be computed by simple scaling arguments [20] or by more sophisticated renormalization group techniques [21, 25, 6].

For the two dimensional electronic system, putting a magnetic field is a simple way of applying a transverse force. If the lattice is sliding at velocity v, it is submitted to a Lorentz transverse force $F_L = evB$. The existence of the transverse critical force F_{tr} thus implies that the channel structure should not slide as long as $F_L < F_{\text{tr}}$. There will thus be no hall voltage generated. On the other hand

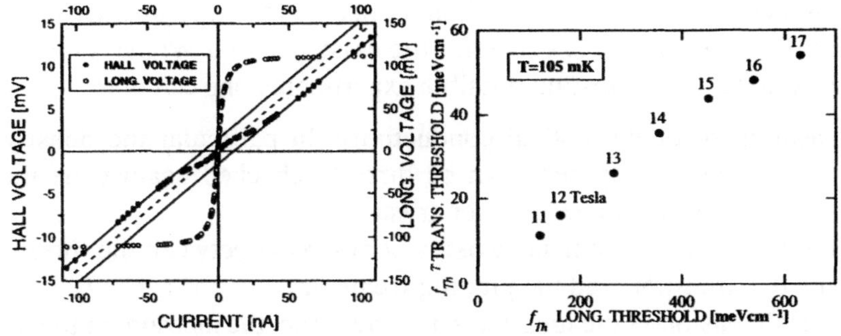

Figure 9. Longitudinal and Hall voltage as a function of the longitudinal current (left). A non-linearity is observed in the Hall voltage. Such a behavior is compatible with the existence of a transverse threshold. This transverse threshold is plotted vs the longitudinal one for different magnetic fields (right) (from [32])

when $F_L > F_{\rm tr}$ the channel structure should slide and a Hall voltage exists. The *periodicity* of the crystalline structure thus implies that one needs a *finite* longitudinal current before a Hall voltage exists.

Such experiment has been performed in the systems under strong magnetic field [32], and the results are shown in Fig. 9. A finite longitudinal current is clearly needed to develop a Hall voltage, in good agreement with the existence of a transverse threshold. The existence of such an effect is a direct probe of the crystalline (existence of a transverse periodicity) nature of the phase.

7. Conclusions and perspecpectives

In these short notes I have discussed how the concepts developed to deal with disordered elastic systems can be fruitfully applied to interacting electrons. They allow to investigate the effect of the disorder on a Wigner crystal.

Both the a.c. transport and thermodynamic quantities such as the compressibility can be reliably computed. There is good agreement with the predicted d.c. transport and the observed behavior of a 2DEG under a strong magnetic field, making a strong case for a Wigner crystal phase in such systems. The compressibility is found to be negative both for a pure Wigner crystal and in the presence of disorder, and detailed comparison with experiments on that point would be clearly fruitful. Even if computing the full d.c. transport is beyond reach at the present, some properties can be obtained. In particular the periodicity of the crystal should lead to the existence of a transverse pinning which should entail a shift in the Hall response, as observed experimentally.

We see that there are thus efficient, if not easy, ways via transport to check for the presence of a crystalline phase. Most of the experiments discussed in the preceding sections were for a 2DEG under a strong magnetic field. It would of course be extremely interesting to use the same techniques to probe the 2DEG in the absence of the magnetic field, and analyze the experiments in the line of what was discussed above to check for the existence of a Wigner crystal in these systems. Among the interesting possible experiments one can note:

- Measurement of the optical conductivity. In particular the density dependence of the pinning peak can be directly checked against the theoretical predictions of the pinned Wigner crystal.
- If the optical measurements exist, a comparison between the threshold field in the d.c. transport and the pinning frequency.
- The Hall tension vs the longitudinal current (i.e. the measure of the transverse pinning force)
- Although not discussed in these notes, noise measurements are also a good way to probe the periodic nature of the systems (see e.g. [34] and references therein).

Acknowledgements. The bulk of the work discussed in these notes results from a fruitful and enjoyable collaboration with R. Chitra and P. Le Doussal, both of whom I would like to specially thank. I would also like to thank E. Abrahams, A. Yacoby and C.M. Varma for many interesting discussions on the compressibility in charged systems and F.I.B. Williams for many enlightening discussions on the Wigner crystal.

References

1. Abrahams, E., P. W. Anderson, D. C. Licciardello, and T. V. Ramakrishnan: 1979. *Phys. Rev. Lett.* **42**, 673.
2. Abrahams, E., S. V. Kravchenko, and M. P. Sarachik: 2001. *Rev. Mod. Phys.* **73**, 251.
3. Abrikosov, A. A. and J. A. Rhyzkin: 1978. *Adv. Phys.* **27**, 147.
4. Altshuler, B. L. and A. G. Aronov: 1985. In: A.L.Efros and M.Pollak (eds.): *Electron-electron interactions in disordered systems*. Amsterdam.
5. Andrei, E. Y. and al.: 1988. *Phys. Rev. Lett.* **60**, 2765.
6. Balents, L., C. Marchetti, and L. Radzihovsky: 1998. *Phys. Rev. B* **57**, 7705.
7. Berezinskii, V. L.: 1974. *Sov. Phys. JETP* **38**, 620.
8. Bonsall, L. and A. A. Maradudin: 1977. *Phys. Rev. B* **15**, 1959.
9. Chitra, R., T. Giamarchi, and P. Le Doussal: 1998. *Phys. Rev. Lett.* **80**, 3827.
10. Chitra, R., T. Giamarchi, and P. Le Doussal: 2001. *Phys. Rev. B* **65**, 035312.
11. Efetov, K. B.: 1983. *Adv. Phys.* **32**, 53.
12. Efetov, K. B., A. I. Larkin, and D. E. Khmel'nitskii: 1980. *Sov. Phys. JETP* **52**, 568.
13. Eisenstein, J. P., L. N. Pfeiffer, and K. W. West: 1978. *Phys. Rev. B* **73**, 535.
14. Finkelstein, A. M.: 1984. *Z. Phys. B* **56**, 189.
15. Fogler, M. M. and D. A. Huse: 2000. *Phys. Rev. B* **62**, 7553.
16. Giamarchi, T. and S. Bhattacharya: 2001. "Vortex phases", cond-mat/0111052; Lecture notes of the 2001 Cargese school on "Trends in high magnetic field science"; to be published by Springer.
17. Giamarchi, T. and P. Le Doussal: 1994. *Phys. Rev. Lett.* **72**, 1530.
18. Giamarchi, T. and P. Le Doussal: 1995. *Phys. Rev. B* **52**, 1242.
19. Giamarchi, T. and P. Le Doussal: 1996a. *Phys. Rev. B* **53**, 15206.

20. Giamarchi, T. and P. Le Doussal: 1996b. *Phys. Rev. Lett.* **76**, 3408.
21. Giamarchi, T. and P. Le Doussal: 1997. *Physica C* **282-287**, 363.
22. Giamarchi, T. and E. Orignac: 2001, 'Disordered Quantum Solids'. In: A. M. Tsvelik (ed.): *New Theoretical Approaches to Strongly Correlated Systems*, Vol. 23 of *NATO SCIENCE SERIES: II: Mathematics, Physics and Chemistry*. Dordrecht. cond-mat/0005220.
23. Ilani, S., A. Yacoby, D. Mahalu, and H. Shtrikman: 1999. cond-mat/9910116.
24. Larkin, A. I. and Y. N. Ovchinnikov: 1979. *J. Low Temp. Phys* **34**, 409.
25. Le Doussal, P. and T. Giamarchi: 1998. *Phys. Rev. B* **57**, 11356.
26. Le Doussal, P. and T. Giamarchi: 2000. *Physica C* **331**, 233.
27. Lee, P. A. and T. V. Ramakhrishnan: 1985. *Rev. Mod. Phys.* **57**, 287.
28. Li, C. C. and al: 1998. cond-mat/9810066.
29. Li, C. C. and *et al*: 2000. *Phys. Rev. B* **61**, 10905.
30. Maki, K. and X. Zotos: 1983. *Phys. Rev. B* **28**, 4349.
31. Nguyen, T. T., A. Y. Grosberg, and B. I. Shklovskii: 2001. cond-mat/0101103.
32. Perruchot, F., F. I. B. Williams, C. J. Mellor, R. Gaal, B. Sas, and M. Henini: 2000. *Physica B* **284**, 1984.
33. Thomas, G. A. and R. F. Milligan: 1985, 'The Metal Insulator Transition'. *Ann. Rev. Phys. Chem.* **36**, 139.
34. Togawa et al., Y.: 2000. *Phys. Rev. Lett.* **85**, 3716.
35. Wegner, F.: 1979. *Z. Phys. B* **35**, 207.
36. Wigner, E.: 1934. *Phys. Rev.* **46**, 1002.
37. Willett, R. L. and *et al.*: 1989. *Phys. Rev. B* **38**, R7881.
38. Williams, F. I. B. and *et al*: 1991. *Phys. Rev. Lett.* **66**, 3285.
39. Zhang, S. C., S. Kivelson, and D. H. Lee: 1992. *Phys. Rev. Lett.* **69**, 1252.

MAGNETO-OPTICS OF COMPOSITE FERMIONS AND SKYRMIONS

I. V. KUKUSHKIN
Max-Planck-Institut für Festkörperforschung, 70569 Stuttgart, Germany
Institute of Solid State Physics, RAS, Chernogolovka, 142432 Russia

1. Introduction

It is well known that in condensed matter physics no single-particle theory can provide us with an exact description of the real experiment due to a huge number of interacting particles involved in the process. However, the main goal for the theory is to catch the basic unusual idea, which would give a clear physical picture and explain essential experimental results. One famous successful example of a many-body theory is the Bardeen-Cooper-Schrieffer model, explaining superconductivity in terms of new composite particles - pairs of electrons which experience a Bose-Einstein condensation. Another successful example is a new theory of Composite Fermions (CFs) [1], explaining the Fractional Quantum Hall Effect [2] in terms of novel composite particles, each consisting of an electron and a number of magnetic flux quanta. The properties of these composite particles are governed by the number of the fluxes attached to the electron - they are bosons for odd numbers of fluxes and fermions for even numbers. The CFs are electrons dressed with two magnetic flux quanta. An attachment of the flux quanta to an electron is a rather natural way to minimize the energy of the system, since these vortices would expel other electrons resulting in a decrease of the repulsion interaction between 2D-electrons. The main consequence of the theory of CFs is that at half filling of the lowest Landau level ($\nu = 1/2$) an external magnetic field is effectively compensated by two flux quanta and a metallic state of these composite particles is formed. This state can be characterized by a Fermi wave-vector and a Fermi-energy [3] and a deviation of the magnetic field from exact half filling results in the appearance of a non-zero effective magnetic field which quantizes the CF motion and discretizes their energy spectrum into Landau levels. In this model, the FQHE is a manifestation of the Landau quantization of CFs and a rich variety of experimental observations can be understood straightforwardly in terms

I.V. Lerner et al. (eds.),
Strongly Correlated Fermions and Bosons in Low-Dimensional Disordered Systems, 185–218.

terms of nearly independent CFs [4, 5]. Recent experiments [6, 7, 8, 9] support the validity of this theoretical concept and moreover demonstrate the semiclassical behavior of these composite particles. One of the main purposes of our investigations was to study magneto-optics of Composite Fermions with the aim to measure their spin polarization and to detect the cyclotron resonance of composite particles.

2. Spin polarization of composite fermions

In spite of above mentioned very strong experimental support of the concept of CFs, there was an apparent inconsistency inside this simple picture, describing the FQHE in terms of CFs. In the case of normal 2D-electrons, their Fermi energy is equal to the cyclotron gap at filling factor $\nu = 2$. If this holds for CFs, their Fermi energy, E_F^{CF}, can be taken as the value of the energy gap measured for $\nu = 1/3$ (for this state, the CF filling factor is $\nu_{CF} = 1$). Since for the magnetic field $B \approx 10T$ this gap is typically 8-12K [5, 10], E_F^{CF} can be estimated as 10K (the same value can be also directly obtained from the published CF mass $m_{CF} \approx 0.6m_0$ [5, 11] at $B \approx 10T$). Due to the small value of the g-factor of electrons in GaAs, the Zeeman energy (E_Z) at $B = 10T$ is only about 3K, which means that $E_F^{CF} >> E_Z$ and, therefore, one may expect that the Fermi gas of CF is practically spin unpolarized at $B \approx 10T$. In contrast, it follows from the above mentioned experiments, in which the CF wave-vector was measured, that the system of CFs is completely spin polarized at $B \approx 10T$. This apparent discrepancy was our main motivation to measure the spin polarization of CFs using a well established optical method [12].

In this section we will consider the magnetic field dependence of the spin polarization of a 2D-system in the $\nu = 1/2$, 3/2 and 1/4 CF states. The CF Fermi energies for each of those states have been determined from the value of the magnetic field B_c at which a system of CFs becomes completely spin polarized ($E_F^{CF} = E_Z$ at $B = B_c$). The CF density of states mass is found to be about 4 times heavier than the previously reported 'activation' mass [5, 11] and more close to the value obtained from the temperature dependence of CFs scattering [13]. Another experimental task solved in the present Letter concerns the phase transitions between differently spin polarized ground states of the FQHE. Several spin transitions between unpolarized, partly polarized and fully polarized spin states, which are accompanied by threshold changes in the electron spin polarization, γ_e, are detected for the $\nu = 2/3$, 3/5, 4/7, 2/5, 3/7, 4/9, 8/5, 4/3 and 7/5 FQHE states and absolute values of γ_e are determined for these partly spin polarized FQHE ground states.

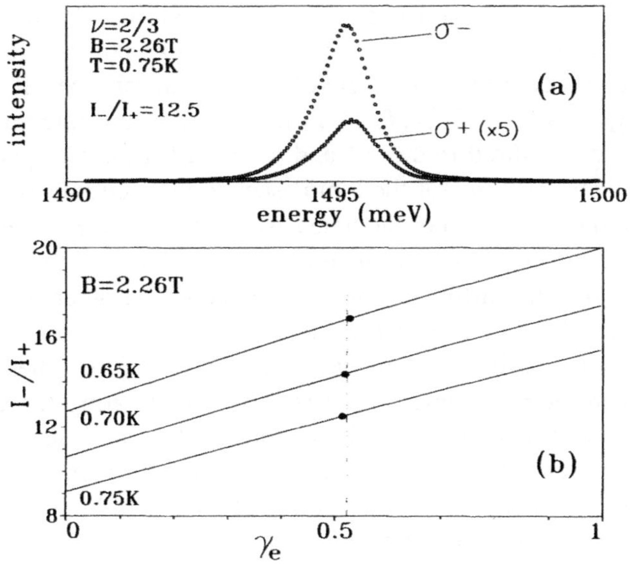

Figure 1. (a) Typical luminescence spectra measured in the σ^- and σ^+ polarizations for $\nu = 2/3$, $B = 2.26T$, $T = 0.75K$ and (b) calibration dependence [12] of the ratio I_-/I_+ on the electron spin polarization γ_e, obtained for B/T =3.01 ($T = 0.75K$), B/T =3.23 ($T = 0.70K$), B/T = 3.48 ($T = 0.65K$). I_- and I_+ are the integrated intensities of the luminescence, measured in the σ^- and σ^+ polarizations, respectively.

2.1. SAMPLES AND METHOD

We studied several low-density ($n_s = (0.36 - 2.4) \times 10^{11}$ cm^{-2}) and high-quality (electron mobility $\mu = (0.9 - 3) \times 10^6$ cm^2/Vs) GaAs/Al$_x$Ga$_{1-x}$As single heterojunctions with a δ-doped monolayer of Be acceptors ($n_A = 2 \times 10^9$ cm^{-2}) located in the wide ($1\mu m$) GaAs buffer layer at a distance of 30 nm from the interface [14]. In all samples a variation of 2D-electron concentration was possible by the use of a top gate. For photoexcitation, we used pulses from a tunable Ti-sapphire laser (the wavelength was close to 780 nm) with a duration of 20 ns, a peak power of 10^{-4} -10^{-2} W/cm^2, and a frequency of $10^4 - 10^6$Hz. Luminescence spectra were detected by a gatable photon counting system with a spectral resolution of 0.03 meV. To analyze the circular polarization of the luminescence signal at low temperatures (down to 300 mK), we used an optical fiber system with a quarter wave plate and a linear polarizer placed in liquid helium just nearby the sample. Measurements were performed in the temperature range of 0.3-1.8K.

The spin polarization of the ground state of 2D-electron system is defined by a competition between the Zeeman and Coulomb (E_C) energies. This was well established both theoretically [15, 16] and experimentally [17, 18, 19] for the ground states of different FQHE states in the extreme quantum limit. In our previous work [12], we demonstrated unambiguously that such spin transitions can be directly detected from the degree of circular polarization γ_L of time-resolved radiative recombination of 2D-electrons with photoexcited holes bound to acceptors ($\gamma_L = (I_- - I_+)/(I_- + I_+)$, I_- and I_+ are integral intensities of the luminescence measured in the σ^- and σ^+ polarizations, respectively). There are two independent reasons for the luminescence to be circularly polarized. One reason is the spin polarization of the hole system due to the Zeeman effect, which depends on the magnetic field and the temperature of the photoexcited holes (i.e. on the population of the different spin-split levels). The other reason is the spin polarization of the 2D-electrons, which depends on filling factor, temperature and magnetic field. It is possible to derive the contribution of the holes to the circular polarization of the luminescence separately by investigating the emission from fully occupied electron Landau levels (at ν=2, 4, 6 ... and also well below the Fermi surface), because in this case both spin up and down states of electrons are completely occupied and the circular polarization of the luminescence is determined only by the spin polarization of the holes [12]. It has been demonstrated experimentally (and in agreement with calculations [12]), that the spin polarization of the hole system is defined by the ratio B/T, so that it can be excluded from the polarization of the luminescence for a fixed value of B/T and a direct correspondence between γ_L and γ_e can be established.

2.2. SPIN TRANSITIONS

Typical luminescence spectra, recorded for both σ^- and σ^+ polarizations and the calibration curves, obtained for different temperatures for a given magnetic field are shown in Fig.1. For a fixed value of B/T, the spin polarization of the holes is also fixed and it determines the interval of variation of γ_L. For B/T = 3 this interval corresponds to the (γ_e=0) low boundary of the ratio I_-/I_+ = 9.1 and to the high (γ_e=1) boundary I_-/I_+ = 15.3. Within this interval, the ratio I_-/I_+ is an almost linear function of γ_e. It was established that γ_e increases slightly with decreasing temperature, but saturates at low temperatures (see Fig.1b).

The value of γ_e found from the extrapolation of $\gamma_e(T)$ to $T \to 0$ was taken as the correct one. The magnetic field dependence of the electron spin polarization $\gamma_e(B)$ measured for two different concentrations of 2D-electrons are plotted in Fig.2. It shows pronounced minima and maxima at several FQHE states. Both of the fundamental sequences of fractional states $n/(2n+1)$ (ν = 2/5, 3/7, 4/9) and $n/(2n-1)$ (ν=2/3, 3/5, 4/7) are present in these data. In addition, the series of fractions 5/3, 8/5, 7/5 and 4/3 around $\nu = 3/2$ is clearly visible in $\gamma_e(B)$ shown in the insets to Fig.2. However, it was rather inconvenient to analyze the magnetic field dependence of γ_e at a fixed electron density since, depending on n_s, the features at fractional filling factors sometimes show up as minima or as maxima, or even disappear. Instead, much more clear behavior was detected for a fixed filling factor value, when the dependence $\gamma_e(B)$ was analyzed at a given ν.

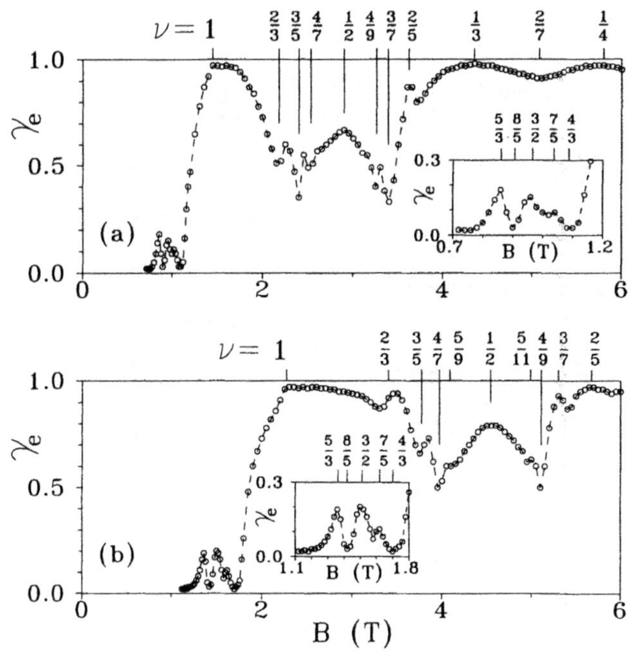

Figure 2. Magnetic field dependence of the electron spin polarization γ_e measured for 2 different 2D-electron concentrations: (a) $n_S = 3.5 \times 10^{10}$ cm^{-2} and (b) $n_S = 5.5 \times 10^{10}$ cm^{-2}. Insets show the details around $\nu = 3/2$.

In Fig.3a we plot the magnetic field dependence of the electron spin polarization measured for several FQHE states from the sequence $\nu = n/(2n-1)$. In order to obtain it, we investigated 6 different samples with top gates, which has allowed us to vary the electron concentration, so that a rather broad interval of magnetic fields was accessible for each fraction. Data measured for different samples are shown by different symbols (for $\nu = 2/3$) and illustrate the consistency of the obtained results. Well defined spin transitions between various differently spin-polarized ground states, governed by the magnetic field are clearly visible from Fig.3a for all studied fractions. For ν=2/3 only one transition (if one neglects the weakly developed feature around B=2.3T) from spin unpolarized to completely spin polarized state was observed at $B \approx 2T$. In contrast, the magnetic field dependence of γ_e obtained for other FQHE states, such as 3/5 and 4/7, definitely indicates a broad region of stability of the partly polarized spin states (γ_e is close to $1/3$ for ν=3/5 and it is about $1/2$ for ν=4/7), and additional transitions between unpolarized, partially polarized and completely spin polarized ground states were detected (for ν=4/7 such transitions take place at $B \approx 1.8T$ and $B \approx 4.8T$). Interestingly, a similar sequence of spin transitions with nearly identical absolute values of γ_e was observed also for FQHE states from the other sequence,

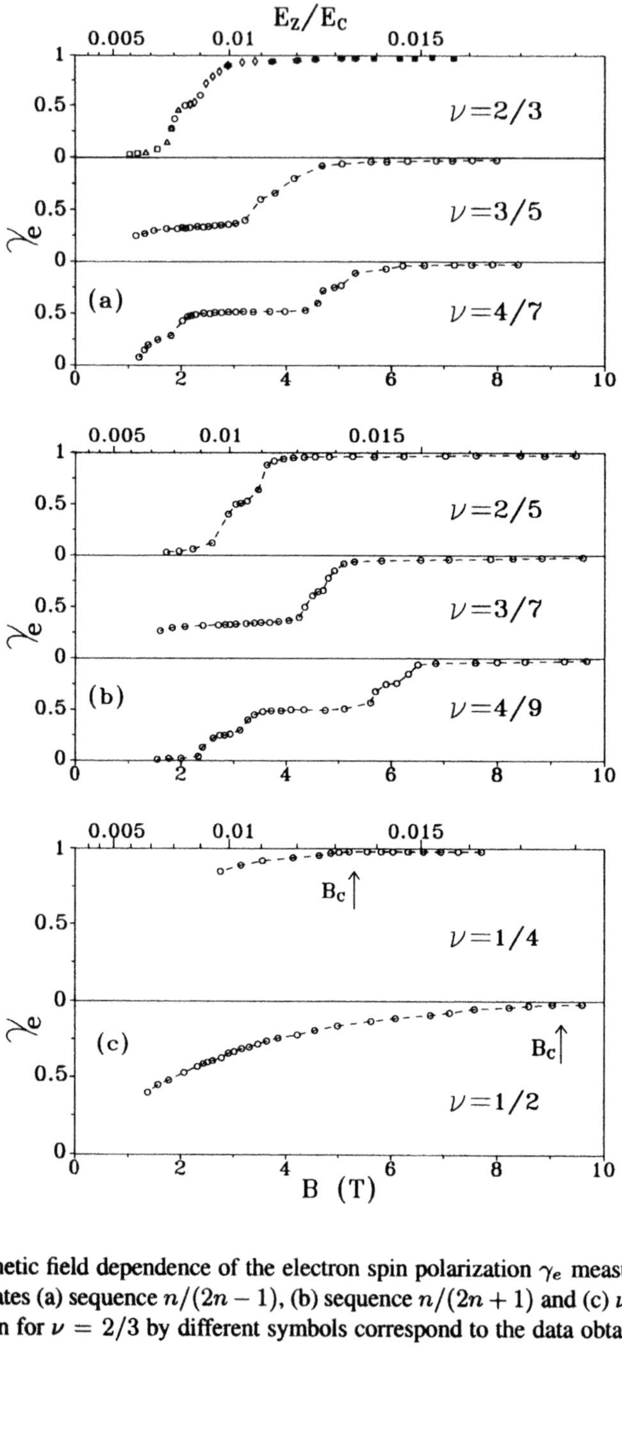

Figure 3. Magnetic field dependence of the electron spin polarization γ_e measured for different FQHE and CF states (a) sequence $n/(2n-1)$, (b) sequence $n/(2n+1)$ and (c) $\nu = 1/2$ and $1/4$. The results shown for $\nu = 2/3$ by different symbols correspond to the data obtained for different gated samples.

$\nu = n/(2n+1)$ (see fig.3b). The correspondence between 2/3, 3/5 and 3/7 states from one side and 2/5, 3/7 and 4/9 states from the other side is apparent from a comparison of figures 3a and 3b. Such a similarity is consistent with the model of Composite Fermions.

Fig.3c shows the magnetic field dependence of the electron spin polarization measured for the CF state at $\nu = 1/2$. First of all, it is clear from the measured dependence that the system of CFs becomes fully spin polarized at $B_c = 9.3T \pm 0.5T$. It means that at $B = B_c$ the CF Fermi energy is equal to the Zeeman splitting, which is equal to $2.8K$ at $9.3T$ (the CF g-factor is the same as that of electrons $g_{CF} = g_e = -0.44$). Note that the Zeeman energy of the CFs can be enhanced by interaction between them. However, this effect is expected for odd CF filling factors (for example, 1/3 FQHE state), but not for the $\nu = 1/2$ state. If we assume a parabolic dispersion law for the CFs at $\nu = 1/2$, then the value of their density of state mass m_{DS} (or 'polarization' mass [20]) can be estimated from the values of their E_F^{CF} and density ($n_{CF} = n_S = 1.26 \times 10^{11}$ cm^{-2} for $B = 9.3$T and $\nu = 1/2$), which yields $m_{DS} = 2.27m_0$ (m_0 is the free electron mass in vacuum). The obtained mass is several times heavier than the previously reported values, measured from the activation energies. However, these masses have different a meaning [20]. Indeed, the filling factor dependence of the ground state energy has a cusp at fractional ν and the activation energy is sensitive to the strength of this cusp (slopes), but not to the absolute value of the minimum, corresponding to the ground state energy. In contrast, the competition between differently spin-polarized ground states, driven by the Zeeman energy, is sensitive only to the energy of this minimum and, therefore, the density of states mass describes the energy splitting between the states with different γ_e. Since there is no energy gap (cusp) at $\nu = 1/2$ the only relevant CF parameters are their Fermi energy and the density of state mass. It is important that if one assumed that CFs are 2D-particles with a parabolic dispersion law (at effective magnetic field $B = 0$ at $\nu = 1/2$), then the magnetic field dependence of γ_e should be a linear function of the parameter E_Z/E_F ($\gamma_e = E_Z/2E_F$), so that $\gamma_e \propto B^{1/2}$ (due to $E_F \propto n_s \propto B$ and the expected increase in the CF mass, as $\sqrt{B}$ [20]). The dependence $\gamma_e(\sqrt{B})$ measured for $\nu = 1/2$ was found to be a slightly sub-linear function of $\sqrt{B}$, which is mainly due to the temperature smearing of the Fermi surface. The dependence of $\gamma_e(B)$ measured for $\nu = 1/4$ is shown in Fig.3c. From the value of $B_c = 5.2T \pm 0.2T$ we derived that for $\nu = 1/4$ and $n_S = 0.32 \times 10^{11}$ cm^{-2}, $m_{DS} = 1.13m_0$. Note that the values of the CF mass obtained for $\nu = 1/2$ and $1/4$ are simply $2\nu m_0/g_e$, which follows from equating the Zeeman and Fermi energies.

In Fig.4a we plot the magnetic field dependence of $\gamma_e(B)$ measured for several FQHE states from the sequences $2 - [n/(2n \pm 1)]$ (ν=8/5, 7/5 and 4/3) and also for $\nu = 3/2$ (the dashed line $\gamma_e = 1/3$ shown on this figure corresponds to the complete spin polarization of the CFs at $\nu = 3/2$). It is clear from these

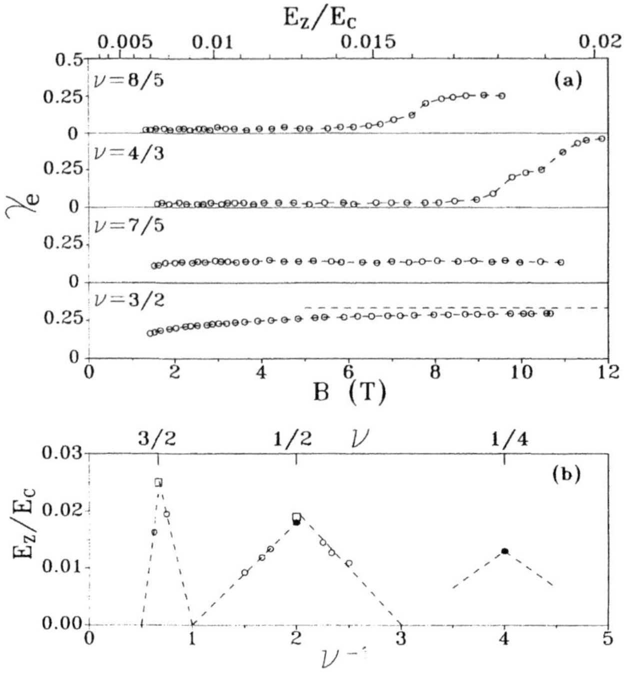

Figure 4. (a) Magnetic field dependence of the electron spin polarization γ_e measured for the $\nu = 8/5, 4/3$ and $7/5$ FQHE states and for the $\nu = 3/2$ CF state. (b) Phase diagram $E_Z/E_C(\nu^{-1})$ illustrating at which magnetic fields the spin-transition into fully polarized states takes place for different fractions (open circles) and also for various CF states ν=3/2, 1/2 and 1/4 (filled circles). Other details are explained in the text.

figures that various spin transitions are also available for these FQHE states, however, in contrast to the fractional states at $\nu < 1$, these transitions are shifted to much higher magnetic fields. Such a breaking of the electron-hole symmetry is in agreement with previously reported transport data [5, 17]. The analysis of the dependence $\gamma_e(B)$ measured for $\nu = 3/2$ shows a very strong sub-linearity of the dependence $\gamma_e(\sqrt{B})$, indicating a considerable nonparabolicity of the CFs dispersion at $\nu = 3/2$ (in contrast to the ν=1/2 state, we find that the CF dispersion law for the $\nu = 3/2$ state is not parabolic, but more close to a linear dependence). In Fig.4b we plot the phase diagram. It illustrates at which magnetic fields (and at what value of $E_Z/E_C \propto \sqrt{B}$) the spin-transition to a fully polarized state takes place for different fractions (open circles) and also for various CF states ν= 1/2 and 1/4 (filled circles). The importance of the parameter E_Z/E_C for the considered problem was already mentioned (also demonstrated in recent theory [20]) and it explains the choice of the coordinates in Fig.4b. The dashed lines on this figure correspond to linear dependence plotted through the experimental points and also through the points at ν = 1, 1/3 and 1/5 since at these filling factors the 2D-electron system is expected to be fully spin polarized even for vanishing Zeeman energy. Empty squares correspond to the extrapolation of these lines to $\nu = 1/2$ and 3/2 (note that for $\nu = 1/2$ the result obtained from the extrapolation is in a good agreement with the directly measured value). The threshold value of E_Z/E_C obtained for $\nu = 3/2$ from the extrapolation procedure corresponds to $B_c = 17.5T \pm 0.5T$, which yields $m_{CF} = 2.27m_0$ at $n_S = 6.3 \times 10^{11}$ cm^{-2} (for $\nu = 3/2, n_{CF} = n_s/3 = 2.1 \times 10^{11}$ cm^{-2}).

3. Spin-wave excitations in the FQHE states and interaction between composite fermions

The attractiveness of the CF-picture is based on the assertion that dressing the electrons with two flux quanta constitutes the main effect of the interaction between the 2D-electrons. The remaining interaction between CFs is weak, so that in many instances the system can be considered as a nearly ideal Fermi-gas of composite particles [1]. The residual interaction between CFs does however exist and plays an essential role in the dispersion of the neutral excitations at FQHE states [21]. No interaction between CFs would imply no dispersion of the neutral excitations, yet it is well known [21, 23] that the dispersion of the CF-exciton is a rather complicated function of the excitonic momentum. Two types of intra-Landau-level neutral excitations for FQHE states below $\nu = 1$ have been recognized. They are neutral charge density (CD) excitations and neutral spin density (SD) excitations, associated with changes of the charge and spin degrees of freedom, respectively. The collective CD-mode has a finite gap at zero wave vector ($k = 0$) and displays a characteristic "magnetoroton" minimum at the inverse magnetic length, $k = 1/l_B$. Its energy approaches in the limit of large k the FQHE energy gap, i.e. the energy to create infinitely separated quasiparticle-quasielectron pairs [21, 24]. The other branch of collective excitations, the SD-mode, takes on the Zeeman energy at zero wave-vector (due to Larmor's theorem) and increases monotonically as a function of momentum until it approaches the exchange energy gap in the limit of infinite wave-vector [22, 23]. Theoretically calculated dispersion laws for the CFs indicate that the interaction between CFs is indeed about an order of magnitude weaker than the inter-electron interaction [21, 22, 24].

In order to measure the interaction energy directly, one thus needs to study the dispersion law of the CD-excitations or, alternatively, measure the spin-wave energy gap at an infinite momentum. The number of experimental possibilities are very limited. Some results were obtained from inelastic light scattering [25, 26] and NMR [27] investigations, however complementary measurements using alternative methods are highly desirable. We have investigated the temperature dependence of the degree of electron spin polarization in the 1/2-CF state. This has allowed us to determine the exchange interaction. Analogue experiments at the 1/3- and 2/3-FQHE-states permit the extraction of the spin wave energy gap. Their results equally illustrate the importance of exchange interaction between CFs.

To this end, we studied several low-density ($n_s = 0.2 - 1.5 \times 10^{11}$ cm^{-2}) and high-quality (electron mobility $\mu = 0.9 - 4 \times 10^6$ cm^2/Vs) GaAs/Al$_x$Ga$_{1-x}$As single heterojunctions with a δ-doped layer of Be-acceptors ($n_A = 2 \times 10^9$ cm^{-2}) located 30 nm from the heterointerface in the wide ($1\mu m$) GaAs buffer layer [28]. In all samples the 2D-electron concentration could be varied by a top gate. Moreover a substrate bias voltage allowed the efficient tuning of the electric field at the heterointerface and thus to modify the width and intersubband splitting of the 2D-channel. For photoexcitation, we used pulses from a tunable Ti-sapphire laser (the wavelength was close to 780 nm) with a duration of 20 ns, a peak power of 10^{-4} -10^{-2} W/cm^2, and a repetition rate of $10^4 - 10^6$Hz. Luminescence spectra were recorded by a gatable photon counting system with a spectral resolution of 0.03 meV. We have measured the degree of the electron spin polarization from an analysis of the degree of circular polarization of the time-resolved radiative recombination of 2D-electrons with photoexcited holes bound to the Be-acceptors. The crux of this method is the full control over the photoexcited hole contribution to the circular polarization of the luminescence. This was achieved by the time-resolved technique to ensure the complete relaxation of the hole system down to the bath temperature [28]. To analyze the circular polarization of the luminescence signal at low temperatures (down to 100 mK), an optical fiber system was used with a quarter wave plate and a linear polarizer placed in liquid helium just nearby the sample. This setup allows the detection of σ^-/σ^+ signal ratios up to 1000 (corresponding to a depolarization coefficient of 0.002). It guarantees a high accuracy (of about 1 percent) measurement of the electron spin polarization. The magnetoresistance was measured with the standard low-frequency (12 Hz), low-current lock-in technique. Temperatures were recorded with a calibrated RuO_2 thermometer. In some special cases, in order to measure the degree of electron spin polarization at very low temperatures more accurately, the temperature of the photoexcited holes (T_h) was effectively increased by a decrease of the delay time. In these cases the calibration of T_h was performed with the use of a completely occupied Landau level below the Fermi surface at fixed magnetic field and bath temperature as described in Ref. [28].

Fig.5a shows the temperature dependence of the electron spin polarization (γ_e), measured at different magnetic fields but fixed filling factor $\nu = 1/2$. Below and above the critical magnetic field $B_c = 9.3T$ [28], for which the CF-system becomes fully spin polarized, the electron spin polarization saturates at low temperatures. In contrast, at $B = B_c$ a well defined linear dependence of γ_e on T is observed in the low T limit. Very similar behaviors at $B = B_c$, $B > B_c$ and $B < B_c$ were predicted theoretically for non-interacting CFs [29] and it was demonstrated that the linear dependence at low temperatures at $B = B_c$ results

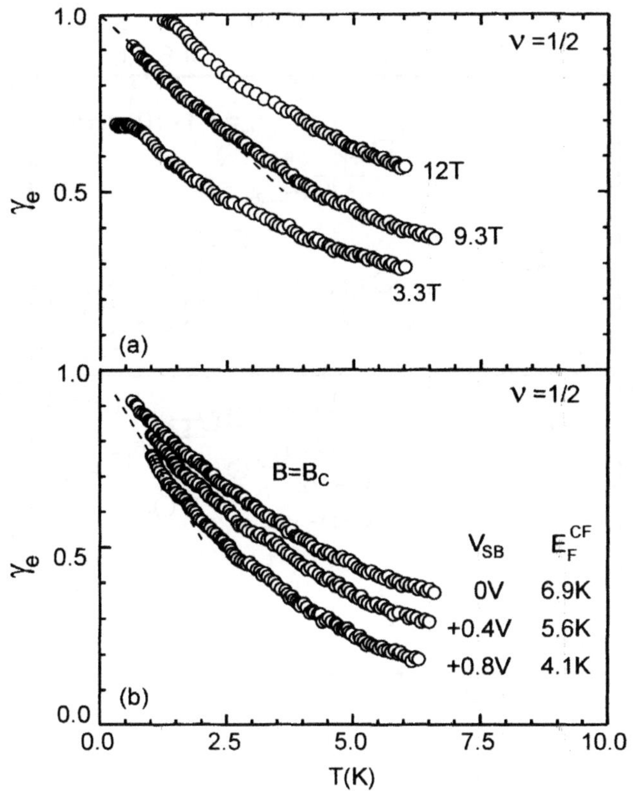

Figure 5. (a) The temperature dependence of the electron spin polarization measured for $\nu = 1/2$ and V_{SB}=0V for different magnetic fields. The dashed line corresponds to the linear extrapolation of the experimental data taken at $B = B_c = 9.3$ T. From its slope the Fermi energy of CF metal was determined. (b) Influence of the substrate-bias voltage on the temperature dependence of the electron spin polarization at $\nu = 1/2$ and $B = B_c$.

from the Fermi statistics (this linear term is expected to be stable for the case of weakly interacting particles). The slope of this dependence is determined by a single parameter, the Fermi energy of CFs: $\gamma_e(T) = 1-(2T/E_F)\cdot ln((1+\sqrt{5})/2)$. The linear extrapolation of the low temperature portion of $\gamma_e(T)$ at $B = B_c$ yields a CF Fermi energy of 6.9K. At $B = B_c$ the Fermi energy equals the Zeeman energy, and one must conclude that the spin splitting of CFs is strongly (by a factor of 2.5) enhanced in comparison with the bare Zeeman energy (2.8K at 9.3T).

Interaction phenomena are quite sensitive to the width of the 2D-channel. They are gradually suppressed upon increasing the channel width. In order to vary this width, we applied a substrate-bias voltage (V_{SB}), while maintaining a fixed carrier concentration through a simultaneous change of the top gate bias. The intersubband splitting E_{10} provides a measure for the finite width w of the 2D-

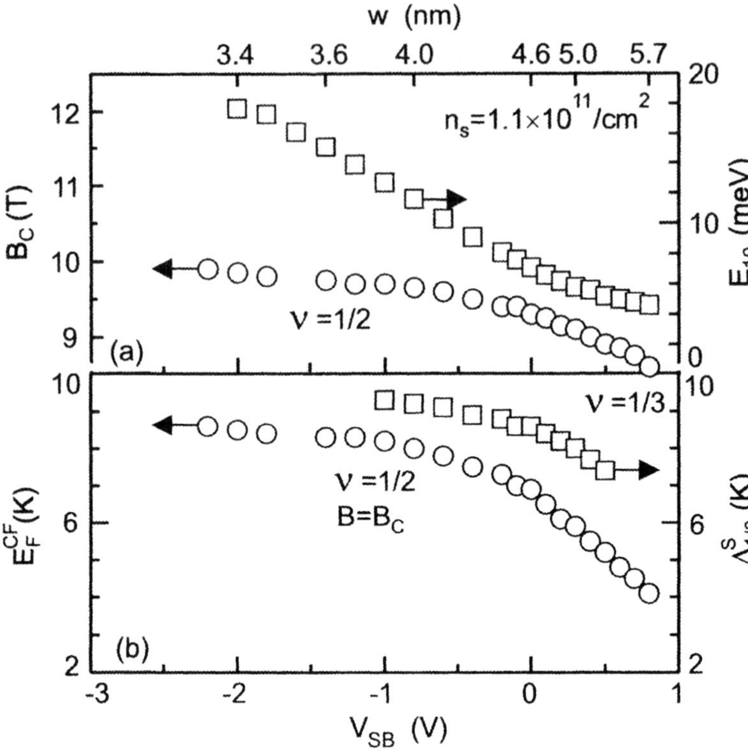

Figure 6. (a) The dependencies of the intersubband energy E_{10} (squares, $n_s = 1.1 \times 10^{11}$ cm^{-2}) and the critical magnetic field B_c (circles, $\nu = 1/2$)) on the substrate-bias voltage V_{SB}. On the top axis the width of wavefunction w, a obtained from a comparison of numerical simulations with the Fang-Howard function is indicated for several values of V_{SB}. (b) The enhanced spin splitting of CFs (circles, $\nu = 1/2$, $B = B_c$) and of the spin-flip gap at $\nu = 1/3$ (squares) as a function of the substrate-bias voltage V_{SB}.

channel and can be obtained directly from luminescence spectra at small delays after the laser pulse, when recombination from both subbands is observable. Alternatively, numerical simulations in the Hartree approximation of the eigenvalues and wave-functions of the ground and first excited subband can serve this purpose. The effective width w can be estimated from a fit of the calculated wave-function to the Fang-Howard function $\psi(z) = z \cdot exp(-bz/2)$, with $w = 1/b$ [30]. Fig. 5(b) depicts the temperature dependence of the electron spin polarization at $B = B_c$ for different V_{SB}. Fig. 6(a) shows the intersubband splitting and the critical magnetic field B_c as a function V_{SB} at a fixed density of $n_s = 1.1 \times 10^{11}$ cm^{-2}. The latter depends only weakly on the substrate bias. On the top axis calculated values for the channel width are indicated for several bias voltages. Fig. 6(b) plots the Zeeman splitting (and Fermi energy) as extracted from the temperature dependent data. Positive (negative) substrate bias voltages correspond to a decrease

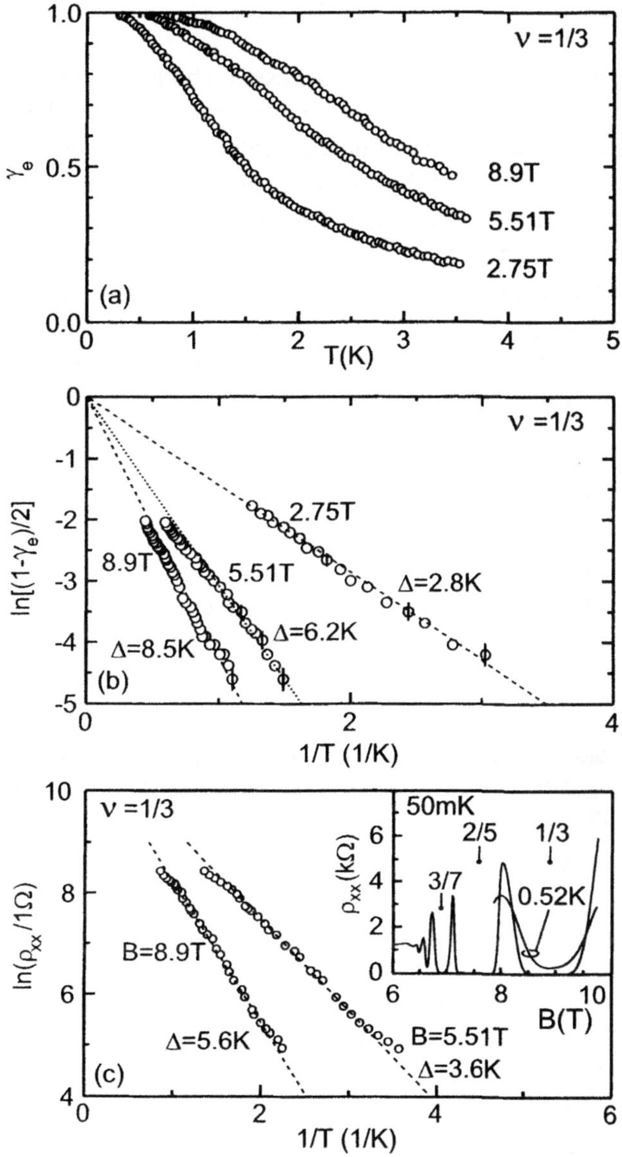

Figure 7. The temperature dependence of the electron spin polarization (a) and Ahrenius plots (b) $ln((1-\gamma_e)/2)$ vs 1/T measured at $\nu = 1/3$ and $V_{SB} = 0$ V for different magnetic fields. (c) Activation behaviour of ρ_{xx}, measured at $B = 5.51$ T and $B = 8.9$ T. The inset depicts the magnetic field dependence of ρ_{xx} near $\nu = 1/3$ at two different temperatures.

(increase) of E_{10} and to an increase (decrease) of the channel width. The enhanced spin splitting of CFs is considerably suppressed for positive V_{SB}, which means that the exchange interaction between CFs drastically drops for wide channels ($w > 6$ nm).

Since the concept of CFs is quite successful (both in numerical calculations and in experiment) in explaining many properties of the FQHE-states, we can proceed in a similar manner and estimate the interaction energy between CFs from the temperature dependence of the electron spin polarization at the 1/3- and 2/3-FQHE states. The analysis of our magneto-optical measurements is very similar to that used for activated magnetotransport investigations. Even if so, the determined energy gap is quite different since it is selectively sensitive to the spin degree of freedom. It therefore provides information about the intra-CF-Landau-level SD-excitation (or CF-spin-waves, CFSW), whereas magnetotransport data delivers the inter-CF-Landau-level CD-excitation gap (or CF-magnetoplasmon, CFMP).

In Fig. 7(a) we show the temperature dependence of γ_e, measured at different magnetic fields, for the $\nu = 1/3$ FQHE state. The presence of a gap in the CFSW mode makes an Ahrenius-type of analysis as shown in Fig. 7(b) reasonable. At low temperatures, the deviation of γ_e from 1 as a function of temperature is well described by a single exponential dependence: $\gamma_e = 1 - 2 \times exp(-\Delta/2k_BT)$. Such a dependence is expected theoretically as described in Ref. [31]. Two possible values for Δ were discussed in this article, because of the presence of two different gaps: Zeeman gap (Δ_Z) and CFSW-gap ($\Delta_{1/3}^{SW}$). In the former case the activated gap Δ equals $2 \cdot \Delta_Z$ ([31]), whereas for the latter case Δ equals $\Delta_{1/3}^{SW}$. The infinite density of states associated with spin wave transitions at large wave-vectors makes it plausible that $\Delta_{1/3}^{SW}$ determines the activated behavior. Indeed, the measured values of Δ are considerably larger than $2 \cdot \Delta_Z$ and exhibit a nonlinear magnetic field dependence. A comparison of the measured CFSW-gap with gaps derived from activated transport under the same conditions revealed that the gap derived from $\gamma_e(T)$ is systematically larger than the transport gaps as shown in Fig. 7(b) and 7(c). We therefore can conclude that in transport a different, smaller, gap, the CFMP gap, is measured. The dependence of the CFSW-gap measured for the 1/3-FQHE state on V_{SB} (and thus the channel width) is illustrated in Fig. 6(b). The interaction energies between CFs, measured for $\nu = 1/2$ and $\nu = 1/3$, as well as their dependence on the channel width are in good agreement with each other.

The $\nu = 2/3$ FQHE state is known for its spin transition at $B = B_{2/3}^C = 2.1T$ [28] from an unpolarized to a completely spin polarized ground state. Because of the close relationship between the exchange interaction energy of CFs and the spin polarization of the system, the spin transition at $B = B_{2/3}^C$ may serve as an excellent illustration of the CF-interaction. The temperature dependence of the electron spin polarization at $\nu = 2/3$ for different magnetic fields both below and above $B = B_{2/3}^C$ is shown in Fig. 8(a). The behavior of $\gamma_e(T)$ at low temperatures

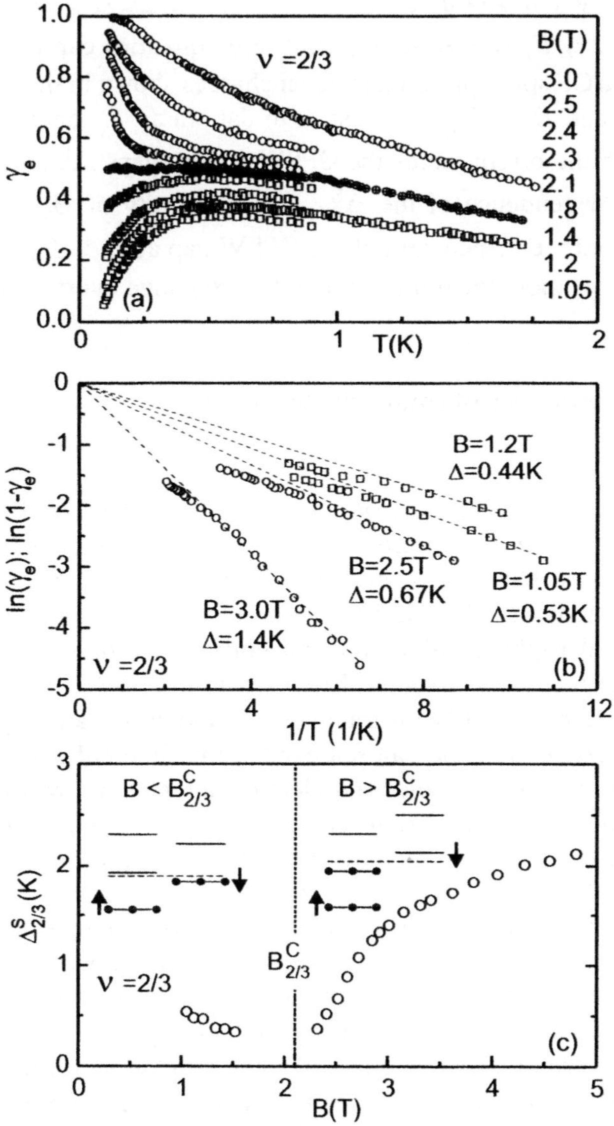

Figure 8. (a) The temperature dependence of the electron spin polarization at $\nu = 2/3$ and V_{SB}=0V for different magnetic field values in the vicinity of the spin-transition $B^{C}_{2/3} = 2.1T$. (b) Ahrenius plots $ln(\gamma_e)$) vs 1/T (for $B < B^{c}_{2/3}$, squares) and $ln(1-\gamma_e)$ vs 1/T (for $B > B^{c}_{2/3}$, circles) at several magnetic fields. (c) Magnetic field dependence of the spin-flip activation energy for $\nu = 2/3$ around $B^{C}_{2/3}$. In the inset the CF spin-split Landau level diagrams are presented.

is qualitatively different for $B > B^{C}_{2/3}$ (circles) and $B < B^{C}_{2/3}$ (squares), even though the temperature dependence in both cases is well described by a single exponential at low temperatures: $\gamma_e = 1 - exp(-\Delta/2k_BT)$ (for $B > B^{C}_{2/3}$) and $\gamma_e = exp(-\Delta/2k_BT)$ (for $B < B^{C}_{2/3}$). These equations can be easily obtained with the aid of a CF spin-split Landau level chart as shown in the inset to Fig. 8(c) for $B > B^{C}_{2/3}$ and $B < B^{C}_{2/3}$. A best fit to data in an Ahrenius plot with a line passing through the origin yields the CFSW energy gaps $\Delta^{S}_{2/3}$ of Fig. 4(b). The magnetic field dependence of the $\Delta^{S}_{2/3}$ in the vicinity of $B^{C}_{2/3}$ is presented in Fig. 8(c). An abrupt enhancement of the CFSW gap at $B > B^{C}_{2/3}$ is obvious and demonstrates once more the importance of the exchange interaction between CFs.

4. Cyclotron resonance of composite fermions

Since the kinetic energy of electrons is entirely quenched in the course of applying a B-field, the CF cyclotron mass is not a renormalized version of the electron conduction band mass, but must be generated entirely from electron-electron interactions [32–35, 3]. The search of the CF cyclotron resonance requires substantial sophistication over conventional methods used to detect the electron cyclotron resonance, since Kohn's theorem [36] must be outwitted. It states that in a translationally invariant system radiation can only couple to the center-of-mass coordinate and can not excite other internal degrees of freedom. Phenomena originating from electron-electron interactions will thus not be reflected in the absorption spectrum. An elegant way to bypass this theorem is to impose a periodic density modulation to break translational invariance. The non-zero wavevectors defined by the appropriately chosen modulation may then offer access to the cyclotron transitions of CFs, even though they are likely to remain very weak. Therefore, the development of an optical detection scheme, that boosts the sensitivity to resonant microwave absorption by up to two orders of magnitude in comparison with traditional techniques, was a prerequisite for our studies. Furthermore, we exploited to our benefit the accidental discovery that microwaves, already incident on the sample, set up a periodic modulation through the excitation of surface acoustic waves (SAW).

The 2DES, patterned into a disk with 1 mm diameter [37], is placed near the end of a 16- or 8-mm short-circuited waveguide in the electric field maximum of the microwave excitation inside a He3-cryostat. At a fixed B-field, luminescence spectra with and without microwave excitation were recorded consecutively. The differential luminescence spectrum is obtained when subtracting both these spectra. To improve signal-to-noise ratio, the same procedure was repeated N times (N = 2 - 20). Subsequently, we integrated the absolute value of the

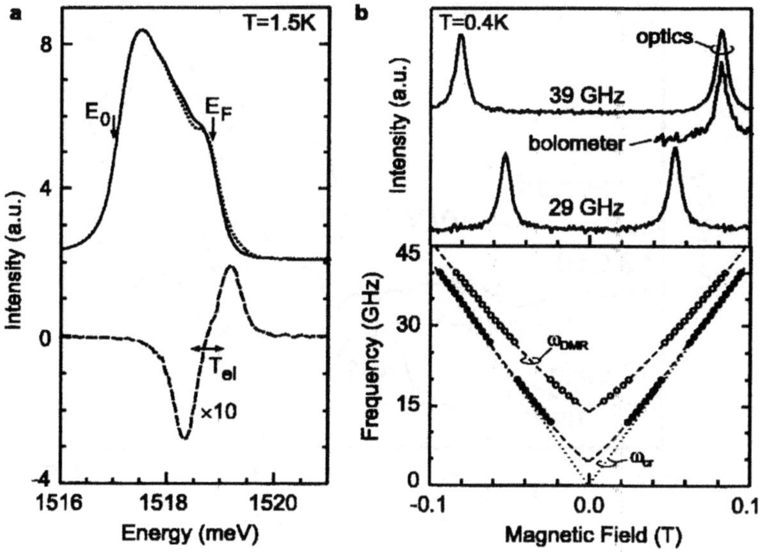

Figure 9. (a) Luminescence spectrum in the presence of (dotted line) and without (solid line) a 0.05 mW microwave excitation of 18 GHz obtained on a disk-shaped 2DES with a diameter of 1 mm and carrier density $n_S = 5.8 \times 10^{10}/cm^2$ at a magnetic field B = 22 mT. (b) Top panel: The microwave absorption amplitude at 29 GHz and 39 GHz as a function of B-field by recording differential luminescence spectra as in a for 1 mT field increments at $n_S = 1.09 \times 10^{11}/cm^2$. The inset shows a conventional bolometer measurement. Bottom panel: resonance position for $n_S = 1.09 \times 10^{11}$ (open circles) and $1.1 \times 10^{10}/cm^2$ (solid circles) as a function of incident microwave frequency. The dashed lines represents the theoretical dependence of the hybrid dimensional magnetoplasma-cyclotron resonance. The dotted line corresponds to the cyclotron mode only.

averaged differential spectrum over the entire spectral range and hereafter refer to the value of this integral as the microwave absorption amplitude. The same procedure is then repeated for different values of B. To establish trustworthiness in this unconventional scheme, we apply it in Fig.9 to the well-known case of the electron cyclotron resonance $\omega_{cr} = eB/m^*$, with m^* the effective mass of GaAs (0.067 m_0). Due to its limited size, the sample also supports a dimensional plasma mode with a frequency ω_p, that depends on both the density n_S and diameter d of the sample, according to $\omega_p^2 = 3\ \pi^2 n_S e^2/(2m^*\epsilon_{eff}d)$. The plasma and cyclotron mode hybridize and the resulting resonance frequency of the upper dimensional magnetoplasma-cyclotron mode ω_{DMR} equals $\omega_{cr}/2 + [\omega_p^2 + (\omega_{cr}/2)^2]^{1/2}$ [37, 38].

The optical method indeed recovers this mode. A comparison with the theoretical expression for ω_{DMR} yields excellent agreement. No fitting is required, since the density can be independently extracted from the luminescence at higher B-fields, where Landau levels can be resolved. At sufficiently low density, the

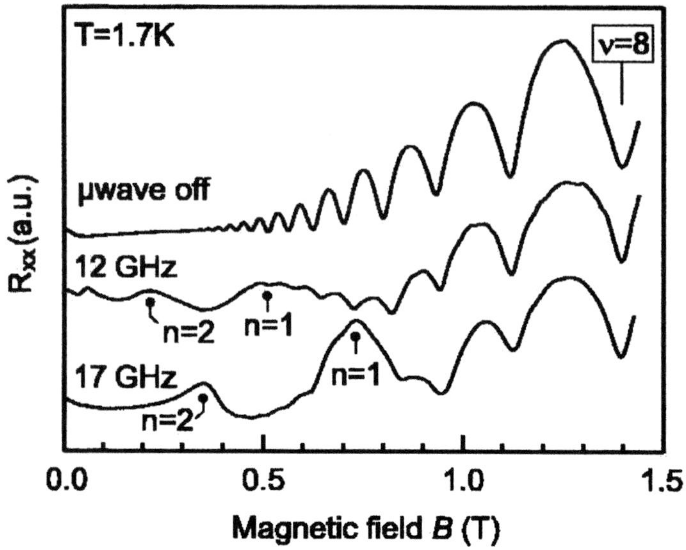

Figure 10. Magnetotransport data without (top curve) and under 0.1mW of microwave radiation at 12 GHz (middle curve) and 17 GHz (bottom curve). Curves are offset for clarity. Besides the well-known Shubnikov-de Haas oscillations, additional magnetoresistance oscillations appear under microwave radiation.

influence of ω_p on the hybrid mode drops and one recovers at large enough B the anticipated $\omega_{cr} = eB/m^*$-dependence. Further details of the electron cyclotron resonance are discussed elsewhere [37]. Additional support for the validity of the detection method comes from a comparison with measurements based on the conventional approach using a bolometer (inset Fig.9b). Not only does one find the same resonance position, but also the same line shape. The only difference is the improved signal to noise ratio (30-100 times) for the optical detection scheme, that enables to observe the electron cyclotron resonance at microwave levels below 10 nW.

Disorder and the finite dimensions of the sample in principle suffice to break translational invariance as attested by the interaction of the cyclotron and dimensional plasma mode. However, they provide access to internal degrees of freedom other than the center-of-mass motion of the electrons either at poorly defined wavevectors or too small a wavevector for appropriate sample sizes. Therefore, the imposition of an additional periodic density modulation, that introduces larger and well-defined wavevectors to circumvent Kohn's theorem, is desirable. Transport experiments in the Hall bar geometry disclosed that additional processing is not required, since the microwaves, already incident on the sample, concomitantly induces a periodic modulation at sufficiently high power. A clear signature is the appearance of commensurability oscillations in the magneto-resistance due

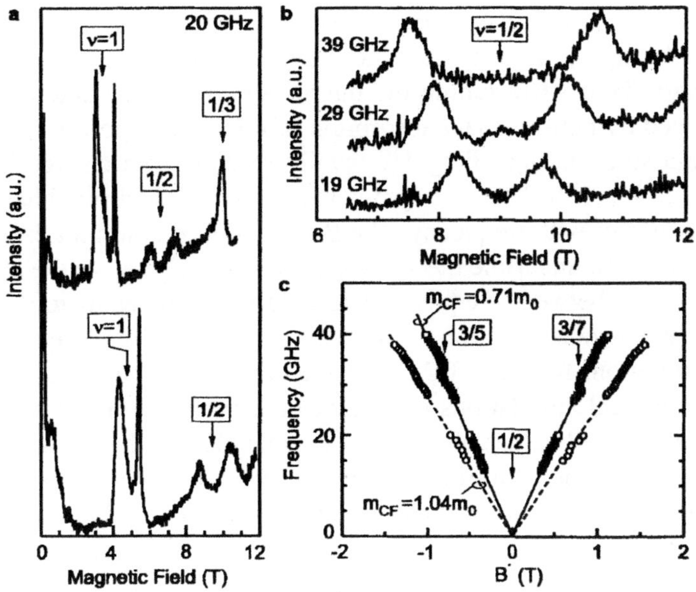

Figure 11. (a) Microwave absorption amplitude measured at high magnetic fields for $n_S = 0.81 \times 10^{11}$ and $1.15 \times 10^{11}/cm^2$ and frequency of 20 GHz ($T = 0.4K$). (b) Microwave absorption amplitude in the vicinity of $\nu = 1/2$ measured at three different frequencies and $n_S = 1.09 \times 10^{11}/cm^2$. (c) Position of the CF cyclotron mode as a function of the effective magnetic field B^* measured for $n_S = 1.09 \times 10^{11}/cm^2$ (circles) and $n_S = 0.59 \times 10^{11}/cm^2$ (squares). The CF effective mass equals 1.04 m_0 and 0.71 m_0 respectively.

to the interplay between the B-dependent cyclotron radius of electrons and the length scale imposed by the modulation. Examples are displayed in Fig.10 and resemble the data in Ref. 15, where the modulation is produced with the help of SAW-transducers. Here, the following scenario is conceivable. Owing to the piezoelectric properties of the AlGaAs-crystal, the radiation is partly transformed into SAW with opposite momentum, so that both energy and momentum are conserved. Reflection from cleaved boundaries of the sample then produces a standing wave with a periodicity determined by the sound wavelength. Photoexcitation creates a very poorly conducting parallel 3D-layer in the Si-doped portion of the AlGaAs-barrier and may enhance the influence of the standing acoustic waves. Carriers are collected in the nodes and affect the local density of the 2DES. The involvement of sound waves can be deduced from transport data, since from the minima we expect the modulation period to be approximately 200 and 250 nm for frequencies of 17 and 12 GHz respectively. The ratio of this period to the sound wavelength at these frequencies is 1.12 and 1.15.

Fig.11a depicts the microwave-absorption amplitude up to high B-fields. Apart from the strong dimensional magnetoplasma-cyclotron resonance signal at low

B-field discussed above, several peaks, that scale with a variation of the density, emerge near filling 1, 1/2 and 1/3. Those peak positions associated with ν = 1 and 1/3 remain fixed when tuning the microwave frequency and are ascribed to heating induced by non-resonant absorption of microwave power. In contrast, the weak maxima surrounding filling 1/2 readily respond to a change in frequency as illustrated in Fig.11b. They are symmetrically arranged around half filling and their splitting grows with frequency. The B-dependence is summarized in Fig.11c for two densities. To underline the symmetry, B^* was chosen as the abscissa. The linear relationship between frequency and field extrapolates to zero at vanishing B^*. We do not expect a deviation at small B^* due to a plasma-like contribution as in Fig.9c. Excitations for the 1/3, 2/5, 3/7 and other fractional quantum Hall states exhibit in numerical simulations no magnetoplasmon-like linear contribution to the dispersion at small values of k [35]. We conclude that the resonance in Fig.11 is the long searched for cyclotron resonance of CFs. Geometric resonances (GR), as they occur in transport at low fields due to the density modulation (Fig.10), are excluded as an alternative interpretation for the observed features on the following grounds: A. In the optical data, only the electron cyclotron resonance peak is observed. Contrary to optical quantities, transport is also sensitive to semi-classical phenomena unrelated to changes in the density of states. B. Even if the 2DES condenses in a FQHE-state and the chemical potential is located in a gap, the resonance peaks surrounding ν = 1/2 occur (Fig.11c). Commensurability effects are not observable in this regime. C. The observation of GRs requires that the density modulation is temporally static on the time scale with which CFs execute their cyclotron orbit. For electrons at low fields this condition is met and accordingly transport displays GRs. For the anticipated enhanced mass of CFs, this condition is violated. D. Analogous resonance peaks were also detected for the higher order CFs around ν = 1/4. Since this CF metallic state is characterized by the same Fermi wavevector GRs would show up at the same distance from ν = 1/4 as they do at ν = 1/2. The observed peaks are located at different positions rendering a commensurability picture untenable (see below Fig.12b).

In contrast to electron cyclotron resonance, the intensity of the CF cyclotron resonance is a strong non-linear function of microwave-power (Fig.12a). Moreover, its observability only at high power correlates with the first appearance of commensurability oscillations. The drop in intensity at even higher power is most probably due to heating. The intensity diminishes to zero at temperatures above 0.7 K, whereas the electron cyclotron resonance persists up to T = 2 K. The slope of the CF cyclotron frequency as a function of B^* in Fig.11c defines the cyclotron mass m_{cr}^{cf}. This mass is set by the electron-electron interaction scale, so that a square root behaviour on density or B-field is forecasted from a straightforward dimensional analysis [3]. Numerical calculations predict $m_{cr}^{cf}/m_0 = 0.079 \times sqrt(B[T])$ for an ideal 2DES, not including Landau level mixing or finite width contributions [20]. The data, shown in Fig.12b, confirm qualitatively the strong

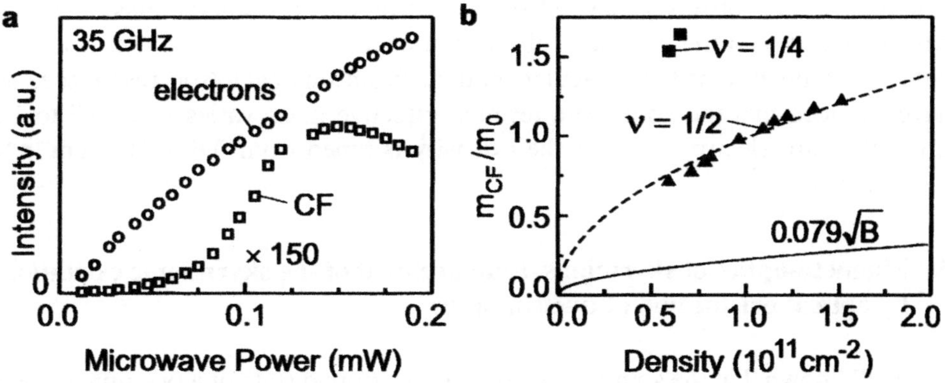

Figure 12. (a) Incident microwave power dependence of the amplitude at the cyclotron resonance (circles for electrons, squares for CFs). (b) Dependence of the CF effective mass near $\nu = 1/2$ on the carrier density n_S (solid triangles). The dashed line is a square root fit to the data. The solid curve is the prediction from theory reported in Ref. 17. Analogous resonance peaks have also been detected around $\nu = 1/4$. The corresponding effective mass values are indicated as solid squares for two different densities.

enhancement in comparison with the electron mass (more than 10 times), however a fit to the square root dependence requires a prefactor that is four times larger. Previously reported mass values based on activation energy gap measurements [5, 41] must be distinguished from the cyclotron mass. The former corresponds to the limit of infinite momentum, whereas here k approaches zero. Moreover, activation gaps can only be extracted at well-developed fractional quantum Hall states and their accurate determination suffers from disorder induced broadening. These and other limitations have been discussed in Ref. 19 for example. The technique discussed here can be performed at arbitrary filling factors.

In summary, the introduction of suitable fictitious entities occasionally allows one to map otherwise difficult strongly interacting many-body systems into a single particle form. One can then take the customary physical approach, using concepts and representations which formerly could only be applied to systems with weak interactions, and yet still capture the essential physics. A most notable recent example refers to the conduction properties of a two dimensional electron system in a strong perpendicular magnetic field B. They are governed by electron-electron interactions, that bring about the fractional quantum Hall effect. Composite fermions, that do not experience the external magnetic field B but a drastically reduced effective magnetic field B^*, were identified as apposite quasi-particles that simplify our understanding of the FQHE. They precess, like electrons, along circular cyclotron orbits, with a diameter determined by B^* rather than B. The frequency of their cyclotron motion remained hitherto enigmatic since their effective mass is no longer related to the band mass of the original electrons and is entirely generated from electron-electron interactions. Here, we have demonstrated the enhanced absorption of a microwave field that resonates with the frequency of their circular motion. From this cyclotron resonance, we have deduced that the composite fermion effective mass varies from 0.7 to 1.2 times the bare electron mass as their density is tuned from 0.6×10^{11} cm^{-2} to 1.2×10^{11} cm^{-2}.

5. Magneto-optics of skyrmions. Enhancement of the skyrmionic excitations by optical orientation of nuclear spins

It is well known that the spin polarization of the ground state of a two dimensional electron system is defined by a competition between the single-particle Zeeman energy (E_Z) and the many-body energy of Coulomb interaction (E_C). This was well established both theoretically [15, 42] and experimentally [17, 18, 19] for the ground states of different Fractional Quantum Hall Effect (FQHE) states in the ultra-quantum limit. The value of the ratio between Zeeman and Coulomb energies E_Z/E_C was especially important for observation of the new collective charged excitations in 2D-electron system, which are called skyrmions (or spin texture excitations) [43, 44]. Skyrmionic excitation describes a macroscopically large distortion of the electronic spins in the vicinity of filling factor $\nu = 1$. According to theory, the ground state of a 2D-electron system is fully spin polarized at $\nu = 1$, however a small deviation from $\nu = 1$ results in a strong spin depolarization of the system. The ground state of the system with one additional (to $\nu = 1$) spin-reversed electron does not correspond to a single flipped spin, but involves a smooth rotation of the spin field from antiparallel to a parallel spin, which is spread over a macroscopic size $K \sim l_B$ (l_B is the magnetic length). Therefore a characteristic feature of skyrmions is a narrow peak in the filling factor dependence of the electron spin polarization in the vicinity of $\nu = 1$. The size of the skyrmion K (measured in units of magnetic length) can be determined from the width (or shape) of this peak. Such dependencies were experimentally observed [45, 46], however some controversy [12] exists due to the results obtained by different methods and further investigations are necessary. It follows from theory, that the size of this spin-texture excitation (skyrmion) is a strong function of the parameter E_Z/E_C and for zero Zeeman energy its size would be infinitely large [43]. Therefore, to enhance skyrmionic excitations and to unambiguously detect them experimentally one would need to use a method of controllable change of Zeeman energy, which allows to suppress it down to zero.

There are several methods to reduce the Zeeman energy of electrons in GaAs [47, 48]. All these methods are based on the influence of the band structure of GaAs on the bare Lande g-factor of electrons. The value of the electronic g-factor in GaAs is $g = -0.44$ [49] and it is defined by the band structure of the material (mainly by the energies of the band gap and of the spin-orbit splitting of the valence band [50]). These energies can be changed by hydrostatic pressure [48] or by quantum confinement [51]. For example, to achieve the limit of $g = 0$, one should apply rather strong hydrostatic pressure of about 20 kbars [48] or to use very narrow quantum wells with a width of 5.5nm [51]. However, a modification of the band structure under the pressure results in a strong decrease of the concentration of 2D-electrons due to corresponding changes of the band-gap and chemical potential level. For example, in our samples with highest concentration of 2D-electrons measured at zero pressure under illumination ($n_s = 6 \times 10^{11}$ cm^{-2}), the electron density decreases 3 times at P=10 kbars and drops to zero at 20 kbars (see also [52]). Such a strong decrease of the electron concentration is accompanied by an even stronger reduction of electron mobility (also several times and more), so that the systems with and without pressure become hardly comparable. Similar problems appear for narrow quantum wells, in which the mobility of 2D-electrons is more than one order of magnitude lower, as compared to conventional single heterojunctions. Since a rather weak disorder is enough to destroy Coulomb correlation effects in the 2D-electron system, the gradual reduction of the quality of a sample is not acceptable to study skyrmionic effects, and an alternative method which would allow us to suppress Zeeman energy without changing the electron density and mobility is necessary.

In this paragraph we consider a method, which is based on compensation of the Zeeman energy by the hyperfine interaction between 2D-electrons and the optically-oriented nuclear spins of Ga and As. A high degree of spin orientation of nuclei is achieved by optical pumping with circularly polarized light. We show that under these conditions the hyperfine energy is comparable with the Zeeman energy (both of about 1-2K in the range of relevant magnetic fields) and therefore an effective compensation, as well as an enhancement, of the Zeeman energy can be obtained. At the same time, we extract the degree of the electron spin polarization from the degree of circular polarization of the light emitted in the recombination between 2D-electrons and photoexcited holes bound to acceptors. Using this technique, we unambiguously demonstrate that a suppression of the Zeeman energy results in a sharp enhancement of the skyrmionic effects, whereas in the opposite case such effects completely disappear. As an *insitu* experimental test, we optically detect nuclear magnetic resonance (NMR), which indicates the importance of the electron-nuclear coupling. Finally, we investigate for the first time a detailed dependence of the skyrmion size as a function of the parameter E_Z/E_C.

The main idea of the optical pumping method developed by Kastler [53] is that optical excitation results in a selective population in one of the degenerate, or nearly degenerate sublevels coupled through an optical transition. The reason for this is the anisotropy of optical excitation, caused by the transversality of the electromagnetic wave. This anisotropy can be enhanced by using for excitation light with linear or circular polarization. For example, in the case of photoexcitation by circularly polarized light, the optical pumping will result in the angular momentum orientation. This is a direct consequence of angular momentum conservation - when polarized photons are absorbed, their momentum is transferred to the medium.

The method of optical pumping is quite universal and for a long time it was used to polarize isolated gas atoms. In solid state physics this method was successfully introduced by Lampel [54]. Optical orientation in semiconductors [55] is based on the fact that in the process of the interband absorption of a photon, an electron in the conduction band and a hole in the valence band are generated, which have the total spin equal to the angular momentum of the absorbed photon. Photons with right (σ^+) or left (σ^-) circular polarization have a projection of the angular momentum on the direction of the wave vector equal to $+1$ or -1, respectively. This angular momentum is distributed between the photoexcited electron and hole in accordance with the selection rules, which are determined by the band structure of the semiconductor. In III-V semiconductors (GaAs type) there are several direct interband optical transitions near the center of the Brillouin zone allowed in the dipole approximation. These are: two dipoles rotating clockwise and counterclockwise in the plane perpendicular to the wave vector k (heavy hole band to conduction band transitions), two dipoles oscillating along the direction of k and two dipoles rotating in the plane perpendicular to k (light hole band to conduction band transitions). It is important that transition matrix elements defining the intensities of the corresponding transitions are different for the optical processes involving heavy holes and light holes. This result coincides with the results obtained by considering the selection rules for optical transitions between two atomic levels with angular momentum $j = 1/2$ and $j = 3/2$. Similar to the atomic physics selection rules, (σ^+) light is emitted by $|+3/2\rangle \rightarrow |+1/2\rangle$ and by $|+1/2\rangle \rightarrow |-1/2\rangle$ transitions, but the intensity of the first transition is 3 times higher than of the second one. Therefore, in the case of excitation by right-hand polarized light, the ratio of the photoexcited electrons with spins $+1/2$ and $-1/2$ is equal to 3, yielding a degree of electron spin polarization of 0.5.

Optical pumping of the nuclear spins is a two-step process [55], which involves in the first step a spin polarization of electrons by circularly polarized light (as described above) and in the second step the electron spin polarization is transferred to the nuclear spin system via the contact hyperfine interaction, orienting the nuclear spins parallel to the optically induced electron spin polarization. Orientation of the nuclear spins results in a static effective nuclear magnetic field (proportional to the degree of nuclear orientation), which in turn acts back on the electron system, changing the Zeeman splitting in the electron spectrum (Overhauser shift). It is important that the energy of the hyperfine interaction, which is usually very small (published maximal value is about 1-2K [55, 56]), may be comparable to a typical electronic Zeeman energy in GaAs (at B=5T E_Z=1.5K). This fact gives an opportunity to control and manipulate the electron

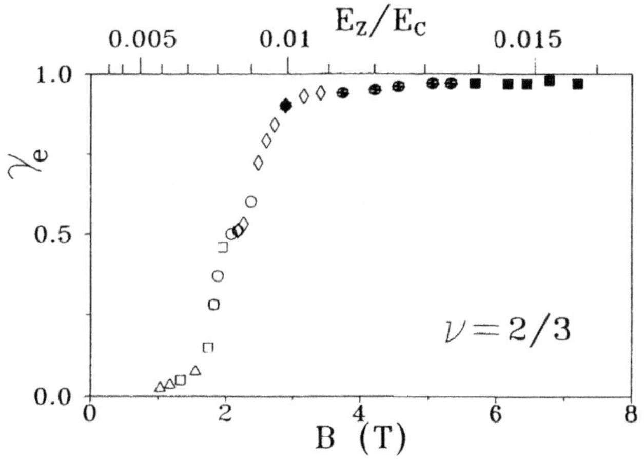

Figure 13. (a) Magnetic field dependence of the spin polarization of 2D-electron system at the $\nu = 2/3$ FQHE state, measured from the analysis of the degree of circular polarization of time-resolved radiative recombination of 2D-electrons with photoexcited holes bound to acceptors. The presented data were obtained from 6 different samples (corresponding data are shown by different symbols) and additional variation of the electron concentration was achieved by use of front gate. For photoexcitation linearly polarized light of very low peak power ($0.1mW/cm^2$) was used. The variation of the parameter E_Z/E_C with magnetic field is also shown.

spin splitting energy at a constant external magnetic field and we use this method both to enhance and to suppress the Zeeman energy.

It has been shown in our previous work [12] that, for the $\nu = 2/3$ FQHE state, an increase of the magnetic field results in a rather sharp transition from a spin-unpolarized state into a spin-aligned ground state. The reason for this effect is a phase transition between two FQHE ground states with different spin configurations, which is very sensitive to the ratio between Zeeman and Coulomb energies. We have studied such spin transitions for all families of FQHE states and these results will be published separately, however, in the present paper we use such transitions to measure the internal nuclear magnetic field, which appears under optical pumping by circularly polarized light. In Fig.13 we show the magnetic field dependence of the electron spin polarization measured for 6 different samples (with additional variation of electron density by top gate) at fixed filling factor $\nu = 2/3$. To measure the electron spin polarization, we used the method which was already described in previous paragraphs [12] and which is based on the analysis of the degree of circular polarization (γ_L) of time-resolved radiative recombination of 2D-electrons with photoexcited holes bound to acceptors.

Fig.13 shows the magnetic field dependence of the electron spin polarization measured under a low excitation intensity (peak power density was less than 10^{-3} W/cm^2). For this power level, we did not find any difference for both circularly and linearly polarized light being used for excitation. However, an increase of

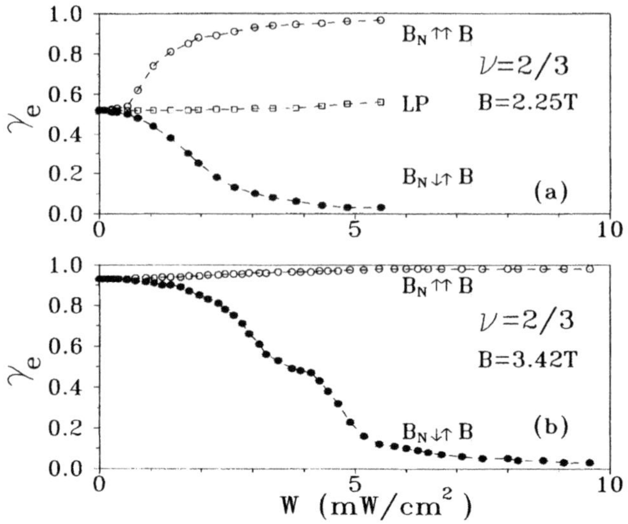

Figure 14. Power density dependence of the degree of electron spin polarization measured for the $\nu = 2/3$ FQHE state in different external magnetic fields (a and b) from the degree of polarization of luminescence under different conditions of photoexcitation. Open (filled) circles correspond to photoexcitation by the circularly polarized light, which orients nuclear spins along (against) the direction of the external magnetic field and open squares correspond to excitation by the linearly polarized light.

the power level results in a dramatic change in the electron spin polarization for the ν=2/3 FQHE state, depending on the polarization of light used for excitation. In Fig.14 we plot the power density dependence of the spin polarization of 2D-electrons, measured for the ν=2/3 FQHE state in different magnetic fields and under various photoexcitation conditions. Circles correspond to excitation by circularly polarized light, whereas squares correspond to excitation by linearly polarized light. One can see from this figure, that in contrast to the case of excitation by linearly polarized light, which was almost ineffective in changing the electron spin polarization, the application of circularly polarized light strongly influences the spin configuration of the 2D-electron system. We attribute this effect to the optical orientation of the nuclear spins by circularly polarized light. This understanding is supported by the fact that the left- and right-hand circularly polarized light produce opposite effects, as they orient nuclear spins in different ways - parallel (open circles in Fig.14) and antiparallel (filled circles in Fig.14) to the external magnetic field.

An effective internal nuclear field is then added or subtracted from the external magnetic field, giving rise to an enhancement or suppression of the effective Zeeman energy, which results in a strong variation of the electron spin polarization. Note that in our system we have two types of electrons - these are the

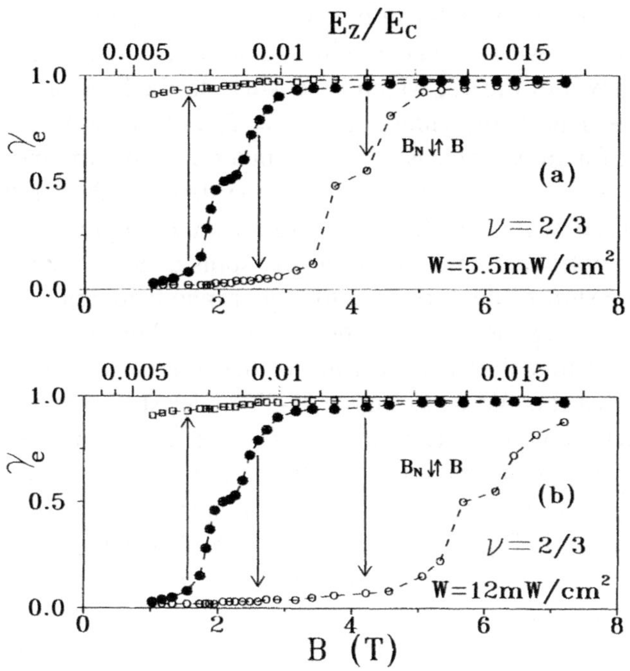

Figure 15. The influence of optical pumping on the threshold dependence γ_e(B), (similar to the one shown on Fig.1), measured for the $\nu = 2/3$ FQHE state under different excitation powers. The variation of electron spin polarization obtained under optical orientation of nuclear spins against (along) external magnetic field is shown by circles (squares).

2D-electrons, which exist without illumination (due to the doping of AlGaAs layer) and are located very close to the interface (typical width is about of 10nm), and electrons which are created in the GaAs buffer layer by absorption of photons within a length of about 1 μm (the penetration length of light) and have rather small life time (1 ns) with respect to the recombination with free photoexcited holes. The optical pumping process (both electronic and nuclear) is only due to the spin orientation of the photoexcited electrons, whereas the studied luminescence is due to the 2D-electrons. The process of nuclear spin orientation by photoexcitation is a rather slow process and requires considerable time (a few seconds) to reach a saturation of the nuclear polarization, however, after being polarized, the nuclear spin system acts on the spins of 2D-electrons, changing the degree of polarization of the corresponding luminescence. The observed increase of variation of γ_e with excitation power is due to an increase of the degree of nuclear polarization, which is defined by the intensity of photoexcitation. As for the results obtained for linearly polarized light, it is clear from Fig.14 that its efficiency for nuclear polarization is very low and gives rise to a small increase of Zeeman energy (in agreement with previous investigations [55]).

To estimate the internal nuclear magnetic field (B_N) created under the optical pumping conditions, we studied the influence of optical pumping on the threshold dependence, shown in Fig.13, which is sensitive to the Zeeman energy. We analyzed such dependencies measured for the $\nu = 2/3$ FQHE state (and also for other incompressible fractions, such as $\nu = 3/5$) under optical pumping conditions and found that, for all samples studied, there is a systematic shift in the dependence γ_e(B), which is shown in Fig.15 for different excitation powers. It is clear from this figure, that optical orientation of nuclear spins against the external magnetic field results in a shift of the threshold to higher magnetic fields, whereas nuclear spin orientation along the magnetic field shifts the dependence to lower fields. The amplitude of this shift is a measure of the internal nuclear magnetic field, and illustrates that it could be as high as 1-3T under strong pumping conditions. However, to determine the exact value of B_N we took into account the change of Coulomb energy as a function of magnetic field and determined an effective Zeeman energy from the value of the electron spin polarization and calibration dependence (shown in Fig.13 for $\nu = 2/3$). Note that this internal nuclear magnetic field B_N acts only on spins of electrons, but not on their orbital motion. As a check of the procedure we have measured the values of B_N using the calibration curves obtained for different fractional states and found that the values of B_N derived from all fractions were in agreement with each other.

An explanation of the observed shifts in the dependencies of $\gamma_e(B)$ under optical pumping conditions by the electron-nuclear interaction requires a more direct experimental verification of the importance of the coupling between 2D-electrons and nuclei. In order to obtain a direct confirmation of such coupling we developed a new method of optical detection of Nuclear Magnetic Resonance (NMR), which is based on analysis of the time-resolved radiative recombination of 2D-electrons with photoexcited holes bound to acceptors. The advantage of this method, in comparison with conventional NMR detection scheme, is that it is selective for nuclei located in the vicinity of the 2D-channel. A depolarization of nuclear spins under conditions of the NMR saturation, destroys the internal nuclear field and results in the variation of the electron spin polarization. As a consequence, the intensity of radiative recombination of 2D-electrons with photoexcited holes measured in a dominating circular polarization will also vary in accordance with a change of electron spin polarization. In Fig. 16a and 16b we show the optically detected NMR, which was measured for different nuclei and isotopes (As^{75}, Ga^{69} and Ga^{71}) in various magnetic fields and sweep rates of the RF field. It is clear from these figures that in order to record a correct NMR curve, it was necessary to use rather slow sweep rates dF/dt, which indicates rather long relaxation times in the nuclear system. Optically detected NMR is not only a direct evidence of a strong electron-nuclear coupling, but also an illustration of a remarkable property of the optical spin orientation method, which gives the possibility to monitor both the electron and the nuclear spin polarization. We were able also to

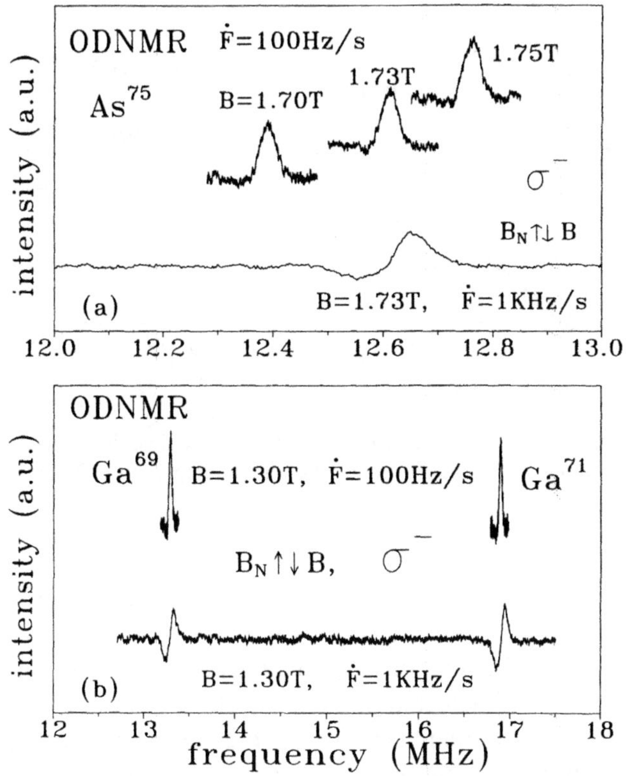

Figure 16. Nuclear Magnetic Resonance (As^{75}(a) and Ga^{69}, Ga^{71} (b)), optically detected as a variation of the luminescence intensity (detection in σ^- polarization) as a function of RF-frequency, measured in different magnetic fields and for different sweep rates (dF/dt) of RF-field under optical pumping, which oriented nuclear spins against the external magnetic field.

measure the Knight shift using the above mentioned possibility to change under different optical pumping conditions (orienting nuclear spins along or against the external magnetic field), from the completely polarized to the unpolarized spin state for $\nu = 2/3$. This shift is a change of the nuclear spin splitting energy due to electron-nuclear coupling and the variation of electron spin polarization. The value of the Knight shift, measured at $B = 4.2T$ for $\nu = 2/3$ from our optically detected NMR was about 17 KHz, which is in reasonable agreement with the data reported in [45]. In addition, NMR was also detected in our samples for luminescence lines corresponding to donor-acceptor recombination in the GaAs buffer layer, which is an indication of a strong nuclear polarization not only at the interface, but also in a rather thick (1 μm) layer of bulk GaAs.

In Fig.17 we plot the dependence of the internal nuclear magnetic field on the excitation power. The data were obtained for the same gated sample at different fixed external magnetic fields under the condition of optical pumping, which ori-

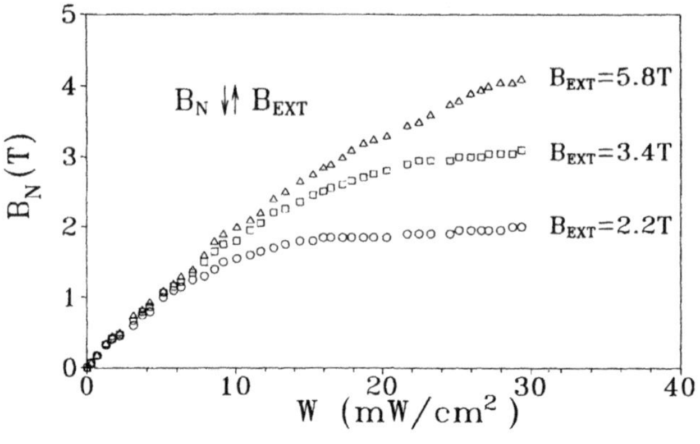

Figure 17. Power dependence of the internal nuclear magnetic field B_N, appearing due to optical pumping by circular light, measured in one of the gated samples in different external magnetic fields.

ents nuclear spins against B_{EXT}. To measure this dependence at fixed B_{EXT} we changed the concentration of 2D-electrons in order to obtain different fractional states at this magnetic field, and, finally, used the corresponding calibration curves for determination of the nuclear field. The results obtained for different FQHE states were in a good agreement with each other. As shown in Fig.17, by using strong optical pumping, we were able to achieve a high level of nuclear spin orientation that creates an internal nuclear magnetic field as high as 4T. Interestingly, we have observed a saturation of the nuclear field as a function of optical power and it was not possible to get B_N higher than B_{EXT}. A possibility to obtain experimentally such a strong B_N is usually provided by the localization of photoexcited electrons on donor impurities, since the nuclear orientation will take place mainly around these donor impurities (within their Bohr radius). In our case, a similar enhancement most probably appears due to the confinement of photoexcited electrons in the potential well near the interface (an electric field in a buffer layer attracts photoexcited electrons to the interface, increasing their local concentration and therefore nuclear orientation efficiency). Additional localization of photoexcited electrons in the plane may arise due to an attractive (for excitons) potential of the neutral acceptor located in the δ-layer near the interface. The binding energy of the exciton to a neutral acceptor is about 2 meV and the corresponding Bohr-radius is about 30 nm. Therefore, in our case the internal nuclear magnetic field can be rather nonuniform, reaching such high values only in the vicinity of acceptor centers. However, since our method is based on locally probing the 2D-electrons exactly around these acceptor centers, it allows polarization measurements with compensation of the Zeeman energy by hyperfine interaction.

The possibility to suppress Zeeman energy by the optical orientation of nu-

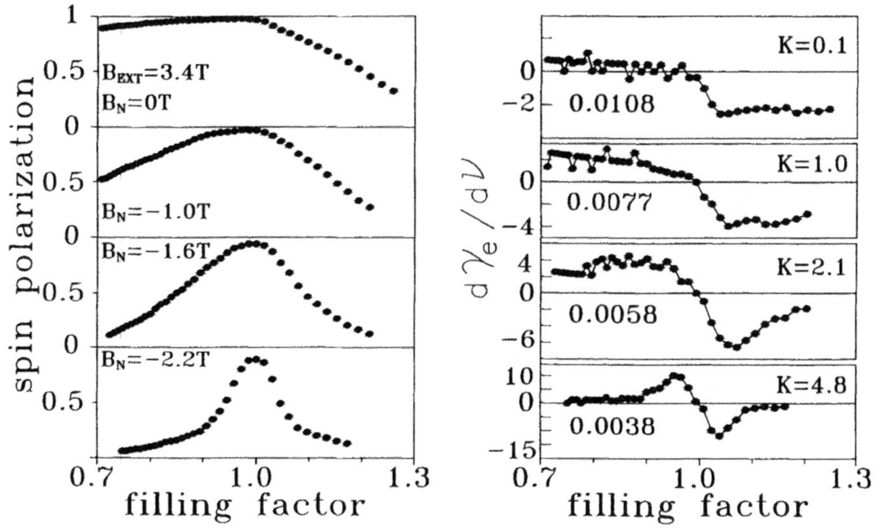

Figure 18. The dependence of the degree of electron spin polarization γ_e (left figure) and its derivative dγ_e/dν (right figure) on filling factor, measured for a sample with electron concentration $8.2 \cdot 10^{10}\ cm^{-2}$ under different conditions of optical pumping which resulted in various internal nuclear fields: B_N=0, -1.0T, -1.6T and -2.2T. Negative values of B_N correspond to nuclear spin orientation against external magnetic field. The size of the skyrmion K was directly measured from the amplitude of discontinuity of dγ_e/dν for different values of the ratio E_Z/E_C (indicated on right figure): $K = 0.1 \pm 0.15$ for E_Z/E_C = 0.0108; $K = 1.0 \pm 0.15$ for E_Z/E_C = 0.0077; $K = 2.1 \pm 0.15$ for E_Z/E_C = 0.0058; $K = 4.8 \pm 0.15$ for E_Z/E_C = 0.0038.

clear spins against the external magnetic field was used to enhance skyrmionic effects in the vicinity of $\nu = 1$. Fig.18 shows the filling factor dependence of the electron spin polarization, measured for the same sample around $\nu = 1$ (B_{EXT}=3.4T) under different conditions of optical pumping. In this figure we present data obtained both without (B_N=0) and with rather strong nuclear optical orientation (B_N=-1.0T, -1.6T and -2.2T), with nuclear spins being aligned against B_{EXT}. It is obvious from the figure that an increase of the strength of the nuclear field results in a decrease of electron spin polarization in the vicinity of $\nu = 1$, and a very narrow peak of spin polarization at $\nu = 1$ was observed at the highest negative B_N. This peak is a characteristic feature of the skyrmion, which gives an opportunity to quantify the parameters of these topological textures. In the theory [44], the size of a skyrmion, K (measured in units of magnetic length) can be measured directly from the filling factor dependence of the electron spin polarization, $\gamma_e(\nu)$. According to [44], for $\nu < 1$ dγ_e/dν = $2K$, whereas for $\nu > 1$, dγ_e/dν = $-2(K+1)$. Fig.18 shows also the filling factor dependence of dγ_e/dν, from which we determine the size of the skyrmion K (the value of K was obtained from the amplitude of discontinuity of dγ_e/dν at ν=1) and its variation under suppression of Zeeman energy (the corresponding values of the parameter E_Z/E_C are indicated

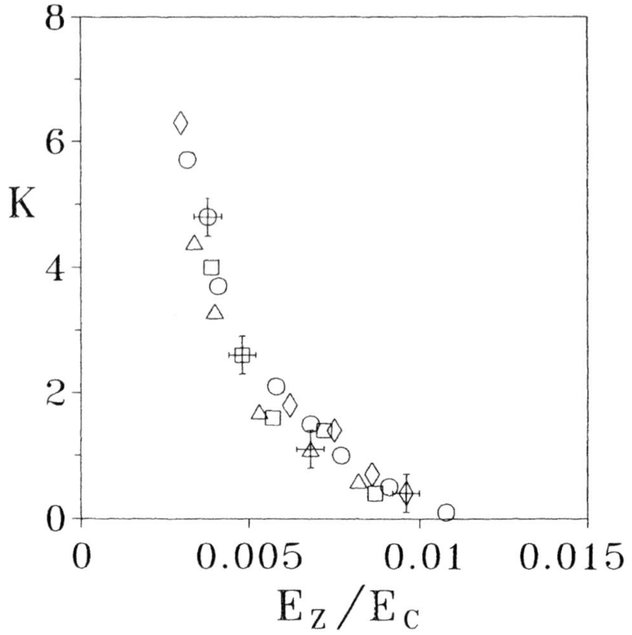

Figure 19. The dependence of the skyrmion size K (in units of magnetic length) on the ratio between Zeeman and Coulomb energies E_Z/E_C, measured for different samples (shown by different symbols) under variation of the Zeeman energy by optical orientation of the nuclear spins. Typical error bars are indicated.

on the right figure). Note that values of K derived from the $\nu > 1$ and $\nu < 1$ regions are close to each other, which illustrates a self-consistency of the obtained results. It is also necessary to mention that the small values of K (like K=0.15) obtained in such a procedure most probably indicate that no skyrmion is formed (since smooth rotation of spins in skyrmion requires rather large K), however, in order to compare results measured in different experimental conditions, we apply the same formal procedure and derive the corresponding values of K.

Finally, we have used the described method to measure the basic dependence of the skyrmion size K as a function of the parameter E_Z/E_C, which was varied by optical orientation of the nuclear spins. In Fig.19 we plot this dependence, $K(E_Z/E_C)$, measured for different samples from the amplitude of discontinuity of $d\gamma_e/d\nu$ around $\nu = 1$. It is clear from this figure, that the size of a skyrmion rapidly grows as the Zeeman energy vanishes and, in contrast, the skyrmion quenches if the parameter E_Z/E_C reaches a value close to 0.011 ± 0.001. These observations are in a contradiction with some previous results [45, 46], which were obtained for the g=-0.44 case (in these publications values of K=2-3 were measured for E_Z/E_C = 0.015-0.018), but they are in a rather good quantitative agreement with the results obtained for the vanishing g-factor [47] (skyrmionic effects in this work were significant only for E_Z/E_C <0.004). A possible explanation of the apparent discrepancy between our data and the results reported in [45, 46] could be due to uncontrollable optically induced nuclear spin polarization, which also compensated Zeeman energy and resulted in the enhancement of the skyrmionic excitation or due to heating effects [12].

Summarizing this paragraph, we have developed an effective experimental method for a compensation of the Zeeman energy of 2D-electrons based on optical pumping of nuclear spins, which was used to enhance skyrmionic excitations. We have demonstrated that photoexcitation by circularly polarized light can result in a strong spin orientation of nuclei both along and against the direction of an external magnetic field and that the energy of the hyperfine interaction between 2D-electrons and optically oriented nuclear spins is close to the electronic Zeeman energy. A compensation of the Zeeman energy by the hyperfine interaction at a fixed magnetic field results in an effective enhancement of the skyrmionic excitations around ν=1, and a dependence of the skyrmionic radius as a function of the ratio between Zeeman and Coulomb energies has been measured. We have demonstrated that the skyrmion quenches in size at E_Z/E_C = 0.011 ± 0.001. A new method of optical detection of NMR is suggested and realized, which is based on the analysis of the circular polarization of time-resolved radiative recombination of 2D-electrons with photoexcited holes bound to acceptors.

Acknowledgments. The financial support from the Max-Planck and Humboldt Research Award, Russian Fund of Fundamental Research and an INTAS grant is gratefully acknowledged.

References

1. J.K.Jain, Phys. Rev. Lett. **63**, 199 (1989); Advances in Physics **41**, 105 (1992)
2. D. C. Tsui *et al.*, Phys. Rev. Lett. **48**, 1559 (1982).
3. B.I.Halperin *et al*, Phys. Rev. **B47**, 7312 (1993)
4. R.L.Willett *et al*, Phys. Rev. Lett. **71**, 3846 (1993)
5. R.R.Du *et al*, Phys. Rev. Lett. **70**, 2944 (1993); Phys. Rev. Lett. **75**, 3926 (1995);
6. W.Kang *et al*, Phys. Rev. Lett. **71**, 3850 (1993)
7. V.J.Goldman *et al*, Phys. Rev. Lett. **72**, 2065 (1994)
8. J.H.Smet *et al*, Phys. Rev. Lett. **77**, 2272 (1996)
9. R.L.Willett *et al*, Surf. Sci. **361/362**, 38 (1996).
10. I.V.Kukushkin *et al*, Phys. Rev. Lett. **72**, 736 (1994).
11. D.R.Leadley *et al*, Phys. Rev. Lett. **72**, 1906 (1994)
12. I.V.Kukushkin *et al*, Phys. Rev. **B55**, 10607 (1997).
13. W.Kang *et al*, Phys. Rev. Lett. **75**, 4106 (1995)
14. I.V.Kukushkin *et al*, Phys. Rev. **B40**, 7788 (1989).
15. B.I.Halperin, Helv. Phys. Acta **56**, 75 (1983).
16. T.Chakraborty *et al*, Phys. Rev. B**29**, 7032 (1984).
17. J.P. Eisenstein *et al*, Phys. Rev. Lett. **62**, 1540 (1989).
18. R.G. Clark *et al*, Phys. Rev. Lett. **62**, 1536 (1989).

19. L.W. Engel *et al*, Phys. Rev. B**45**, 3418 (1992).
20. K.Park and J.K.Jain, Phys. Rev. Lett. **80**, 4237 (1998).
21. J. K. Jain and R. K. Kamilla in 'Composite Fermions' p.3 (1998).
22. T. Nakajima and H. Aoki, Phys. Rev. Lett. **73**, 3568 (1994).
23. G. Murthy, Phys. Rev. B **60**, 13702 (1999).
24. S. M. Girvin *et al.*, Phys. Rev. Lett. **54**, 581 (1985).
25. A. Pinczuk *et al.*, Phys. Rev. Lett. **70**, 3983 (1993); Phys. Rev. Lett. **68**, 3623 (1992).
26. H. D. M.Davies *et al.*, Phys. Rev. Lett. **78**, 4095 (1997).
27. P. Khandelwal *et al.*, Phys. Rev. Lett. **81**, 673 (1998).
28. I. V. Kukushkin *et al.*, Phys. Rev. B **40**, 7788 (1989); Phys. Rev. B **55**, 10607 (1997); Phys. Rev. B **60**, 2554 (1999); Phys. Rev. Lett. **82**, 3665 (1999).
29. I. V. Kukushkin *et al.*, JETP Lett. **70**, 722 (1999).
30. F. C. Zhang and S. Das Sarma, Phys. Rev. B **33**, 2903 (1986).
31. A. H. MacDonald and J. J. Palacios, Phys. Rev. B **58**, R10171 (1998).
32. S. Das Sarma, A. Pinczuk. (eds.), Perspectives on quantum Hall effects, Wiley, N.Y., (1996).
33. J.K. Jain. Physics Today, 39-45 (April 2000).
34. J.K. Jain, Phys. Rev. Lett. **63**, 199 (1989).
35. O. Heinonen, (ed.), Composite Fermions, World Scientific Publishing, (1998).
36. W. Kohn, Phys. Rev. **123**, 1242 (1961).
37. S.I. Gubarev *et al*, JETP Lett. **72**, 324 (2000).
38. S.J. Allen *et al*, Phys. Rev. **B 28**, 4875 (1983).
39. R.R. Gerhardts *et al*, Phys. Rev. Lett. **62**, 1173 (1989).
40. J.M. Shilton *et al*, Phys. Rev. **B 51**, 14770 (1995).
41. R.L. Willett, Composite Fermions, Ed. O. Heinonen, World Scientific Publishing, 349 (1998)
42. T.Chakraborty and F.C.Zhang, Phys. Rev. B**29**, 7032 (1984).
43. S.L.Sondhi *et al*, Phys. Rev. B**47**, 16419 (1993).
44. H.A. Fertig *et al*, Phys. Rev. B**50**, 11018 (1994); Phys. Rev. B**55**, 10671 (1997).
45. S.E.Barrett *et al*, Phys. Rev. Lett. **74**, 5112 (1995).
46. E.H.Aifer *et al*, Phys. Rev. Lett. **76**, 680 (1996).
47. D.K.Maude *et al*, Phys. Rev. Lett. **77**, 4604 (1996).
48. S.Holmes *et al*, Semicond. Sci. Technol. **9**, 1549 (1994).
49. C.Weisbuch and C.Hermann, Phys. Rev. **15**, 816 (1977).
50. C.Hermann and C.Weisbuch, Phys. Rev. **15**, 823 (1977).
51. M.J.Snelling *et al*, Phys. Rev. **B44**, 11345 (1991).
52. S.Ernst *et al*, Phys. Rev. Lett. **72**, 4029 (1994).
53. A.Kastler, Sciences, **158**, 214 (1967).
54. G.Lampel, Phys. Rev. Lett., **20**, 491 (1968).
55. F.Meier and B.Zakharchenya, *Optical Orientation*, North Holland (1984).
56. S.W.Brown *et al*, Phys. Rev. **B54**, R17339 (1996).

METAL-INSULATOR TRANSITION IN DILUTE 2D ELECTRON AND HOLE GASES

A.K. SAVCHENKO
School of Physics, University of Exeter, Exeter, EX4 4QL, U.K.

1. Introduction

The existence of a critical metal-insulator transition in two-dimensional (2D) systems has been a subject of great debate over the last few decades. The scaling theory of electron localisation [1] predicts a critical metal-to-insulator transition with increasing disorder only in 3D, and no such transition in 2D. According to this theory, the resistance of a 2D sample should monotonically increase with increasing disorder or decreasing temperature, so that with any small disorder the 2D sample will become non-conducting at $T = 0$. There has been vast experimental support of such behaviour of resistance of 2D systems, thin films and channels in transistor structures. Transistors offer a great opportunity for metal-to-insulator transition studies as the effective disorder is easily controlled by changing the carrier density with gate voltage [2, 3]. As an example, Fig. 1 shows the resistance of a low-mobility channel of a n-GaAs transistor, which monotonically increases with decreasing temperature at all electron densities n [4]. At high n the weak temperature dependence is described by weak localisation (WL) [5], but with decreasing n and increasing disorder parameter $k_F l$ (k_F is the Fermi wave vector and l is the mean free path), the localisation radius decreases and becomes smaller than the sample size. In this case the current is carried by electron transport via localised states [6], variable range hopping in this example. In this case the change in the character of electron transport is a non-critical metal-to-insulator transition which is simply a transition from weak to strong localisation.

The scaling theory, however, does not take into account electron-electron interactions, which can change the sign of the $R(T)$ dependence from 'insulating', when $dR/dT < 0$, to 'metallic', when $dR/dT > 0$ [7]. In the theory of quantum corrections the total correction to the classical Drude conductivity comes from a WL contribution (normally negative) and a correction from electron-electron interaction (EEI), which can be either negative or positive, depending on the value

I.V. Lerner et al. (eds.),
Strongly Correlated Fermions and Bosons in Low-Dimensional Disordered Systems, 219–239.

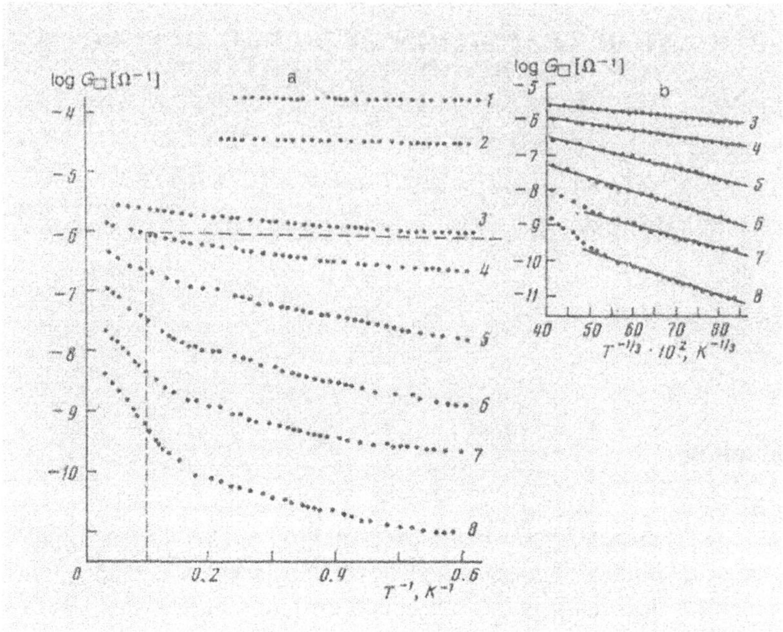

Figure 1. (a) Temperature dependence of the conductance of a low-mobility channel in a n-GaAs MESFET at different electron densities, which decrease in the figure from 1 to 8, showing increasing resistance with decreasing T at all electron densities [4]. (b) Hopping regime (2D VRH) at low densities.

of the interaction constant F [5, 7]:

$$\Delta\sigma(T) = \Delta\sigma_{WL} + \Delta\sigma_{EEI}(F), \tag{1}$$

Although one can, in principle, expect from Eq. (1) the positive sign of $dR(T)/dT$ caused by EEI, the WL and EEI theories were developed for small r_s (the ratio of the Coulomb and Fermi energies), and for dilute 2D systems with large r_s they cannot be applicable and the interaction constant F is not known.

The possibility of suppression of localisation by electron-electron interactions has been considered over the years in many theories (e.g.[8–10]). Recently, the problem of the co-existence of disorder and interactions has been given a boost by observations of a crossover in the sign of dR/dT in some 2D systems [11–16]. Fig. 2 shows an example of such a crossover with varying 2D electron density in a Si MOSFET [11], where the 'metallic' $R(T)$ is in obvious contradiction with the scaling theory for non-interacting electrons. A crossover at $R_\square \sim h/e^2$ has been seen in (100) Si [11, 12], Si/SiGe [13, 14] and GaAs structures [15, 16]. The effect was observed in high-mobility structures and at low electron and hole densities, which suggests that Coulomb interactions can play a significant role in this phenomenon. Indeed, at low densities $r_s = U/E_F \propto m^*/p^{1/2}$ becomes large.

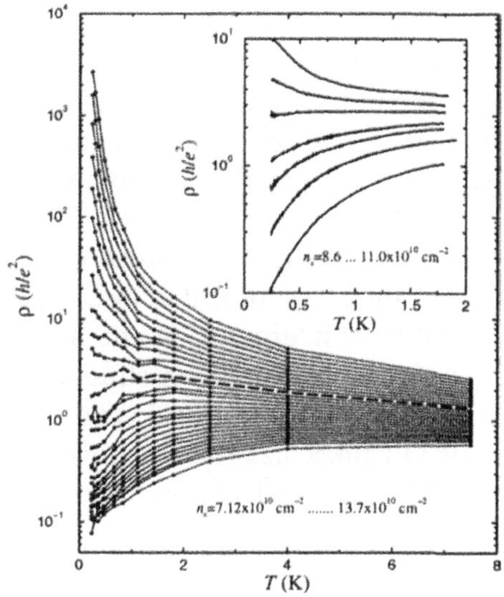

Figure 2. Temperature dependence of the resistivity of a 2DEG in Si MOSFET [11] with a crossover in the sign of dR/dT at some electron density.

The new observations have raised several questions: What is the origin of the 'metallic' behaviour of the resistance? Is the crossover a critical metal-to-insulator transition? How universal is this phenomenon in different low-density high-mobility 2D systems? Is the Fermi-liquid description still valid in these systems with large r_s?

There are different opinions about the origin of the crossover and the 'metal' with $dR/dT > 0$. For Si MOSFET structures these effects have been interpreted as a result of a critical metal-to-insulator transition [17]. The opinion that in these systems the Fermi-liquid approach is not valid stimulated new theories [18–22]. On the contrary, there have been several models proposed where the 'metallic' behaviour is attributed to conventional, although non-trivial low-temperature transport in low-density structures, with the features not discussed earlier [23–28]. These models explain the new effect by the following: the specifics of a transistor structure in which the 2D channel is formed [23], the facts that impurity scattering of electrons can be temperature dependent even in degenerate electron gas [23–25] and that a low-density 2D channel can be inhomogeneous [26]. For 2DHGs, the unusual behaviour can also be ascribed to a complex band structure of holes in GaAs and the importance of the interband scattering at large enough hole densities [27, 28].

The results of our work will be relevant to the first three models which consider electron scattering by impurities. Model [23] suggests that the 'metallic' state in Si MOSFETs can be caused by an impurity band (IB) in the oxide. Theories [25] and [24] consider the effect of temperature dependent impurity scattering, caused by the change in the screening of the impurity potential.

On the experimental side, there are several facts which have been widely reproduced and are the signature of the new effect. These are the existence of the crossover itself (although it has now been seen not only in extremely high-mobility structures) and a positive parallel magnetoresistance, seen both in Si- and GaAs-based structures. A characteristic effect of the parallel field is that it produces a transition from the 'metallic' to 'insulating' $R(T)$. In addition, there have also been several interesting experimental observations which, in my opinion, require further work on their interpretation and understanding of how these effects depend on the specifics of the transistor structures where the 2D channel is formed. Examples of these could be observations of an enhancement of the g-factor [29] and increase of $1/f$ noise near the crossover in Si MOSFETs [32]; an anomaly in the compressibility [30], and a Wigner crystal formation at $r_s \simeq 35$ in 2DHGs in GaAs [31].

In the results presented here we examine the behaviour near the crossover of two high-mobility 2D systems: a) electrons in Si MOSFETs and b) holes in GaAs/AlGaAs heterostructures. Our aim is to see whether their properties can be described by conventional models, and if not, see the deviations from Fermi-liquid approaches.

This work was done in collaboration of the University of Exeter (UE) with the Institute of Semiconductor Physics, Novosibirsk (ISP), and Cavendish Laboratory, Cambridge (CL). I would like to acknowledge the contributions of S.S. Safonov and S.H. Roshko (UE), A.G. Pogosov and Z.D. Kvon (ISP) to the study of the 2DEG in Si MOSFETs; Y.Y. Proskuryakov and S.S. Safonov (UE), M. Pepper, M.Y. Simmons and D.A. Ritchie (CL) to the work on the 2DHG in GaAs.

2. Impurity scattering as the origin of the crossover in a 2DEG on vicinal Si

All previous studies of the crossover in the 2DEG in Si MOSFETs were performed on electrons on the (100) Si surface. To see whether the crossover is a genuine property of the 2DEG or depends on the surface orientation of the MOSFET, we have studied the 2DEG on a *vicinal* Si surface, which is tilted from the (100) surface around the [011] direction by an angle of 9.5°. One would not expect that such a small difference in the surface orientation is important for the manifestation of the crossover in the 2DEG. The high mobility of electrons in our structures, peak mobility 2×10^4 cm^{2}V^{-1}s^{-1} at $T = 4.2$ K, is comparable to the highest mobility of the 2DEGs on (100)Si. The oxide thickness is 120 nm. Fig. 3 shows the resistance as a function of V_g for vicinal sample Si-4.1 in the temperature range below 1 K, with the crossover near $\rho_{xx} \sim 1$ kOhm $\sim 0.04 \times h/e^2$, which is much smaller than that seen in (100) Si structures. Furthermore, at $T \leq 4.2$ K two distinct groups of curves are seen for opposite directions of V_g-sweeps, with different values of the crossover resistance.

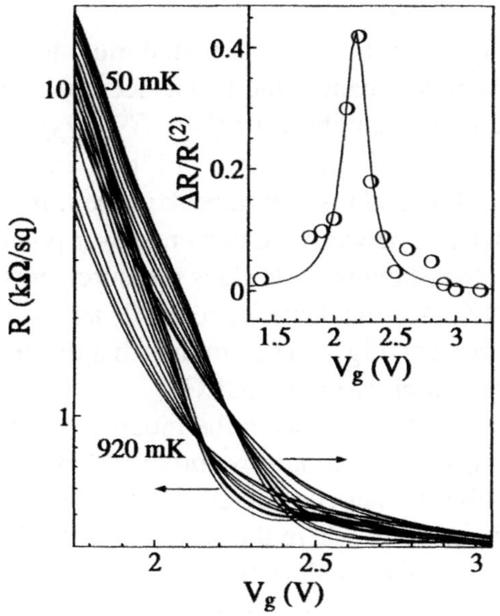

Figure 3. Resistance as a function of the gate voltage for sample Si-4.1 at $T = 50 - 920$ mK. The arrows show different directions of the V_g sweeps. Inset: the resistance difference for the opposite directions of the gate voltage sweep.

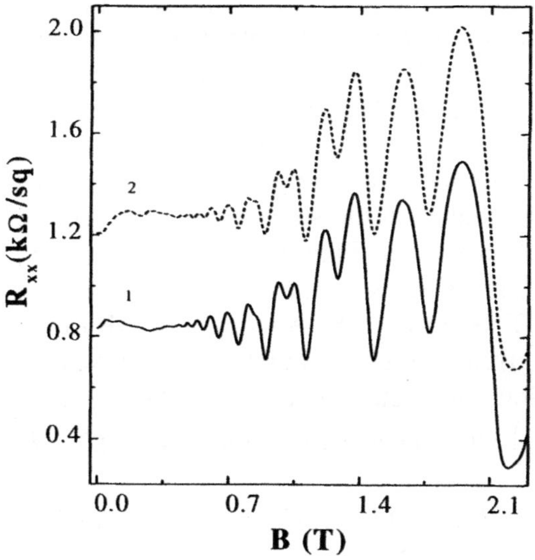

Figure 4. SdH oscillations measured at $V_g = 2.2V$, after this gate voltage was approached from two opposite directions: 1 - $V_g = 9.5V$, 2 - $V_g = 0.5V$; $T = 50$ mK.

To study the hysteresis in more detail, we have performed an experiment at T=50 mK where a particular V_g-point in the studied interval of gate voltages is approached, by a jump, from the opposite boundaries: $V_g^{(1)} = 0.5$ V and $V_g^{(2)} =$ 9 V. After an initially rapid change, the resistance comes to saturation, and the measurements of the quasi-equilibrium values $R^{(1)}(V_g)$ and $R^{(2)}(V_g)$ reveals a difference $\Delta R =| R^{(1)} - R^{(2)} |$ presented in Fig. 3, inset. The hysteresis exists at temperatures below 4 K, and it shows a peak in the difference between the two resistance values exactly in between the two crossover points.

At the same time, the Shubnikov-de Haas measurements show that for a resistance difference of 40%, electron densities $n^{(1)}(V_g)$ and $n^{(2)}(V_g)$ are equal within an accuracy of a few percent, Fig. 4. This implies that the difference ΔR is caused by the difference in the mobility of the 2DEG.

The presence of a sharp hysteresis in the mobility of the 2DEG suggests that the crossover can be caused by a narrow impurity band (IB) at the Si/Si-oxide interface which is similar to that considered in [23]. The hysteresis peak appears in the range of the Fermi level where the IB is positioned, and the IB width can roughly be estimated from the width of the hysteresis peak: $W \leq 0.3$ meV.

The IB scatters electrons when it is positively charged. With rising Fermi level in the 2DEG the number N_+ of ionised impurities in it decreases and this increases the mobility. The hysteresis in the mobility at lowest temperatures is caused by a slow electron exchange between the 2DEG and the IB which are separated by a barrier.

Let us discuss how the IB scattering of 2D electrons produces a crossover in the $R(T)$. When the Fermi level E_F is in the upper part of the IB (inset to Fig. 5), smearing the Fermi distribution with increasing temperature decreases the number of electrons in the IB and increases the number of ionised scatterers N_+, therefore the resistance of the 2DEG rises ('metallic' behaviour). On the contrary, when the Fermi level is in the lower part of the IB, the number of scatterers decreases with increasing T ('insulating' behaviour). For an IB with symmetric density of states, the change in the sign of dR/dT occurs when the Fermi level is at the IB centre (crossover). Figure 5 shows the temperature dependence of the resistance $R^{(2)}$, presented as $\mu^{-1}(T)$, near the crossover at $n_c \simeq 4.18 \times 10^{11}$ cm^{-2}, in the range T=50 mK -4 K. Assuming the IB to be centered at the level E_i, the mobility is given by the total number of positive charges in it. This is the integral of the distribution function of unoccupied states in it $f(E, E_F - E_i, T) = 1/\left(1 + \exp \frac{E_F - E_i}{k_B T}\right)$ with the density of states in the IB $g(E, E_i)$:

$$N_+ \propto \int g(E, E_i)\ f(E, E_F, T)\ dE \tag{2}$$

To calculate the IB contribution, we have subtracted the value of μ^{-1} at $V_g = 2.4$ V when the IB is full. The value $E_F - E_i$ is calculated from the known electron density and the Fermi energy. In this analysis the IB is taken as a step with width

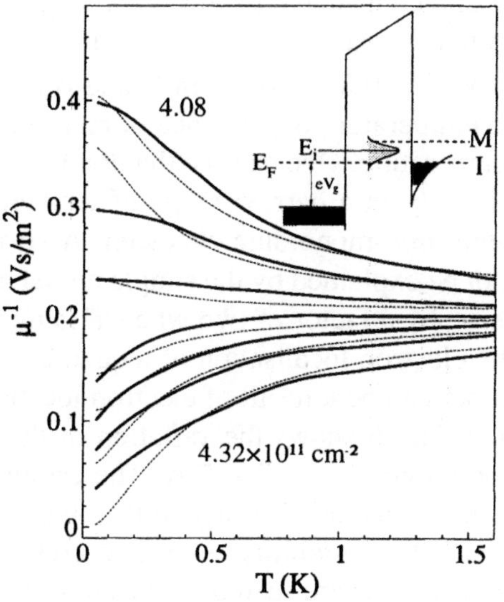

Figure 5. Temperature dependence of the mobility near the crossover at different electron densities, $n = 4.08 - 4.32 \times 10^{11}$ cm^{-2} and $V_g = 2.1 - 2.24$ V, from top to bottom. Dashed lines show the theoretical fit. Inset: a diagram showing a narrow impurity band which scatters electrons in the 2DEG. The number of ionised impurities in the IB increases (M) or decreases (I) depending on the position of the Fermi level.

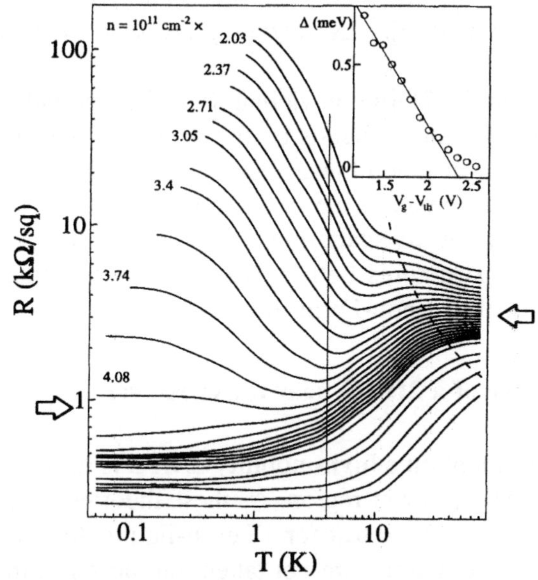

Figure 6. $R(T)$ in the full range of temperatures. Low- and high-T crossovers are indicated by the arrows. The dashed line shows the Fermi temperature T_F. Inset: the variation of the activation energy in the insulator with gate voltage, with extrapolation to zero value where the SdH oscillations disappear with decreasing n (V_{th} is the threshold gate voltage).

$W \simeq 0.08$ meV found as an adjustable parameter. Using the formula for the mobility due to ionised impurity scattering [25] we estimate the total concentration of scatterers in the IB: $N_+^{(1)} = 0.34 \times 10^{11}$ cm^{-2} and $N_+^{(2)} = 0.23 \times 10^{11}$ cm^{-2}.

Figure 6 shows the temperature dependence of resistance in a broad range of electron density and temperature. One can see that the low-temperature crossover at $R \sim 1$ kOhm exists only in a narrow range of temperatures, and $R(T)$ has in general a complicated, non-monotonic character. A strong insulator at small electron densities cannot be explained by the simple model of the IB as it can only give a resistance decrease by a factor of two when T is increased. This insulator cannot occur due to the electron localisation of a quantum nature because of the low sample resistance, but can be a result of electron localisation due to classical percolation. The inset in Fig. 6 shows the variation of the slope of $R(T)$ using an exponential fit in the range $T \sim 1.5 - 7$ K. The estimated activation energy $\Delta = E_c - E_F$ is seen to increase linearly with lowering of the Fermi level, and it extrapolates to zero at the mobility edge E_c corresponding to $n = 3.74 \times 10^{11}$ cm^{-2}. The SdH oscillations are seen on both sides of the crossover, but they disappear when n approaches the value found by the extrapolation of Δ to zero.

Let us briefly discuss $R(T)$ with increasing temperature. Surprisingly, at a higher T another crossover in dR/dT is seen at $\rho_{xx} \sim 3$ kOhm. We note that in the temperature range $T > 4$ K the system experiences a transition from degenerate to nondegenerate state. The variation of the Fermi temperature T_F corresponds to shift of the hump in Fig. 6. The main features of the model [24, 25] for the temperature dependent ionised impurity scattering are seen in the high-temperature crossover.

The experiments on the 2DEG on vicinal Si [33] show that the crossover phenomenon is not universal and depends on the details of the MOSFET structure. The crossover can occur at much lower resistance than h/e^2, and there can even be two crossovers if one considers a broad temperature range. These experiments show that the interface impurity scattering has to be taken into account in Si MOSFET structures. In our structures both crossovers can be explained by impurity scattering, without involving non-Fermi-liquid ideas.

3. Magnetoresistance of a 2DHG near the crossover

The aim of this experiment on a high-mobility 2DHG is to try and apply WL and EEI descriptions to the crossover region. As these theories have been derived for $r_s \ll 1$ [5], they provide a good test for a Fermi-liquid. Athough in the WL theory electron-electron interactions are not taken into account, the phase-breaking rate, which determines the WL correction, presents valuable information about interaction between the carriers.

There have already been indications that WL and weak hole-hole interaction (HHI) effects do not show anomalies up to $r_s \sim 10 - 14$ [34, 35]. With increasing

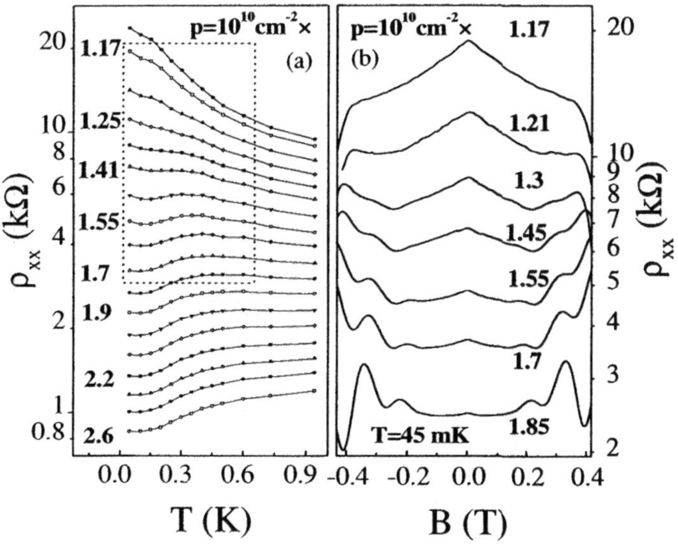

Figure 7. (a) Temperature dependence of the resistivity at different hole densities. The dashed box encloses the range of NMR study. (b) NMR at $T = 45$ mK, for different hole densities.

the mobility, the crossover in 2DHGs shifts towards smaller hole densities. In our structures with higher mobility than in [34], the crossover region occurs at lower densities ($p \sim 1.17 \times 10^{10}$ cm^{-2}, $r_s \sim 25$), so that we have more reason to expect deviations from Fermi-liquid behavior. This range of r_s approaches $r_s \sim 35$ where a Wigner crystal formation is expected [36] and was claimed to be observed in a low-density 2DHG [31].

The experiments have been performed on a heterostructure with a peak mobility of $\mu_p = 6.5 \times 10^5$ cm ^{2}V^{-1}s^{-1} at $T = 45$ mK, formed on a (311)A GaAs substrate, where the 2DHG is separated from the Si doped layer by a 500Å spacer. The hole density is varied by the front gate voltage in the range of r_s from 10 to 29 (with effective mass m^* taken as 0.38 m_0).

Fig. 7 shows a typical temperature dependence of the longitudinal resistivity. In the crossover region, where r_s varies from 23 to 29, negative perpendicular magnetoresistance (NMR) is seen, Fig. 7(b), which increases with decreasing p (determined from the Shubnikov-de Haas effect) and can be attributed to WL. The SdH oscillations are seen on both sides of the crossover, which is a clear indication of the presence of the Fermi surface in the whole range of hole densities studied.

Analysis of the WL magnetoresistance in high-mobility structures is not that straightforward. The conventional theory is based on the diffusion approximations and restricted by the range of magnetic fields $B < B_{tr}$, where $B_{tr} = \hbar/4De\tau$ is the 'transport' magnetic field, D is diffusion coefficient and τ is momentum relaxation time. Physically, this means that no negative magnetoresistance is expected for $B > B_{tr}$ when the magnetic length L_B becomes smaller than the

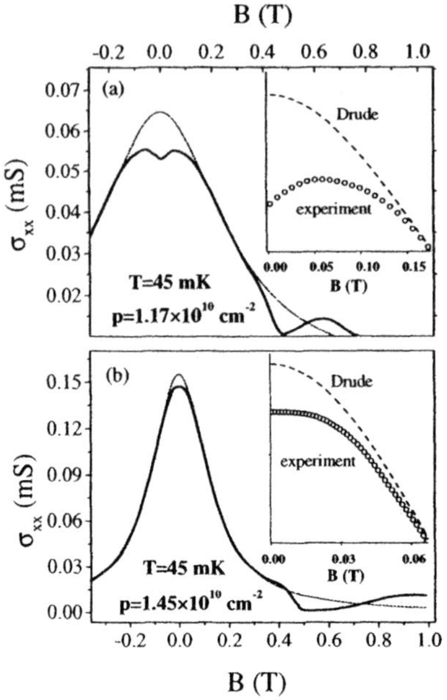

Figure 8. (a,b) Experimental conductivity $\sigma_{xx}(B)$ for two hole densities shown together with Drude conductivity (dotted line). The insets present zoomed-in regions.

mean free path l. In our samples B_{tr} ranges from 0.003 to 0.08 T, however the NMR is seen up to $B \sim 0.2$ T where SdH oscillations start. The theory of WL for $B > B_{tr}$ has also been developed [38], although no experimental test of it has yet been performed. Theories of WL discuss the positive magnetoconductivity $\Delta\sigma^{WL}(B) = \delta\sigma_{xx}(T,B) - \delta\sigma_{xx}(T,0)$, where $\delta\sigma_{xx}(T,B)$ is a negative correction to the classical (Drude) conductivity. To analyse $\Delta\sigma^{WL}(B)$ in a high-mobility system, one should take into account the magnetic field dependence of the Drude conductivity itself: $\sigma_{xx}^D(B) = \sigma_0/(1+(\mu B)^2)$. Figure 8 shows the total conductivity as a function of magnetic field obtained by inversion of the resistivity tensor, $\sigma_{xx}^{tot}(B) = (1/\rho_{xx})/(1+\rho_{xy}^2/\rho_{xx}^2)$. The dotted lines are plotted using the classical expression with $\mu B = \rho_{xy}(B)\sigma_0$. The zero-field Drude conductivity σ_0 is found as an adjustable parameter to make the best fit in the higher field region of $\sigma_{xx}^{tot}(B)$ where WL is expected to be totally suppressed. The difference between the solid and dotted lines gives the WL correction $\delta\sigma_{xx}(T,B)$.

We analyse the WL magnetoconductivity using theory [38] which is valid for an arbitrary ratio B/B_{tr}, provided $\tau < \tau_\varphi$ is satisfied:

$$\Delta\sigma^{WL}(B) = \frac{-e^2}{\pi h(1+\gamma)^2}\left[\sum_{n=0}^{N}\left(\frac{b\cdot\psi_n^3(b)}{1+\gamma-\psi_n(b)}\right) - \ln\frac{1+\gamma}{\gamma}\right] \tag{3}$$

where $\gamma = \tau/\tau_\varphi$, τ_φ is the dephasing time, $b = \frac{1}{(1+\gamma)^2}\frac{B}{B_{tr}}$, $\psi_n(b) = \int_0^\infty d\xi\,\cdot$

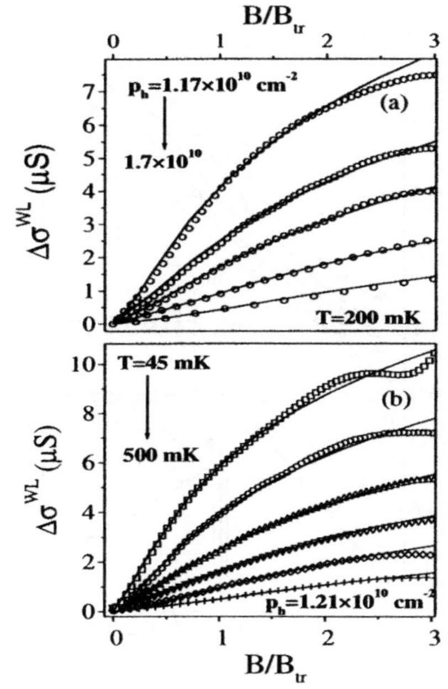

Figure 9. Magnetoconductivity $\Delta\sigma_{xx}^{WL}$ as a function of dimensionless magnetic field ($B_{tr} = \hbar/4De\tau$), for different densities (a) and temperatures (b). Solid lines are the fit to Eq. (3).

$e^{-\xi-b\xi^2/4}L_n(b\xi^2/2)$, and L_n are the Laguerre polynomials. In Fig. 9 we show representative data, plotted against dimensionless magnetic field B/B_{tr}, with B_{tr} found as $(4\pi\hbar\mu\sigma_0/e^2)^{-1}$. Solid lines in Fig. 9 are obtained from Eq. (3), where γ is used as an adjustable parameter. The γ-values are obtained with an error of less than 10%, they satisfy the condition $\tau < \tau_\varphi$ and range from 0.04 to 0.4.

Good agreement of the negative magnetoresistance with WL theory speaks in favour of a Fermi liquid. Further evidence for the existence of a Fermi liquid comes from the temperature dependence of the dephasing rate $\tau_\varphi^{-1}(T)$. Estimations show that the contribution to $\tau_\varphi^{-1}(T)$ from electron-phonon scattering is negligible. According to the Fermi-liquid theory [5, 39], the dephasing rate due to electron-electron scattering is dominated either by a linear or quadratic term, depending on the parameter $k_BT\tau/\hbar$. The corresponding relations can be presented as follows:

$$\tau_\varphi^{-1}\cdot p = \frac{m^*}{\pi\hbar^3}\left[\alpha k_B^2T^2\ln\left(\frac{4E_F}{k_BT}\right)\right], \qquad k_BT\tau/\hbar \gg 1 \tag{4}$$

$$\tau_\varphi^{-1}\cdot p = \frac{m^*}{\pi\hbar^3}\left[k_BT\frac{\hbar}{2\tau}\ln\left(\frac{2E_F\tau}{\hbar}\right)\right], \qquad k_BT\tau/\hbar \ll 1 \tag{5}$$

where $\alpha = \pi/8$ [39]. The quadratic term, Eq. (4), is due to Landau-Baber scattering in a clean Fermi-liquid associated with collisions with large momentum

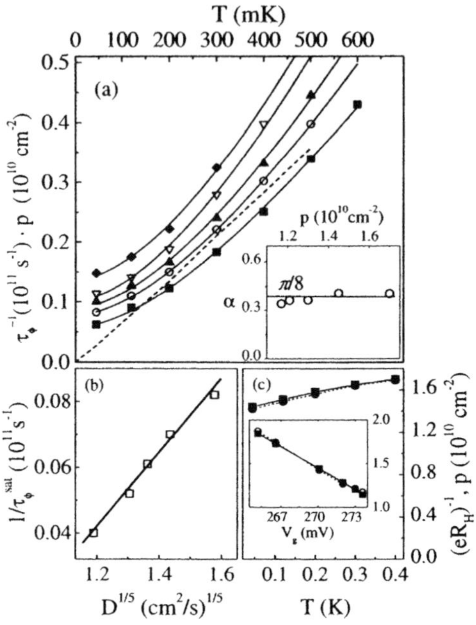

Figure 10. (a) Temperature dependence of the dephasing rate at different gate voltages, $p = 1.17$ - 1.7×10^{10} cm^{-2} (top). Solid lines are fits to Eq. (3). (b) The saturation value of the dephasing rate at $T = 0$ against diffusion coefficient. (c) Comparison of the hole densities determined from the Hall coefficient and SdH effect. Inset: Hole density measured by the Hall and SdH effect at different V_g, T= 45 mK.

transfer, and the linear term, Eq. (5), corresponds to interactions with small energy transfer in a disordered conductor. (To take into account a slight ($\sim$ 10%) increase of the hole density with T in the sample, we analyse the product $\tau_\varphi^{-1} \cdot p$, with density p measured directly at each temperature by the SdH effect.)

We have established that in the entire range of p and T, the data are well described by the quadratic term with a zero-temperature saturation value $1/\tau_\varphi^{sat}$ added. Coefficient α, found as an adjustable parameter, agrees within 20% accuracy with the value $\pi/8$, inset to Fig. 10(a). The results show that the linear term is suppressed by more than an order of magnitude, its value estimated using Eq. (5) for $p = 1.3 \times 10^{10}$ cm^{-2} is shown in Fig. 10(a) as a straight line.

Usually, WL in a disordered system is accompanied by the HHI effect, seen as a quantum correction to the conductivity and Hall coefficient. At small r_s, the two corrections are related as $\delta R_H(T)/R_H = -2\delta\sigma(T)/\sigma$ [5]. In Fig. 10(c) the Hall coefficient is presented as $(R_H e)^{-1}$ for different temperatures (solid squares). From [5] one can expect a variation of R_H of about 10% in the studied temperature range. However, the experimental temperature dependence of R_H agrees, within 2%, with the temperature dependence of the hole density measured by the SdH effect. We suggest that the suppression of the temperature dependence $\delta R_H(T)$, when compared with estimations based on the diffusive model [5] (valid at $k_B T\tau/\hbar \ll 1$), is linked to the observed absence of the linear term in the dephasing rate.

Let us briefly discuss the saturation of τ_φ^{-1}. In our case, the effect has one characteristic feature - the saturation becomes more pronounced with increasing p. In the experiment with a constant current, this excludes overheating as origin of the saturation and agrees qualitatively with the suggestion that it can be related to an external noise, Fig. 10(b). According to [40], the expected saturation is then $1/\tau_\varphi^{sat} \simeq D^{1/5}(\Omega e E_{ac})^{2/5}$, where Ω is the radiation frequency and E_{ac} is amplitude of an external ac field.

These experiments [41] have demonstrated that, surprisingly, the WL description is applicable to the description of the negative magnetoresistance of the 2DHG even at very large r_s. The phase breaking rate determined from the WL magnetoresistance agrees well with the Fermi-liquid T^2-dependence. At the same time the contribution from HHE effect to the Hall coefficient and the phase-braking time appears to be suppressed. A possible reason for this will be discussed at the end of the next section.

4. 'Metallic' regime of the 2DHG

Let us discuss the origin of the 'metallic' behaviour at larger hole densities in Fig. 11, where $r_s \sim 10 - 17$. The densities studied in our work are lower than in [27], so that the interband scattering can be neglected. Our analysis shows that at the highest densities ($p \sim 9 \cdot 10^{10}\mathrm{cm}^{-2}$) the increase of resistance with temperature can be explained simply by piezo-electric phonon scattering. However, with decreasing p another contribution becomes visible and at the lowest densities ($p \sim 2 \cdot 10^{10}\mathrm{cm}^{-2}$) the phonon scattering becomes altogether negligible. In Fig. 11(c) we plot the results of mobility calculations for GaAs structures [42, 43] (the phonon contribution for each density is added to the value of the resistance due to impurity scattering at $T = 0$). Fig. 12(a) shows the additional contribution at lower hole densities. It has a peak-like shape, with the maximum at $T \simeq 0.3T_F$ for different hole densities, and can be attributed to impurity scattering with temperature dependent screening [24, 25]. The presence of such a peak in $\rho(T)$ is in qualitative agreement with numerical calculations in [44] where it is explained by the transition to the non-degenerate regime at $T > 0.3T_F$. (Similar $\rho(T)$ with a peak has also been seen in [45].) At $T < T_{\max}$, a linear dependence is apparent and described by $\rho(T) = \rho_0 + C(p) \cdot T/T_F$. Coefficient $C(p)$ increases with decreasing density in accordance with [25].

There has recently been a model proposed in [46–48], which also predicts a linear dependence of resistivity of a form which is very similar to the result in [25]. However, it explains this behaviour in terms of elastic (coherent) electron scattering on the modulated density of other electrons (Friedel oscillation) caused by an impurity with a short-range potential. The phase of the Friedel oscillation,

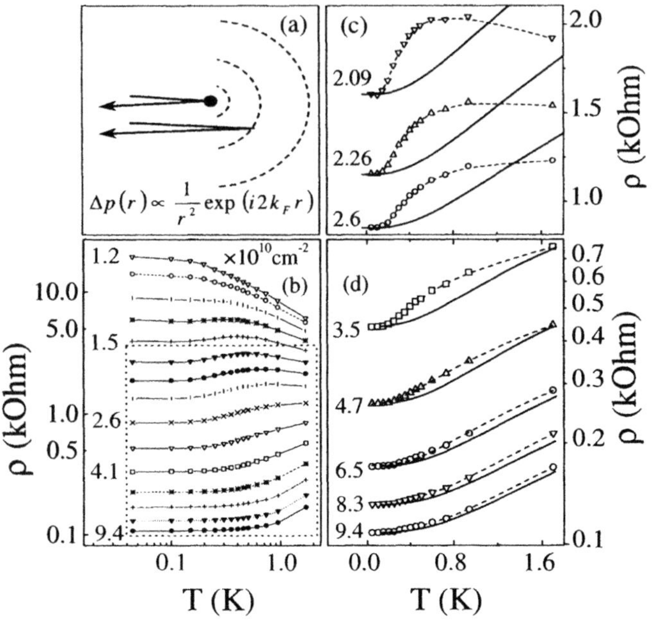

Figure 11. (a) Diagram of electron scattering by an impurity and the Friedel oscillation it produced. (b) Temperature dependence of the resistivity at different hole densities near the crossover in the sign of $d\rho/dT$. (c,d) Resistivity on the 'metallic' side of the crossover (symbols), and contribution to $\rho(T)$ due to phonon scattering (solid lines).

$\Delta p \propto \exp(i2k_F r)$, is such that the wave scattered from the impurity interferes constructively with the wave scattered from the oscillation, Fig. 11(a), leading to the quantum correction to the Drude conductivity σ_0. This correction has the same physical nature as the logarithmic interaction correction in [5], which exists in the diffusive regime, $k_B T\tau/\hbar < 1$. The new theory considers EEI in the ballistic, $k_B T\tau/\hbar > 1$, and in the intermediate regimes. It important to note that in the high-mobility structures where the crossover is studied the latter condition is usually satisfied even at low temperatures.

The model [46] gives several predictions to be tested experimentally.

Firstly, the logarithmic correction in the regime of multiple impurity scattering [5] becomes a linear temperature dependence for a single scatterer at $k_B T\tau/\hbar > 1$:

$$\begin{aligned} \delta\sigma(T) &= \frac{e^2}{\pi\hbar}\frac{k_B T\tau}{\hbar}\left(1 + \frac{3F_0^\sigma}{1+F_0^\sigma}\right) \\ &= \sigma_0\left(1 + \frac{3F_0^\sigma}{1+F_0^\sigma}\right)\frac{k_B T}{E_F} \end{aligned} \tag{6}$$

where F_0^σ is the Fermi-liquid interaction parameter in the triplet channel. The coefficient in the temperature dependence originates from two contributions: the

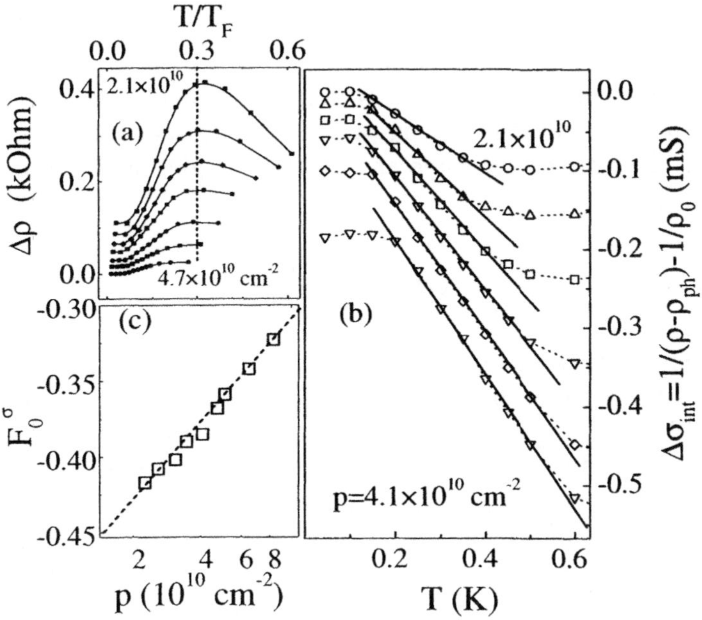

Figure 12. (a) Impurity scattering contribution $\Delta\rho$ against dimensionless temperature at different p. For clarity, curves in (a) and (b) are offset vertically from zero value at $T = 0$. (b) The same data as in (a) but in conductivity form, with linear fitting. (c) Fermi liquid parameter versus hole density.

first term is due to exchange processes (Fock) and the second due to direct interaction (Hartree). Similar to the diffusion regime, the sign of $d\rho/dT$ depends on the constant F_0^σ. It is interesting to note that for a given F_0^σ, dependence $\rho(T)$ can be 'metallic' even when the logarithmic correction at smaller $k_B T\tau/\hbar$ is 'insulating'. It is important that according to [46] the actual transition to the ballistic regime occurs at $k_B T\tau/\hbar \sim 0.1$, so that experiments on high-mobility structures can be easily driven into the ballistic regime.

Secondly, for a wide range of parameter F_0^σ the model allows the change of the sign of $d\rho/dT$ with parallel magnetic field - the effect seen in the experiments on the crossover. Magnetic field suppresses the Hartree correction in the triplet channel in Eq. (6), resulting in a universal, positive correction to the Drude conductivity in magnetic field, σ_0^B, and hence the 'insulating' behaviour of $\rho(T)$:

$$\delta\sigma = \sigma_0^B \frac{T}{T_F} \mathrm{at} B \geq B_S. \tag{7}$$

Here B_S is the field corresponding to the full spin polarisation of the 2D system, $B_S = 2E_F/g^*\mu_B$, where g^* is the Lande g-factor, μ_B is the Bohr magneton, and T_F is the Fermi temperature.

The authors of [46] note that the model [25], where the same functional dependence as in Eq. (6) is derived in terms of the temperature dependent screening, considers only the Hartree potential of interacting electrons and ignores the Fock contribution. As a result, it only predicts the positive sign of $d\rho/dT$ and does not allow the change in the sign of $d\rho/dT$ with magnetic field.

The Fermi-liquid constant F_0^σ has a significant physical meaning. It can be considered as the ratio of exchange and kinetic energies, and it also comes into the magnetic susceptibility [49]:

$$\chi(n) = \frac{\chi_0}{1 + F_0^\sigma} \tag{8}$$

Recently, there have been reports that in the 'metallic' state of the 2DEG in Si-MOSFETs the g-factor diverges when approaching the 'metal'-to-'insulator' transition [29], indicating the ferromagnetic (Stoner) instability expected in low-density 2D systems [2]. In our 2DHG the parameter r_s is twice as large as in [29], $r_s \sim 20$ near the crossover, and we can also expect a manifestation of the Stoner instability. Therefore, our analysis of the conductance in terms of Eq. (6) can give the value of the interaction parameter F_0^σ and show how close to the Stoner instability the system is - from Eq. (8) the instability is expected to occur at $F_0^\sigma = -1$.

In order to compare the results in the low-temperature range of $\rho(T)$ with Eq. (6), we present the data in Fig. 12(b) in the conductivity form : $\Delta\sigma(T) = (\rho(T) - \rho_{ph}(T))^{-1} - \rho_0^{-1}$. The condition for the ballistic regime $k_B T\tau/\hbar \geq 1$ is satisfied in our structure at $T > 50 - 100$ mK, and a linear fit of $\Delta\sigma(T)$ gives the value of parameter F_0^σ, which is presented in Fig. 12(c) for different hole densities. The following comments can be made on this result. Firstly, the interaction constant is negative and this provides the 'metallic' slope in $\rho\,(T)$. Secondly, its absolute value decreases with increasing density, which is in agreement with the expectation that the ratio of the exchange to kinetic energy of quasi-particles decreases to zero at large densities. Thirdly, one can see that the measured value does not exceed 0.42, and when extrapolated to the density of the crossover from 'metal' to 'insulator' ($p \sim 1.5 \times 10$ cm^{-2}), is much smaller than the value of $|F_o^\sigma| = 1$ expected for the Stoner instability. This implies that our description of the 'metallic' system as a weakly interacting Fermi-liquid is self-consistent.

Let us now turn to the increase of resistance with parallel field shown in Fig. 13(a), which is similar to that observed earlier on a 2DHG [50]. There is a characteristic feature in the data - a bend, which shifts towards smaller fields as the density is decreased. It was recently shown that this hump corresponds to the magnetic field B_S of full spin polarisation of the 2DHG [51]. However, there has been no quantitative analysis of the magnetoresistance in the 2DHG, which we now attempt. For an overall view of the magnetoresistance, we will first analyse it as is done for 2D electrons in a Si MOSFET [29] and GaAs heterostructure [52]. This analysis is based on the model [53] of positive magnetoresistance at $T = 0$, which considers the effect of parallel field on the screening of the impurity poten-

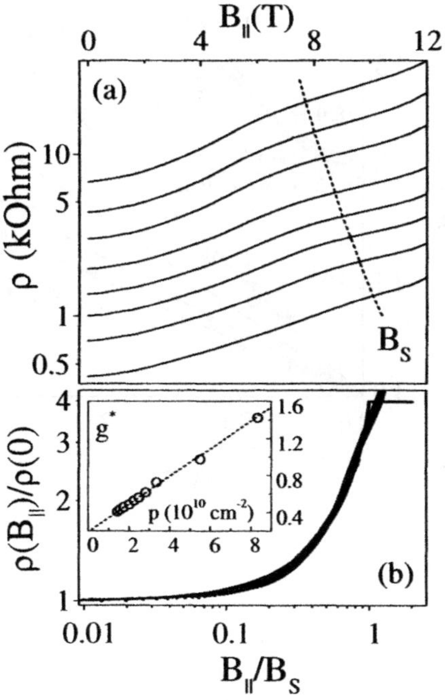

Figure 13. (a) Dependence of the resistivity on parallel magnetic field at $T = 50$ mK and $p = 1.43$; 1.57; 1.75; 2.03; 2.26; 2.49; 2.83; 3.36 $\times$ 10^{10} cm^{-2}, from bottom to top. (b) Scaled data, with an added curve $\rho(B_{||})$ for $p = 8.34 \times 10^{10}$ cm^{-2}; solid line is the result of the model [53]. Inset: dependence of the effective g-factor on the hole density, obtained from the value of B_S.

tial, affected by the presence of two spin subbands with different Fermi momenta k_F.

Fig. 13(b) shows $\rho(B_{||})/\rho(B_{||} = 0)$ as a function of dimensionless magnetic field B/B_S, with B_S found as a fitting parameter. Its value is shown by the dashed line in Fig. 13(a) and corresponds to the position of the hump. In accordance with [53], all the data in the density range $p = 1.43 - 8.34 \times 10^{10}$ cm^{-2} collapse on one curve which is close to the theoretical dependence, apart from the region near B_S where one can expect a contribution from another mechanism taking place [54]. Using the value of B_S one can obtain the effective g-factor, using the relation $g^* = 2E_F/\mu_B B_S$, whose dependence on the density is shown in the inset to Fig. 13(b). Note that contrary to [29], where similar analysis on a 2DEG in Si structures showed a rapid increase of the g-factor with decreasing density near the 'metal'-to-'insulator' transition, the g-factor in our case decreases with decreasing density. Similar behaviour was recently observed for 2D electrons in GaAs [55].

Fig. 14(inset) shows the temperature dependence of the magnetoresistance, where one can see that $B_{||}$ drives the 'metallic' state into 'insulator'. To compare this result with the prediction given by Eq. (7) we analyse the temperature dependence of the resistivity at field B_S. (In this analysis we neglected altogether the

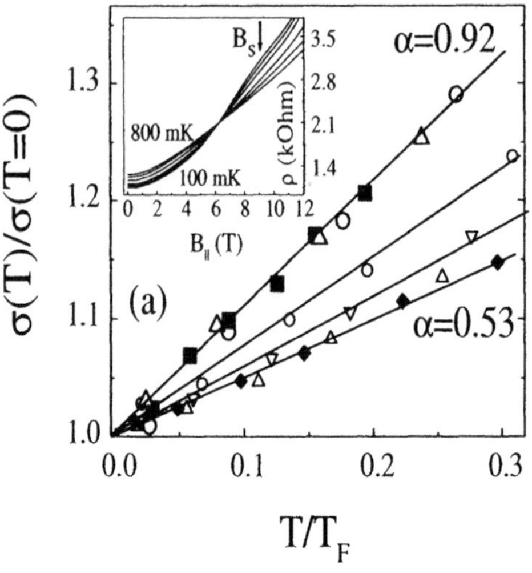

Figure 14. (a) Temperature dependence of the conductivity at $B_{||} = B_S$ for different hole densities. Coefficient $\alpha = 0.53$ is obtained for $p = 2.49$, 2.83×10^{10} cm^{-2} ; $\alpha = 0.62$ for $p = 2.26 \times 10^{10}$ cm^{-2} ; $\alpha = 0.74$ for $p = 2.03 \times 10^{10}$ cm^{-2}; and $\alpha = 0.92$ for $p = 1.43$, 1.57, 1.75×10^{10} cm^{-2}. Inset: $\rho(B_{||})$ for $p = 2.26\times 10^{10}$ cm^{-2}, at different temperatures: $T = 0.1$, 0.2, 0.3, 0.45, 0.6, 0.8 K. The crossover from 'insulator' to 'metal' is clearly seen.

phonon contribution, because at $T < 0.5$ K it becomes less than 5%, due to the four-fold resistance increase at $B_{||} = B_S$.) The resulting dependences, Fig. 14(a), are indeed linear, in accordance with Eq. (7). By extrapolation to $T = 0$, we find the value of the Drude conductivity σ_0^B and determine the slope α of the straight lines. Its value is close to the expected one, α=1, at all studied densities, although we find that agreement is better for smaller p, where α increases to 0.92. Such a behaviour of α can be attributed to the fact that in the real system the scatterers are not exactly point-like, as assumed in the theory. However, with decreasing density and increasing Fermi wavelength, $\lambda_F \propto p^{-1/2}$, the approximation of short-range scatterers becomes more applicable.

It is interesting to note that in the previous study of the crossover region on similar structures [41], i.e. in the range of hole densities from 1.2 to 1.7×10^{10} cm^{-2} in Fig. 1(b), only the weak localisation effect was seen. Surprisingly, the interaction correction, estimated from the diffusion theory [5], appeared to be suppressed both in the Hall coefficient and the phase breaking rate. The origin of this can be understood if the theory [46] is taken into account. According to this theory, the transition between the diffusion and ballistic regimes is realised at $k_B T\tau/\hbar = 0.1 - 1$ where the interaction correction is minimal, which corresponds to the conditions of the NMR experiments. The temperature dependence of the Hall coefficient is also expected to be significantly suppressed when $k_B T \tau/\hbar > 0.1$ [47]. Therefore in the intermediate regime studied earlier the interaction correction can become negligible compared with that of weak localisation. In the present section we have discussed the results at larger hole densities where the parameter $k_B T \tau/\hbar$ is $1 - 10$, hence the interaction correction develops and dominates the weak localisation contribution.

In this part we have studied the temperature dependence of the conductivity and the magnetoresistance in parallel field of a low-density (large r_s) 2D hole gas in the 'metallic' phase, near the crossover in the sign of $d\rho/dT$ [56]. We have demonstrated that the 'metallic' character of $\rho(T)$ and the positive magnetoresistance are caused by the hole-hole interaction in the ballistic limit $k_B T \tau/\hbar > 1$. The Fermi-liquid constant F_0^σ, which determines the sign of $d\rho/dT$, has been found. Its value near the crossover appears to be significantly smaller than expected for the ferromagnetic instability.

5. Conclusion

Our results on high-mobility 2D electron and hole gases show that the properties of the studied systems near the crossover can be explained within Fermi-liquid approaches. At the same time, it is clear that conventional approaches do not exclude some exciting physics which the new low-density high-mobility 2D systems distinctly demonstrate.

References

1. E. Abrahams, P. W. Anderson, D. C. Licciardello, and T. V. Ramakrishnan, Phys. Rev. Lett. **42**, 673 (1979).
2. T. Ando, A. B. Fowler, and F. Stern, Rev. Mod. Phys. **54**, 347 (1982).
3. D. J. Newson, M. Pepper, and T. J. Thornton, Phil.Mag. B **56**, 775 (1987).
4. E. I. Laiko, A. O. Orlov, A. K. Savchenko, E. A. Il'ichev, and E. A. Poltoratskii, Sov. Phys. JETP **66**, 1258 (1987).
5. B. L. Altshuler and A. G. Aronov, in *Electron -Electron Interaction in Disordered Systems*, edited by A. L. Efros and M. Pollak, North Holland, Amsterdam (1985).
6. B. I. Shklovskii and A. L. Efros, *Electronic Properties of Doped Semiconductors*, Springer Verlag, Berlin (1984).
7. A.M. Finkel'stein, Sov. Phys. JETP **57**, 97 (1983).
8. D. L. Shepelyansky, Phys. Rev. Lett. **73**, 2607 (1994).
9. F. von Oppen, T. Wettig, and J. Muller, Phys. Rev. Lett. **76**, 491 (1996).
10. D. Weimann and J.-L. Pichard, Phys. Rev. Lett. **77**, 1556 (1996); G. Benenti, X.Waintal, and J.-L. Pichard, Phys. Rev. Lett. **83**, 1826 (1999).
11. S. V. Kravchenko, G. V. Kravchenko, J. E. Furneaux, V. M. Pudalov and M. DIorio, Phys. Rev. B **50**, 8039 (1994); S. V. Kravchenko, D. Simonian, M. P. Sarachik, W. Mason and J. E. Furneaux, Phys. Rev. Lett. **77**, 4938 (1996); D. Simonian, S. V. Kravchenko, M. P. Sarachik, and V. M. Pudalov, Phys. Rev. Lett. **79**, 2304 (1997).
12. D. Popovic, A. B. Fowler, and S. Washburn, Phys. Rev. Lett. **79**, 1543 (1997).

13. P. T. Coleridge, R. L. Williams, Y. Feng, and P. Zawadzki, Phys. Rev. B **56,** R12764 (1997).
14. V. Senz, T. Heinzel, T. Ihn, K. Ensslin, G. Dehlinger, D. Grtzmacher, and U. Gennser, Phys. Rev. B **61,** 5082 (2000).
15. Y. Hanein, U. Meirav, D. Shahar, C. C. Li, D. C. Tsui, and H. Shtrikman, Phys. Rev. Lett. **80,** 1288 (1998).
16. M. Y. Simmons, A. R. Hamilton, M. Pepper, E. H. Linfield, P. D. Rose, D. A. Ritchie, A. K. Savchenko and T. G. Griffiths, Phys. Rev. Lett. **80,** 1292 (1998).
17. E. Abrahams, S. V. Kravchenko, and M. P. Sarachik, Rev. Mod. Phys. **73**, 251 (2001).
18. V. Dobrosavljevic, E. Abrahams, E. Miranda, and S. Chakravarty, Phys. Rev. Lett. **79,** 455 (1997).
19. C. Castellani, C. Di Castro, and P. A. Lee, Phys. Rev. B **57,** 9381 (1998).
20. S. He and X. C. Xie, Phys. Rev. Lett. **80,** 3324 (1998).
21. D. Belitz and T. R. Kirkpatrick, Phys. Rev. B **58,** 8214 (1998).
22. Q. Si and C. M. Varma, Phys. Rev. Lett. **81,** 4951 (1998).
23. B. L. Altshuler and D. L. Maslov, Phys. Rev. Lett. **82,** 145 (1999).
24. S. Das Sarma and E. H. Hwang, Phys. Rev. Lett. **83,** 164 (1999).
25. A. Gold and V. T. Dolgopolov, Phys. Rev. B, **33,** 1076 (1986).
26. Y. Meir, Phys. Rev. Lett. **83,** 3506 (1999).
27. Y. Yaish, O. Prus, E. Buchstab, S. Shapira, G. Ben Yoseph, U. Sivan, and A. Stern, Phys. Rev. Lett. **84**, 4954 (2000).
28. S.S. Murzin, S. I. Dorozhkin, G. Landwehr, and A. C. Gossard, JETP Lett. **67**, 113 (1998).
29. A. A. Shashkin, S. V. Kravchenko, V. T. Dolgopolov, and T. M. Klapwijk, Phys. Rev. Lett. **87**, 086801 (2001).
30. S. C. Dultz, and H. W. Jiang, Phys. Rev. Lett. **84,** 4689 (2000).
31. J. Yoon, C. C. Li, D. Shahar, D. C. Tsui, and M. Shayegan, Phys. Rev. Lett. **82,** 1744 (1999).
32. S. Bogdanovich and D. Popovic, cond-mat/0106545 (2001).
33. S.S. Safonov, S.H.Roshko, A.K.Savchenko, A.G.Pogosov, and Z.D.Kvon, Phys. Rev. Lett. **86,** 272 (2001).
34. M. Y. Simmons, A. R. Hamilton, M. Pepper, E. H. Linfield, P. D. Rose, and D. A. Ritchie, Phys. Rev. Lett. **84**, 2489 (2000).
35. V. Senz, T. Ihn, T. Heinzel, K. Ensslin, G. Dehlinger, D. Grtzmacher, and U. Gennser, Phys. Rev. Lett. **85**, 4357 (2000).
36. B. Tanatar and D. M. Ceperley, Phys. Rev. B **39,** 5005 (1989).
37. K. Hirakawa, Y. Zhao, M. B. Santos, M. Shayegan, and D. C. Tsui, Phys. Rev. B **47,** 4076 (1993); H. L. Stormer, Z. Schlesinger, A. Chang, D. C. Tsui, A. C. Gossard, and W. Wiegmann, Phys. Rev. Lett. **51**, 126 (1983).
38. H.-P. Wittman and A. Schmid, J. Low Temp. Phys. **69**, 131 (1987).
39. M. Reizer and J. W. Wilkins, Phys. Rev. B **55** , R7363 (1997).
40. I. L. Aleiner, B. L. Altshuler, and M. E. Gershenson, Wave Random Media **9** , 201 (1999) [cond-mat/9808053].
41. Y.Y. Proskuryakov, A.K. Savchenko, S.S.Safonov, M. Pepper, M.Y. Simmons, and D.A. Ritchie, Phys. Rev. Lett. **86**, 4895 (2001).
42. V. Karpus, Sem. Sci. Tech. **5**, 691 (1990).
43. V. Karpus, Sov. Phys. Semicond. **21**, 1180 (1987).
44. S. Das Sarma and E. H. Hwang, Phys. Rev. B **61**, R7838 (2000).
45. A. P. Mills, Jr., A. P. Ramirez, L. N. Pfeiffer, and K. W. West, Phys. Rev. Lett. **83**, 2805 (1999).
46. G. Zala, B. N. Narozhny, and I. L. Aleiner, Phys. Rev. B **64**, 214204 (2001).
47. G. Zala, B. N. Narozhny, and I. L. Aleiner, Phys. Rev. B **64**, 201201 (2001).
48. G. Zala, B. N. Narozhny, and I. L. Aleiner, cond-mat/0109531.

49. E. M. Lifshitz and L. P. Pitaevskii, *Landau and Lifshitz Course of Theoretical Physics*, Vol. 9, Pergamon Press (1980).
50. J. Yoon, C. C. Li, D. Shahar, D. C. Tsui, and M. Shayegan, Phys. Rev. Lett. **84**, 4421 (2000).
51. E. Tutuc, E. P. De Poortere, S. J. Papadakis, and M. Shayegan, Phys. Rev. Lett. **86**, 2858 (2001).
52. V. S. Khrapai, E. V. Deviatov, A. A. Shashkin, V. T. Dolgopolov, cond-mat/0005377.
53. V. T. Dolgopolov and A. Gold, JETP Lett. **71**, 27 (2000).
54. S. Das Sarma and E. H. Hwang, Phys. Rev. Lett. **84**, 5596 (2000).
55. E. Tutuc and M. Shayegan, cond-mat/0107145.
56. Y. Y. Proskuryakov, A. K. Savchenko, S. S. Safonov, M. Pepper, M. Y. Simmons, and D. A. Ritchie, cond-mat/0109261 (2001).

SPECTRAL DECOMPOSITION OF GEODESIC FLOWS ON CONSTANT CURVATURE SURFACES

BORIS MUZYKANTSKII
University of Warwick, Coventry CV4 7AL, UK
STEPHEN ROBERTS
University of Cambridge, Cambridge CB3 0HE, UK

1. Introduction

The dynamics of a Hamiltonian system can either be described in terms of the trajectories $x_0 \to x(t) = U_t x_0$ in the phase space, or by specifying the laws of evolution of a function $\varphi(x)$ on the phase space: $\varphi \to \hat{U}_t \varphi$. The evolution operator $\hat{U}_t$ which pulls the function backwards along the trajectories is defined by

$$\hat{U}_t \varphi(x) = \varphi(U_t x) = \varphi(x(t)).$$

Quantum mechanics has a natural relation to the trajectory based approach through Feynman's Path Integral. In the semiclassical limit the path integral can be approximated by the saddle point contributions (which turn out to be the classical orbits) and leads to the Gutzwiller trace formula [1] for the Green's function of the quantum mechanical Hamiltonian. This is a useful tool for studying such problems as the quantum energy level correlations but it requires a knowledge of the long periodic orbits in order to obtain the correlations at small energy differences [2, 3]. The situation is particularly bad in chaotic systems, where the periodic orbits proliferate exponentially with length. In practice some uncontrolled approximations about correlations of actions for different periodic orbits are made to get the analytical results.

In contrast, the flow based approach seems suitable for studying the behaviour of chaotic systems at long times when due to the decay of correlations the dynamics becomes trivial. To use this approach in quantum chaos two problems however need to be overcome. First, it is unclear how to relate the quantum mechanics to the evolution operator since there is no analogue of the Feynman path integral. Second, one needs to be able to calculate various properties of the evolution operator which are naturally formulated in terms of the spectral decomposition of $\hat{U}_t$ in decaying eigenmodes.

I.V. Lerner et al. (eds.),
***Strongly Correlated Fermions and Bosons in Low-Dimensional Disordered Systems*, 241–258.**

The first problem has been addressed in [4, 5] where it has been conjectured that the (suitably averaged) correlation functions of the quantum energy levels and/or quantum eigenfunctions can be generated from an effective action of non-linear σ-model type – the so called "ballistic" σ-model. This action involves the Liouville operator $\hat{L} = \frac{d}{dt}\Big|_{t=0} \hat{U}_t$, and for example, the correlation function of the energy levels is given by the sum over eigenvalues of $\hat{L}$ (see Eqs. (50,51)). The inspiration for the ballistic σ-model comes from the well developed theory of weakly disordered metals, where the disorder averaged properties of an ensemble of macroscopically identical systems are calculated using a similar σ-model with a diffusion operator instead of the Liouvillian one. (See [6, 7] for a detailed discussion of the diffusive case and [8] for a recent overview of ballistic σ-model applicability).

This paper is devoted to the second problem. We first consider the spectral decomposition of $\hat{U}_t$ in the space of square integrable functions on the energy shell Γ_E

$$\hat{U}_t = \sum_{\lambda} e^{\lambda t} |f_\lambda\rangle\langle g_\lambda|, \tag{1}$$

where f_λ, g_λ are right and left eigenfunction of $\hat{U}_t$ respectively

$$\hat{U}_t f_\lambda = e^{\lambda t} f_\lambda, \tag{2}$$

$$g_\lambda \hat{U}_t = e^{\lambda t} g_\lambda. \tag{3}$$

It turns out that for chaotic dynamics such a decomposition does not exist and the functional space needs to be enlarged (see below). We will always assume that the energy shell Γ_E is compact.

The evolution operator $\hat{U}_t$ preserves the Liouvillian measure μ on Γ_E and thus is unitary with respect to the scalar product

$$\langle\phi|\psi\rangle = \int_{\Gamma_E} \phi(x)\psi^*(x)d\mu. \tag{4}$$

As a consequence all eigenvalues λ in decomposition (1) are purely imaginary[1] and right and left eigenfunctions are related by complex conjugation $f_\lambda(x) = g^*_\lambda(x)$.

Along a trajectory $x(t) = U_t x_0$ an eigenfunction f_λ obeys

$$f_\lambda(x(t)) = e^{\lambda t} f_\lambda(x), \tag{5}$$

which for a periodic trajectory with period T means that either f_λ is constant on $x(t)$ (and $\lambda =$) or

$$\lambda = \frac{2\pi i}{T} n \qquad (n \text{ is integer}). \tag{6}$$

[1] strictly speaking the λ-s are the eigenvalues of the Liouvillian operator $\hat{L}$. For brevity we will call them the eigenvalues of the flow.

In a system with only finite number of different periods of primary periodic orbits condition (6) completely determines the spectrum of $\hat{U}_t$. For example in section 3 we consider geodesic motion on a sphere, where all primary periodic orbits have the same period $T = 2\pi R/v$, where R is the radius of the sphere and v is the velocity on Γ_E. In this system the eigenvalues are given by

$$\lambda_n^{\text{sphere}} = i\frac{v}{R}n. \tag{7}$$

In a chaotic system the periodic orbits are dense and (in general) have incommensurate periods, so an eigenfunction from L_2 will be non-zero on infinitely many periodic trajectories and conditions (6) on the corresponding eigenvalue become incompatible. The only exception is a constant eigenfunction with eigenvalue $\lambda = 0$. For a different (and more rigorous) proof of this fact see [9]. Spectral decomposition (1) with just one term clearly does not make sense and to fix it for chaotic dynamics we need to extend the functional space. A non-trivial solution of Eq. (5) outside L_2 can be constructed with the help of the Dirac δ-function. For example the integral

$$f_\lambda^\delta(x) = \int dte^{\lambda t}\delta(x - x(t)), \tag{8}$$

is a formal solution of Eq. (2) corresponding to eigenvalue λ. Below we call these solutions eigenfunctionals[2]. The condition (6) still holds if the trajectory $x(t)$ is periodic. Any complex values of λ are allowed if $x(t)$ is not periodic. This is not surprising because $\hat{U}_t$ is no longer unitary once we extended the functional space and the invariant scalar product is lost (Eq. (4) does not make sense if ϕ and ψ are both δ-functions).

To summarise, for the chaotic dynamics the functional space for decomposition (1) is either too small and does not contain non-trivial eigenfunctions (in case of L_2) or too large and contains eigenfunctions with all possible complex eigenvalues (in case when δ-functions are allowed). Ideally, one would hope to find an intermediate functional space that would yield an infinite but countable set of eigenfunctions. We are not aware of such choice. Instead we propose a concrete procedure for picking up a subset of eigenvalues that is in some sense complete. The corresponding eigenfunctionals are, in general, quite non-trivial and, in particular, are not given by construction (8). All selected eigenvalues have non-positive real parts, so Eq. (1) is a decomposition in decaying eigenmodes. For a chaotic system the set of selected eigenvalues is countable and consists of Ruelle resonances [10] (see section 2).

The general procedure is outlined in section 2. In section 3 we apply it to the case of geodesic motion on a two dimensional sphere. This system is integrable and all eigenvalues end up being purely imaginary and infinitely degenerate (see Eq. 7). In section 4 we extend the calculation of section 3 to the geodesic motion

[2] The discussion in section 2 after Eq (15) explains this naming convention

on a constant negative curvature surface (CNCS). These systems are strongly chaotic, and the resulting non-trivial eigenvalues are non-degenerate and have non-vanishing negative real parts. Finally, in section 5 we discuss the relevance of the spectral decomposition (1) to the ballistic σ-model.

Throughout this paper we try to introduce the relevant techniques without paying much attention to the mathematical rigour. In particular, the conditions of convergence of the spectral decompositions are not discussed at all. For a more formal presentation covering these issues we refer the reader to our paper [11].

2. Spectral decomposition of the resolvent of $\hat{L}$

Decomposition (1) would imply that the correlation function

$$\langle\xi|\hat{U}_t|\varphi\rangle = \int_{\Gamma_E} d\mu\xi(x)\hat{U}_t\varphi^*(x) = \int_{\Gamma_E} d\mu\,\xi(x)\phi^*(U_t(x)),$$

considered as a function of t (with φ and ξ being some fixed C^∞ functions) can be decomposed into the infinite sum

$$\langle\xi|\hat{U}_t|\varphi\rangle = \sum_\lambda e^{\lambda t}\langle g_\lambda|\varphi\rangle\langle\xi|f_\lambda\rangle. \tag{9}$$

Taking the Laplace transform of the both sides and using $\hat{U}_t = e^{\hat{L}t}$ we introduce the resolvent of $\hat{L}$

$$\hat{R}(z) = -\int_0^\infty e^{-zt}\hat{U}_t dt = (\hat{L}-z)^{-1} \tag{10}$$

and obtain for its matrix element

$$F_{\xi,\varphi}(z) \equiv \langle\xi|\hat{R}(z)|\varphi\rangle = \sum_\lambda \frac{\langle\xi|f_\lambda\rangle\langle g_\lambda|\varphi\rangle}{\lambda - z}. \tag{11}$$

Conversely we can obtain decomposition (11) (and therefore Eq.1) by treating the integral

$$F_{\xi,\varphi}(z) = -\int_0^\infty e^{-zt}dt\int_{\Gamma_E} d\mu\,\xi(x)\phi^*(U_t(x)) \tag{12}$$

as a function of the complex variable z and analysing its singularities. Since the functions ϕ and ξ are smooth and Γ_E is compact the correlation function $\langle\xi|\hat{U}_t|\varphi\rangle$ is bounded and the integral (12) converges in the half plane $\mathrm{Re}\, z > 0$. The analytical continuation of $F(z)$ to the half plane $\mathrm{Re}\, z < 0$ will have some poles and possibly branch cuts. The position of poles should be identified with the eigenvalues of $\hat{U}_t$, while the the branch cuts (if any) correspond to the continues parts of the spectra of $\hat{U}_t$.

The poles of the resolvent $\hat{R}(z)$ for a chaotic flow were considered by Ruelle [10] and are called Ruelle resonances. He also conjectured that for a chaotic flow the only singularities in the analytic continuation of $F_{\xi,\varphi}(z)$ are poles, and their positions do not depend on the choice of the functions φ and ξ. An integrable flow will (in general) have branch cuts along the imaginary axis.

To obtain the eigenfunctions of $\hat{U}_t$ corresponding to the eigenvalue λ we consider the pole of the function $F_{\xi,\varphi}(z)$ at $z = \lambda$. The residue at this pole is a linear functional of ξ which depends anti-linearly on φ

$$\operatorname*{Res}_{z=\lambda} F_{\xi,\phi}(z) = K_{\varphi}[\xi].$$

Here and below the arguments of a linear functional are given in square brackets. Consider the linear space $\mathcal{K}$ spanned by the functionals $K_{\varphi}[\,]$ for all possible φ. In the non-degenerate case, this space is one dimensional, i.e., different choices of φ yield proportional functionals $K_{\varphi}[\,]$

$$K_{\varphi}[\xi] = g[\varphi]^* f[\xi], \tag{13}$$

where $f[\,]$ and $g[\,]$ are some linear functionals on C^{∞}. In the degenerate case when $\dim \mathcal{K} > 1$ we select a basis $f_k[\,] \in \mathcal{K}$ where $1 \leq k \leq \dim \mathcal{K}$. Then decomposition (13) contains many terms:

$$K_{\varphi}[\xi] = \sum_{k=1}^{\dim \mathcal{K}} g_k[\varphi]^* f_k[\xi],$$

and the eigenvalue λ has degeneracy $\dim \mathcal{K}$. To simplify the notation we consider the non-degenerate case below.

When the flow can be decomposed over the space of square integrable functions, the functionals $f[\,]$ and $g[\,]$ correspond to the scalar product with some L_2 functions

$$f[\xi] = \langle \xi | f_{\lambda} \rangle; \qquad g[\varphi] = \langle \varphi | g_{\lambda} \rangle, \tag{14}$$

where f_{λ}, g_{λ} are now the "normal" (i.e., from L_2) right and left eigenfunctions of $\hat{U}_t$ obeying Eq. 2. In a more general (and in particular chaotic) case the decomposition (9) still holds if the scalar products $\langle \xi | f_{\lambda} \rangle$ and $\langle \varphi | g_{\lambda} \rangle$ are replaced by the result of application of the linear eigenfunctionals $f_{\lambda}[\,], g_{\lambda}[\,]$ to the functions ξ, ϕ

$$\langle \xi | \hat{U}_t | \varphi \rangle = \sum_{\lambda} e^{\lambda t} g_{\lambda}[\varphi]^* f_{\lambda}[\xi]. \tag{15}$$

This decomposition is found to be valid for just positive times $t > 0$ [11]. By allowing generic linear functionals we drastically extend the functional space. For example, the Dirac δ-function (as well as its derivatives) can be viewed as a linear functionals on C^{∞}. Applying this functional to $\phi(x) \in C^{\infty}$ gives the value of ϕ at zero

$$\delta(x)[\varphi] = \varphi(0).$$

Clearly such a functional cannot be represented as a scalar product with a function from L_2.

We define the action of $\hat{U}_t$ on a functional $f[\,]$ by the pull-back

$$(\hat{U}_t f)[\varphi] = f[\hat{U}_t \varphi], \tag{16}$$

where $\varphi \in C^\infty$. Then the eigenfunctional $f_\lambda[\,]$ still obeys Eq. (2). Note, however, that contrary to the usual decomposition over L_2, Eq. (2) does not define the eigenfunctionals and eigenvalues uniquely. In fact, as we mentioned in the introduction, in the space of linear functionals Eq (2) admits solutions for arbitrary complex λ. To select the subset of λ entering decomposition (15) we need to find the positions of the poles in the analytical continuation of the matrix element of the resolvent. To find the corresponding right and left eigenfunctionals we need to decompose the residue at $z = \lambda$ according to Eq. (13).

Using the unitarity of $\hat{U}_t$ on C^∞ functions a decomposition for negative times $t < 0$ can be derived [11] from (15).

$$\langle \xi | \hat{U}_t | \varphi \rangle = \sum_\lambda e^{-(\lambda)^* t} f_\lambda[\varphi]^* \, g_\lambda[\xi] \tag{17}$$

For integrable dynamical systems, where the eigenvalues λ are purely imaginary and $f_\lambda = g_\lambda$, the positive and negative time descriptions (15) and (17) are the same (see e.g. the decomposition for the geodesic motion on the sphere (34)). By contrast, for chaotic systems, where λ is complex, we require two different decompositions to describe the future and past evolutions. The complete sets of eigenfunctions f_λ used to characterise the forward time evolution differ from the eigenfunctions g_λ for the evolution backwards in time. The need for two separate descriptions depending on the direction of time reflects the irreversible nature of these dynamical systems.

3. Example of Integrable dynamics – motion on a sphere

We now consider the geodesic motion on a two dimensional sphere. We use this example to illustrate the general method for constructing spectral decompositions of the previous section.

The 2-sphere may be embedded in 3-dimensional Euclidean space as the surface

$$|\mathbf{r}|^2 = x^2 + y^2 + z^2 = R^2. \tag{18}$$

We denote the position of the particle by $\mathbf{r}$ and its velocity by $\mathbf{v}$. The velocity is tangential to the sphere, lying in the plane $\mathbf{r} \cdot \mathbf{v} = 0$. Under the motion the particle moves along the great circle passing through $\mathbf{r}$ in the direction of $\mathbf{v}$ with the speed v. The trajectories of this motion (arcs of great circles) are geodesics of the sphere (see Fig. 1).

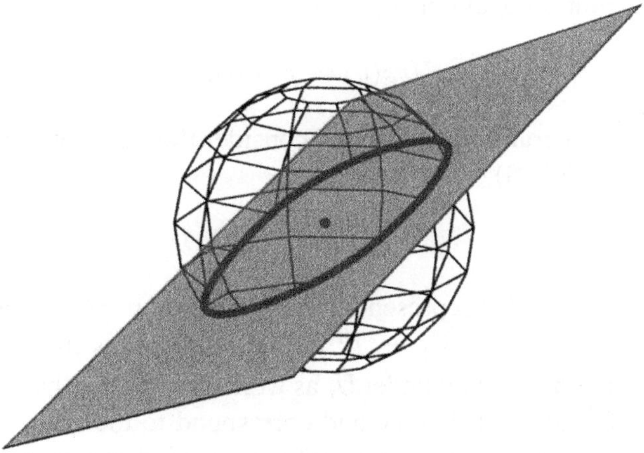

Figure 1. Trajectories for the geodesic motion on the sphere are great circles (intersections of the sphere with a plane going through the centre)

On an energy shell the modulus v of the velocity is constant. Therefore a point on the energy shell may be specified by its position $\mathbf{r}$ on the sphere and the direction of its velocity $\mathbf{v}$. We take O to be the point on the energy shell with position $(R, 0, 0)$ and velocity $(0, 0, v)$. The isometries of the sphere are given by the group of transformations $SO(3)$. For each point on the energy shell there is a unique transformation from the group $SO(3)$ mapping O to that point. In this way the energy shell may be identified with the group $SO(3)$. This identification is a diffeomorphism mapping the point gO to the group element $g \in SO(3)$.

The elements g can be written as a product of three rotations.

$$g(\phi, \theta, \psi) = R^x_\phi R^y_\theta R^x_\psi \tag{19}$$

The rotation R^x is about the x-axis and R^z rotates about the z-axis. The angles θ, ϕ, ψ of rotation are known as the Euler angles. The angles ϕ and θ of g give the coordinates of the position on the sphere of gO

$$\mathbf{r} = R(\cos\theta, -\sin\phi\sin\theta, \cos\phi\cos\theta).$$

The angle ψ specifies the direction of the velocity corresponding to gO.

From now until the end of this section we shall measure time in units R/v and distance in units of R. After a time t the geodesic motion moves the point O to $h_t O$ where

$$h_t = R^y_t, \tag{20}$$

By applying the transformation g we see that the point gO moves to gh_tO. In terms of the group $SO(3)$ the geodesic motion on the sphere is given by a right shift by h_t

$$U_t g = g h_t \tag{21}$$

and hence the evolution operator is given by

$$\hat{U}_t\varphi(g) = \varphi(gh_t). \tag{22}$$

It is useful to generalise Eq. (22) and consider the right multiplication by any group element $h \in SO(3)$

$$T^R(h)\varphi(g) = \varphi(gh). \tag{23}$$

$T^R(h)$ is the right regular representation of the group $SO(3)$ on the space of functions $L_2(SO(3))$. Since $\hat{U}_t = T^R(h_t)$ the subspaces that remain invariant under T^R will remain invariant under $\hat{U}_t$ as well. On the other hand, the subspaces invariant under T^R are well known and correspond to the decomposition of the right regular representation of $SO(3)$ into irreducible representations (see e.g. [12]). The irreducible representations of $SO(3)$ are classified by the total angular momentum $l = 0, 1, \ldots$ which takes integer values. The irreducible representation T^l has dimension $2l + 1$. It turns out that the right regular representation T^R contains $2l + 1$ copies of T^l for each allowed value of l (see e.g., [12])[3]

$$T^R = \bigoplus_{l=0}^{\infty} \bigoplus_{m=-l}^{l} T^{lm} \tag{24}$$

The corresponding invariant subspaces H^{lm} are spanned by the orthonormal functions

$$|l, m, n\rangle = e^{im\phi} e^{in\psi} P^l_{mn}(\cos\theta) \quad n = -l, -l+1, \ldots l-1, l, \tag{25}$$

where $P^l_{mn}(\cos\theta)$ are the Legendre polynomials.

The Casimir operator Ω is defined on the space $L_2(SO(3))$ by

$$\Omega = \frac{1}{\sin\theta}\frac{\partial}{\partial\theta}\sin\theta\frac{\partial}{\partial\theta} + \frac{1}{\sin^2\theta}\left(\frac{\partial^2}{\partial\phi^2} + \cos\theta\frac{\partial^2}{\partial\phi\partial\psi} + \frac{\partial^2}{\partial\psi^2}\right).$$

It can be verified that the Casimir operator commutes with the regular representation. Therefore by Schur's Lemma Ω is constant on each of the invariant subspaces H^{lm}. As seen from Eq. (25) the functions $|l, m, 0\rangle$ do not depend on the angle of the velocity ψ and therefore can be viewed as functions on the sphere itself. On the space of such functions the Casimir operator coincides with the Laplacian operator Δ (also known as the operator of total angular momentum). The eigenvalues of Δ are well known

$$\Delta|l, m, 0\rangle = l(l+1)|l, m, 0\rangle, \tag{26}$$

[3] in fact this is always the case for a regular representation, it contains all the irreducible representations and the number of copies of each representation is equal to its dimension

from which we deduce that on the space H^{lm} the Casimir operator acts as the multiplication by $l(l+1)$

$$\Omega|l,m,n\rangle = l(l+1)|l,m,n\rangle. \tag{27}$$

In fact, for the geodesic motion on any constant curvature surface the spectrum of the Casimir operator on the constant energy shell co-insides with the spectrum of the Laplacian on the configuration space (although the multiplicities are different).

Since $\hat{U}_t$ leaves each of the spaces H^{lm} invariant, we decompose the flow on each of these spaces separately. On a particular subspace H^{lm} it is sufficient to find the matrix elements of $\hat{U}_t$ between the basis functions (25). Using $\hat{U}_t = T^{lm}(h_t)$ together with the expressions for the matrix elements of $T^{lm}(h)$ (see [12]) we get

$$\langle l,m,n|\hat{U}_t|l,m,n'\rangle = P^l_{nn'}(\cos t). \tag{28}$$

Combining (28) with (10) we see that the matrix elements of the resolvent $F_{nn'}(z)$ are given by the Laplace transform

$$F_{nn'}(z) = \langle n|R(z)|n'\rangle = -\int e^{-zt} P^l_{nn'}(\cos t), \tag{29}$$

where we omitted the common indexes l, m from the notations for basis vectors in H^{lm}. According to our general procedure, the eigenvalues of $\hat{U}_t$ are given by the positions of the singularities of the analytical continuation of $F_{nn'}(z)$. This procedure is particularly simple in the present case because $P^l_{nn'}$ are polynomials of degree n. So $P^l_{nn'}(\cos t)$ can be written as finite sums of exponential functions:

$$P^l_{nn'}(\cos t) = \sum_{p=-l}^{l} C^l_{nn'}(p) e^{ipt}, \tag{30}$$

where $C^l_{nn'}(p)$ are some coefficients. As a result

$$F_{nn'}(z) = \sum_{p=-l}^{l} \frac{C^l_{nn'}(p)}{ip - z}. \tag{31}$$

From Eq. (31) and restoring the correct units we immediately see that the eigenvalues are given by (6). The eigenfunctions are found in the appendix where we show that the coefficients $C^l_{nn'}(p)$ are given by the product

$$C^l_{nn'}(p) = a^l_{np}\left(a^l_{n'p}\right)^* \tag{32}$$

Therefore the eigenvectors corresponding to the eigenvalue $\lambda = ip$ have the form

$$|f^{lm}_p\rangle = \sum_n a^l_{np}|l,m,n\rangle. \tag{33}$$

On each of the invariant subspaces H^{lm} the flow $\hat{U}_t$ is non-degenerate, but on the whole functional space $L_2(SO(3))$ the eigenvalues are infinitely degenerate. The eigenspace corresponding to the eigenvalue $\lambda = ip$ is spanned by the vectors $|f_p^{lm}\rangle$ with $l \geq p$ and $-l \leq m \leq l$.

Combining contributions from the eigenfunctions in each of the subspaces H^{lm} of the irreducible representations (24) we obtain a decomposition of the evolution operator on the whole space $L_2(SO(3))$.

$$\langle\xi|\hat{U}_t|\varphi\rangle = \sum_{l=0}^{\infty} \sum_{m=-l}^{l} \sum_{p=-l}^{l} e^{ipt} \langle f_p^{lm}|\varphi\rangle\langle\xi|f_p^{lm}\rangle \tag{34}$$

The relative weightings of the eigenfunctions appearing in this expansion oscillate with frequencies p determined by their eigenvalues.

4. Example of Chaotic dynamics – motion on a constant negative curvature surface

In this section we consider a model chaotic system – geodesic motion on a constant negative curvature surface (CNCS). The distance between two neighbouring geodesics on a constant curvature (K) Riemann manifold changes with time according to

$$\delta\mathbf{r} \sim \mathrm{Re}\left(e^{(\sqrt{-K})t}\right).$$

On a positive curvature manifold (e.g., sphere) the distance between close geodesics oscillates with time. On the contrary, on a CNCS the close trajectories diverge exponentially. The resulting motion is chaotic when the surface is compact.

A two-dimensional CNCS cannot be embedded in the 3-dimensional Euclidean space. The easy way of seeing this is by replacing R with iR in Eq. (18). The solution of the resulting equation (if it existed) would have described a surface with constant negative curvature $-1/R^2$. On the other hand, a CNCS can be embedded into Minkovskii space where the length is given by $|\mathbf{r}|^2 = x^2 + y^2 - z^2$ as the solution of the equation

$$|\mathbf{r}|^2 = x^2 + y^2 - z^2 = -R^2. \tag{35}$$

The surface given by Eq. (35) is called the Lobachevskii plane (Fig. 2). We parameterise it with the two angles $0 < \theta < \infty$ and $0 \leq \phi < 2\pi$.

$$\begin{pmatrix} x \\ y \\ z \end{pmatrix} = R \begin{pmatrix} \sinh\theta\cos\phi \\ \sinh\theta\sin\phi \\ \cosh\theta \end{pmatrix} \tag{36}$$

A geodesic on the Lobachevskii plane is given by the intersections between a plane going through the origin and the hyperboloid specified by Eq. (35) (see

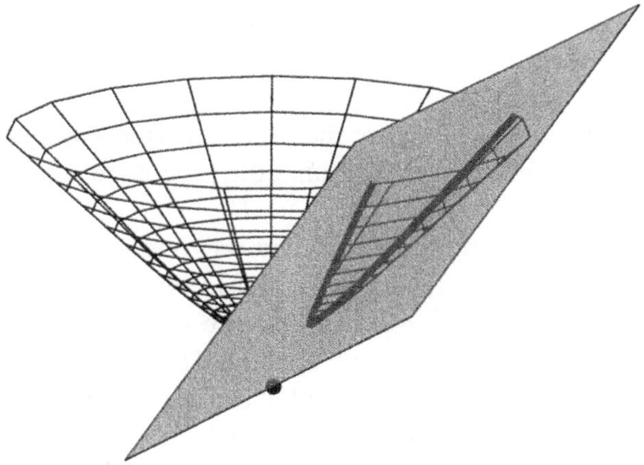

Figure 2. Trajectories of geodesic motion on the Lobachevskii plane are intersections of the hyperboloid (35) with a plane going through the origin.

Fig. 2). A particle moves along a geodesic with constant magnitude velocity (in the Minkowskii metric) $v = |\mathbf{v}| = \text{const.}$

Similarly to the constant positive curvature case (i.e., motion on the sphere considered in the previous section) the energy shell Γ_E can be mapped onto the isometry group of the Lobachevskii plane. From Eq. (35) we conclude that this group is $SO(2,1)$. The geodesic motion is again given by the right multiplication by an element $h_t = e^{Kt}$ from some one-dimensional subgroup

$$U_t g = g h_t, \quad h_t = e^{Kt} \in SO(2,1), \quad K = \begin{pmatrix} 0 & 1 & 0 \\ 1 & 0 & 0 \\ 0 & 0 & 0 \end{pmatrix} \tag{37}$$

The Lobachevskii plane is not compact and the geodesic motion on it is integrable. To make the motion chaotic we need to "compactify" the surface. To achieve this we select a discrete subgroup $\Gamma \subset SO(2,1)$ and identify the pairs of points on the Lobachevskii plane if they can be obtained from each other by an element from Γ. As a result the constant energy shell is reduced from $SO(2,1)$ (which is not compact) to the factor group $\Gamma \backslash SO(2,1)$ (which is compact for some choices of the subgroup Γ)

$$\Gamma_E \equiv \Gamma \backslash SO(2,1). \tag{38}$$

The geodesic motion is still given by the right multiplication by h_t

$$U_t \Gamma g = \Gamma g h_t, \tag{39}$$

where Γg denotes the left coset of the element g. The choice of the subgroup Γ is constrained by the requirement that the left factor Γ_E given by Eq. (38) is compact. There are many subgroups satisfying this condition giving rise to many compact constant negative curvature surfaces with inequivalent dynamics. Our method allows us to treat all such surfaces in a uniform way. An excellent introduction to the geodesic motion on the Lobachevskii plane and its compactifications is given in [13].

Eqs. (37) and (39) allow us to use the representation theory of $SO(2,1)$ following (with some important modifications) the procedure from section 3. In the rest of this section we choose the units where $R = v = 1$.

The group $SO(2,1)$ is non-compact and its irreducible representations are quite different from the representations of $SO(3)$. We are interested in the unitary irreducible representations of $SO(2,1)$ which are infinitely dimensional. Analogously to the $SO(3)$ case, the Casimir operator Ω for $SO(2,1)$ acts as a multiplication by a constant on the spaces of the irreducible representations. It is convenient to denote this constant by $l(l+1)$ where l is some complex number uniquely identifying the irreducible representation T^l. The representation T^l has a "natural" orthonormal basis $|l,n\rangle$ labelled by the integer n satisfying $-\infty \leq n \leq \infty$, so we have

$$\Omega|l,n\rangle = l(l+1)|l,n\rangle. \tag{40}$$

The right regular representation of $SO(2,1)$ contains infinitely many copies of each T^l (see [12]), where l is given by

$$l = -1/2 + i\rho \text{ and } \rho \in \mathbb{R}, \quad \rho > 0 \tag{41}$$

with positive real ρ. Due to compactification (38) we need to consider the functions invariant under left multiplication by Γ. Let us denote the right action of $SO(2,1)$ on this space by T^Γ. It turns out [14] that decomposition of T^Γ into irreducible representations contains only representations T^l corresponding to some discreet values of l (denoted by l_k below). To find out which l_k are allowed we consider the Laplacian operator on the Lobachevskii plane

$$\Delta = \frac{1}{\sinh\theta}\frac{\partial}{\partial\theta}\sinh\theta\frac{\partial}{\partial\theta} + \frac{1}{\sinh^2\theta}\frac{\partial^2}{\partial\phi^2}$$

By imposing the periodicity condition

$$\psi(\Gamma\mathbf{r}) = \psi(\mathbf{r}).$$

we make its eigenvalues discrete. Since the Casimir operator Ω has the same spectrum as Δ the values l_k are given by

$$l_k(l_k+1) = -\epsilon_k, \tag{42}$$

where $\epsilon_k \geq 0$ is an eigenvalue of $-\Delta$

$$\Delta\psi_k = -\epsilon_k\psi_k \text{ and } \psi_k(\Gamma\mathbf{r}) = \psi_k(\mathbf{r}).$$

The eigenvalue $\epsilon_0 = 0$ comes from the constant eigenfunction $\psi_0 = 1$ and is always present. It corresponds to the trivial representation I which enters T^Γ

once. For every other ϵ_k we pick up one of the two solutions of Eq (42) according to the following rule. If ϵ_k is less then $1/4$ then we choose the solutions $l^k = -1/2 + \sqrt{1/4 - \epsilon_k}$ which is real and obeys $-1/2 < l_k < 0$. This corresponds to the irreducible representation T^l from the so called complimentary series. If $\epsilon_k > 1/4$ we pick up the solution $l_k = -1/2 + i\sqrt{\epsilon_k - 1/4}$, which has the form (41). This solution labels the representation T^l from the principle series. In a non-degenerate case all eigenvalues of Δ are simple, so each T^l enters T^Γ once. The corresponding invariant subspace is denoted H^l. We do not specify the explicit coordinates on Γ_E, and do not quote the explicit forms of the basis functions $|l, n\rangle \in H^l$. This information can be found in [11]. Here we only mention that the function $|l_k, 0\rangle$ does not depend on the direction of the velocity $\mathbf{v}$ and is proportional to the eigenfunction $\psi_k(\mathbf{r})$ of Δ.

The rest of the computation is completely analogous to section 3. The evolution operator $\hat{U}_t$ can be decomposed separately on each irreducible representation. The trivial representation yields the eigenvalue $\lambda = 0$ and corresponds to the constant eigenfunction. On the space H^l the matrix element of $\hat{U}_t$ between the basis functions $|l, n\rangle$ is given by

$$\langle l, n|\hat{U}_t|l, n'\rangle = B^l_{nn'}(\cosh t), \tag{43}$$

where $B^l_{nn'}$ is the Jacobi function (see [12]). It can be represented as an infinite sum [11]

$$B^l_{nn'} = \sum_{p=0}^{\infty} \sum_{\pm} f^{l_\pm}_{np} g^{l_\pm}_{n'p} e^{(l_\pm - p)t}, \tag{44}$$

where $l_+ = l$ and l_- is the other solution of Eq (42). Using this representation we see that the spectrum of the decomposition of the flow over H^l contains infinitely many non-degenerate eigenvalues with negative real parts $\lambda = l_\pm - p$, where $p \geq 0$ is an arbitrary non-negative integer. Contrary to the sphere case the eigenvalues coming from different H^l are not equal, so the resulting spectrum of U_t is not degenerate (see Fig 3). The eigenvalues appearing in this decomposition coincide with the Ruelle resonances calculated using zeta-functions (see e.g. [15]).

The left and right eigenvectors corresponding to the eigenvalue $\lambda = l - p$ are given by

$$|f^\lambda\rangle = \sum_n f^l_{np}|l, n\rangle, \qquad \langle g^\lambda| = \sum_n g^l_{np}\langle l, n| \tag{45}$$

where the coefficients f^l_{np}, g^l_{np} come from Eq (44). They are computed in [11]. The modulus of the coefficients does not decrease with n, so the series (45) does not formally converge. In line with our general considerations in section 2 the eigenvectors (45) should be treated as eigenfunctionals acting on the space of C^∞ functions

$$f^\lambda[\xi] = \sum_n f^l_{np}\langle \xi|l, n\rangle, \qquad g^\lambda[\varphi] = \left(\sum_n g^l_{np}\langle l, n|\varphi\rangle\right)^*, \tag{46}$$

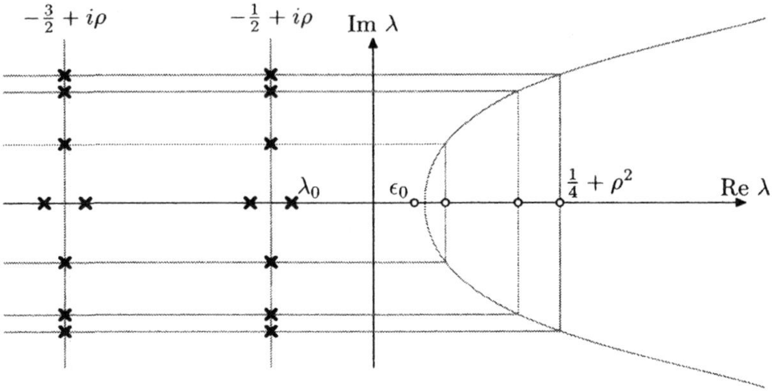

Figure 3. The eigenvalues λ (crosses) of the evolution operator and the eigenvalues ϵ of the Laplacian operator $-\Delta$ (open circles). The correlation function decays exponentially at long times $\langle\phi|U_t|\xi\rangle - \langle\phi|1\rangle\langle 1|\xi\rangle \to e^{\mathrm{Re}(\lambda_0)t}$ with the exponent λ_0 determined by the eigenvalue λ_0 with the least negative real part.

where $\lambda = l - p$. Putting together the contributions from each of the irreducible representations we obtain the spectral decomposition for the evolution operator

$$\langle\xi|\hat{U}_t|\varphi\rangle = \langle 1|\varphi\rangle\langle\xi|1\rangle + \sum_{\lambda=l_k^{\pm}-p} g^{\lambda}[\varphi]^* f^{\lambda}[\xi]e^{\lambda t}, \tag{47}$$

where the first term on the right hand side comes from the trivial representation.

Since all eigenvalues λ has negative real parts the correlation function at the long times decays away to leave just the constant term

$$\lim_{t\to\infty}\langle\xi|\hat{U}_t|\varphi\rangle = \langle 1|\varphi\rangle\langle\xi|1\rangle.$$

The rate of the decay at sufficiently long times t is controlled by the eigenvalue λ_0 with the least negative real part

$$|\langle\xi|\hat{U}_t|\varphi\rangle - \langle 1|\varphi\rangle\langle\xi|1\rangle| < Ce^{\mathrm{Re}(\lambda_0)t}. \tag{48}$$

By keeping more terms in (47) we can get a more accurate prediction of the long time behaviour of the correlation function.

5. Regularisation and relation to ballistic σ-model

The statistical properties of energy levels of a quantum mechanical system can be characterised by the two point correlation function

$$R_2(\omega) = L^2\langle d(E+\hbar\omega)d(E)\rangle - 1 \tag{49}$$

where $d(E)$ is the density of states, L is the average level spacing and the average $\langle\rangle$ is taken over an ensemble of quantum systems with similar classical dynamics. The nonlinear ballistic σ-model [4] relates R_2 with the eigenvalues λ of the classical evolution operator. The expression for $R_2(\omega)$ at high frequencies $\omega \gg L$ involves the spectral function $S(\omega)$ which is given by

$$S(\omega) = \sum_{\lambda} \frac{1}{(\lambda - i\omega)^2}. \tag{50}$$

In particular, the smooth part of R_2 is proportional to the real part of $S(\omega)$

$$R_2^{sm}(\omega) = C \operatorname{Re} S(\omega), \tag{51}$$

while the remaining part of R_2 oscillates on the scale of the level spacing and can also be expressed through some integrals involving $S(\omega)$.

In order to make use of these expressions we need to decide which eigenvalues of $\hat{L}$ should be used in the sum (50). The eigenvalues appearing in our decompositions provide a set of eigenvalues which characterise the dynamics and are a natural choice. In the particular case of the geodesic motion on a CNCS the spectral function (50) will be given by

$$S(\omega) = \sum_{p=0}^{\infty} \sum_{k=0}^{\infty} \sum_{\pm} \frac{1}{(l_k^{\pm} - p - i\omega)^2}, \tag{52}$$

where $l_k^{\pm}$ are the two solutions of Eq (42) corresponding to the eigenvalue ϵ_k of the Laplacian operator. At large k the eigenvalue ϵ_k grows like k^2, so

$$l_k \sim k \text{ at } k \gg 1.$$

The above means that the double sum (52) diverges logarithmically. In fact, because the terms tend to cancel each other the finite result can be obtained but will depend on the order in which the summation is done.

A regularisation of the sum leading to a finite answer can be achieved by perturbing the Liouville operator $\hat{L}$ of the dynamical system by some operator $\hat{X}$ to give a new operator $\hat{L}(\epsilon) = \hat{L} + \epsilon\hat{X}$. The eigenvalues $\lambda(\epsilon)$ of this operator can be used in (50) instead of λ. The limiting value of the sum as $\epsilon \to 0$ gives the regularised spectral function $S_{reg}(\omega)$.

$$S_{reg}(\omega) = \lim_{\epsilon \to 0} \sum_{\lambda(\epsilon)} \frac{1}{(\lambda(\epsilon) - i\omega)^2} \tag{53}$$

The choice of the operator $\hat{X}$ should be fixed by the σ-model and, in general, may depend on the details of the averaging procedure employed in (49). So far this relation is unclear.

In the rest of the paper we report the result of our investigation of a more simple question: the behaviour of individual eigenvalues $\lambda(\epsilon)$ as $\epsilon \to 0$. The geodesic motion on a CNCS was perturbed by the angular diffusion operator $\hat{D}$. The perturbed Liouville operator $\hat{L}(\epsilon) = \hat{L} + \epsilon\hat{D}$ still commutes with with T^{Γ} and therefore preserves the invariant subspaces H^l. The action of $L(\epsilon)$ on H^l is given by

$$\hat{L}(\epsilon)|l,n\rangle = \frac{1}{2}(l-n)|l,n-1\rangle + \frac{1}{2}(l+n)|l,n-1\rangle + \epsilon n^2|l,n\rangle$$

At any finite $\epsilon > 0$ the operator $\hat{L}(\epsilon)$ has square integrable eigenfunctions $|f^{\lambda(\epsilon)}\rangle = \sum_n C_n|l,n\rangle$ which were found numerically by solving the difference equation

$$(l-n+1)C_{n-1} + (l+n+1)C_{n+1} = 2(\lambda(\epsilon) + \epsilon n^2)C_n, \tag{54}$$

with the boundary condition $C_n \to 0$ as $n \to \pm\infty$. The results are presented on fig 4 where the few eigenvalues $\lambda(\epsilon)$ are plotted for the subspace H^l with $l = -1/2 + 0.6i$. As ϵ gets smaller the eigenvalues converge to the eigenvalues of the flow defined from the poles in the resolvent. This convergence is non-uniform in a sense that for arbitrary small ϵ the eigenvalues $\lambda(\epsilon)$ with sufficiently large modulus will be very different from the unperturbed ones. We also found the the coefficients C_n converge (non uniformly coefficient by coefficient) to the coefficients f^l_{np} from Eq. (44),(45).

Appendix

6. Eigenfunctions on the sphere

In this appendix we study the eigenfunctions f^{lm}_p of the sphere. Our starting point will be Eq. 28 for the correlation function of the evolution operator $\hat{U}_t = T^{lm}(h_t)$ on the space H^{lm}.

$$\langle n|\hat{U}_t|n'\rangle = \langle n|\hat{T}^{lm}(h_t)|n'\rangle = P^l_{nn'}(\cos t) \tag{55}$$

We note that conjugating by $R^z_{\pi/2}$ rotates the axis of a rotation in $SO(3)$ by $\pi/2$ about the z-axis. In particular conjugating the element h_t gives

$$R^z_{-\pi/2}h_t R^z_{\pi/2} = R^z_{-\pi/2}R^y_t R^z_{\pi/2} = R^x_t \equiv k_t$$

This means that

$$\begin{aligned}\langle n|T^{lm}(h_t)|n'\rangle &= \langle n|T^{lm}(R^z_{\pi/2}k_t R^z_{-\pi/2})|n'\rangle \\ &= \langle n|T^{lm}(R^z_{\pi/2})T^{lm}(k_t)T^{lm}(R^z_{\pi/2})^{-1}|n'\rangle\end{aligned} \tag{56}$$

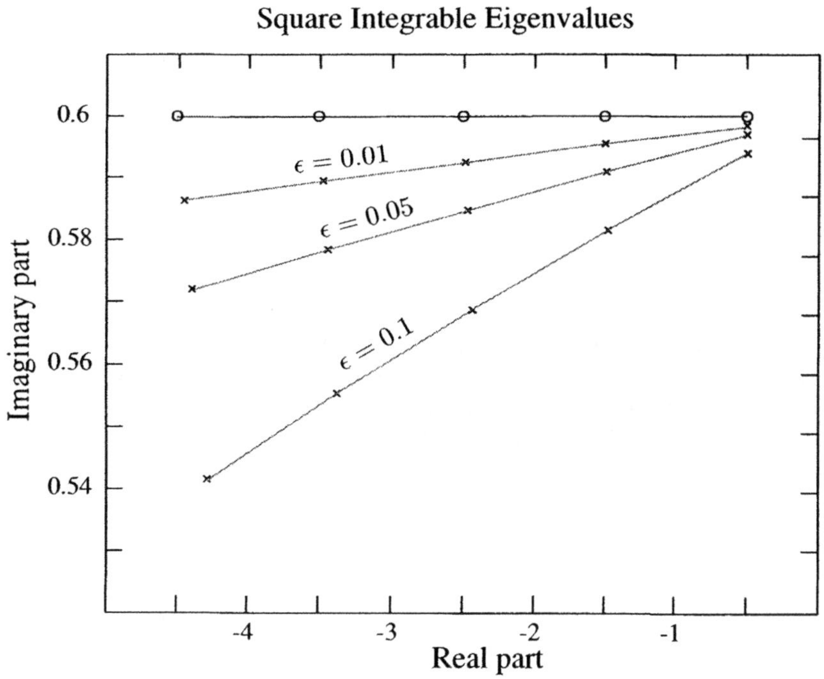

Figure 4. The square integrable eigenvalues $\lambda(\epsilon)$ (crosses) are shown for different values of ϵ. As $\epsilon \longrightarrow 0$ they converge to the eigenvalues λ (shown as circles) in decomposition (47). The irreducible representation is specified by $l = -\frac{1}{2} + 0.6i$.

We now evaluate the product of matrices 56.

$$\langle n|T^{lm}(h_t)|n'\rangle = \sum_{p,p'=-l}^{l} \langle n|T^{lm}(R^z_{\pi/2})|p'\rangle\langle p|T^{lm}(k_t)|p'\rangle\langle p'|T^{lm}(R^z_{\pi/2})^{-1}|n'\rangle \tag{57}$$

Using the matrix elements

$$\langle p|T^{lm}(k_t)|p'\rangle = \langle p|e^{ip't}|p'\rangle = \delta_{pp'}e^{ipt}$$

we can write the product as

$$\langle n|T^{lm}(h_t)|n'\rangle = \sum_{p=-l}^{l} \langle n|T^{lm}(R^z_{\pi/2})|p\rangle\langle p|T^{lm}(R^z_{\pi/2})^{-1}|n'\rangle e^{ipt} \tag{58}$$

Let us denote the matrix entries $\langle n|T^{lm}(R^z_{\pi/2})|p\rangle$ by a^l_{np}. Since the representation T^{lm} is unitary the matrix entry $\langle p|T^{lm}(R^z_{\pi/2})^{-1}|n'\rangle$ will be $\left(a^l_{n'p}\right)^*$. Therefore

Eq. 58 becomes

$$\langle l,m,n|\hat{U}_t|l,m,n\rangle = \langle n|T^{lm}(h_t)|n'\rangle = \sum_{p=-l}^{l} a^l_{np}\left(a^l_{n'p}\right)^* e^{ipt} \qquad (59)$$

Combining (28), (30) with (59) proves Eq. (32).

References

1. M. C. Gutzwiller, Journal of Mathematical Physics **12**, 343 (1971).
2. M. V. Berry, Proceedings of the Royal Society of London A **413**, 183–198 (1987).
3. E. B. Bogomolny and J. P. Keating, Physical Review Letters **77**, 1472–1475 (1996).
4. B. A. Muzykantskii and D. E. Khmelnitskii, JETP letters **62**, 76-82 (1995).
5. A. V. Andreev, O. Agam, B. D. Simons, and B. L. Altshuler, Physical Review Letters **76**, 3947 (1996).
6. K. B. Efetov, Advances in Physics **32**, 53–127 (1983).
7. K. B. Efetov, *Supersymmetry in Disorder and Chaos*, Cambridge University Press, New York (1997).
8. I.V.Gornyi and A.D.Mirlin, condmat/0107552 (2001).
9. Arnold, *Ergodic Theory of Classical Mechanics*, W.A. Benjamin Inc.,New York (1968).
10. R. D., Physical Review Letters **56**, 405–407 (1986).
11. S. Roberts and B. Muzykantskii, Journal of physics A-mathematical and general **33**, 8953 – 8972 (2000).
12. N. Vilenkin, *Special Functions and the Theory of Group Representations*, American Mathematical Society (1968).
13. N. Balazs and A. Voros, PHYSICS REPORTS (Review Section of Physics Letters) **143**, 109–240 (1986).
14. S. Lang, $SL_2(\mathbb{R})$, Springer-Verlag (1985).
15. H. Rugh, Ergodic Theory and Dynamical Systems **16**, 805-819 (1996).

PHASE COHERENCE PHENOMENA IN DISORDERED SUPERCONDUCTORS

A. LAMACRAFT AND B. D. SIMONS
Cavendish Laboratory, Madingley Road, Cambridge CB3 OHE, UK

1. Introduction

Mechanisms of quantum phase coherence heavily influence spectral and transport properties of weakly disordered normal conductors. Such effects are manifest in weak and strong localization effects, and characteristic fluctuation phenomena. Over the past thirty years, theoretical progress in elucidating the mechanisms of quantum phase coherence in weakly disordered structures has been substantial: By now a consistent theory of weakly interacting disordered structures has been developed (For a review, see e.g., Refs. [1–3]).

At the same time, considerable experimental effort has been directed towards the exploration of the influence of phase coherence effects on the quasi-particle properties of disordered superconductors. Again, attempts to develop a consistent theory have enjoyed great success. By now a reliable theory of the weakly interacting superconducting system has been formulated. Yet, a complete description of the phenomenology of the disordered superconductor in the presence of strong interaction effects has yet to be established. The continuing developments and refinements of experimental techniques continue to present fresh challenges to theoretical investigations.

On this background, the aim of these lecture notes is to selectively review the recent development of a quasi-classical field theoretic framework to describe phase coherence phenomena in disordered superconductors. This approach, which is motivated by the parallel formulation of the theory of the normal disordered system, presents average properties of the superconductor as a quantum field theory with an action of non-linear σ-model type. The limited scope of these lectures does not permit an extensive review the many applications of this technique. Instead, to illustrate the impact of quantum phase coherence phenomena on the quasi-particle properties of the disordered superconducting system, and the practical application of the field theoretic scheme, the final part of these notes will be devoted to a study of the magnetic impurity system.

I. V. Lerner et al. (eds.),
Strongly Correlated Fermions and Bosons in Low-Dimensional Disordered Systems, 259–312.

Before turning to the construction of the field theoretic scheme, we will begin these notes with a qualitative discussion of phase coherence phenomena in the superconducting environment placing emphasis on the importance of fundamental symmetries. To close the introductory section, we will outline the quasi-classical theory which forms the basis of the field theoretic scheme. In section 2 we will develop a quantum field theory of the weakly disordered non-interacting superconducting system (i.e. in the mean-field BCS approximation). To illustrate a simple application of this technique, we will explore the spectral properties of a normal quantum dot contacted to a superconducting terminal. Finally, in section 3, we will present a detailed study of the influence of magnetic impurities in the disordered superconducting system. This single application will emphasize a number of generic features of the phase coherent superconducting system including unusual spectral and localization properties and the importance of effects non-perturbative in the disorder.

To orient our discussion, however, let us first briefly recapitulate the BCS mean-field theory of superconductivity in order to establish some notations and definitions.

1.1. THE BCS THEORY

In the mean-field approximation, the second quantized BCS Hamiltonian of a weakly disordered superconductor is defined by

$$\hat{H}_{\rm bcs} = \int d\mathbf{r}\Big[\sum_{\sigma=\uparrow,\downarrow} \psi^\dagger_\sigma(\mathbf{r})\left(\frac{1}{2m}(\hat{\mathbf{p}} - e\mathbf{A}/c)^2 + W(\mathbf{r}) - \epsilon_F\right)\psi_\sigma \qquad (1)$$
$$+\Delta(\mathbf{r})\psi^\dagger_\uparrow(\mathbf{r})\psi^\dagger_\downarrow(\mathbf{r}) + \Delta^*(\mathbf{r})\psi_\downarrow(\mathbf{r})\psi_\uparrow(\mathbf{r})\Big]$$

where $\psi^\dagger_\sigma(\mathbf{r})$ creates an electron of spin σ at position $\mathbf{r}$, ϵ_F denotes the Fermi energy, $\mathbf{A}$ represents the vector potential of an external electromagnetic field, and $W(\mathbf{r})$ an impurity scattering potential. The order parameter is determined self-consistently from the condition $\Delta(\mathbf{r}) = -(\lambda/\nu)\langle\psi_\uparrow(\mathbf{r})\psi_\downarrow(\mathbf{r})\rangle$, where λ is the (dimensionless) BCS coupling constant and ν represents the average density of states (DoS) per spin of the normal system.[1] Defining the Bogoliubov transform

$$\psi_\uparrow(\mathbf{r}) = \sum_i\left[\gamma_{i\uparrow}u_i(\mathbf{r}) - \gamma^\dagger_{i\downarrow}v^*_i(\mathbf{r})\right], \qquad \psi_\downarrow(\mathbf{r}) = \sum_i\left[\gamma_{i\downarrow}u_i(\mathbf{r}) + \gamma^\dagger_{i\uparrow}v^*_i(\mathbf{r})\right]$$

the Hamiltonian can be brought to a diagonal form by choosing the spinor elements $u_\alpha(\mathbf{r})$ and $v_\alpha(\mathbf{r})$ to satisfy the coupled Bogoliubov-de Gennes (BdG)

[1] To avoid ambiguity, this is be the density of states per d-dimensional volume, for an effectively d-dimensional system

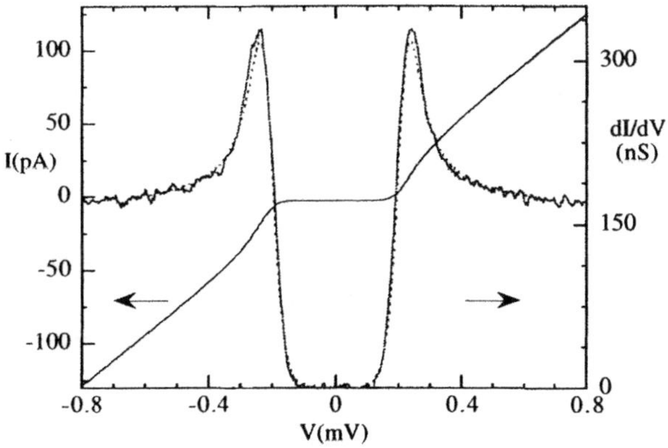

Figure 1. I-V characteristic and differential conductance measured by scanning tunneling microscopy on a superconducting layer of Al at 60mK. The dashed line is a fit using a BCS density of states ($\Delta_{\rm Al} = 210\mu$eV) convoluted with a thermal Fermi distribution (at $T = 210$mK). Taken from Ref. [6].

presence of localization provided that $|\Delta|\nu L_{\rm loc}^d \gg 1$, where $L_{\rm loc}$ is the localization length and d the dimensionality [7]. In fact, the destruction of superconductivity can occur in far more metallic samples due to the dramatic effects of disorder combined with the residual Coulomb interaction. The mean-field treatment of this physics is due to Finkelstein (see e.g. [8]) — but the effects of the Coulomb interaction in dirty superconductors are only well understood in certain limits and not at all generally. Even more surprising is that the BCS model in section 1.1 is compatible with a huge variety of unusual spectral and transport behaviour enabled by novel mesoscopic phase coherence mechanisms.

1.2.1. *Evading the Anderson Theorem*

Thermodynamic properties have not historically been the best place to start looking for mesoscopic effects (it was, for example, a long time before attention was focussed on the persistent currents in normal metals). Spectral properties are the domain of mesoscopics, but the conclusion drawn from Anderson's theorem about the quasi-particle spectrum may appear to preclude any new effects particular to superconducting systems.

In fact the assumptions of Anderson's theorem seem more restrictive today than at the time. The investigation of hybrid electronic devices containing both superconducting (S) and normal (N) metallic elements is an extremely active field of research. Here the order parameter is not constant throughout the system and Anderson's theorem does not apply. At the very least one needs a formulation of the Gor'kov theory capable of handling this spatial inhomogeneity. We will come to this *quasi-classical* description presently. Beyond this description — which dates back to the late 60s — SN systems do in fact exhibit a wide range of novel

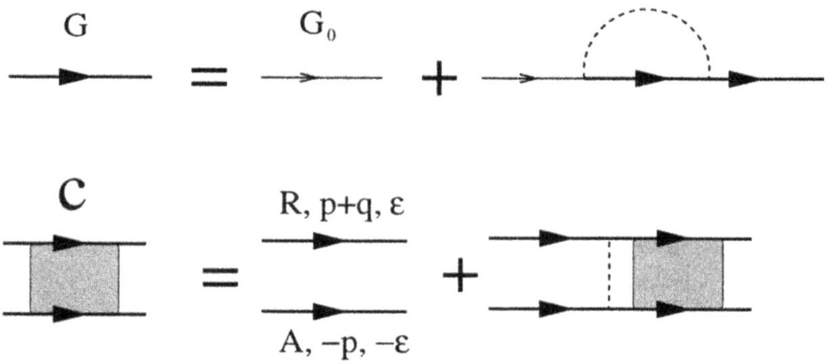

Figure 2. Diagrams for the evaluation of the Cooperon.

mesoscopic phenomena. These are mediated by Andreev [9] reflection — the phase coherent inter-conversion of electrons and holes at the SN interface due to the spectral gap of the bulk superconductor.

We will be concerned only tangentially with hybrid structures in later chapters, so a qualitative description of these effects here is not appropriate (for a discussion, see [10]). There are many other ways, however, to avoid Anderson's conclusion even in a 'bulk' superconductor (including thin films and wires). An important second strand of experimental evidence discussed in Anderson's paper relates to the deleterious effect of *magnetic* impurities on superconductivity. Unconventional superconductors with non s-wave pairing (the high-T_c materials being the most prominent examples) are likewise affected by normal disorder. All these counter-examples have very recently been shown to display dramatic mesoscopic behaviour. We will come to this through a fuller explanation of the robustness to disorder in the conventional s-wave case.

1.3. PAIR PROPAGATION AND THE COOPERON

Within the Gor'kov formalism outlined in section 1.1, an estimate for T_c can be determined by linearizing the self-consistent equation (4) in Δ

$$\begin{aligned}\Delta(\mathbf{r}) &= -\frac{\lambda}{\nu}T\sum_{\epsilon_n}\int d\mathbf{r}'\Delta(\mathbf{r}')\hat{G}_{i\epsilon_n}(\mathbf{r},\mathbf{r}')\hat{G}_{-i\epsilon_n}(\mathbf{r},\mathbf{r}') \\ &= -\frac{\lambda}{\nu}\int d\mathbf{r}'\int\frac{d\epsilon}{2\pi}\tanh\left(\frac{\epsilon}{2T}\right)\mathrm{Im}\,\hat{G}^{\mathrm{r}}_{\epsilon}(\mathbf{r},\mathbf{r}')\hat{G}^{\mathrm{a}}_{-\epsilon}(\mathbf{r},\mathbf{r}')\,,\end{aligned} \tag{7}$$

where $\hat{G}_{\epsilon_n}$ is the Green's function corresponding to the single-particle Hamiltonian $\hat{H}$ at imaginary frequency and $\hat{G}^{\mathrm{r,a}}_{\epsilon}$ the real frequency advanced and retarded counterparts. Taking Δ to be constant as before we average over disorder configurations to find $\langle\hat{G}^{\mathrm{r}}_{\epsilon}(\mathbf{r},\mathbf{r}')\hat{G}^{\mathrm{a}}_{-\epsilon}(\mathbf{r},\mathbf{r}')\rangle$. The evaluation may be performed using

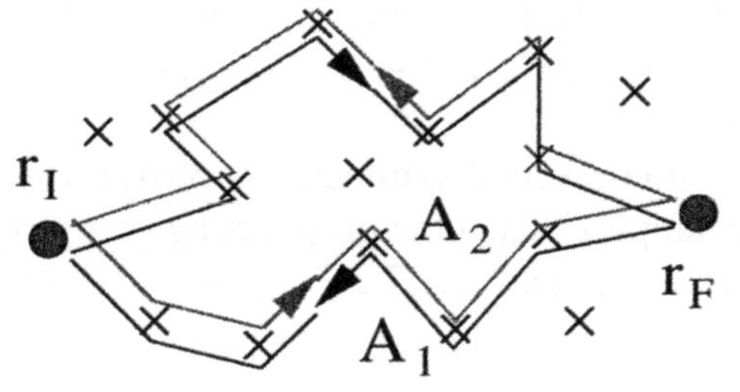

Figure 3. Dominant contributions to time-reversed pair propagation in the Feynman picture. The phase of the amplitude A_1 is the opposite of A_2 if time-reversal symmetry is preserved.

the standard 'cross' technique [11] based on a Gaussian δ-correlated impurity distribution,

$$\langle W(\mathbf{r})\rangle = 0, \qquad \langle W(\mathbf{r})W(\mathbf{r}')\rangle = \frac{1}{2\pi\nu\tau}\delta^d(\mathbf{r}-\mathbf{r}')\,, \tag{8}$$

and is illustrated in Fig. 2. The result is [12]

$$\langle \hat{G}^{\mathrm{r}}_{\epsilon}(\mathbf{r},\mathbf{r}')\hat{G}^{\mathrm{a}}_{-\epsilon}(\mathbf{r},\mathbf{r}')\rangle = \left(\frac{2\pi\nu}{Dq^2-2i\epsilon}\right)_{\mathbf{rr}'}\,. \tag{9}$$

Here $D = v_F^2\tau/d$ is the diffusion constant, where $v_F = p_F/m$ denotes the Fermi velocity. The two-particle quantity under consideration evidently relates to the propagation of a pair of electrons between two points in opposite directions. The diffusion pole structure of the average signals the presence of a hydrodynamic mode of pair propagation known as the *Cooperon*. In the language of the Feynman path integral, this is because the dominant trajectories for the propagation of the pair through a given disorder realization come from the the electrons tracing out precisely time-reversed paths, so that the phase accumulated in the overall amplitude in propagation is completely canceled (see Fig. 3). The phase of of a single propagating electron is scrambled after a time $\sim\tau$, but two particle averages like the above depend on the 'bulk' property D. Their inclusion in diagrammatic calculations typically leads to anomalously large contributions from long wavelengths due to their diffusive structure.

Returning to the matter of determining T_c, from the result above, the self-consistency condition (7) takes the form

$$1 = -\lambda\int d\epsilon\tanh\left(\frac{\epsilon}{2T}\right)\frac{1}{2\epsilon}\,, \tag{10}$$

independent of disorder, yielding $T_c \sim \omega_D\exp(1/\lambda)$, with ω_D the Debye frequency at which the interaction is cut off. The multiple scattering between time-reversed electrons summarized by (7) is absolutely indifferent to the disorder potential through which they propagate. Thus we see the intimate connection between time-reversal invariance in the original single-particle Hamiltonian and Anderson's theorem.

What happens if time-reversal symmetry is broken (by the application of a magnetic field, for example)? Then the propagating pair progressively loses relative phase coherence as time passes. The Cooperon ceases to be a hydrodynamic mode

$$\langle G_{i\epsilon_n}(\mathbf{r},\mathbf{r}')G_{-i\epsilon_n}(\mathbf{r},\mathbf{r}')\rangle = \left(\frac{2\pi\nu}{Dq^2+2|\epsilon_n|+1/\tau_\varphi}\right)_{\mathbf{rr}'} .$$

Here $1/\tau_\varphi$ represents some rate characteristic of the symmetry-breaking perturbation. Substituting this into (7) one obtains the celebrated result obtained by Abrikosov and Gor'kov [13],

$$\ln\left(\frac{T_{c\,0}}{T_c}\right) = \psi\left(\frac{1}{4\pi\tau_\varphi T_c}+\frac{1}{2}\right) - \psi\left(\frac{1}{2}\right) , \tag{11}$$

where $T_{c\,0}$ is the critical temperature at $1/\tau_\varphi = 0$. The complete destruction of T_c is predicted at $1/\tau_\varphi = 1.76T_{c\,0}$ (see Fig. 4).

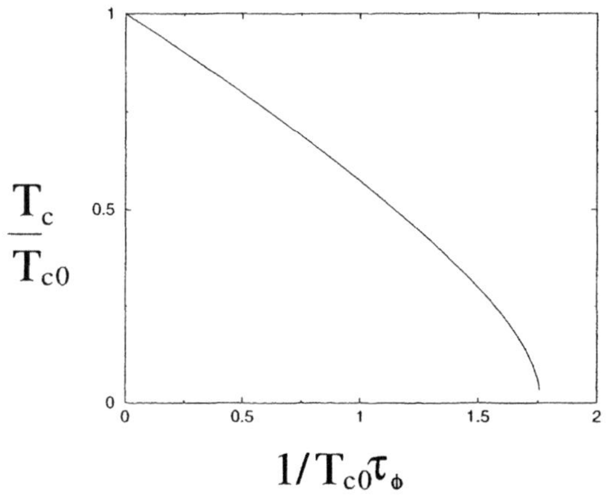

Figure 4. Suppression of T_c predicted by the Abrikosov-Gor'kov theory

One of the main themes in the following chapters will be the mesoscopic nature of various processes that impinge on the coherent pair propagation responsible for superconductivity. In this context, we should note that, in addition to the time-reversal symmetry breaking perturbations discussed here, these include both the static and dynamic parts of the Coulomb interaction. While the static part acts like the BCS interaction, the dynamic part like a pair-breaking perturbation. Before we can begin, there is one more subject to introduce.

1.4. SYMMETRIES OF THE HAMILTONIAN AND RANDOM MATRIX THEORY

In the previous section we encountered an important theme in mesoscopics; the central role played by the basic symmetries of the Hamiltonian. In fact there is a limiting sense in which a mesoscopic system is entirely characterized by its symmetries.[4] Let us first focus on the normal system. From the conductivity σ, we can define the conductance $G = \sigma L^{d-2}$ which, making use of the Einstein relation $\sigma = e^2 \nu D$ can be expressed as

$$G = \frac{e^2}{\hbar} \nu L^d \frac{\hbar D}{L^2} = \frac{e^2}{\hbar} g, \qquad g \equiv \frac{E_{\mathrm{T}}}{\delta} \tag{12}$$

where $\delta = 1/\nu L^d$ denotes the average energy level spacing of the normal system, and $E_{\mathrm{T}} = \hbar D/L^2$ represents the typical inverse diffusion time for an electron to cross a sample of dimension L^d — the 'Thouless energy'. This result shows that the conductance of a metallic sample can be expressed as the product of the quantum unit of conductance $e^2/\hbar = (4.1k\Omega)^{-1}$, and a *dimensionless conductance* g equal to the number of levels inside an energy interval E_c. In a good metallic sample, the dimensionless conductance is large, $g \gg 1$.

One of the central tenets of mesoscopic physics is that the spectral properties of Hamiltonian of a disordered electronic system can be modeled as a random matrix of the appropriate symmetry. This remarkable correspondence holds if we are concerned only with energies within E_{T} of the Fermi surface, or equivalently, with times longer than the transport time $t_{\mathrm{D}} = L^2/D$ across the system. Crudely speaking, this is due to the existence of an ergodic regime at these scales when the entire phase space has been explored. If we are only concerned with this regime it is appropriate to take the 'universal' $g \to \infty$ limit. Within the σ-model formalism that will be developed later, the emergence of the random matrix description is very natural.

The random matrix description is formalized by defining a statistical ensemble $P(H)\, dH$ from which the Hamiltonian which models our system will be drawn. The choice encountered most frequently in the literature is the Gaussian ensemble

$$P(H)\, dH = \exp\left[-\frac{1}{v^2} \operatorname{tr} H^2\right] dH\ . \tag{13}$$

Restricting the discussion to ordinary normal metals, three principal universality classes of the Random Matrix Theory (RMT) description can be identified [15] according to whether the matrix H is constrained to be real symmetric ($\beta = 1$, *Orthogonal*), complex Hermitian ($\beta = 2$, *Unitary*), or real quaternion ($\beta = 4$,

[4] In this section we discuss only non-interacting systems (including the mean-field treatment of interactions represented by the Gor'kov Hamiltonian (3)). Recently this has been extended to the interacting case [14]

Symplectic). Hamiltonians invariant under time-reversal belong to the orthogonal ensemble, while those which are not belong to the unitary ensemble. Time-reversal invariant systems with half-integer spin and broken rotational symmetry belong to the third symplectic ensemble.

Expressed in the basis of eigenstates $H = U^\dagger \Lambda U$, where Λ denotes the matrix of eigenvalues, the probability distribution (13) can be recast in the form

$$P(\{\epsilon\})\, d[\{\epsilon\}] = \prod_{i<j} |\epsilon_i - \epsilon_j|^\beta \prod_k e^{-\epsilon_k^2/v^2}\, d\epsilon_k$$

where the invariant measure reveals the characteristic repulsion of the energy levels.

The Dyson classification is made on the basis of the symmetries of time reversal $\mathcal{T}$ and spin rotation $\mathcal{S}$:

$$\mathcal{T} : H = \sigma_2^{\rm sp} H^T \sigma_2^{\rm sp}, \qquad \mathcal{S} : [H, \sigma^{\rm sp}] = 0,$$

where $\sigma_i^{\rm sp}$ are Pauli matrices acting on spin.

In the present context it is natural to ask what happens when we extend the discussion to superconducting systems described by the Gor'kov Hamiltonian. Altland and Zirnbauer [16] have provided the answer, introducing a further *seven* symmetry classes, exhausting the Cartan classification of symmetric spaces upon which they turn out to be based. Their analysis was technical, but we can see the idea through a simple example. As a prototype of the superconducting system let us consider the example of a $2N \times 2N$ matrix Hamiltonian with a particle/hole structure. The simplest case corresponds to $\mathcal{S}$ preserved and $\mathcal{T}$ broken. The Hamiltonian

$$H = \begin{pmatrix} h & \Delta \\ \Delta^\dagger & -h^T \end{pmatrix}, \tag{14}$$

where the block diagonal elements are complex Hermitian, $h^\dagger = h$, and the off-diagonal blocks are symmetric, $\Delta^T = \Delta$, exhibits the ph-symmetry

$$H = -\sigma_2^{\rm ph} H^T \sigma_2^{\rm ph}. \tag{15}$$

In this case, according to the Cartan classification scheme, the Hamiltonian (14) belongs to the symmetry class C. Taking the elements to be drawn from a Gaussian ensemble $P(H)\, dH = \exp[-{\rm tr}\, H^2/2v^2]\, dH$, the distribution function takes the general form

$$P(\{\epsilon\}) d[\{\epsilon\}] = \prod_{i<j} |\epsilon_i^2 - \epsilon_j^2|^\beta \prod_k |\epsilon_k|^\alpha e^{-\epsilon_k^2/v^2} d\epsilon_k$$

where $\beta = 2$ and $\alpha = 2$ [16]. The repulsion that the levels feel from $\epsilon = 0$ follows from the privileged place that energy possesses in the Gor'kov Hamiltonian. By imposing the further symmetry of time-reversal (i.e. $h^* = h$ and $\Delta^* = \Delta$), the symmetry is raised to class CI with $\beta = 1$ and $\alpha = 1$. Once again, an extension to a spinful structure identifies two more symmetry classes [17].

Why is the classification scheme useful? In fact, the low-energy, long-ranged properties of the disordered superconducting system are heavily constrained by the fundamental symmetries of the Hamiltonian. We will see that the localization properties of the low-energy quasi-particle states can typically be immediately inferred from the symmetry classification alone.[5]

We saw that the existence of a hydrodynamic Cooperon mode was a fundamental consequence of time-reversal symmetry in a ordinary (non-Gor'kov) Hamiltonian. Therefore the Cooperon should be viewed as a perturbative, finite g, counterpart of the universal RMT description of the orthogonal class. In the same way we can expect that new soft modes will appear as signatures of the new symmetry classes. As their very existence depends on the Gor'kov structure of the Hamiltonian, it is not surprising that the effects of these new modes are singular at low energies. Crudely speaking, the order parameter can be viewed as a potential scattering particle excitations of energy ϵ to hole excitations of energy $-\epsilon$. It is evident that these processes, like the Cooperon, are coherent as $\epsilon \to 0$. Hence the existence of *low energy quasi-particle states* is absolutely necessary for the new channels of interference to be effective. All the aforementioned examples of superconducting systems that evade Anderson's theorem have this property for some parameter ranges and, as such, are candidates for the observation of new mesoscopic effects. For instance, systems of class C symmetry will presumably display some precursor of the level repulsion from $\epsilon = 0$ in the averaged density of states before the universal limit is reached. The possibility of observing dramatic behaviour in *single* quasi-particle properties instead of two-particle properties is an exciting prospect.

This completes our discussion of the phenomenology of the weakly disordered superconducting system. In the following we will develop and apply a field theoretic framework which captures both the perturbative and non-perturbative effects of quantum interference on the quasi-particle properties of the system. However, to prepare our discussion of the field theoretic scheme we begin with a brief review of the quasi-classical theory of superconductivity which forms the basis of this approach.

1.5. THE QUASI-CLASSICAL THEORY

Typically, it is found experimentally that the Fermi energy ϵ_F of a superconductor is always greatly in excess of the order parameter, Δ. In conventional 'low-temperature' superconductors, the ratio ϵ_F/Δ is often as much as 10^3. From

[5] There are rare cases — such as the disordered d-wave superconductor [18, 19] — where the particular nature of the disorder is important.

this fact we can infer that the description of the superconductor in terms of the exact Green function carries with it a certain amount of redundant information. The quasi-classical method exploits this redundancy to develop a simplified theory describing the variation of the Green function on length scales comparable with the coherence length (which, in the clean system, is given by $\xi = v_F/\Delta \gg \lambda_F$). This makes the quasi-classical method ideal for the description of inhomogeneous situations (like the hybrid devices mentioned before).

In the BCS mean-field approximation, the single quasi-particle properties of the superconductor are contained within the equation (of motion) for the advanced Gor'kov Green function (3)

$$\left[\epsilon_- - \hat{\zeta}\sigma_3^{\rm ph} - \hat{\Delta}\right] \hat{G}^{\rm a}_{\rm Gorkov}(\mathbf{r}_1 - \mathbf{r}_2) = \delta^d(\mathbf{r}_1 - \mathbf{r}_2)$$

where $\epsilon_- = \epsilon - i0$, $\hat{\zeta} = \hat{\mathbf{p}}^2/2m - \epsilon_F$, and $\hat{\Delta} = |\Delta|\sigma_1^{\rm ph} e^{-i\varphi\sigma_3^{\rm ph}}$.

In the quasi-classical limit, $\epsilon_F \gg |\Delta|$, fast fluctuations of the Gor'kov Green function (i.e. those at the Fermi wavelength $\lambda_F = 1/p_F$) are modulated by slow variations at the scale of the coherence length $\xi = v_F/\Delta$ of the clean system. In this limit, the important long-ranged information contained within the slow variations of the Gor'kov Green function can be exposed by averaging over the fast fluctuations. Following the procedure outlined in the seminal work of Eilenberger [21], and later by Larkin and Ovchinnikov [22, 23], the resulting equation of motion for the average Green function assumes the form of a kinetic equation

$$v_F \mathbf{n} \cdot \nabla \hat{g}(\mathbf{r}, \mathbf{n}) - i\left[\hat{g}(\mathbf{r}, \mathbf{n}), (\epsilon_- + \hat{\Delta})\sigma_3^{\rm ph}\right] = 0$$

where, defining $\mathbf{r} = (\mathbf{r}_1 + \mathbf{r}_2)/2$, $\zeta = v_F(p - p_F)$, and $\mathbf{n} = \mathbf{p}/p_F$,

$$\hat{g}(\mathbf{r}, \mathbf{n}) = \frac{i}{\pi}\sigma_3^{\rm ph} \int d\zeta \overbrace{\int d(\mathbf{r}_1 - \mathbf{r}_2) \hat{G}^{-}_{\rm Gorkov}(\mathbf{r}_1, \mathbf{r}_2) e^{i\mathbf{p}\cdot(\mathbf{r}_1 - \mathbf{r}_2)}}^{\hat{G}^{-}_{\rm Gorkov}(\mathbf{r}, \mathbf{p})} .$$

This Boltzmann-like equation of motion, known as the Eilenberger equation, represents an expansion to leading order in the ratio of λ_F to the scale of spatial variation of the slow modes of the Gor'kov Green function. The Eilenberger Green's function satisfies the non-linear constraint: $\hat{g}(\mathbf{r}, \mathbf{n})^2 = \mathbb{1}$, fixed in the usual formulation by the homogeneous BCS solution discussed below [24] (for reasons which will become clear later, we will not dwell here upon the origin of this condition).

In the presence of weak impurity scattering (i.e. $\ell \equiv v_F\tau \gg \lambda_F$), the Eilenberger equation must be supplemented by an additional term which, in the language of the kinetic theory, takes the form of a collision integral. In the Born scattering approximation, the corresponding equation of motion for the average

Green function assumes the form

$$v_F \mathbf{n}\cdot\nabla \hat{g}(\mathbf{r},\mathbf{n}) - i\left[\hat{g}(\mathbf{r},\mathbf{n}), (\epsilon_- + \hat{\Delta})\sigma_3^{\mathrm{ph}}\right] = -\frac{1}{2\tau}\left[\hat{g}(\mathbf{r},\mathbf{n}), \int d\mathbf{n}'\, \hat{g}(\mathbf{r}',\mathbf{n}')\right] .$$

Now, in the dirty limit $\ell \ll \xi$, where $\xi = (D/\Delta)^{1/2}$ represents the superconducting coherence length in the dirty limit, the Eilenberger equation can be simplified further. In this regime the dominant transport mechanism is diffusion. Under these conditions, the dependence of the Green function on the momentum direction ($\mathbf{n} = \mathbf{p}/p_F$) is weak, justifying a moment expansion: $\hat{g}(\mathbf{r},\mathbf{n}) = \hat{g}_0(\mathbf{r}) + \mathbf{n}\cdot\hat{\mathbf{g}}_1(\mathbf{r}) + \ldots$, where $\hat{g}_0(\mathbf{r}) \gg \mathbf{n}\cdot\hat{\mathbf{g}}_1(\mathbf{r})$. A systematic expansion of the Eilenberger equation in terms of $\hat{\mathbf{g}}_1$ then leads to a nonlinear second-order differential equation — the Usadel equation — for the isotropic component [25],

$$D\nabla\left(g_0(\mathbf{r})\nabla\hat{g}_0(\mathbf{r})\right) + i\left[\hat{g}_0(\mathbf{r}), (\epsilon_- + \hat{\Delta})\sigma_3^{\mathrm{ph}}\right] = 0\,. \qquad (16)$$

As in the parent Eilenberger case, the matrix field obeys the non-linear constraint $\hat{g}_0(\mathbf{r})^2 = \mathbb{1}$. Finally, when supplemented by the self-consistent equation for the order parameter,

$$|\Delta(\mathbf{r})| = -\frac{\lambda\pi}{2}T\sum_{\epsilon_n}\mathrm{tr}\left[\sigma_2^{\mathrm{ph}} e^{-i\varphi\sigma_3^{\mathrm{ph}}}\,\hat{g}_0(\mathbf{r})\right]_{\epsilon=i\epsilon_n}, \qquad (17)$$

where the trace runs over the particle/hole degrees of freedom, this equation describes at the mean-field level the quasi-classical properties of the disordered superconducting system. By averaging over the fast fluctuations at the scale of the Fermi wavelength, the long-range properties of the average quasi-classical Green function are expressed as the solution to a non-linear equation of motion.

Let us illustrate the quasi-classical Usadel theory for a weakly disordered bulk singlet superconducting system. In this case, the solution of the mean-field equation can be obtained by adopting the homogeneous parameterization

$$\hat{g}_{\mathrm{bcs}} = \cosh\theta\;\sigma_3^{\mathrm{ph}} - i\sinh\theta\;\sigma_2^{\mathrm{ph}} e^{-i\varphi\sigma_3^{\mathrm{ph}}}\,. \qquad (18)$$

When substituted into Eq. (16), one obtains the homogeneous solution

$$\cosh\theta_s = \frac{\epsilon_-}{E}, \qquad \sinh\theta_s = \frac{|\Delta|}{E} \qquad (19)$$

where $E = (\epsilon_-^2 - |\Delta|^2)^{1/2}$. Here the root is taken in such a way that $\lim_{\epsilon\to\infty} E \to \epsilon_-$, i.e. $\theta = 0$. Finally, when the solution (19) is substituted back into the self-consistent equation (17), one obtains the BCS equation for the order parameter,

$$|\Delta| = -\lambda\pi T\sum_{\epsilon_n}\frac{|\Delta|}{(\epsilon_n^2 + |\Delta|^2)^{1/2}}.$$

i.e. at the level of mean-field, the average quasi-classical Green function is insensitive to the random impurity potential — a result compatible with the Anderson theorem.

This concludes our introductory discussion of the disordered superconducting system. The quasi-classical theory (and it's extension to the non-equilibrium systems) has proved to be remarkably successful in explaining mechanisms of phase coherent transport observed in hybrid superconducting/normal compounds. However, as a comprehensive theory, the quasi-classical scheme alone is incomplete: In such environments, low-energy quasi-particle properties become heavily influenced by quantum phase coherence effects not accommodated by the present theory. In the following section, we will develop a description of the superconducting system within the framework of a quantum field theory. Here we will find that the quasi-classical theory above represents the saddle-point of an effective action whose fluctuations encode the missing mechanisms of quantum phase coherence.

2. Field theory of the disordered superconductor

The development of a statistical field theory of the weakly disordered superconductor closely mirrors the formulation of the quasi-classical theory outlined in section 1. However, the benefits of the field theoretic scheme are considerable:

1. Firstly, the field theoretical approach provides a consistent method to explore the influence of mesoscopic fluctuation phenomena both in the "particle/hole" and "advanced/retarded" channels. As discussed above, such effects become pronounced when low-energy quasi-particle states persist. Indeed, such quantum interference effects can be explored even in situations where the mean-field structure is spatially non-trivial such as that encountered with hybrid superconducting/normal structures.
2. Secondly, and more importantly, it provides a secure platform for the further development and analysis of Coulomb interaction effects and non-equilibrium phenomena through straightforward refinements of the field theoretic scheme.
3. Finally, the field theoretic approach has great aesthetic appeal: it's content is largely constrained by the fundamental symmetries of the disordered superconducting system. Within this formulation, the soft low-energy modes responsible for the long-ranged phase coherence properties described in the previous section are exposed.

For these reasons, we will provide a detailed exposition of the field theoretic method from formulation to application. The starting point will be an exact functional integral representation of the generating function of the electron Green function. The latter must be normalized *independently of the disorder*. This can be achieved via the supersymmetry, replica, or Keldysh methods. Since we will restrict attention to the non-interacting system, we will focus on the supersymmetry technique (which extends to the mean-field treatment of superconductivity). In the semi-classical approximation, we will use the intuition afforded by the quasi-classical scheme to identify the low-energy content of the theory of the ensemble averaged system. As a result, we will show that the low-energy, long-ranged properties of the disordered superconductor can be presented as a supersymmetric non-linear σ model.

In the remainder of the chapter we will apply the supersymmetric scheme to analyze the spectral properties of a hybrid superconductor/normal quantum dot device. Later, in the subsequent chapter we will see how this scheme presents a method to explore non-perturbative effects in the magnetic impurity system.

2.1. FUNCTIONAL METHOD

2.1.1. *Generating functional*

To compute the disorder averaged Green function, we will use Efetov's supersymmetry method [26, 27] tailored to the description of the superconducting system [28, 29, 10]. The analysis (and notation) adopted here is based on a pedagogical exposition of the method by Bundschuh, Cassanello, Serban and Zirnbauer [30]. Within the supersymmetric approach, the Gor'kov Green function is obtained from the generating functional[6]

$$\mathcal{Z}[j] = \int D[\bar{\psi}, \psi] \exp\left[\int d\mathbf{r} \left(i\bar{\psi}(\hat{H}_{\text{Gorkov}} - \epsilon_-)\psi + \bar{\psi}j + \bar{j}\psi\right)\right],$$

where, as usual, $\epsilon_- \equiv \epsilon - i0$ and, in the mean-field approximation, $\hat{H}_{\text{Gorkov}}$ denotes the Gor'kov Hamiltonian (3). For the moment we ignore the spin structure and retain only the Nambu space. Formally, the infinitesimal, which provides convergence of the field integral, imposes the analytical structure of the Green function. The functional integral is over supervector fields $\psi(\mathbf{r})$ and $\bar{\psi}(\mathbf{r})$, whose components are commuting and anticommuting (i.e. Grassmann) fields [26]. Introducing both commuting and anticommuting elements ensures the normalization of the field integral, $\mathcal{Z}[0] = 1$ — a trick clearly limited to the mean-field (single quasi-particle) approximation. Thus, in addition to the (physical) particle-hole (ph) or Nambu structure, the fields are endowed with an auxiliary "boson-fermion" (bf) structure. A generalization to averages over products of Green functions follows straightforwardly by introducing further copies of the field space.

To capture all possible channels of quantum interference in the effective theory is is necessary to further double the field space [27]. This "charge conjugation" (or cc) space, is introduced by rearranging the quadratic form of the generating functional as follows:

[6] Historically the field-theoretic approach to disordered electron problems is due to Wegner [31] who used the replica formalism for the derivation of the nonlinear sigma model.

$$\begin{aligned} &2\bar{\psi}(\hat{H}_{\text{Gorkov}} - \epsilon_-)\psi \\ &\quad = \bar{\psi}(\hat{H}_{\text{Gorkov}} - \epsilon_-)\psi + \psi^T(\hat{H}^T_{\text{Gorkov}} - \epsilon_-)\bar{\psi}^T \\ &\quad = \bar{\psi}(\hat{H}_{\text{Gorkov}} - \epsilon_-)\psi + \psi^T(-\sigma_2^{\text{ph}}\hat{H}_{\text{Gorkov}}\sigma_2^{\text{ph}} - \epsilon_-)\bar{\psi}^T \\ &\quad = \bar{\Psi}(\hat{H}_{\text{Gorkov}} - \epsilon_-\sigma_3^{\text{cc}})\Psi \end{aligned}$$

where

$$\bar{\Psi} = \frac{1}{\sqrt{2}}\begin{pmatrix} \bar{\psi} & -\psi^T\sigma_2^{\text{ph}} \end{pmatrix}, \qquad \Psi = \frac{1}{\sqrt{2}}\begin{pmatrix} \psi \\ \sigma_2^{\text{ph}}\bar{\psi}^T \end{pmatrix}.$$

Here the superscript T denotes the supertransposition operation,[7] and σ_i^{cc} represent Pauli matrices acting in the charge conjugation space. As a consequence, the two supervector fields $\bar{\Psi}$, and Ψ are not independent but obey the symmetry relations

$$\Psi = \sigma_2^{\text{ph}}\gamma\,\bar{\Psi}^T, \qquad \bar{\Psi} = -\Psi^T\,\sigma_2^{\text{ph}}\gamma^{-1}, \tag{20}$$

where

$$\gamma = \mathbb{1}^{\text{ph}} \otimes \begin{pmatrix} \sigma_1^{\text{cc}} & \\ & -i\sigma_2^{\text{cc}} \end{pmatrix}_{\text{bf}} \tag{21}$$

To summarize, the generating functional for averages of products of Green functions can be written as

$$\mathcal{Z}[0] = \int D[\bar{\Psi}, \Psi]\,\exp\left[i\int d\mathbf{r}\;\bar{\Psi}(\hat{H}_{\text{Gorkov}} - \epsilon_-\sigma_3^{\text{cc}})\Psi\right].$$

For clarity, explicit reference to the structure of the source term has been suspended. The latter can be restored when necessary.

[7] In the following it will be important to note that the transformation rules for supervectors and supermatrices differ from those of conventional vectors and matrices. In particular, if we define a pair of supervectors

$$\psi = \begin{pmatrix} S \\ \chi \end{pmatrix}, \qquad \bar{\psi} = \begin{pmatrix} \bar{S} & \bar{\chi} \end{pmatrix}$$

with commuting and anticommuting elements S, $\bar{S}$ and χ, $\bar{\chi}$ respectively, the supertransposition operation is defined according to

$$\psi^T = \begin{pmatrix} S & -\chi \end{pmatrix}, \qquad \bar{\psi}^T = \begin{pmatrix} \bar{S} \\ \bar{\chi} \end{pmatrix}.$$

Similarly, under a supertransposition, a supermatrix transforms as

$$F = \begin{pmatrix} S_1 & \chi_1 \\ \chi_2 & S_2 \end{pmatrix}, \qquad F^T = \begin{pmatrix} S_1 & -\chi_2 \\ \chi_1 & S_2 \end{pmatrix}, \qquad \text{i. e. } F \neq (F^T)^T.$$

2.1.2. *Impurity averaging*

To develop the low-energy theory of the disordered superconductor, the first step in the program is to implement the impurity average. The result will be to transform the free theory into an interacting theory. Separating the Gor'kov Hamiltonian into regular and stochastic parts as $\hat{H}_{\text{Gorkov}} = \hat{H}^{(0)}_{\text{Gorkov}} + W(\mathbf{r})\sigma_3^{\text{ph}}$ and subjecting the generating function to an ensemble average over a Gaussian δ-correlated impurity distribution (8),

$$P(W)DW = \frac{e^{-\pi\nu\tau\int d\mathbf{r}\, W^2(\mathbf{r})}DW}{\int DWe^{-\pi\nu\tau\int d\mathbf{r}\, W^2(\mathbf{r})}}$$

one obtains

$$\langle \mathcal{Z}[0]\rangle_W = \int D[\bar{\Psi},\Psi]\exp\left[\int d\mathbf{r}\ \left(i\bar{\Psi}(\hat{H}^{(0)}_{\text{Gorkov}} - \epsilon_-\sigma_3^{\text{cc}})\Psi \right.\right.$$
$$\left.\left. -\frac{1}{4\pi\nu\tau}(\bar{\Psi}\sigma_3^{\text{ph}}\Psi)^2\right)\right].$$

In this form we can proceed in two ways: firstly, we could undertake a perturbative expansion in the interaction. Indeed, an appropriate rearrangement of the resulting series recovers the diagrammatic diffusion mode expansion. A second, and more profitable route, is to seek an appropriate mean-field decomposition of the interaction. Specifically, we are interested in identifying the diffusive modes discussed in chapter 1, *i.e.* two-particle channels arising from multiple scattering with momentum difference smaller than the inverse of the elastic mean free path, $\ell = v_F\tau$.

2.1.3. *Slow mode decoupling*

Isolating these modes is a standard, if technical, procedure [27] which is conveniently performed in Fourier space. Let us then focus on the quartic interaction generated by the impurity average:

$$\frac{1}{4\pi\nu\tau}\int d\mathbf{r}\ \left(\bar{\Psi}(\mathbf{r})\sigma_3^{\text{ph}}\Psi(\mathbf{r})\right)^2.$$

From this term, we want to isolate within it the collective modes involving small momentum transfer, $|\mathbf{q}| < q_0 \sim 1/\ell$, which are to be decoupled by a Hubbard-Stratonovich transformation — these represent the soft modes identified in section 1.4. To achieve this, following Ref. [30], we present the interaction in the Fourier representation, viz.

$$\int d\mathbf{r}\ \left(\bar{\Psi}(\mathbf{r})\sigma_3^{\text{ph}}\Psi(\mathbf{r})\right)^2$$
$$= \sum_{\mathbf{k}_1,\mathbf{k}_2,\mathbf{k}_3} \bar{\Psi}(\mathbf{k}_1)\sigma_3^{\text{ph}}\Psi(\mathbf{k}_2)\ \bar{\Psi}(\mathbf{k}_3)\sigma_3^{\text{ph}}\Psi(-\mathbf{k}_1-\mathbf{k}_2-\mathbf{k}_3).$$

Now there are three independent ways of pairing two fast single-particle momenta to form a slow two-particle momentum $\mathbf{q}$:

	$\bar{\Psi}(\mathbf{k}_1)$	$\Psi(\mathbf{k}_2)$	$\bar{\Psi}(\mathbf{k}_3)$	$\Psi(-\mathbf{k}_1-\mathbf{k}_2-\mathbf{k}_3)$
(a)	$\mathbf{k}$	$-\mathbf{k}+\mathbf{q}$	$\mathbf{k}'$	$-\mathbf{k}'-\mathbf{q}$
(b)	$\mathbf{k}$	$-\mathbf{k}'-\mathbf{q}$	$-\mathbf{k}+\mathbf{q}$	$\mathbf{k}'$
(c)	$\mathbf{k}$	$\mathbf{k}'$	$-\mathbf{k}'-\mathbf{q}$	$-\mathbf{k}+\mathbf{q}$

Term (a) can be decoupled trivially, producing no more than energy shifts that can be absorbed by a redefinition of the chemical potential. The other two terms can be rearranged in the following way. For term (b) we have

$$\begin{aligned}
&\sum_{\mathbf{k},\mathbf{k}',\mathbf{q}} \bar{\Psi}(\mathbf{k})\sigma_3^{\mathrm{ph}}\Psi(-\mathbf{k}'-\mathbf{q})\,\bar{\Psi}(-\mathbf{k}+\mathbf{q})\sigma_3^{\mathrm{ph}}\Psi(\mathbf{k}')\\
&= \sum_{\mathbf{k},\mathbf{k}',\mathbf{q}} \bar{\Psi}(\mathbf{k})\sigma_3^{\mathrm{ph}}\Psi(-\mathbf{k}'-\mathbf{q})\,\Psi^T(\mathbf{k}')\sigma_3^{\mathrm{ph}}\bar{\Psi}^T(-\mathbf{k}+\mathbf{q})\\
&= \sum_{\mathbf{k},\mathbf{k}',\mathbf{q}} \bar{\Psi}(\mathbf{k})\sigma_3^{\mathrm{ph}}\Psi(-\mathbf{k}'-\mathbf{q})\left(-\bar{\Psi}(\mathbf{k}')\gamma^{-1}\sigma_2^{\mathrm{ph}}\right)\sigma_3^{\mathrm{ph}}\left(\gamma\sigma_2^{\mathrm{ph}}\Psi(-\mathbf{k}+\mathbf{q})\right)\\
&= \sum_{\mathbf{q}} \mathrm{str}\left[\left(\sum_{\mathbf{k}'}\Psi(-\mathbf{k}'-\mathbf{q})\otimes\bar{\Psi}(\mathbf{k}')\sigma_3^{\mathrm{ph}}\right)\left(\sum_{\mathbf{k}}\Psi(-\mathbf{k}+\mathbf{q})\otimes\bar{\Psi}(\mathbf{k})\sigma_3^{\mathrm{ph}}\right)\right].
\end{aligned}$$

Here we have introduced the supertrace operation which acts on a supermatrix M according to $\mathrm{str}\ M = \mathrm{tr}\ M_{\mathrm{bb}} - \mathrm{tr}\ M_{\mathrm{ff}}$. Moreover, we have made use of the symmetry relations $\bar{\Psi}^T = \gamma\sigma_2^{\mathrm{ph}}\Psi$, and $\Psi^T = -\bar{\Psi}\gamma^{-1}\sigma_2^{\mathrm{ph}}$, which follow from Eq. (20). Finally, the term (c) is easily brought to the same form by using the cyclic invariance of the supertrace. Therefore, to assimilate the soft degrees of freedom, we may affect the replacement

$$\frac{1}{4\pi\nu\tau}\int d\mathbf{r}\,\left(\bar{\Psi}(\mathbf{r})\sigma_3^{\mathrm{ph}}\Psi(\mathbf{r})\right)^2 \simeq 2\times\frac{1}{4\pi\nu\tau}\sum_{|\mathbf{q}|<q_0}\mathrm{str}\left[\Gamma(-\mathbf{q})\Gamma(\mathbf{q})\right],$$

where the factor of 2 reflects the two channels of decoupling (b) and (c), and Γ is given by a sum of dyadic products of the fields Ψ and $\bar{\Psi}$

$$\Gamma(\mathbf{q}) = \sum_{\mathbf{k}}\Psi(-\mathbf{k}+\mathbf{q})\otimes\bar{\Psi}(\mathbf{k})\sigma_3^{\mathrm{ph}}.$$

(Note that, if the summation over $\mathbf{q}$ was unrestricted, the Hubbard-Stratonovich transformation would involve an overcounting by a factor of 2.)

With this definition, we can now implement a Hubbard-Stratonovich decoupling with the introduction of 8×8 supermatrix fields, Q,

$$\exp\left[-\frac{1}{2\pi\nu\tau}\sum_{\mathbf{q}}\mathrm{str}\left(\Gamma(\mathbf{q})\Gamma(-\mathbf{q})\right)\right]$$

$$= \int DQ \exp\left[\frac{1}{2\tau}\sum_{\mathbf{q}} \mathrm{str}\left(\frac{\pi\nu}{4}Q(\mathbf{q})Q(-\mathbf{q}) - Q(\mathbf{q})\Gamma(-\mathbf{q})\right)\right].$$

The symmetry properties of Q reflect those of the dyadic product $\Gamma(\mathbf{q})$. In particular, the symmetry relation

$$\begin{aligned}
\mathrm{str}\left[Q\Psi\otimes\bar{\Psi}\sigma_3^{\mathrm{ph}}\right] &= \mathrm{str}\left[\sigma_3^{\mathrm{ph}}\bar{\Psi}^T\otimes\Psi^T Q^T\right]\\
&= \mathrm{str}\left[\sigma_3^{\mathrm{ph}}(\gamma^{-1}\sigma_2^{\mathrm{ph}}\Psi)\otimes(-\bar{\Psi}\gamma\sigma_2^{\mathrm{ph}})Q^T\right]\\
&= \mathrm{str}\left[\sigma_2^{\mathrm{ph}}\gamma Q^T\gamma^{-1}\sigma_3^{\mathrm{ph}}\sigma_2^{\mathrm{ph}}\Psi\otimes\bar{\Psi}\right]\\
&= \mathrm{str}\left[\sigma_1^{\mathrm{ph}}\gamma Q^T\gamma^{-1}\sigma_1^{\mathrm{ph}}\Psi\otimes\bar{\Psi}\sigma_3^{\mathrm{ph}}\right],
\end{aligned}$$

is accounted for by subjecting the supermatrix Q to the linear condition

$$Q = \sigma_1^{\mathrm{ph}}\gamma\, Q^T\gamma^{-1}\sigma_1^{\mathrm{ph}}. \tag{22}$$

Finally, integrating out the fields Ψ, and $\bar{\Psi}$, and switching back to the coordinate representation, we obtain $\langle\mathcal{Z}[0]\rangle = \int DQ\ \exp\left[-S[Q]\right]$, where

$$S[Q] = -\int d\mathbf{r}\left[\frac{\pi\nu}{8\tau}\mathrm{str}\ Q^2 - \frac{1}{2}\ \mathrm{str}\ \ln\hat{\mathcal{G}}^{-1}\right]. \tag{23}$$

Here

$$\hat{\mathcal{G}}^{-1} = \hat{\zeta} + \sigma_3^{\mathrm{ph}}\hat{\Delta} - \epsilon_-\sigma_3^{\mathrm{cc}}\otimes\sigma_3^{\mathrm{ph}} + \frac{i}{2\tau}Q \tag{24}$$

represents the 'supermatrix' Green function with $\hat{\Delta} = |\Delta|\sigma_1^{\mathrm{ph}}e^{-i\varphi\sigma_3^{\mathrm{ph}}}$.

The domain of integration of the Hubbard-Stratonovich field Q is important. It is fixed by the requirement of convergence (in the boson-boson block), and this ultimately determines the structure of the saddle-point manifold of the σ-model. Historically, the first careful analysis of this issue is due to Weidenmüller, Verbaarschot and Zirnbauer [32] for the normal case. Later, Zirnbauer [17] provided a construction for each of the ten universality classes that emphasizes the algebraic aspects in ensuring convergence. In chapter 3 the integration manifold will be vital in our analysis of instanton saddle-points: we will specify the required contours there and refer to the literature for the details.

The problem of computing the disorder averaged Green function (and, if necessary, its higher moments) has been reduced to considering an effective field theory with the action $S[Q]$. Further progress is possible only within a saddle-point approximation.

2.1.4. *Saddle-point approximation and the σ-model*

The next step in deriving the low-energy theory is to explore the saddle-point structure of the effective action (23), and to classify and incorporate fluctuations by means of a gradient expansion. This is most straightforwardly achieved by implementing a two-step procedure devised in Ref. [10]. For in the dirty limit $\Delta \ll 1/\tau$, the scales set by the disorder and by the superconducting order parameter are well separated, so that one can perform two minimizations in sequence.

The strategy adopted in Ref. [10] is as follows: at first, one neglects the order parameter Δ and the deviation of the energy from the Fermi level, ϵ. By varying the resulting effective action, one finds the corresponding saddle-point manifold stabilized by the semi-classical parameter $\epsilon_F\tau \gg 1$. Then, fluctuations inside this manifold are considered; they couple to the order parameter and to the energy ϵ. The resulting low-energy effective action is varied once again inside the first (high-energy) saddle-point manifold. We will find that the corresponding low-energy saddle-point equation coincides with the Usadel equation (16) for the average quasi-classical Gor'kov Green function in the dirty limit.

In the absence of the order parameter, a variation of the action functional at the Fermi energy $S[Q]$ yields the saddle point equation:

$$Q(\mathbf{r}) = \frac{i}{\pi\nu}\mathcal{G}(\mathbf{r},\mathbf{r})$$

Taking the solution $Q_{\rm sp}$ to be spatially homogeneous, and setting $\int d\mathbf{p}/(2\pi)^d = \int \nu(\zeta)d\zeta \simeq \nu(0)\int d\zeta$, the saddle-point equation can be recast as

$$Q_{\rm sp} = \frac{i}{\pi}\int \frac{d\zeta}{\zeta - \epsilon_- \sigma_3^{\rm ph}\otimes\sigma_3^{\rm ph} + iQ_{\rm sp}/2\tau}, \tag{25}$$

where the positive infinitesimal 0^+ allows a distinction to be drawn between the physical and unphysical solutions. For $\epsilon_F\tau \gg 1$ the integral (25) may be evaluated in the pole approximation from which one obtains the diagonal matrix solution $Q = \mathrm{diag}(q_1, q_2, ...)$, with $q_i = \pm 1$. To choose the signs correctly, we note that the expression on the right-hand side of the saddle-point equation relates to the Green function of the disordered normal system evaluated in the self-consistent Born approximation. The disorder preserves the causal (i.e. retarded versus advanced) character of the Green function, and therefore the sign of q_i must coincide with the sign of the imaginary part of the energy. This singles out the particular solution

$$Q_{\rm sp} = \sigma_3^{\rm ph}\otimes\sigma_3^{\rm cc}.$$

As anticipated, however, this solution is not unique for $\epsilon \to 0$. Dividing out rotations that leave $\sigma_3^{\rm cc}\otimes\sigma_3^{\rm ph}$ invariant, the degeneracy of the manifold spanned by $Q = TQ_{\rm sp}T^{-1}$ is specified by the coset space $\mathrm{SU}(2,2|4)/\mathrm{SU}(2|2)\otimes\mathrm{SU}(2|2)$. The above form of $Q_{\rm sp}$ means that the manifold may also be defined by the non-linear condition $Q^2 = \mathbb{1}$.

Fluctuations transverse to this manifold are integrated out using the saddle-point parameter $\nu L^d/\tau \gg 1$. In the Gaussian approximation they do not couple to fluctuations on the saddle-point. Furthermore, the integration yields a factor of unity by supersymmetry [27] (for a more complete discussion see, e.g., Ref. [30]). With the saddle point approximation understood, it is straightforward to derive the σ-model action from (23) by inserting $Q(\mathbf{r}) = T(\mathbf{r})Q_{\text{sp}}T^{-1}(\mathbf{r})$ into the expression (23) for $S[Q]$ and expanding in $\hat{\Delta}$ and ϵ, and up to second order in gradients of $Q(\mathbf{r})$, neglecting higher-order derivatives.

$$S[Q] = -\frac{\pi\nu}{8}\int d\mathbf{r}\ \text{str}\left[D(\nabla Q)^2 - 4i(\hat{\Delta} + \epsilon_-\sigma_3^{\text{cc}})\sigma_3^{\text{ph}}Q\right], \tag{26}$$

where $D = v_F^2\tau/d$ denotes the classical diffusion constant of the normal metal. The effect of a vector potential $\mathbf{A}$ is included by the replacement $\nabla \to \tilde{\nabla} \equiv \nabla - ie\mathbf{A}[\sigma_3^{\text{ph}}, \ ]$ [27].

Let us emphasize the approximations used in the derivation of (26). Besides the quasi-classical ($\epsilon_F\tau \gg 1$) and saddle-point ($\nu L^d/\tau \gg 1$) parameters, one requires that all energies left are small compared to $1/\tau$, which allows us to truncate the expansion. Thus the action applies to $(Dq^2, \epsilon, \Delta) \ll 1/\tau$, where q is a wavevector characterizing the scale of variation of Q. This includes the usual dirty limit. We stress again that the completeness of the description provided by the action (26) within these approximations means that all physics at these energies should be contained.

This completes the derivation of the intermediate energy scale action. However, even on the soft manifold $Q^2(\mathbf{r}) = \mathbb{1}$, the majority of degrees of freedom are rendered massive by the order parameter and energy. To explore the structure of the low-energy action it is necessary to implement a further saddle-point analysis of (26) taking into account the influence of the superconducting order parameter.

2.1.5. *Low-energy saddle-point and soft modes*

To identify the low-energy saddle-point it is necessary to seek the optimal energy configuration of the supermatrix field Q for a non-vanishing order parameter $\hat{\Delta}$ and, in principle, a non-vanishing magnetic vector potential $\mathbf{A}$. We therefore require $S[Q]$ to be stationary with respect to variations of $Q(\mathbf{r})$ that preserve the non-linear constraint $Q^2(\mathbf{r}) = \mathbb{1}$. Following Ref. [30], such variations can be parametrized by transformations

$$\delta Q(\mathbf{r}) = \eta\left[X(\mathbf{r}), Q(\mathbf{r})\right],$$

where $X = -\sigma_1^{\text{ph}}\gamma X^T\gamma^{-1}\sigma_1^{\text{ph}}$ so as to preserve the symmetry (22). Subjecting the action to this variation, and linearizing in X, the stationarity condition $\delta S = 0$ translates to the equation of motion

$$D\tilde{\nabla}\left(Q\tilde{\nabla}Q\right) + i\left[Q, (\epsilon_-\sigma_3^{\text{cc}} + \hat{\Delta})\sigma_3^{\text{ph}}\right] = 0. \tag{27}$$

Associating $Q(\mathbf{r})$ with the average quasi-classical Gorkov Green function $g_0(\mathbf{r})$, the saddle-point equation is identified as the mean-field Usadel equation (16) derived in section 1.5. In hindsight the coincidence should not be surprising. At each stage of this calculation we have implemented approximations consistent with the quasi-classical scheme. With this understanding, we will tend to refer to the above as the Usadel equation.

Although the solution of this equation constrains many of the degrees of freedom to a single saddle-point, in the limit $\epsilon = 0$ several degrees of freedom remain massless for any value of the order parameter $\hat{\Delta}$. Specifically, the action functional $S[Q]$ is invariant under transformations

$$Q(\mathbf{r}) \mapsto TQ(\mathbf{r})T^{-1}, \quad \text{if} \quad T = 1\!\!1^{\text{ph}} \otimes t \tag{28}$$

with $t = \gamma(t^{-1})^T\gamma^{-1}$ constant in space. The latter condition means that t runs through an orthosymplectic Lie supergroup OSp(2|2). According to the classification scheme discussed in section 1.4, this defines the symmetry class CI. In presence of a magnetic field, the space of massless fluctuations is further diminished to the coset manifold $\mathrm{OSp}(2|2)/\mathrm{GL}(1|1)$ characterizing the symmetry class C. Not all of the classes are available to us in the present formulation. The classes designated D and DI require the introduction of spin degrees of freedom. This will be done in the next chapter, where we will encounter a realization of class D and the associated novel phase coherent phenomena.

This completes the formal construction of the low-energy statistical field theory of the weakly disordered superconductor. At the level of the mean-field of saddle-point, an application of this theory reproduces the results of the quasi-classical scheme. The role of fluctuations around the mean-field impacts most strongly on situations where low-energy quasi-particles are allowed to exist, e.g. quasi-particle states trapped around a vortex in the mixed phase [30], bulk superconductors driven into a gapless phase by a parallel magnetic field or magnetic impurities (see section 3), or hybrid superconductor/normal structures. To explore the impact of these novel mechanisms of quantum interference, in the following section we will explore the phenomenology of the magnetic impurity system. However, before doing so, let us first explore the mean-field structure of the action focusing on two simple examples: the bulk s-wave superconductor (and the restoration of the Anderson theorem), and the case of a quantum dot contacted to a superconducting terminal. Indeed, the latter solution will be needed in section 3.

2.2. DISORDERED BULK SUPERCONDUCTOR

In the absence of a magnetic field, taking the order parameter to be spatially homogeneous and specifying the gauge $\varphi = 0$ (i.e. $\hat{\Delta} = \sigma_1^{\text{ph}}|\Delta|$), the saddle-point equation for Q can be solved straightforwardly. With the ansatz

$$Q_{\text{sp}} = \sigma_3^{\text{cc}} \otimes \sigma_3^{\text{ph}} \cosh\hat{\theta} - i\sigma_2^{\text{ph}} \sinh\hat{\theta} \tag{29}$$

where the matrix $\hat{\theta}$ is diagonal in the superspace with elements $\hat{\theta} = \text{diag}(\theta_{\text{b}}, \theta_{\text{f}})$, the saddle-point or Usadel equation assumes the form

$$D\nabla^2\theta + 2\left(|\Delta|\cosh\hat{\theta} - \epsilon_- \sinh\hat{\theta}\right) = 0 \tag{30}$$

Taking $\hat{\theta}$ to be homogeneous with $\theta_{\text{b}} = \theta_{\text{f}}$, we obtain the BCS solution $\theta = \theta_{\text{s}}$ (19).

Having obtained the quasi-classical Green's function we can impose the self-consistency condition $\Delta = -(\lambda/\nu)\langle\psi_\downarrow\psi_\uparrow\rangle$ as usual. We obtain the gap equation

$$|\Delta| = i\lambda\pi T \sum_{\epsilon_n} \sinh\theta|_{\epsilon_- = i\epsilon_n} \tag{31}$$

where the summation is taken over fermionic Matsubara frequencies $\epsilon_n = \pi T(2n + 1)$. Similarly, from the saddle-point solution, we obtain the quasi-particle DoS

$$\begin{aligned}\nu(\epsilon) &= \frac{1}{\pi}\text{tr Im }\hat{G}^-(\epsilon) = -\frac{1}{4\pi}\text{tr Im }\hat{\mathcal{G}}\sigma_3^{\text{ph}} \otimes \sigma_3^{\text{cc}} \\ &= \frac{\nu_{\text{n}}}{4}\text{Re str}\left[\sigma_3^{\text{bf}} \otimes \sigma_3^{\text{cc}} \otimes \sigma_3^{\text{ph}}\, Q\right] = 2\nu_{\text{n}}\text{Re }\cos\theta(\epsilon)\ ,\end{aligned}$$

just as in the usual quasi-classical theory.

We finish this first example with an important technical comment. In the present formalism the above result follows from a saddle-point approximation. Yet normally any quantity calculated in this way is weighted by a factor $e^{-S[Q_{\text{sp}}]}$. To complete the correspondence with the usual quasi-classical theory, we note that the saddle-point Q_{sp} should be chosen proportional to unity in the boson-fermion space. Through the definition of the supertrace, this ensures that $S[Q_{\text{sp}}] = 0$. In the same way, any fluctuation corrections to the saddle-point action vanish by supersymmetry [27].[8] Saddle-point configurations that are not 'supersymmetric' in this sense can be important and we will discuss such a case in the next chapter.

2.3. HYBRID SN-STRUCTURES

With the Usadel equation in hand, one can proceed (once the correct boundary conditions are known) to find solutions in more complex geometries, that describe hybrid superconductor-normal systems [10]. In chapter 3 we will need the mean-field result for a geometry that cannot in fact be described by the Usadel equation as it stands. This is the case of a quantum dot contacted to a superconductor.

2.3.1. *Quantum dot contacted to a superconductor*

The case of a normal quantum dot coupled to superconducting lead through a contact of arbitrary transparency (see Fig. 5) presents us with a dilemma. The

[8] Of course, fluctuations are important in the calculation of non-supersymmetric source terms used to extract physical quantities from the action.

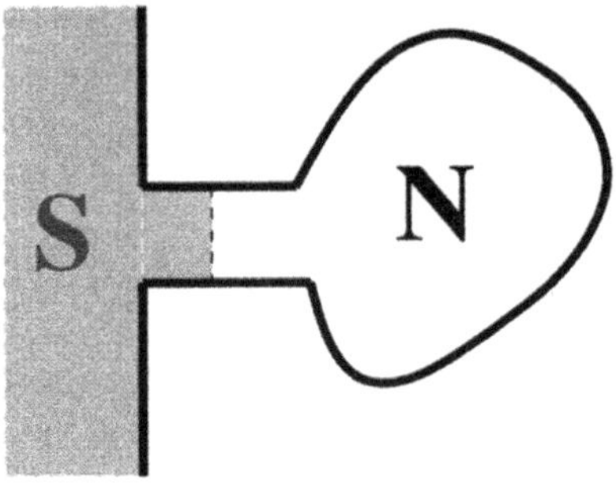

Figure 5. Metallic quantum dot coupled to a superconducting lead.

lead has N propagating modes. The quantum dot is a small metallic region with $D/L^2 \gg N\delta$.[9] The energy scale that determines the influence of the contact on the properties of the dot is the inverse of the time taken for an electron in the dot to feel the contact. This defines the generalized Thouless energy [33], and for the quantum dot, this scale is set by $N\delta$ (modulo factors relating to the transparency of the lead). In a large dot with $D/L^2 \ll \delta$ the diffusive motion of the electrons would set this scale.

A naive expectation is that this problem should involve the solution of the Usadel equation as before, with the right boundary conditions. The above considerations show this not to be the case. With D/L^2 the largest energy scale in the problem, gradients of Q are frozen out of the action. One must explicitly include the coupling to the leads from the outset, as the saddle point will be determined by the competition between the energy ϵ and this coupling (of order $N\delta$) in the action. D/L^2 will appear nowhere. Put simply, the gradient expansion is not the true low-energy action in such a confined geometry.

Unfortunately, a fully microscopic derivation of the correct form of the zero-dimensional (that is, containing no spatial gradients) action is laborious [27]. We can get to the answer more directly by using the general principle that the zero-dimensional limit of the action describes the appropriate random matrix model, or equivalently, that the quantum dot system in the limit $D/L^2 \gg \delta$ may be modeled by random matrix theory with matrices of size $M \to \infty$, as described in section 1.4. The random matrix model for the dot is simply a Gor'kov Hamiltonian (3) with $\Delta = 0$ — the dot is normal — and $\hat{H}$ given by an appropriate random Hamiltonian with mean level spacing δ from the orthogonal symmetry class. The non-trivial element is the coupling to the leads. The standard approach [34] is to

[9] This includes the case of a ballistic chaotic quantum dot, provided the ergodic time (the time required for an electron to explore the available phase space) is much longer than the dwell time of the electrons in the dot.

write the lead-dot coupling as

$$\hat{H}_{\rm LD} = \sum_{j,\alpha} \int \frac{dk}{2\pi} (W_{\alpha j}(|\alpha, p\rangle\langle j, k, p| - |\alpha, h\rangle\langle j, k, h|) + \text{h.c.}) \,. \tag{32}$$

In this expression $|\alpha, n\rangle$ with $n = p, h$ denotes a basis of the random matrix model for the dot, and $|j, k, n\rangle$ is the obvious basis for the $j = 1 \ldots N$ propagating modes of the lead. Though this coupling is formally the same as a tunneling Hamiltonian it is capable of describing contacts of arbitrary transparency with proper interpretation of the couplings $W_{\alpha j}$. It is possible to show that the dot can be described by the 'effective Hamiltonian',[10]

$$\hat{H}_{\rm eff} \equiv \hat{H}\sigma_3^{\rm ph} - i\pi\nu W W^\dagger \hat{g}_{\rm bcs}(\epsilon) \,,$$

where $g_{\rm bcs}$ is defined in Eq. (18). It is this structure that is needed in the derivation of the zero-dimensional σ-model. By expanding only in ϵ in the 'str ln' form of the action (23) one arrives at

$$S[Q] = \frac{i\pi\epsilon_-}{2\delta} \text{str} \left[\sigma_3^{\rm cc} \otimes \sigma_3^{\rm ph} Q\right] - \frac{1}{2} \sum_j \text{str} \left[\ln(1 + \alpha_j Q_{\rm bcs} Q)\right] , \tag{33}$$

where $Q_{\rm bcs}$ is used to denote the bulk BCS saddle-point found in the previous section. In the above we have taken $WW^\dagger$ to be the $M \times M$ diagonal matrix $\text{diag}\{\alpha_1, \ldots, \alpha_N, 0, \ldots, 0\}$. (33) is the proper form of the σ-model for a quantum dot with superconducting leads. It was first used by [35] in their investigation of the class C spectral statistics of such a device. Since we are typically interested in energies of the order of the level spacing, the order parameter may be taken to infinity so that $Q_{\rm bcs} = \sigma_1^{\rm ph}$. We will specialize at this stage to the case of N perfectly ballistic contacts, so that all $\alpha_j = 1$.

As before, to obtain a mean-field expression for the DoS it is necessary to minimize the action with respect to variations in Q. Doing so, one obtains the saddle-point equation

$$-\frac{i\pi\epsilon_-}{2\delta}[Q, \sigma_3^{\rm cc} \otimes \sigma_3^{\rm ph}] + \frac{N}{2}[Q, (1 + Q_{\rm bcs}Q)^{-1} Q_{\rm bcs}] = 0$$

Applying the ansatz that the saddle-point solution is contained within the diagonal parameterization (29), the saddle-point equation takes the form

$$-\frac{\pi\epsilon_-}{\delta} \sinh\hat{\theta} + \frac{N}{2} \frac{\cosh\hat{\theta}}{1 + i\sinh\hat{\theta}} = 0. \tag{34}$$

[10] This has a well-defined meaning only within the context of a scattering approach [34]. For an informal derivation, write down the BdG equations (2) for the whole system and eliminate states from outside the dot.

We can straightforwardly determine that there is a 'minigap' $E_{\rm gap}$ in the DoS by setting $\cosh\theta_{\rm sp}$ to be imaginary. Thus $\sinh\theta_{\rm s} \equiv -ib$ for real b and (34) gives

$$\epsilon(b) = \frac{N\delta}{2\pi}\frac{1}{b}\sqrt{\frac{b-1}{b+1}} .$$

The extremum of this function gives the largest energy corresponding to a real value of b. This occurs at $b = (1+\sqrt{5})/2 = 1+\gamma$, where γ is the golden mean, and yields $E_{\rm gap} = (N\delta/2\pi)\gamma^{5/2} \approx 0.048N\delta$. With a bit more effort, one can expand in the vicinity of $E_{\rm gap}$ to obtain

$$\nu(\epsilon) \simeq \begin{array}{ll} 0 & \epsilon < E_{\rm gap}, \\ \frac{1}{\pi L^d}\sqrt{\frac{\epsilon - E_{\rm gap}}{\Delta_g^3}} & \epsilon < E_{\rm gap}, \end{array} \tag{35}$$

where $\Delta_g \approx 0.068N^{1/3}\delta$.

Finally, we note that, in the opposite case of α_j small, one can expand the logarithm in α_j. In the first order the action is just the same as for a BCS superconductor with gap $(\delta/\pi)\sum_j \alpha_j$. The formation of the minigap is a highly non-trivial effect. Indeed, in Ref. [33], the integrity of the gap is proposed as a signature of irregular or chaotic dynamics inside the dot. A dot with integrable dynamics appears to possess only a 'soft' gap in the DoS, with the DoS going to zero at zero energy. It is no surprise that 'diffusive' SN structures, where the gradient action and Usadel equation are the appropriate description, also display a minigap. For a modern theoretical review of minigap structures in superconductor/normal compounds, see Ref. [36].

This completes our study of the mean-field spectral properties of the hybrid superconducting/normal system. In principle, these results could have been recovered without resort to the field theoretic scheme. To address the importance of mesoscopic fluctuations on the coherence properties of the superconducting system, we now turn to a bulk system which exhibits low-energy quasi-particle excitations. Here we will require the full machinery of the non-linear σ-model.

3. Superconductors with magnetic impurities: instantons and sub-gap states

3.1. INTRODUCTION

In section 1.2 we discussed Anderson's observation that the thermodynamic properties of an s-wave superconductor in the dirty limit are independent of the amount of normal (non-magnetic) impurities added to the system. In the argument the time-reversal symmetry of the single-particle Hamiltonian plays a prominent role: pairing occurs between degenerate time-reversed eigenstates. When time-reversal symmetry is broken we expect pairing to be disrupted and superconductivity suppressed. This can be achieved by applying a magnetic field or by adding magnetic impurities. The effect is described by the classic theory of Abrikosov and Gor'kov [13] (AG), who considered the magnetic impurity case, though the description has a high degree of universality [37].

It is easy to see the importance of time-reversal symmetry from the Gor'kov Hamiltonian

$$\hat{H}_{\text{Gorkov}} = \begin{pmatrix} \hat{H} & \Delta\sigma_2^{\text{sp}} \\ \Delta^*\sigma_2^{\text{sp}} & -\hat{H}_0^T \end{pmatrix}_{\text{ph}} . \tag{36}$$

This differs from Eq. (3) through the introduction of the spin space (with Pauli matrices denoted σ_i^{sp}). The Pauli matrix σ_2^{sp} in the off-diagonal particle-hole block reflects singlet pairing. We introduce scattering by normal and magnetic impurities through the simple model

$$\hat{H} = \frac{\hat{\mathbf{p}}^2}{2m} - \epsilon_F + W(\mathbf{r}) + J\mathbf{S}(\mathbf{r}) \cdot \sigma^{\text{sp}} , \tag{37}$$

In addition to the weak potential impurity distribution $W(\mathbf{r})$, the particles experience a quenched random magnetic impurity distribution $J\mathbf{S}(\mathbf{r})$ where J represents the exchange coupling. The inclusion of $J\mathbf{S}(\mathbf{r})$ evidently prevents the simple diagonalization of (36) in terms of the single-particle eigen-energies as before.

AG solved the model defined by Eq. (36) together with the self-consistent equation for the order parameter (4) in the self-consistent Born approximation. Their results are expressed in terms of the spin-flip scattering rate $1/\tau_s$ through the natural dimensionless parameter

$$\zeta \equiv \frac{1}{\tau_s|\Delta|} . \tag{38}$$

The relation between $1/\tau_s$ and $J\mathbf{S}(\mathbf{r})$ will be given shortly. In section 1.3 we explained how a time-reversal symmetry breaking perturbation leads to the suppression of superconductivity (in the present model $1/\tau_\varphi = 2/\tau_s$). This certainly has the flavour of a mesoscopic effect: it depends on the loss of phase rigidity in the single-particle wavefunctions as the time-reversal symmetry is broken.[11] It is, however, of a 'mean-field' character. In this chapter we will see that a complete description of the DoS within the model defined by Eq. (37) necessitates the inclusion of non-perturbative effects as well as the novel channels of quantum phase coherence discussed in the introduction.

[11] This notion of phase rigidity can be made precise. In Ref. [38] the 'order parameter' $\rho \equiv \langle |\int d\mathbf{r}\phi_\alpha^2|\rangle$ is calculated for the crossover from the orthogonal ($\rho = 1$) to the unitary ($\rho = 0$) symmetry classes.

3.1.1. *Density of states*

We saw that AG's formula (11) followed from general considerations and it is indeed universal [37, 4]. Quantities such as the quasi-particle DoS are more model dependent. In the present model AG found that, remarkably, the suppression of the gap in the DoS is more rapid than that of the superconducting order parameter (Fig. 6). They found a narrow 'gapless' superconducting phase in which the quasi-particle energy gap is destroyed while the superconducting order parameter remains non-zero. This prediction was soon confirmed experimentally.

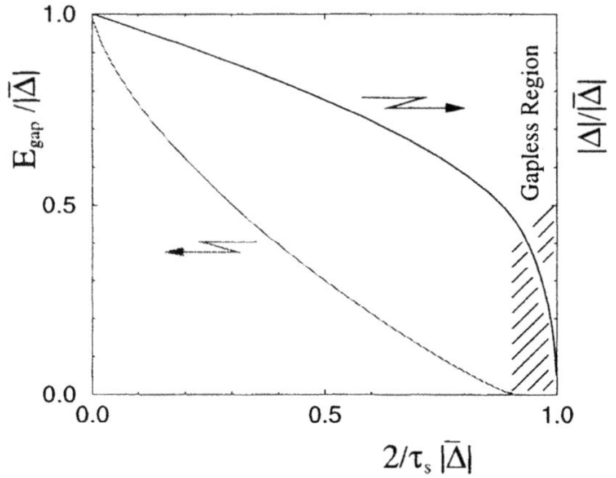

Figure 6. Variation of the energy gap $E_{\rm gap}$ and the self-consistent order parameter $|\Delta|$ as a function of (normalized) scattering rate $2/\tau_s|\bar{\Delta}|$. $|\bar{\Delta}|$ is the order parameter at $1/\tau_s = 0$.

This immediately presents two questions:

1. According to AG, the gap is maintained up to a critical concentration of magnetic impurities (at $T = 0$, 91% of the critical concentration at which superconductivity is destroyed). Yet, being unprotected by the Anderson theorem, it seems likely that the gap structure predicted by the mean-field theory is untenable and must be subject to non-perturbative corrections. What is the structure of the resulting 'sub-gap' states?
2. The gapless superconducting phase has quasi-particle states all the way down to zero energy. These low energy states should be strongly affected by channels of quantum interference discussed in section 1.4. Where does the gapless system fit into this classification and what are the consequences for the spectral and transport properties?

Once identified, the answer to the second question can be straightforwardly inferred from existing studies of the relevant universality class. Here we will be more concerned with answering the first question.

Sub-gap states in the magnetic impurity system have been discussed before. Strong magnetic impurities [39–41] evidently lie outside the Born approximation used by AG. In particular it was shown that, in the unitarity limit, a single magnetic impurity leads to the local suppression of the order parameter and creates a bound sub-gap quasi-particle state [39]. For a finite impurity concentration, these intra-gap states broaden into a band [40] merging smoothly with the continuum bulk states.

We will argue that there is a mesoscopic view of this problem which is more universal. Sub-gap states are those which are anomalously lacking in phase rigidity in the presence of a time-reversal symmetry breaking perturbation. This could be either an extrinsic or intrinsic effect. By intrinsic we mean that this is simply what happens to some proportion of states of this random Hamiltonian when we switch on such a perturbation. Alternatively, one can conceive of an extrinsic mechanism: The AG theory shows the gap to follow the relation

$$E_{\text{gap}}(\tau_s) = |\Delta| \left(1 - \zeta^{2/3}\right)^{3/2} \tag{39}$$

showing an onset of the gapless region at $\zeta = 1$ (note $\hbar = 1$ throughout). Even for weak disorder, however, it is apparent that optimal fluctuations of the random potential must generate sub-gap states in the interval $0 < \zeta < 1$, thus providing non-perturbative corrections to the self-consistent Born approximation used by AG. A fluctuation of the random potential which leads to an effective Born scattering rate $1/\tau_s'$ in excess of $1/\tau_s$ over a range set by the superconducting coherence length,

$$\xi = \left(\frac{D}{|\Delta|}\right)^{1/2}, \tag{40}$$

induces quasi-particle states down to energies $E_{\text{gap}}(\tau_s')$.[12] These sub-gap states are localized, being bound to the region where the scattering rate is large, see Fig. 7. We will return to this picture later.

The situation bears comparison with band tail states in semi-conductors. In this instance, rare or optimal configurations of the random impurity potential generate bound states, known as Lifshitz tail states [43], which extend below the band edge. The correspondence is, however, somewhat superficial: band tail states in semi-conductors are typically associated with smoothly varying, nodeless wavefunctions. By contrast, the tail states below the superconducting gap involve the superposition of states around the Fermi level. As such, one expects these states to be rapidly oscillating on the scale of the Fermi wavelength λ_F, but modulated by an envelope which is localized on the scale of the coherence length ξ. This difference is not incidental. Firstly, unlike the semi-conductor, one expects the energy dependence of the density of states in the tail region below the mean-field gap edge to be 'universal', independent of the nature of the weak impurity distribution

[12] Similar arguments have been made by Balatsky and Trugman [42].

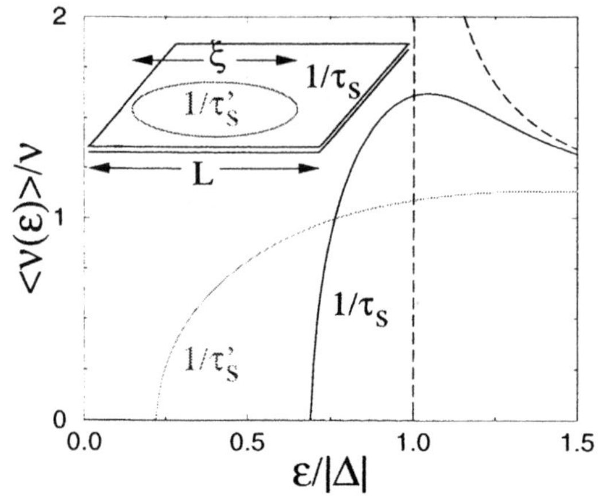

Figure 7. Mechanism of extrinsic sub-gap state formation.

but dependent only on the pair-breaking parameter ζ. Secondly, as we will see, one can not expect a straightforward extension of existing theories [43, 44] of the Lifshitz tails to describe the profile of tail states in the superconductor.

3.1.2. *Outline*

In this chapter, following Refs. [45], we will first show how to extend the statistical field theory described in chapter 2 to incorporate scattering by magnetic impurities. As anticipated in the previous chapter, a saddle-point approximation recovers the mean-field theory of AG. We discuss the soft-modes of the action that exist in the gapless phase and determine the consequences of these new channels of interference. In section 3.4, with the field theory in hand, we turn to problem of the sub-gap states. We find that these are described by instantons of the field theory; we identify the profile of the instanton with the envelope modulating the quasi-classical sub-gap states. A careful analysis allows us to evaluate the sub-gap density of states with exponential accuracy. In section 3.5 we examine the zero dimensional limit and prove a recent universality conjecture [46]. We next discuss the universality of the $d > 0$ problem in the context of other realizations of gapless superconductivity.

3.2. FIELD THEORY OF THE MAGNETIC IMPURITY PROBLEM

Incorporating the additional structure of (36) into the field theoretic description obtained in the previous chapter is straightforward. As before one starts from the generating functional

$$\mathcal{Z}[J] = \int D(\bar{\psi}, \psi) e^{\int d\mathbf{r} \left(i\bar{\psi}(\hat{H}_{\text{Gorkov}} - \epsilon_-)\psi + \bar{\psi}J + \bar{J}\psi \right)}, \tag{41}$$

where $\epsilon_- \equiv \epsilon - i0$ and the supervector fields have the internal structure $\bar{\psi} = (\bar{\psi}_\uparrow\, \bar{\psi}_\downarrow\, \psi_\uparrow\, \psi_\downarrow)$, $\psi^T = (\psi_\uparrow\, \psi_\downarrow\, \bar{\psi}_\uparrow\, \bar{\psi}_\downarrow)$. As in chapter 2 we will only be concerned with the average of a single Green's function.

3.2.1. *σ-model action*

For clarity it is desirable to remove the $\sigma_2^{\rm sp}$ from the off-diagonal terms in Eq. (36). To do this, we perform the rotation $\psi \mapsto \psi' = U\psi$, $\bar{\psi} \mapsto \bar{\psi}' = \bar{\psi}U^\dagger$ with

$$U = \begin{pmatrix} 1 & 0 \\ 0 & i\sigma_2^{\rm sp} \end{pmatrix}_{\rm ph},$$

after which the Gor'kov Hamiltonian takes the form

$$\hat{H}_{\rm Gorkov} = \left(\frac{\hat{\mathbf{p}}^2}{2m} + W(\mathbf{r}) - \epsilon_F\right) \otimes \sigma_3^{\rm ph} + J\mathbf{S}(\mathbf{r}) \cdot \sigma^{\rm sp} + |\Delta|\sigma_2^{\rm ph}.$$

Since, in the following, the global phase can be chosen arbitrarily, the order parameter can be chosen to be real. The unusual phase coherence properties of the superconducting system rely on the particle/hole or charge conjugation symmetry

$$\hat{H}_{\rm Gorkov} = -\sigma_2^{\rm ph} \otimes \sigma_2^{\rm sp} \hat{H}^T_{\rm Gorkov} \sigma_2^{\rm sp} \otimes \sigma_2^{\rm ph}. \tag{42}$$

As before, one can include all channels of interference by further doubling the field space as in chapter 2. Rather than present all the intermediate steps, we give only the symmetry relation on Q, the Hubbard-Stratonovich field introduced to decouple the average over W. In this case

$$Q = \sigma_1^{\rm ph} \otimes \sigma_2^{\rm sp} \gamma Q^T \gamma^{-1} \sigma_1^{\rm ph} \otimes \sigma_2^{\rm sp}, \tag{43}$$

where now, in contrast to Eq. 21, we have defined

$$\gamma = 1\!\!1^{\rm ph} \otimes \begin{pmatrix} i\sigma_2^{\rm cc} & \\ & \sigma_1^{\rm cc} \end{pmatrix}_{\rm bf}.$$

We will see presently that, when there are quasi-particle states at low energy in the present system, their localization properties are radically different to those of systems in the previous chapter. It is through this new γ that the distinction enters the present formalism.

Turning to the magnetic impurity scattering due to the $J\mathbf{S}(\mathbf{r}) \cdot \sigma^{\rm sp}$ term, we use the Gaussian model specified by zero mean and variance

$$\langle JS_\alpha(\mathbf{r}) JS_\beta(\mathbf{r}')\rangle_S = \frac{1}{6\pi\nu\tau_s}\delta^d(\mathbf{r} - \mathbf{r}')\delta_{\alpha\beta}, \tag{44}$$

where $1/\tau_s$ is the spin flip scattering rate introduced earlier.[13] Averaging yields the term in the Ψ field action

$$\left\langle \exp\left[i\int d\mathbf{r}\Psi J\mathbf{S}(\mathbf{r})\cdot\sigma^{\rm sp}\bar{\Psi}\right]\right\rangle_{JS} = \exp\left[-\frac{1}{12\pi\nu\tau_s}\int d\mathbf{r}(\bar{\Psi}\sigma^{\rm sp}\Psi)^2\right]. \quad (45)$$

The interaction generated by the magnetic impurity averaging can be treated [27] by performing all possible pairings and making use of the saddle-point approximation $Q(\mathbf{r}) = 2\langle\Psi(\mathbf{r})\otimes\bar{\Psi}(\mathbf{r})\sigma_3^{\rm ph}\rangle_\Psi/\pi\nu$. This leads to the replacement

$$\frac{1}{12\pi\nu\tau_s}\int d\mathbf{r}\ (\bar{\Psi}\sigma^{\rm sp}\Psi)^2 \mapsto \frac{\pi\nu}{24\tau_s}\int d\mathbf{r}\ {\rm str}\left(Q\sigma_3^{\rm ph}\otimes\sigma^{\rm sp}\right)^2.$$

Such an approximation, which neglects pairings at non-coincident points is allowed by the strong inequality $(\ell/\xi)^d \ll 1$. In addition we discard the contraction $\langle\bar{\Psi}\sigma^{\rm sp}\Psi\rangle_\Psi$. The term generated by this procedure could in any case be decoupled by a slow bosonic field $\mathbf{S}(\mathbf{r})$ which would immediately be set to zero for the singlet saddle-points that will be the basis of this section.

Gaussian in the fields Ψ and $\bar{\Psi}$, the functional integration can be performed explicitly after which one obtains $\langle\mathcal{Z}[0]\rangle_{V,S} = \int DQ\exp(-S[Q])$ where

$$S[Q] = -\int d\mathbf{r}\left[\frac{\pi\nu}{8\tau}{\rm str}\ Q^2 - \frac{1}{2}{\rm str}\ \ln\left(\sigma_3^{\rm ph}(\hat{\mathcal{H}}_0 - \epsilon_-\sigma_3^{\rm cc}) + \frac{i}{2\tau}Q\right)\right.$$
$$\left.-\frac{\pi\nu}{24\tau_s}{\rm str}\ (Q\sigma_3^{\rm ph}\otimes\sigma^{\rm sp})^2\right].$$

From this point, the σ-model follows precisely as before

$$S[Q] = -\frac{\pi\nu}{8}\int d\mathbf{r}\ {\rm str}\ \left[D(\nabla Q)^2 - 4i\left(\epsilon_-\sigma_3^{\rm cc} + |\Delta|\sigma_2^{\rm ph}\right)\sigma_3^{\rm ph}Q\right.$$
$$\left.-\frac{1}{3\tau_s}\left(Q\sigma_3^{\rm ph}\otimes\sigma^{\rm sp}\right)^2\right]. \quad (46)$$

The saddle point manifold is given by $Q = TQ_{\rm sp}T^{-1}$, with $Q_{\rm sp} = \sigma_3^{\rm ph}\otimes\sigma_3^{\rm cc}$ and T chosen to be consistent with (43). The quasi-particle DoS is obtained from the functional integral

$$\langle\nu(\epsilon,\mathbf{r})\rangle_{V,S} = \frac{\nu}{4}{\rm Re}\left\langle{\rm str}\left(\sigma_3^{\rm bf}\otimes\sigma_3^{\rm ph}\otimes\sigma_3^{\rm cc}Q(\mathbf{r})\right)\right\rangle_Q. \quad (47)$$

The numerical factor leads to a DoS of 4ν for the system as $|\epsilon| \to \infty$. This is because both the particle-hole structure of the original Bogoliubov Hamiltonian

[13] Following AG, we take the quenched distribution of magnetic impurities to be 'classical' and non-interacting throughout — indeed, otherwise our method would not apply in its present formulation. For practical purposes, this entails the consideration of structures where both the Kondo temperature [47] and, more significantly, the RKKY induced spin glass temperature [48] are smaller than the relevant energy scales of the superconductor.

and the spin each cause a doubling of the DoS. With the appropriate extension of the σ-model in hand, we should now check that the mean-field description of AG is recovered at the saddle-point level, as anticipated.

3.2.2. *AG Mean-Field Theory*

Variation of the σ-model action with respect to fluctuations of Q obtains the Usadel equation

$$D\nabla\left(Q\nabla Q\right)+i\left[Q,\epsilon_{-}\sigma_3^{\rm cc}\otimes\sigma_3^{\rm ph}+i|\Delta|\sigma_1^{\rm ph}\right] \\ +\frac{1}{6\tau_s}\left[Q,\sigma_3^{\rm ph}\otimes\sigma^{\rm sp}Q\sigma_3^{\rm ph}\otimes\sigma^{\rm sp}\right]=0. \qquad (48)$$

With the ansatz

$$Q_{\rm sp}=\left[\sigma_3^{\rm cc}\otimes\sigma_3^{\rm ph}\cosh\hat\theta+i\sigma_1^{\rm ph}\sinh\hat\theta\right]\otimes 1\!\!1^{\rm sp}, \qquad (49)$$

where the elements $\hat\theta={\rm diag}(\theta_1,i\theta)_{\rm bf}$ are diagonal in the superspace, the saddle-point equation decouples into boson-boson and fermion-fermion sectors, and takes the form

$$\nabla_{{\bf r}/\xi}^2\hat\theta+2i\left(\cosh\hat\theta-\frac{\epsilon}{|\Delta|}\sinh\hat\theta\right)-\zeta\sinh(2\hat\theta)=0\ . \qquad (50)$$

As explained in section 2.2 we take $\hat\theta=\theta_1 1\!\!1^{\rm bf}$ and spatially constant to recover the results of the usual Usadel theory [24] for this problem. Together with the self-consistency equation (31) we have[14]

$$0=\epsilon\sinh\theta_1-|\Delta|\cosh\theta_1-\frac{i}{\tau_s}\sinh(2\theta_1), \\ |\Delta|=-i\pi\lambda\int d\epsilon\ \sinh\theta_1(\epsilon). \qquad (51)$$

The saddle-point equations (51) can be solved self-consistently following the procedure outlined, for example, in Ref. [37]. Setting $\epsilon=\tilde\epsilon-(1/2\tau_s)\cosh\theta_1$ and $|\Delta|=|\tilde\Delta|+(1/2\tau_s)\sinh\theta_1$, the saddle-point equation for each energy ϵ takes the form $\tilde\epsilon\sinh\theta_1=|\tilde\Delta|\cosh\theta_1=0$. Setting $\tilde v\equiv\tilde\epsilon/|\tilde\Delta|$ and recalling the definition $\zeta=1/\tau_s|\Delta|$, one obtains

$$v\equiv\frac{\epsilon}{|\Delta|}=\tilde v\left(1-\zeta\frac{1}{\sqrt{1-\tilde v^2}}\right).$$

To reiterate, the latter equation should be regarded as a self-consistent solution for $\tilde v$ from which one can obtain $\theta=\arcsin(1/\sqrt{1-\tilde v^2})$. The corresponding

[14] Here we work at zero temperature.

self-consistent equation for the gap parameter then takes the form

$$|\Delta| = -\pi\lambda \int d\epsilon \, \frac{1}{\sqrt{1-\widetilde{\upsilon}^2}}.$$

Although there is no simple closed analytic expression for the solution of the mean-field equation, much is known about its form. In particular, the system exhibits a transition at $\zeta = 1$ from a gapped to a gapless phase. In the gapped phase, i.e. for $\zeta < 1$, the gap edge is fixed by the solution $\widetilde{\upsilon}_{\text{gap}} = \left(1 - \zeta^{2/3}\right)^{1/2}$, from which one obtains $E_{\text{gap}} = \Delta\left(1 - \zeta^{2/3}\right)^{3/2}$. A numerical solution for the AG DoS for various values of the dimensionless parameter ζ is shown in Fig. 8.

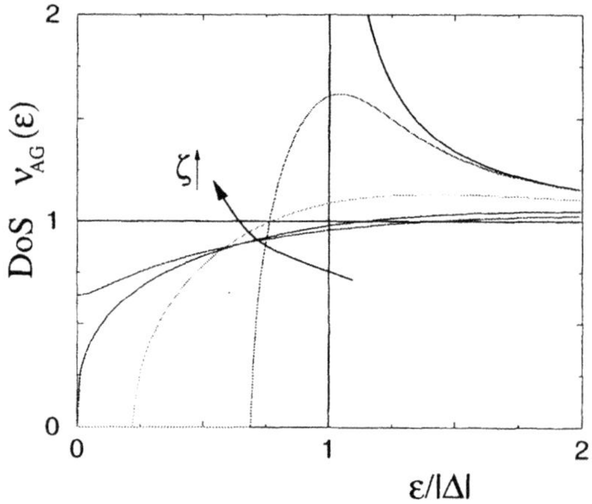

Figure 8. Average DoS as obtained from the Abrikosov-Gor'kov mean-field theory for $\zeta = 0$, 0.1, 0.5, 1, 1.3 and ∞. Note that for $\zeta > 1$, the system enters the gapless phase with the DoS at $\epsilon = 0$ non-vanishing.

Turning to the self-consistent equation, at $T = 0$, the gap equation can be written in the form,

$$1 = -\lambda \int_0^{\omega_D} \frac{d\epsilon}{|\Delta|} \frac{1}{(1+\widetilde{\upsilon}^2)^{1/2}}$$

Changing the integration variable from ϵ to $\widetilde{\upsilon}$, the gap equation assumes the form,

$$1 = -\lambda \int_{\widetilde{\upsilon}_l}^{\omega_D/\Delta} d\widetilde{\upsilon} \left[1 - \zeta \frac{1}{(1+\widetilde{\upsilon}^2)^{3/2}}\right] \frac{1}{(1+\widetilde{\upsilon}^2)^{1/2}}$$

where the lower limit is defined by

$$\widetilde{\upsilon}_l = \begin{matrix} 0 & \zeta \le 1, \\ (\zeta^2 - 1)^{1/2} & \zeta > 1. \end{matrix}$$

The integration can be performed analytically and obtains the solution,

$$\ln\left(\frac{\Delta}{\bar{\Delta}}\right) = \begin{array}{ll} -\frac{\pi}{4}\zeta & \zeta \leq 1 \\ -\text{arccosh}\zeta - \frac{1}{2}\left(\zeta\arcsin(1/\zeta) - (1 - 1/\zeta^2)^{1/2}\right) & \zeta > 1 \end{array}$$

where $\bar{\Delta}$ is the order parameter for $\zeta = 0$. For small ζ, Δ decreases linearly,

$$\Delta - \bar{\Delta} = -\frac{\pi}{4\tau_s}.$$

The onset of the gapless region occurs when $\zeta = 1$. At this point, $\Delta = \bar{\Delta}e^{-\pi/4}$, from which one obtains $1/\tau_s = \bar{\Delta}e^{-\pi/4}$. Using the fact that superconductivity completely disappears when $1/\tau_s = \bar{\Delta}/2$, one finds that the gapless region arises at 91% of the critical impurity concentration [13]. The variation of the self-consistent order parameter and quasi-particle energy with $2/\tau_s|\bar{\Delta}|$ is shown in shown in Fig. 6.

3.3. PHASE COHERENCE EFFECTS IN THE GAPLESS PHASE

Having determined the homogeneous solution of the mean-field equation we now turn attention to influence of fluctuations. As in the time-reversal invariant bulk s-wave superconductor, in the gapped region of the phase diagram, fluctuations are rendered massive by the energy ϵ. Here the fluctuations serve only to provide a small renormalization of the mean-field DoS above the gap. However, in the gapless phase, quasi-particle states persist to zero energy. In this limit, some fluctuations become soft.

More precisely, in the limit $\epsilon \to 0$, the mean-field solution (49) to the saddle-point equation (48) is not unique: here the saddle-point equation admits an entire manifold of homogeneous solutions parameterized by the transformations $Q = TQ_{\text{sp}}T^{-1}$ where $T = 1\!\!1_{\text{ph}} \otimes 1\!\!1_{\text{sp}} \otimes t$ and $t = \gamma(t^{-1})^T\gamma^{-1}$: soft fluctuations of the fields, which are controlled by a non-linear σ-model defined on the group manifold $T \in \text{OSp}(2|2)/\text{GL}(1|1)$. This corresponds to symmetry class D in the Altland-Zirnbauer classification scheme discussed in section 1.4. This is no surprise as this class corresponds to Gor'kov Hamiltonians with broken time-reversal and spin rotation symmetry. The massless fluctuations control the low-energy, long-range properties of the gapless system giving rise to unusual localization and spectral properties.

Taking into account slow spatial fluctuations of the fields

$$Q(\mathbf{r}) = T(\mathbf{r})Q_{\text{sp}}T^{-1}(\mathbf{r}) = (\sinh\theta_1\sigma_1^{\text{ph}} + \cosh\theta_1\,\sigma_3^{\text{ph}}Q_s(\mathbf{r})) \otimes 1\!\!1^{\text{sp}},$$

where $Q_s(\mathbf{r}) = T(\mathbf{r})\sigma_3^{\text{cc}}T^{-1}(\mathbf{r})$, one obtains the soft mode non-linear σ-model action

$$S_{Q_s} = -\frac{\pi\nu}{8}\int d\mathbf{r}\,\text{str}\left[D_s(\nabla Q_s)^2 - 4i\epsilon_s\sigma_3^{\text{cc}}Q_s\right], \tag{52}$$

where $D_s = D\cos^2\theta_1$ denotes the effective diffusion constant, and $\epsilon_s = \epsilon\cosh\theta_1$ represents the (potentially complex) energy source. The mean-field solution therefore brings about an energy dependent renormalization of the coupling constants in the non-linear σ-model action. We emphasize that, these constants aside, the form of the soft-mode action (by which we mean the structure of the saddle-point manifold) — and therefore low-energy behaviour — is entirely determined by the symmetry of the original random Hamiltonian. In the present case, the Q-matrix relation (43) follows from the symmetry (42) of the Hamiltonian. This point of view is most elegantly put by Zirnbauer [17]. Fortunately extensive analysis of class D through the action (52) exists in the literature [49–52], and we can just quote the results. These fall into two categories

Thermal transport: Since neither charge nor spin are conserved in the original Hamiltonian, novel effects with be present only in *thermal* transport by quasiparticles. A standard perturbative RG analysis of (52) yields for 2D the flow

$$\frac{dg(L)}{d(\ln L/\ell)} = \frac{1}{\pi^2},$$

g now gives the thermal conductivity, and we see that transport is metallic. Indeed, the differing behaviours of classes C and D can be attributed to the following observation [50]. Friedan's [53] general result on the one-loop β-function of a σ-model in 2D shows that the flow is determined by the curvature of the field space. The different γ matrices used in this chapter and the last indicate that moving from class C to D involves switching the manifolds of the boson-boson and fermion-fermion sectors, which reverses the curvature (in the appropriate superspace generalization) and leads to *anti-localization* in the class D case.

Spectral properties: One of the common features of the superconductor universality classes [16] is non-stationary behaviour of the DoS due to the distinguished position of $\epsilon = 0$. The correction to the mean-field (AG in our case) DoS can be calculated in perturbation theory for $|\epsilon_s| \gg E_c = |D_s|/L^2$. In the 2D case we have

$$\begin{aligned}\nu(\epsilon) &= \mathrm{Re}\,\nu\cosh\theta_1\left(1 + \frac{1}{\pi\nu}\int\frac{d^2q}{(2\pi)^2}\frac{1}{D_s q^2 - 2i\epsilon_s}\right)\\ &= \mathrm{Re}\,\nu\cosh\theta_1\left(1 + \frac{1}{8\pi^2 D_s\cosh^2\theta_1}\ln\left(1 + \left(\frac{D_s}{2\epsilon_s\ell^2}\right)^2\right)\right). \qquad (53)\end{aligned}$$

For $|\epsilon_s| < E_c$, the action is dominated by the zero spatial mode, which must be integrated exactly. This is the universal limit, and the result should coincide with the average DoS of a random matrix ensemble with the class D symmetry (see section 1.4). The result is [16, 50]

$$\nu(\epsilon) = \nu(E_c)\left[1 + \frac{\sin(2\pi\epsilon/\delta)}{2\pi\epsilon/\delta}\right]$$

where $\delta = 1/\nu(E_c)L^2$ represents the average energy level spacing at energy E_c. This accounts for the RG scaling from the microscopic scale to the Thouless scale E_c suggested by (53).

Thus, according to the class D field theory, the DoS obtained from the AG mean-field should be modulated by a logarithmic divergence at low energies which is cut on the scale E_c. Below E_c, the modulation should be universal corresponding to the random matrix result. In particular, as $\epsilon/\delta \to 0$, the DoS should exhibit a jump by a factor of two.

This completes the formal description of the bulk superconducting phase. The solution of the AG mean-field equation provides an adequate description of the bulk extended states. New channels of quantum interference are described by soft modes in the gapless phase with dramatic consequences. This answers the second of the two questions posed in the introduction. Now we move onto the first: the sub-gap structure in the gapped phase.

3.4. INSTANTONS AND SUB-GAP STATES

Although the reduction and eventual destruction of the quasi-particle energy gap predicted by the AG mean-field theory can be reasonably justified on purely physical grounds, the integrity of the gap of the range $0 < \zeta < 1$ is less credible. Once time-reversal symmetry is broken and the protection of Anderson's theorem is lost, there remains no reason why a sharp gap should persist. Add to this the observation that the spin scattering rate must be subject to spatial fluctuations from the average value $1/\tau_s$, and one concludes that corrections to the DoS predicted by the AG theory must lead to the appearance of sub-gap states analogous to "band tails" in a disordered semiconductor [43, 44].

This analogy is of course not new [42, 54] nor, as far as practical calculation in the present formulation is concerned, is it particularly deep. This is because all averages have already been taken, so we can not look for an optimal fluctuation of some potential, as in the classic approaches to the study of band tail states in disordered semi-conductors [44]. However, these studies hint at how one can proceed.

Band tail states in semi-conductors can be studied within the same functional integral formulation. In particular, the generating function of the single-particle Green function of a normal disordered conductor can be presented in the form of a supersymmetric field integral

$$\mathcal{Z}[0] = \int D(\Psi, \bar{\Psi}) \exp\left[i \int d\mathbf{r} \bar{\Psi} \left(\epsilon_+ - \frac{\hat{\mathbf{p}}^2}{2m} - W(\mathbf{r}) \right) \Psi \right],$$

where, once again, the random impurity distribution is drawn from a Gaussian δ-correlated white-noise impurity potential. The optimal fluctuation method involves minimizing the action with respect to fluctuations in the fields Ψ and potential W. This involves seeking inhomogeneous solutions of the non-linear Schrödinger equation

$$\left(\epsilon - \frac{\hat{\mathbf{p}}^2}{2m} - W(\mathbf{r})\right)\Psi = 0,$$

where the corresponding optimal potential is determined self-consistently by the relation $W(\mathbf{r}) = -|\Psi(\mathbf{r})|^2/2\pi\nu\tau$. In the supersymmetric formulation, band tail states are identified with 'supersymmetry broken' inhomogeneous solutions of the saddle-point equation (see Cardy [55] and Affleck [56]). Indeed, the anticipated exponential suppression of the DoS necessitates a breaking of supersymmetry to support a finite action. Here the phrase "supersymmetry breaking" is potentially misleading. We use it only to refer to *field configurations*, ubiquitous in the problems under discussion here, that do not respect the parity between Bose and Fermi degrees of freedom. However, any such configuration is just one member of a degenerate manifold differing by supersymmetric transformations. The latter maintain the invariance of the generating functional $\mathcal{Z}[0]$ under global supersymmetric transformations.

What does this tell us about the identification of optimal fluctuations and sub-gap states in the superconductor? Following the analysis above, one might guess that sub-gap states are associated with inhomogeneous configurations of the Ψ field action. However, we anticipate that optimal solutions corresponding to sub-gap states are localized on a length scale in excess of the superconducting coherence length. In the dirty limit, $\xi \gg \ell \gg \lambda_F$, this implies that the localized sub-gap states are quasi-classical in nature. Their existence on the level of the Ψ field action will be buried in the fast λ_F oscillations of the wavefunction. To reveal the sub-gap states, we must first remove the fast short length scale fluctuations of the quasi-classical Green function and look for an equation of motion for the slowly varying envelope of the wavefunction. But this is just the program of the usual quasi-classical scheme.

The term "sub-gap states" is a little misleading in this context. Band tails are bound states of some rare potential that sit by themselves below the bulk of the spectrum. Each rare configuration that make the gap soft in the present case will give rise to many states beneath the AG gap. Thus the term "gap fluctuation", used in Ref. [46] to describe the zero dimensional SN system, may be more appropriate.

As well as being quasi-classical in nature, the existence of sub-gap states is not affected by working in the dirty limit. As such, their existence must be accommodated in the non-linear σ-model functional (46) since the validity of this description relied only on the quasi-classical parameter $\epsilon_F\tau \gg 1$ and the dirty limit assumption. To identify sub-gap states in the present formalism, we should therefore investigate inhomogeneous solutions of the low-energy saddle-point equation in Q — the Usadel equation [25]. Such a solution should be thought of as defining an envelope for the quasi-classical sub-gap states.

Therefore, let us revisit the mean-field equation and look for inhomogeneous solutions at energies $\epsilon < E_{\rm gap}$. To focus our discussion, let us begin by restricting attention to the quasi one-dimensional geometry. To stay firmly within the diffusive regime, we therefore impose the requirement that the system size L be much smaller than the localization length of the normal system $L_{\rm loc.} \simeq \nu d_\perp^2 D$, where $d_\perp$ is the wire diameter. Later, in section 3.4.3, we will generalize our discussion to encompass systems of higher dimension. Furthermore, since, over the interval $0 < \zeta < 1$, the quasi-particle energy gap varies more rapidly than the superconducting order parameter, we will neglect self-consistency of the order parameter. Taking self-consistency into account will not alter our qualitative findings, and will only weakly affect the quantitative results.

3.4.1. *Instantons in the Quasi One-dimensional Geometry*

To investigate inhomogeneous solutions of the mean-field equation (50) it is convenient to recast the equation in terms of its first integral

$$(\partial_{x/\xi}\hat{\theta})^2 + V(\hat{\theta}) = \text{const}, \tag{54}$$

where

$$V(\hat{\theta}) = 4i\left(\sinh\hat{\theta} - \frac{\epsilon}{\Delta}\cosh\hat{\theta}\right) - \zeta\cosh 2\hat{\theta}$$

denotes the complex potential. Let us denote by $\theta_{\rm AG}$ the values of θ_1 and $i\theta$ at the conventional saddle point, and focus on an energy ϵ below the gap predicted by the AG theory. Here $\text{Im}\ \theta_{\rm AG} = \pi/2$ such that the mean-field DoS $\nu_{\rm AG}(\epsilon) = 4\nu\text{Re}\ \cosh\theta_{\rm AG}$ vanishes. The corresponding value of Re $\theta_{\rm AG}$ depends sensitively on the energy, with Re $\theta_{\rm AG} = 0$ for $\epsilon = 0$.

Considering the boson-boson sector only, if we require that $\theta_1(x \to \pm\infty) = \theta_{\rm AG}$, what kind of inhomogeneous solution is possible? The values of θ_1 at which $\partial_x\theta_1 = 0$ can be identified by considering the complex (dimensionless) potential function $V(\theta_1)$ from which we can determine the endpoints of the 'motion' in the complex plane, just as one would use a real potential normally. By inspection one may see that, on the line Im $\theta_1 = \pi/2$, the potential is purely real. This is not the only contour where Im $V = 0$, but, by considering forces, it is not hard to see that either Im $\theta_1 = \pi/2$ always during the motion, or θ_1 follows a trajectory with an endpoint at Im $\theta_1 < 0$. For reasons outlined below, we will discount this latter possibility. The former case amounts to considering "bounce" trajectories in the *real* potential $V(i\pi/2 + \phi) = V_{\rm r}(\phi)$ where

$$V_{\rm r}(\phi) \equiv -4\left(\cosh\phi - \frac{\epsilon}{|\Delta|}\sinh\phi\right) + \zeta\cosh 2\phi. \tag{55}$$

A typical potential is shown in Fig. 9.

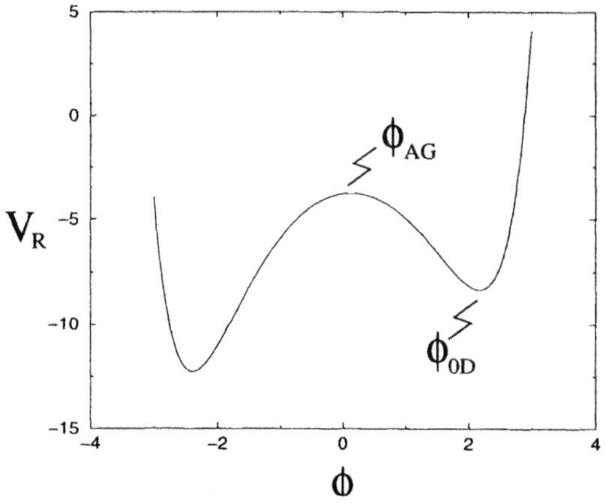

Figure 9. Potential $V_r(\phi) = V(i\pi/2 + \phi)$ for $\epsilon/|\Delta| = 0.1$ and $\zeta = 0.2$. The AG saddle point corresponds to the central maximum. The saddle point marked ϕ_{0d} is used in the analysis of the zero-dimensional problem (section 3.5).

Now integration over the angles $\hat{\theta}$ is constrained to certain contours [27]. Is the bounce solution accessible to both? As usual, the contour of integration over the boson-boson field θ_1 includes the entire real axis, while for the fermion-fermion field, $i\theta$ runs along the imaginary axis from 0 to $i\pi$. With a smooth deformation of the integration contours, the AG saddle-point is accessible to both the angles $\hat{\theta}$ [10]. By contrast, the bounce solution *and* the AG solution can be reached simultaneously by a smooth deformation of the integration contour *only* for the boson-boson field θ_1 (see Fig. 10). The bounce solution is therefore associated with a *breaking of supersymmetry* at the level of the saddle point.

Thus we have identified an inhomogeneous saddle-point configuration for which the supersymmetry is broken: θ_1 executes a bounce whilst $i\theta$ remains at the mean-field value $\theta_{\rm AG}$. The symmetry broken solution then incurs the (finite) real action

$$S = 4\pi\nu L_W (D|\Delta|)^{1/2} S_\phi(\epsilon/|\Delta|, \zeta)$$

where, defining ϕ' as the endpoint of the motion,

$$S_\phi \equiv \int_{\phi_{\rm AG}}^{\phi'} d\phi \sqrt{V_r(\phi_{\rm AG}) - V_r(\phi)}. \tag{56}$$

Now, as mentioned above, there exists a second possibility for a bounce solution in which one moves away from $\theta_{\rm AG}$ parallel to the imaginary axes. Indeed, such a solution would seem to be a natural candidate for the fermion-fermion field $i\theta$. However, since the endpoint for this trajectory lies at Re $\theta < 0$ outside the integration domain which runs from 0 to π, this would seem to be excluded.

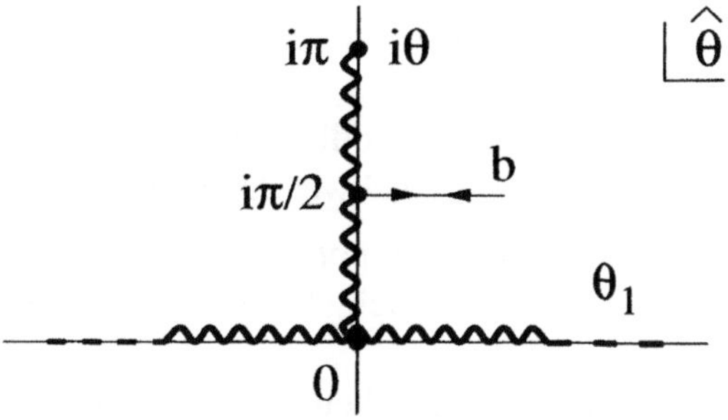

Figure 10. Integration contours for boson-boson and fermion-fermion fields in the complex $\hat{\theta}$ plane. The bounce solution for $\epsilon = 0$ (labelled as 'b') is shown schematically.

As ϵ approaches E_{gap} from below, the potential (55) becomes more shallow, with the maximum merging with one of the minima when we reach the gap. Near the edge, up to an irrelevant constant, an expansion of the potential in powers of $(\phi - \phi_{\text{AG}})$ leads to the cubic form

$$V_{\text{r}}[\phi] \simeq -\alpha \left(\frac{E_{\text{gap}} - \epsilon}{|\Delta|} \right)^{1/2} (\phi - \phi_{\text{AG}})^2 + \beta(\phi - \phi_{\text{AG}})^3 \qquad (57)$$

where the dimensionless coefficients are specified by

$$\alpha = 6\sqrt{\frac{2}{3}} \left(\frac{E_{\text{gap}}}{|\Delta|} \right)^{1/6}, \qquad \beta = 2 \left(\frac{\zeta E_{\text{gap}}}{|\Delta|} \right)^{1/3}. \qquad (58)$$

Note that, making use of Eq. (39), both of these coefficients depend solely on the dimensionless parameter ζ. From this expansion, one can obtain an analytic solution for S_ϕ. To leading order in $(E_{\text{gap}} - \epsilon)/|\Delta|$ one finds

$$S_\phi = \frac{4}{15} \frac{\alpha^{5/2}}{\beta^2} \left(\frac{E_{\text{gap}} - \epsilon}{|\Delta|} \right)^{5/4}. \qquad (59)$$

Note that the action vanishes exactly at the gap. For completeness we give the explicit form of the bounce solution

$$\phi(x) - \phi_{\text{AG}} = \frac{\alpha}{\beta} \frac{1}{\cosh^2(x/2r_0)},$$

where the extent of the instanton is set by

$$r_0(\epsilon) = \frac{\xi}{\alpha^{1/2}} \left(\frac{|\Delta|}{E_{\text{gap}} - \epsilon} \right)^{1/4}. \qquad (60)$$

Indeed the size of the instanton is easily understood from the quadratic "stiffness" term in Eq. (57). Thus one finds that, while the overall scale is set by the superconducting coherence length ξ, the size of the localized region diverges both as

ϵ approaches E_{gap} and, noting that $\alpha \sim (1 - \zeta^{2/3})^{1/4}$, as one approaches the gapless phase $\zeta \to 1$.

This completes the analysis of the saddle-point solution together with the corresponding action. However, as this level we are presented with two problems:

1. The contribution of a second saddle point would seem to spoil the normalization condition $\langle \mathcal{Z}[0] \rangle_{V,S} = 1$, which should be preserved within the saddle point approximation;
2. Confined to the line $\text{Im}\,\theta = \pi/2$, when substituted into the DoS source (47), the bounce configuration does not appear to generate states!

The resolution of both problems lies in the nature of the fluctuations around the symmetry broken mean-field solution. These field fluctuations can be separated into "radial" and "angular" contributions. The former involve fluctuations of the diagonal elements $\hat{\theta}$, while the latter describe rotations including those Grassmann transformations which mix the bf sector. Both classes of fluctuations play a crucial role.

3.4.2. *Fluctuations*

Let us outline qualitatively the influence of the fluctuations around the mean-field. As usual, associated with radial fluctuations around the bounce, there exists a zero mode (due to translational invariance of the solution), and a negative energy mode. The latter, which necessitates a $\pi/2$ rotation of the corresponding integration contour to follow the line of steepest descent (c.f. Ref. [57]), has two effects: firstly it ensures that the non-perturbative contributions to the local DoS are non-vanishing, and secondly, that they are positive. Turning to the angular fluctuations, the breaking of supersymmetry is accompanied by the appearance of a Grassmann zero mode separated by a gap from higher excitations which restores the global supersymmetry (c.f. spin symmetry breaking in a ferromagnet of finite extent). The zero mode ensures that the symmetry broken inhomogeneous saddle-point configurations respect the normalization condition $\langle \mathcal{Z}[0] \rangle_{V,S} = 1$.

A careful analysis of the fluctuations to formally check all these features is contained in Ref. [45]. The result is that, taking into account Gaussian fluctuations and zero modes, one obtains the non-perturbative, one instanton contribution to the sub-gap DoS:

$$\frac{\langle \nu(\epsilon) \rangle_{V,S}}{4\nu} \sim (-i|K|) \int dx\, i(\sinh\phi(x) - \sinh\phi_{\text{AG}})\, |\varphi_0^-(x)|^2 \sqrt{\frac{LS_\phi}{\xi}} \times \exp\left[-4\pi\nu L_{\text{W}} \sqrt{D|\Delta|S_\phi}\right], \tag{61}$$

where the factor $\sqrt{LS_\phi/\xi}$ represents the Jacobian associated with the introduction of the collective coordinate [57], $-i|K|$ is the overall factor arising from the non-zero modes, and the Grassmann zero mode wavefunction φ_0^- is normalized such that $\int dx|\varphi_0^-|^2 = 1$. Eq. (61) is the main result of this section. Note the non-perturbative nature of the result, both in the coupling constant g^{-1} of the σ-model, and the (dimensionless) spin scattering rate ζ.

3.4.3. *Sub-gap States in Dimensions* $1 < d < 6$

The calculation above was tailored to the consideration of the quasi one-dimensional geometry. The generalization to higher dimensions follows straightforwardly. In particular, it is necessary to seek inhomogeneous solutions of the saddle-point equation (54) where the gradient operator must be interpreted as the higher dimensional generalization. Generally, this equation must be solved numerically. However, for energies ϵ in the vicinity of the gap $E_{\rm gap}$, an analytic expression for the energy scaling can be obtained for $d < 6$.[15]

Using the approximation to $V_{\rm r}[\phi]$ (57) valid when $(E_{\rm gap} - \epsilon)/|\Delta| \ll 1$, the exponential dependence of the sub-gap DoS can be deduced in higher dimension. In this limit, dimensional analysis of the cubic equation of motion yields the scaling form

$$\phi(\mathbf{r}) - \phi_{\rm AG}(\epsilon) = \frac{\alpha}{\beta} f(\mathbf{r}/r_0) \, ,$$

where r_0 is the characteristic length defined by Eq. (60). When substituted back into the action, one finds that the DoS depends exponentially on the parameter $4\pi g(\xi/L)^{d-2} S_\phi$

$$S_\phi = a_d \, \zeta^{-2/3}(1 - \zeta^{2/3})^{-(2+d)/8} \left(\frac{E_{\rm gap} - \epsilon}{|\Delta|} \right)^{(6-d)/4} . \tag{62}$$

Here $g = \nu D L^{d-2}$ denotes the bare dimensionless conductance of the normal system, and a_d is a numerical constant ($a_1 = 8\,^4\sqrt{24}/5$). In particular, the exponent depends linearly on the energy separation from the gap in two dimensions.

Having completed this calculation, it is interesting to explore the connection of the results presented above to related problems in the literature. The resulting expression for the DoS (62) is non-perturbative in the σ-model coupling $1/g$, which measures the strength of non-magnetic disorder. We note that other non-perturbative results in disordered systems have been obtained by related instanton calculations. As well as the investigation of tail states in semi-conductors [55], a

[15] The importance of $d = 6$ is the following: the intuition from the quasi one-dimensional problem is that since the potential $V_{\rm r}$ becomes more shallow as $\epsilon \to E_{\rm gap}^-$ the action for the instanton is well approximated by retaining only those terms due to the cubic potential (57). This is correct in $d < 6$ because the instanton found using the truncated potential have an action that vanishes as $\epsilon \to E_{\rm gap}^-$, and it is clear that the 'remainder' of the action evaluated on this instanton vanishes even more rapidly. For $d > 6$ this procedure fails and no truncation of the potential is possible. Effectively the potential is $V_{\rm r} \sim \zeta \exp(2|\phi|)$ and no instanton solutions exist. Thus for $d > 6$ the mean-field gap is hard.

supersymmetric field theory was developed by Affleck [56] (see also Refs. [58]) to investigate tail states in the lowest Landau level. There it was shown that tail states correspond to instanton configurations of the Ψ*-field action* (c.f. Ref. [55]). It is also interesting to compare the present scheme with the study of 'anomalously localized states' [59] (see also, Ref. [60]). There one finds that long-time current relaxation in a disordered wire is also associated with instanton configurations of the σ-model action. Finally, a Lifshitz argument has been applied on the level of the Usadel equation in the study of gap fluctuations due to inhomogeneities of the BCS interaction [64].

3.5. ZERO DIMENSIONAL PROBLEMS AND UNIVERSALITY

In the previous section we considered the instanton contribution to the sub-gap DoS in the infinite system. For completeness let us now consider the zero dimensional case that obtains when r_0, the size of the instanton, exceeds the system size L, which will happen when ϵ approaches close enough to E_{gap} from below in any finite system. In this limit one can clearly not fit an instanton inside the system. Leaving aside the practical relevance of this situation, theoretical motivation is provided by a recent paper [54] that explored this regime using a random matrix analysis.

However, before turning to the consideration of the zero-dimensional limit of the present problem let us first try to draw some intuition from a system that turns out to be closely related. In section 2.3.1 we discussed the properties of a normal quantum dot contacted to a superconducting terminal (Fig. 5). There we saw that near the gap edge the DoS of the dot takes the singular form

$$\nu(\epsilon > E_{\mathrm{gap}}) \simeq \frac{1}{\pi L^d}\sqrt{\frac{\epsilon - E_{\mathrm{gap}}}{\Delta_{\mathrm{g}}^3}}\,, \tag{63}$$

However, the location of the gap edge relies on a mean-field analysis of the coupled system. In Ref.[46] Vavilov *et al.* have argued that optimal fluctuations of the impurity potential give rise to gap fluctuations. The hypothesis introduced in Ref. [46] is that the spectral statistics near a gap edge are universal. This allows a random matrix theory analysis of gap fluctuations and leads to the following expression for the sub-gap DoS,

$$\frac{\nu(\epsilon < E_{\mathrm{gap}})}{\nu} \sim \exp\left[-\frac{2}{3}\left(\frac{E_{\mathrm{gap}} - \epsilon}{\Delta_{\mathrm{g}}}\right)^{3/2}\right]. \tag{64}$$

Now the AG mean-field solution for a superconducting quantum dot with magnetic impurities also predicts the existence of a square root edge (see Eq. (35)). Then, when recast in the form of Eq. (63), it is pertinent to ask whether the expression for the sub-gap DoS coincides with Eq. (64) in the zero dimensional limit. This is the situation addressed in Ref.[54].

In the following we will show that the universal result (64) is explicitly recovered by the present theory. Moreover, in doing so, we will expose the origin of the universal structure reported in Ref. [46] and describe its implications for universality of the $d > 0$ problem.

3.5.1. *Superconducting dot with magnetic impurities*

Let us therefore consider explicitly the action of a superconducting grain in the presence of a weak magnetic impurity potential. When $r_0 \gg L$ only the zero spatial mode contributes significantly to the action (46). In this limit, the action assumes the zero dimensional form

$$S[Q] = \frac{i\pi}{2\delta}\mathrm{str}\left[\left(\epsilon_{-}\sigma_3^{\mathrm{cc}} + |\Delta|\sigma_2^{\mathrm{ph}}\right)\sigma_3^{\mathrm{ph}}Q\right] + \frac{\pi}{24\tau_s\delta}\mathrm{str}\left(Q\sigma_3^{\mathrm{ph}}\otimes\sigma^{\mathrm{sp}}\right)^2, \quad (65)$$

where, as usual, δ denotes the single-particle level-spacing. As in the higher dimensional problem, a variation of the action with respect to Q obtains a mean-field equation, now without spatial variation. Parameterizing the saddle-point equation as in section 3.2.2, we obtain the zero-dimensional Usadel or AG mean-field equation (50)

$$2i\left(\cosh\hat{\theta} - \frac{\epsilon}{|\Delta|}\sinh\hat{\theta}\right) - \zeta\sinh(2\hat{\theta}) = 0.$$

From this equation, we can identify the usual AG solution which in turn recovers the AG phenomenology.

The inclusion of bounce configurations in the previous calculations was based upon the observation that, although the contribution they make is exponentially small, they are the least action configurations on the part of the contour that gives a finite sub-gap DoS. In the zero-dimensional case we are spared having to think about the problem in function space. The action is proportional to the potential of Fig. 9. The correct contour thus passes through the maximum of the potential (minimum of the action) corresponding to the usual AG saddle point, and turns away from the real ϕ axes (i.e. the line Im $\theta_1 = \pi/2$) at the minimum of the potential (marked in Fig. 9 as ϕ_{0d}). This part of the contour, parallel to the imaginary axes, gives a contribution to the DoS, and the second saddle point is in fact a *maximum* on this portion by analyticity. Following the same arguments as in section 3.4.1, this solution is inaccessible to the fermionic contour. Thus we find that the sub-gap DoS in the zero dimensional case is given by

$$\frac{\nu(\epsilon < E_{\mathrm{gap}})}{\nu} \sim \exp\left[-\pi\frac{|\Delta|}{\delta}(V_{\mathrm{r}}(\phi_{\mathrm{AG}}) - V_{\mathrm{r}}(\phi_{\mathrm{0d}}))\right]. \quad (66)$$

As before analytic results may be obtained near the gap edge using the cubic potential (57). We find

$$\phi_{\mathrm{0d}}(\epsilon) \sim \phi_{\mathrm{AG}}(\epsilon) + \frac{2\alpha}{3\beta}\sqrt{\frac{E_{\mathrm{gap}} - \epsilon}{|\Delta|}},$$

which leads to the sub-gap DoS near $E_{\rm gap}$ scaling as

$$\frac{\nu(\epsilon < E_{\rm gap})}{\nu} \sim \exp\left[-\frac{4\pi}{27}\frac{|\Delta|}{\delta}\frac{\alpha^3}{\beta^2}\left(\frac{E_{\rm gap}-\epsilon}{|\Delta|}\right)^{3/2}\right] .$$

We note that the general result for the energy dependence of the exponent written down in section 3.4.3 for dimensions $d \geq 1$ applies also for $d = 0$.

To establish contact with the universal result given in Eq. (63) it is helpful to recast the result in a modified form. To do this we note that, in the vicinity of the mean-field gap edge, the DoS can be expanded as [37]

$$\frac{\nu_{\rm AG}(\epsilon > E_{\rm gap})}{4\nu} \simeq \sqrt{\frac{2}{3}}\,\zeta^{-2/3}(1-\zeta^{2/3})^{-1/4}\left(\frac{\epsilon - E_{\rm gap}}{|\Delta|}\right)^{1/2} .$$

Then, if we define

$$\Delta_{\rm g}^{-3/2} \equiv \frac{4\pi}{\delta}\sqrt{\frac{2}{3|\Delta|}}\zeta^{-2/3}(1-\zeta^{2/3})^{-1/4} ,$$

the mean-field DoS can be brought to the form of Eq. (63), and the sub-gap DoS takes the universal form

$$\frac{\nu(\epsilon < E_{\rm gap})}{\nu} \sim \exp\left[-\frac{4}{3}\left(\frac{E_{\rm gap}-\epsilon}{\Delta_{\rm g}}\right)^{3/2}\right] . \tag{67}$$

Taking into account the particular convention for the definition of the DoS[16], the sub-gap DoS obtained above coincides with the universal expression shown in Eq. (64).

Away from the mean-field gap edge, we can in principle obtain an exact expression for the exponential dependence of the sub-gap DoS by solving the saddle-point equation for $V_{\rm r}(\phi_{\rm AG})$ and $V_{\rm r}(\phi_{\rm 0d})$ explicitly. However, such a program can not be performed analytically. However, the asymptotic dependence of the DoS tail far from the mean-field gap can be obtained by developing an asymptotic expansion in ζ. In doing so, we can establish contact with the results of Ref. [54]. Taking $\zeta \ll 1$ and keeping terms at order ζ^{-1} and ζ^0 in the potential $V_{\rm r}$ one finds that $\phi_{\rm AG}$ coincides with the BCS solution at leading order and $\phi_{\rm 0d}$ moves to the larger value

$$\phi_{\rm 0d}(\epsilon) \sim \ln\left(2\frac{|\Delta|-\epsilon}{|\Delta|\zeta}\right) .$$

[16] We note that the factor of 2 discrepancy between the result here and that presented in Ref. [46] can be straightforwardly accommodated into a redefinition of $\Delta_g^{-3/2}$, or, equivalently, a rescaling of the mean-field DoS. For reasons outlined in the text, we believe that the convention adopted here is the consistent one.

Evaluating $V_r(\phi_{AG})$ and $V_r(\phi_{0d})$ and substituting in Eq. (66) gives the result

$$\frac{\nu(\epsilon < E_{gap})}{\nu} \sim \exp\left[-\frac{2\pi\tau_s}{\delta}(|\Delta| - \epsilon)^2 + \frac{4\pi}{\delta}(|\Delta|^2 - \epsilon^2)^{1/2}\right] .$$

In particular, we note that the two terms in this expansion have the same functional form as those obtained in Ref. [54], the coefficients differ. The reason for this difference is unclear.

The rescaling of the DoS above and the appearance of the universal form suggests that we should revisit the d-dimensional result and look for a similar rescaling. From Eq. (62) it is straightforward to verify that in this case

$$\frac{\nu(\epsilon < E_{gap})}{\nu} \sim \exp\left[-\tilde{a}_d \left(\frac{r_0(\epsilon)}{L}\right)^d \left(\frac{E_{gap} - \epsilon}{\Delta_g}\right)^{3/2}\right] ,$$

where $\tilde{a}_d$ represents some numerical constant, and r_0 is the characteristic length defined by Eq. (60). Finally, by defining

$$\tilde{\Delta}_g^{-3/2}(\epsilon) \equiv \frac{\delta}{\tilde{\delta}(\epsilon)} \Delta_g^{-3/2} ,$$

where $\tilde{\delta}(\epsilon) = 1/(\nu r_0^d(\epsilon))$ is the level spacing inside a region of size r_0, the volume dependent prefactor can be absorbed into the expression and DoS can be brought to the form

$$\frac{\nu(\epsilon < E_{gap})}{\nu} \sim \exp\left[-\tilde{a}_d \left(\frac{E_{gap} - \epsilon}{\tilde{\Delta}_g}\right)^{3/2}\right] ,$$

revealing a simple relation between the $d = 0$ and $d > 0$ problems.

3.5.2. *Universalities*

The coincidence of Eqs. (64) and (67) indeed suggests that gap fluctuations are universal. To understand why, let us consider the following: At first glance the actions (65) and (33), used in section 2.3.1 to describe the quantum dot problem, would seem not to have much in common. However, a simple and general argument may be established to reveal the universal character. As before, defining $\theta_{mf}(\epsilon) = i\pi/2 + \phi_{mf}(\epsilon)$, the mean-field DoS for the SN device is given by $\nu_{mf}(\epsilon) = 2\nu \;\mathrm{Im}\; \sinh\phi_{mf}(\epsilon)$, where ϕ_{mf} is determined by the condition $\delta S/\delta\phi[\phi = \phi_{mf}] = 0$. Since the DoS displays a square root singularity described by Eq. (63), the (saddle-point) action near the edge is constrained to be of the form

$$S[\hat{\phi}] = -k \;\mathrm{str}\left[\frac{1}{3}\hat{s}^3 + \left(\frac{\delta}{2\pi}\right)^2 \left(\frac{\epsilon_+ - E_{gap}}{\Delta_g^3}\right)\hat{s}\right] ,$$

where $\hat{s}(\epsilon) = \sinh\hat{\phi}(\epsilon) - \sinh\hat{\phi}_{\rm mf}(E_{\rm gap})$. Here, the elements $\hat{\phi} = {\rm diag}(\phi_{\rm bb}, \phi_{\rm ff})$ and $\hat{s} = {\rm diag}(s_{\rm bb}, s_{\rm ff})$ are diagonal in the superspace. (As one may check, a variation of the action for $\epsilon > E_{\rm gap}$ obtains the symmetric mean-field solution

$$s_{\rm bb} = s_{\rm ff} = i\frac{\delta}{2\pi}\sqrt{\frac{\epsilon - E_{\rm gap}}{\Delta_g^3}}$$

which in turn recovers the expression (63) for $\nu(\epsilon)$.) Moreover, since the term containing $\hat{s}$ is linear in the energy, we can determine the value of k from the knowledge that ϵ appears in the action as $(2\pi\epsilon_+/\delta)\sinh\hat{\phi}$. (It is this term that can more generally contain the Dyson index 'β', which therefore appears in the general expression for gap fluctuations described in Ref. [46].) In the present case, we thus have $k = (2\pi\Delta_{\rm g}/\delta)^3$.

Now, as discussed in the previous section, when $\epsilon < E_{\rm gap}$ there exists two saddle-point solutions at

$$s_\pm = \pm\frac{\delta}{2\pi}\sqrt{\frac{E_{\rm gap} - \epsilon}{\Delta_g^3}}.$$

As before, one of these solutions ($s_-(\epsilon) \rightsquigarrow \phi_{\rm mf}(\epsilon)$) is associated with the conventional symmetric mean-field solution while the other represents a second saddle-point accessible only to the bosonic contour. Taking this second, symmetry broken saddle-point into account (i.e. setting $s_{\rm bb} = s_+$ and $s_{\rm ff} = s_-$), one obtains the saddle-point action

$$S[\hat{\phi}] = \frac{4}{3}\left(\frac{E_{\rm gap} - \epsilon}{\Delta_{\rm g}}\right)^{3/2}.$$

It is this symmetry broken saddle-point which controls the sub-gap DoS and leads to the universal scaling form proposed in Ref. [46]. This generalizes the arguments applied to the superconducting dot with magnetic impurities.

3.5.3. *Discussion*

Following on from this discussion, to conclude this section, let us make two remarks which bear on the universality of the general scheme. The first of these remarks concerns the integrity of the scaling of the sub-gap DoS when different impurity distributions are taken into account. The second remark concerns the extension of the ideas above to the consideration of the hybrid SN system beyond the zero-dimensional regime.

Firstly, for the superconductor with magnetic impurities, one can generalize the arguments above to show that the energy scaling of the sub-gap DoS even in the d-dimensional case is insensitive to the nature of the random impurity distribution. This is in contrast to Lifshitz band tail states in semi-conductors where the energy scaling depends sensitively on this distribution. To understand this, let us suppose that the distribution of magnetic impurities $J\mathbf{S}(\mathbf{r})$ is not Gaussian δ-correlated, as we assumed throughout, but obeys some arbitrary statistics defined by a probability functional $P[J\mathbf{S}(\mathbf{r})]$. When the ensemble average over $J\mathbf{S}(\mathbf{r})$ is performed one would obtain in the Ψ-field action a contribution of the form

$$\ln\left\langle \exp\left[-i\int d\mathbf{r}\ \bar{\Psi}J\mathbf{S}\cdot\sigma^{\mathrm{sp}}\Psi\right]\right\rangle_P \equiv C[\bar{\Psi}\sigma^{\mathrm{sp}}\Psi(\mathbf{r})]\,,$$

which defines $C[\cdots]$, the generating functional of connected correlators of $J\mathbf{S}(\mathbf{r})$. Though this is in general a very complicated and indeed non-local functional of $\bar{\Psi}\sigma\Psi(\mathbf{r})$, one can in principle find a *local* Q-field action by including pairings only at coincident points, justified by the assumption $(\ell/\xi)^d \ll 1$ about the non-magnetic disorder. The mean-field description of this system then follows from the homogeneous solution of the saddle-point equation, an Usadel equation like (50) with some potential. Generally this potential will have the same characteristics as the real potential of (55) plotted in Fig. 9 on the line Im $\theta_1 = \pi/2$. The central maximum is due to the $|\Delta|$ term; the upturn at large ϕ arises from the small pair-breaking part, and the asymmetry comes from the ϵ term. Now, if mean-field theory leads to a square-root singularity in the DoS (a circumstance which can be avoided only by a special tuning of parameters), one can expect that increasing the energy leads to the maximum merging with one of the minima according to

$$V_{\mathrm{r}}[\phi] \simeq -\alpha\left(\frac{E_{\mathrm{gap}}-\epsilon}{|\Delta|}\right)^{1/2}(\phi-\phi_{\mathrm{mf}})^2+\beta(\phi-\phi_{\mathrm{mf}})^3$$

with α and β chosen appropriately. Then the analysis of section 3.4 applies. In particular the scaling of the exponent with $((E_{\mathrm{gap}}-\epsilon)/|\Delta|)^{(6-d)/4}$ is expected to be *universal and independent of the details of the magnetic impurity potential.*

Now let us turn to the generality of the present scheme in describing 'gap fluctuations' in extended hybrid superconductor/normal systems. The latter has been discussed in a very recent paper by Ostrovsky *et al.* [61]. In this work, the authors developed an instanton approach analogous to that employed in the magnetic impurity system here to estimate the profile of gap fluctuations in the d-dimensional SNS system. Now from the discussion above, it is possible to expose the relation between these two works: in the SNS system, the energy gap induced in the normal region due to the proximity effect is determined by the Thouless energy defined as $E_{\mathrm{T}} \sim 1/\tau_{\mathrm{dwell}}$, where τ_{dwell} is the time required for electrons in the normal region to feel the presence of the superconductor [33]. The Thouless energy is determined by $E_{\mathrm{T}} \sim \min\{D/L^2, \Gamma N\delta\}$ where Γ is the transparency of the contact to the superconductor ($\Gamma = 1$ in the analysis of the zero-dimensional system above).

In the diffusive limit $D/L^2 \ll \Gamma N\delta$, at the mean-field level, the position of the quasi-particle energy gap is found by solving the Usadel equation with the appropriate boundary conditions [62]. As a result one obtains a square root singularity in the DoS. In this case the mean-field solution is itself inhomogeneous. The sub-gap correction is found by identifying a second inhomogeneous instanton configuration that breaks supersymmetry at the level of the action [61]. Both solutions merge at the mean-field gap. Now, following the arguments above, it is simple to see how the phenomenology of Ref. [61] fits into the same general scheme: in this case the relevant coordinate of Q interpolates between the inhomogeneous mean-field solution and the instanton. The result is a sub-gap DoS which assumes the familiar form of Eq. (67), with appropriately defined geometry dependent parameters E_{gap} and Δ_{g}. Naturally the introduction of $d_\perp$ transverse dimensions gives the expected energy dependence of $(E_{\text{gap}} - \epsilon)^{(6-d_\perp)/4}$ in the exponent.

In the opposite limit $D/L^2 \gg \Gamma N \delta$ (not considered in Ref. [61]), gradients of Q are heavily penalized and the coupling to the lead *must* be retained in its 'logarithmic form', with Q being taken as constant in the dot. (Indeed, the logarithm is crucial to reproduce even the mean-field expression for the DoS (63) with the correct coefficients.) This is the true zero-dimensional limit treated above. As we have seen, with this action, one recovers the known universal expression for the spectrum of gap fluctuations below the mean-field edge. Finally, it is interesting to note that the parallel (at the mean field level) between the magnetic impurity problem and proximity effect situations was noted a long time ago [37, 63].

So far we have uncovered a high degree of universality in our discussion of sub-gap states. This parallels the well-appreciated fact that the AG theory recurs in a great many pair-breaking scenarios (see the 'equivalence theorems' of Ref. [37] and Ref. [64] for a more unusual context). In the next section we will discuss two of the classical realizations of AG and ask whether the equivalence extends to description of sub-gap states

3.6. SUB-GAP STATES IN OTHER REALIZATIONS OF THE AG THEORY

Evidently, time-reversal symmetry can be broken by an external perturbation in a number of ways. Soon after the seminal work of AG it was understood [65, 66] that the single quasi-particle Green's function of a thin dirty film in a parallel field is described by the AG theory. Thus the predictions for both T_c and the DoS are the same as before. In fact, the DoS of the parallel field system is experimentally better described by the AG theory than the magnetic impurity system. This has been blamed on weaknesses of the model (37) in the magnetic impurity case [4]. We now analyze the parallel field system to address the universality of the sub-gap structure.

3.6.1. *Thin Film Superconductor in an in-plane Magnetic Field*

Consider a thin film of thickness $d_\perp$ in the $x - y$ plane and in the limit where $d_\perp \ll \xi, \lambda_L$. Here λ_L is the London penetration depth. The effective action for this system reads

$$S = -\frac{\pi\nu}{8}\int \mathrm{str}\left[D(\tilde{\nabla}Q)^2 - 4i(\epsilon\sigma_3^{\mathrm{cc}} + |\Delta|\sigma_2^{\mathrm{ph}})\sigma_3^{\mathrm{ph}}Q\right] .$$

Recall $\tilde{\nabla} \equiv \nabla - ie\mathbf{A}[\sigma_3^{\mathrm{ph}}, \,]$ is the covariant derivative. To analyze the saddle-point structure, note that in London gauge we have $A_y = Hz$, and this also implies that Q_{sp} is constant across the film. Since

$$\frac{1}{d_\perp}\int dz\, A(z) = 0, \qquad \frac{1}{d_\perp}\int dz\, A^2(z) = \frac{1}{12}(Hd_\perp)^2 ,$$

we see that the paramagnetic piece must vanish from the saddle-point equation and we are left with

$$D\nabla(Q\nabla Q) + i\left[Q, \epsilon\sigma_3^{\mathrm{ph}} \otimes \sigma_3^{\mathrm{cc}} + i\Delta\sigma_1^{\mathrm{ph}}\right] + \frac{1}{2\tau_H}\left[Q, \sigma_3^{\mathrm{ph}}Q\sigma_3^{\mathrm{ph}}\right] = 0 ,$$

where $1/\tau_H = D(Hd_\perp)^2/6$, and we understand that the spatial derivatives are in the plane. We note that this coincides with the Usadel equation (48) for the magnetic impurity problem, once the spin singlet ansatz is made there. Evidently the results of the AG theory are carried over wholesale with τ_H for τ_s. Furthermore, it is not difficult to see that the description of the sub-gap states is also identical, with the instantons living in the $x - y$ plane, where they are not disturbed by the z-dependence of $\mathbf{A}$.

In the introduction to this chapter we referred to intrinsic and extrinsic pictures of tail state formation. It is clear that there is no extrinsic picture for the present situation like the 'droplet' view of the magnetic impurity system. Tail state formation is more likely due to a proportion of the spectrum being anomalously lacking in phase rigidity.

Before moving on to an example where this universality is not repeated, it should be noticed that the parallel field and magnetic impurity problems are not equivalent as far as spin related physical quantities (like the spin susceptibility) are concerned [37], for obvious reasons. In the same way, if we drive a thin film into the gapless phase by applying a parallel field, it is clear on symmetry grounds alone that the spectral and transport properties of the low-energy quasi-particles are those of class C. This is in fact a thermal (and spin) *insulator* — see [35].

3.6.2. *Pair breaking by supercurrent*

Up to now we have not needed to discuss the phase of the order parameter. To describe the situation where a supercurrent flows we will need to take this into account. Varying the action with respect to Q, an applying the Ansatz,

$$Q_{\rm sp} = \left[\sigma_3^{\rm cc} \otimes \sigma_3^{\rm ph} \cosh \hat{\theta} + i \sinh \hat{\theta} (\sigma_1^{\rm ph} \cos \phi - \sigma_2^{\rm ph} \sin \phi)\right] \otimes 1\!\!1^{\rm sp} ,$$

where ϕ will coincide with the phase of the order parameter $\Delta = |\Delta| e^{i\phi}$ at a saddle-point, the saddle-point equations obeyed by $\hat{\theta}$ and ϕ divide into two. With $\mathbf{A} = 0$,

$$\begin{aligned} \nabla^2_{\mathbf{r}/\xi} \hat{\theta} + 2i \left(\cosh \hat{\theta} - \frac{\epsilon}{|\Delta|} \sinh \hat{\theta} \right) - \frac{D}{2} (\nabla_{\mathbf{r}/\xi} \phi)^2 \sinh(2\hat{\theta}) = 0 \, , \\ \nabla_{\mathbf{r}/\xi} (\sinh \hat{\theta} \nabla_{\mathbf{r}/\xi} \phi) = 0 \, . \end{aligned} \tag{68}$$

The second equation describes the conservation of (spectral) current. Evidently a constant phase gradient (that arises when we apply a phase difference across a piece of uniform superconducting wire, say) acts as a pair-breaking perturbation and the AG mean-field result holds as before with $\zeta = D(\nabla \phi)^2 / 2|\Delta|$.

Do localized tail states form in this situation? On the basis of the first equation (68) we would be inclined to think so. But the current conservation law adds a complication. This can be dealt with in 1D by substituting

$$\partial_x \phi = \frac{\rm const}{\sinh^2 \hat{\theta}} \, . \tag{69}$$

It is straightforward to see that the resulting potential does not allow for the existence of bounce solutions that we discussed before: there are no localized sub-gap states! On reflection this makes sense: a localized state can not carry a supercurrent, so there would be no pair-breaking effect.

Thus we see that the description of sub-gap states may not have quite the universality of the AG mean-field solution for the extended part of the spectrum. To finish we mention one slightly unusual example where the tail state formation *does* proceed in the same way. This is the case of gap fluctuations in superconductors with a quenched inhomogeneous distribution of the BCS coupling constant [67], where a mean-field description identical to AG was given in [64]

References

1. B. L. Altshuler and A. G. Aronov, in *Electron-Electron Interactions in Disordered Conductors*, ed. A. J. Efros and M. Pollack, Elsevier Science Publishers, North-Holland, 1985.
2. B. L. Altshuler, A. G. Aronov, M. E. Gershenson, and Y. V. Sharvin, in: Physics Reviews, ed., I. M. Khalatnikov (Harwood Academic Publishers, Switzerland, 1987), p. 225.
3. B. L. Altshuler, P. A. Lee, and R. A. Webb, editors, *Mesoscopic Phenomena in Solids* (North-Holland, Amsterdam): p. 1, 1991.

4. M. Tinkham, *Introduction to Superconductivity*, McGraw-Hill, New York, 1996.
5. P. W. Anderson, J. Phys. Chem. Sol. **11**, 26 (1959).
6. N. Moussy, H. Courtois, and B. Pannetier, preprint cond-mat/0106299.
7. M. Ma and P. A. Lee, Phys. Rev. B **22**, 5658 (1985).
8. A. M. Finkelstein, Physica B, **197**, 636 (1994).
9. A. F. Andreev, Sov. Phys. JETP **19**, 1228 (1964).
10. A. Altland, B. D. Simons and J. P. D. Taras-Semchuk, JETP Lett. **67**, 22 (1998); Adv. in Phys. **49**, 321 (2000).
11. A. A. Abrikosov, L. P. Gor'kov, and I. E. Dzyaloshinski, *Methods of quantum field theory in statistical physics*, Dover, New York, 1975.
12. L. P. Gor'kov, A. I. Larkin, and D. E. Khmel'nitskii, JETP Lett. **30**, 228 (1979).
13. A. A. Abrikosov and L. P. Gor'kov, Sov. Phys. JETP **12**, 1243 (1961).
14. I. L. Kurland, I. L. Aleiner, and B. L. Altshuler, Phys. Rev. B, **62**, 14886, (2000).
15. F. J. Dyson, J. Math. Phys. **3**, 140, 157, 166 (1962); F. J. Dyson and M. L. Mehta, J. Math. Phys. **4**, 701 (1963).
16. A. Altland and M. R. Zirnbauer, Phys. Rev. Lett. **76**, 3420 (1996); Phys. Rev. B **55**, 1142 (1997).
17. M. R. Zirnbauer, J. Math. Phys. **37**, 4986 (1996).
18. W. A. Atkinson, P. J. Hirschfeld, A. H. MacDonald, preprint cond-mat/0002333; W. A. Atkinson, P. J. Hirschfeld, A. H. MacDonald, and K. Ziegler, preprint cond-mat/0005487;
19. A. Altland, B. D. Simons, and M. R. Zirnbauer, Phys. Rep. in press.
20. T. Senthil, M. P. A. Fisher, L. Balents, and C. Nayak, Phys. Rev. Lett, **81**, 4704 (1998).
21. G. Eilenberger, Z. Phys. **182**, 427 (1965); *ibid.* **214**, 195 (1968).
22. A. I. Larkin and Yu N. Ovchinnikov, Sov. Phys. JETP **28**, 1200 (1969).
23. A. I. Larkin and Yu. N. Ovchinnikov, in *Nonequilibrium Superconductivity*, eds. D. N. Langenberg and A. I. Larkin (Elsevier, New York, 1986).
24. N. Kopnin *Theory of Non-equilibrium Superconductivity*, Oxford Science Publications, 2001.
25. K.D. Usadel, Phys. Rev. Lett. **25**, 507 (1970).
26. K. B. Efetov, Adv. Phys. **32**, 53 (1983).
27. K. B. Efetov, *Supersymmetry in Disorder and Chaos*, Cambridge University Press, 1997.
28. R. Oppermann, Nuclear Phys. B **280**, 753 (1987).
29. V. E. Kravtsov and R. Oppermann, Phys. Rev. B **43**, 10865 (1991).
30. R. Bundschuh, C. Cassanello, D. Serban, and M. R. Zirnbauer, Nucl. Phys. B **532**, 689 (1998).
31. F.Wegner, Z. Phys. B. **35**, 207 (1979)
32. J. J. M. Verbaarschot, H. A. Weidenmüller, and M. R. Zirnbauer, Phys. Rep. **129**, 367 (1985).
33. J. A. Melsen, P. W. Brouwer, K. M Frahm, and C. W. J Beenakker, Europhys. Lett. **35**, 7 (1997); Physica Scripta **69**, 223 (1997).
34. See, for e.g., C. W. J. Beenakker, Rev. Mod. Phys. **69**, 731 (1997).
35. K. M Frahm, P. W. Brouwer, J. A. Melsen, and C. W. J. Beenakker, Phys. Rev. Lett. **76**, 2981 (1996).
36. J. P. D. Taras-Semchuk and A. Altland, preprint cond-mat/0010413
37. K. Maki, in *Superconductivity*, edited by R. D. Parks (Dekker, New York, 1969), Vol. 2, p. 1035.
38. S. A. van Langen, P. W. Brouwer, and C. W. J. Beenakker, Phys. Rev. E **55**, R1 (1997).
39. Lu Yu, Physica Sinica **21**, 75 (1965).
40. H. Shiba, Prog. Theo. Phys. **40**, 435 (1968).
41. A. I. Rusinov, JETP Lett. **9**, 85 (1969).
42. A. V. Balatsky and S. A. Trugman, Phys. Rev. Lett. **79**, 3767 (1997).
43. I. M. Lifshitz, Sov. Phys. Usp. **7**, 549 (1965); Usp. Fiz. Nauk. **83**, 617 (1964).

44. B. I. Halperin and M. Lax, Phys. Rev. **148**, 722 (1966); J. Zittartz and J. S. Langer, Phys. Rev. **148**, 741 (1966).
45. A. Lamacraft and B. D. Simons, Phys. Rev. B **64**, 014514 (2001).
46. M. G. Vavilov, P. W. Brouwer, V. Ambegaokar, and C. W. J Beenakker, Phys. Rev. Lett. **86**, 874 (2001).
47. K. Maki, Phys. Rev. **153**, 428 (1967).
48. A. I. Larkin, V. I. Melnikov and D. E. Khmel'nitskii, Sov. Phys. JETP **33**, 458 (1971) [ZhETP **60**, 846 (1971)].
49. R. Bundschuh, C. Cassanello, D. Serban, and M. R. Zirnbauer, Phys. Rev. B **59**, 4382 (1999).
50. M. Bocquet, D. Serban and M. R. Zirnbauer, Nucl. Phys. B **578**, 628 (2000).
51. T. Senthil and M. P. A. Fisher, Phys. Rev. B **61**, 9690 (2000).
52. N. Read and D. Green, Phys. Rev. B **61**, 10267 (2000).
53. D. Friedan, Phys. Rev. Lett. **45**, 1057 date
54. I. S. Beloborodov, B. N. Narozhny and I. L. Aleiner, Phys. Rev. Lett. **85**, 816 (2000).
55. J. Cardy, J. Phys. C **11**, L321 (1978).
56. I. Affleck, J. Phys. C **17**, 2323 (1984).
57. See, e.g., S. Coleman, in *Aspects of Symmetry* (CUP) (1985).
58. K. Benedict, Nucl. Phys. B **280**, 549 (1987); K. B. Efetov and V. G. Marikhin, Phys. Rev. B **40**, 12126 (1989); O. Viehweger and K. B. Efetov, Phys. Rev. B **44**, 1168 (1991).
59. B. A. Muzykantskii and D. E. Khmel'nitskii, Phys. Rev. B **51**, 5840 (1995).
60. V. I. Fal'ko and K. B. Efetov, Europhys. Lett. **32**, 627 (1995); A. D. Mirlin, Phys. Rev. B **53**, 1186 (1996).
61. P. M. Ostrovsky, M. A. Skvortsov and M. V. Feigel'man, preprint cond-mat/0012478
62. A. A. Golubov and M. Yu. Kupriyanov, Sov. Phys. JETP **69**, 805 (1989).
63. P. G. De Gennes, *Superconductivity of metals and alloys*, Addison-Wesley, Redwood City, 1989.
64. A. I. Larkin and Y. N. Ovchinnikov, Sov. Phys. JETP **34**, 1144 (1972).
65. K. Maki, Progr. Theoret. Phys. **29**, 333 (1963).
66. P. G. de Gennes and M. Tinkham, Physics **1**, 107 (1964).
67. J. S. Meyer and B. D. Simons, Phys. Rev. B **64**, 134516 (2001).

KELDYSH AND DOI–PELITI TECHNIQUES FOR OUT–OF–EQUILIBRIUM SYSTEMS

ALEX KAMENEV
Department of Physics, Technion, Haifa 32000, Israel

1. Quantum Keldysh formalism

1.1. A BIT OF HISTORY

The name of the technique takes its origin in 1964 paper of L. V. Keldysh [1]. Among the earlier closely related approaches, one should mention J. Schwinger [2] and R. P. Feynman and F. L. Vernon [3]. Classical counterparts of the Keldysh technique are extremely useful and interesting on their own. I shall mention Wyld diagrammatic technique [4], which is the basis of the modern theory of turbulence; Martin–Siggia–Rose–DeDominics method [5] for stochastic systems; and Doi–Peliti technique [6, 7] for reaction–diffusion systems and cellular automata. Despite the unique power and wide popularity of the technique its pedagogical presentations are rare. One may find some in books [8, 9] or in a review [10]. This lecture is not an alternative for the existing literature. Using the simplest possible example, I try to explain the general structure of the theory, its relation to other approaches, its potential for novel applications and open problems. Technicalities of the numerous practical applications may be found in the literature and left outside the scope of the lecture. I tried hard, however, to mention possible traps and silent points which are rarely exposed explicitly.

1.2. MOTIVATION

The Keldysh formulation of the many–body theory is useful in the following cases:

- For treatment of systems not in the thermal equilibrium [1].
- For calculation of the full counting statistics of a quantum mechanical observable (as opposed to an average value or correlators) [11, 12].
- As an alternative to the replica and supersymmetry methods in the theory of disordered and glassy systems [13–17].

I. V. Lerner et al. (eds.),
Strongly Correlated Fermions and Bosons in Low-Dimensional Disordered Systems, 313–340.

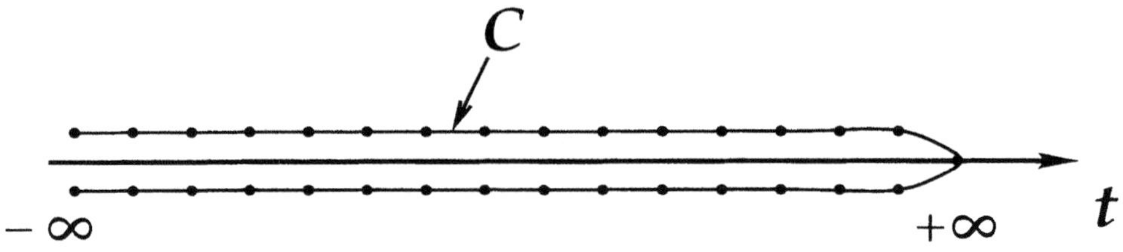

Figure 1. The closed Keldysh time contour C. Dots on the forward and the backward branches of the contour denote the discretized time points.

- For equilibrium problems where the Matsubara analytical continuation may prove to be too sophisticated [10].

1.3. CLOSED TIME CONTOUR

The standard construction of the zero temperature (or equilibrium) many–body theory (see e.g. [18, 9]) involves the procedure of adiabatic switching interactions on at distant past and then switching it off at distant future. The crucial assumption here is that starting from the ground (or equilibrium) state of the system at $t = -\infty$ one arrives to the same state at $t = +\infty$ (acquiring some phase factor along the way). This is clearly not the case out of the equilibrium. Starting from some arbitrary distribution and then switching interactions on and off, one is going to find the system in some unpredictable state. The latter depends, in general, on the peculiarities of the switching procedure. The absence of knowledge about the finite state spoils completely the entire construction [18, 9]. One would like thus to build a theory which avoids reference to the state at $t = +\infty$. We still need to know a finite state (since we calculate traces). The Schwinger suggestion is to make the finite state to be exactly the initial one. The central idea is to let a quantum system to evolve first in the forward direction in time and then to "rewind" its evolution back, playing the movie in the backward direction. One ends up thus with the need to construct a theory with the time evolution along the two–branch contour depicted on Fig. 1.3. Then, no matter what was the state at $t = +\infty$, after the backward evolution it will return back to the known initial state. In this construction there is no switching of interactions in the future. Both switchings on and off take place in the past: on – on the forward branch and off – on the backward. How to construct such a theory and how to use it – is the subject of this chapter.

1.4. FIELD THEORY

To be specific, let me consider the simplest possible many–body system. It consists of bosons living in a single quantum state with the energy ω_0:

$$H = \omega_0 a^\dagger a\,; \tag{1}$$

here $a^\dagger$ and a are bosonic creation and annihilation operators. Let us consider the "partition function" defined as

$$Z = \mathrm{Tr}\{\rho_0 U_C\}/\mathrm{Tr}\{\rho_0\}\,, \tag{2}$$

where U_C is the evolution operator along the closed contour C. If we assume (as we shall do for a while) that all the external fields are exactly the same on the forward and backward branches of the contour, then $U_C = 1$ and therefore $Z = 1$. In Eq. (2) $\rho_0 = \rho(H)$ is some [1] density matrix operator defined at $t = -\infty$. In our example $\mathrm{Tr}\{\rho_0\} = [1 - \rho(\omega_0)]^{-1}$. An important point is that in general $\mathrm{Tr}\{\rho_0\}$ is an interaction– and disorder–*independent* constant. Indeed both interactions and disorder are supposed to be switched on (and off) on the forward (backward) parts of the contour sometime after (before) $t = -\infty$. I shall therefore frequently omit this constant or refer to it as being unity – it never causes a confusion.

The next step is to divide the C contour into $2N+1$ time steps of the length δ_t, such that $t_1 = t_{2N+1} = -\infty$ and $t_{N+1} = +\infty$ as shown on Fig. 1.3. Following the standard route [19], we obtain the coherent state functional integral, by introducing a resolution of unity at each time step. Taking the $N \to \infty$ limit we obtain for the partition function

$$Z = \frac{1}{\mathrm{Tr}\{\rho_0\}} \int \mathcal{D}\bar\phi\phi \, \exp\{iS[\bar\phi(t), \phi(t)]\}\,, \tag{3}$$

where the action is given by

$$S[\bar\phi(t), \phi(t)] = \int\limits_C dt\, \bar\phi(t) \mathcal{D}^{-1} \phi(t)\,, \tag{4}$$

[1] To accommodate the equilibrium initial density matrix, $\rho_0 = \exp\{-\beta(H - \mu N)\}$, it is sometimes suggested to add an imaginary (vertical) part to the Keldysh contour at $t = -\infty$. To my opinion this procedure only obscures the structure of the theory. Indeed, the technique is not restricted in any way to the initial distribution being the equilibrium one. Therefore it is much preferred to keep the structure general and not commit oneself with unnecessary assumptions.

and $\mathcal{D}^{-1} = [i\partial_t - \omega_0]$. It is important to remember that this continuous notation is only a short way to represent the $2N \times 2N$ matrix $\mathcal{D}_{ij}^{-1}$

$$-i\mathcal{D}_{ij}^{-1} = \begin{bmatrix} 1 & 0 & & & & & & a_+\rho(\omega_0) \\ a_+ & 1 & 0 & & & & & \\ & a_+ & 1 & 0 & & & & \\ & & a_+ & 1 & 0 & & & \\ & & & \ddots & \ddots & \ddots & & \\ & & & & & & & \\ & & & & & a_- & 1 & 0 \\ & & & & & & a_- & 1 \end{bmatrix}, \tag{5}$$

where $a_\pm = -1 \pm \delta_t i\omega_0$. The change of sign $+\delta_t \to -\delta_t$ takes place in the center of the matrix at the point where the time axis changes its direction. It is straightforward to evaluate the determinant of such matrix

$$\det[-i\mathcal{D}^{-1}] = 1 - \rho(\omega_0)(-1 + \delta_t i\omega_0)^N(-1 - \delta_t i\omega_0)^N \to 1 - \rho(\omega_0)\,; \tag{6}$$

Recalling that $\rho(\omega_0) < 1$, one may convince himself that all eigenvalues of the matrix $-i\mathcal{D}_{ij}^{-1}$ have positive real part. This fact secures convergence of the functional integral. Performing the integration, one not–surprisingly obtains for the partition function

$$Z = \frac{\det^{-1}[-i\mathcal{D}^{-1}]}{\mathrm{Tr}\{\rho_0\}} = 1\,. \tag{7}$$

Seemingly absent in the continious notations – the upper right element of the matrix is extremely important to obtain the correct normalization. It is not less important to find the proper correlation functions.

We divide next the bosonic field $\phi(t)$ onto the two components $\phi_f(t)$ and $\phi_b(t)$ which reside on the forward and the backward parts of the time contour correspondingly. The action may be now rewritten as $S[\phi_f] - S[\phi_b]$, with the time integration running along the real axis (the minus sign comes from reversing direction of the time integration on the backward part of the contour).

1.5. CORRELATION FUNCTIONS

Now comes the most unpleasant part: to invert the matrix and find the correlation (or Green) functions. We may save ourselves a lot of efforts by recalling that the functional integral always produces the time–ordered Green functions [19] and that the time arguments on the backward branch are always *after* those on the forward. This way one finds [2]

$$-i\langle\phi_f(t)\bar{\phi}_f(t')\rangle = \mathcal{D}^T(t,t') = \theta(t-t')\mathcal{D}^>(t,t') + \theta(t'-t)\mathcal{D}^<(t,t')\,;$$

[2] The more familiar definition of the Green functions is $\mathcal{D} = -i\langle\phi\bar{\phi}\rangle/Z$, since in our case $Z = 1$ – we are doing OK.

$$
\begin{aligned}
-i\langle\phi_f(t)\bar{\phi}_b(t')\rangle &= \mathcal{D}^<(t,t') = \left(-ine^{-i\omega_0(t-t')}\right); \\
-i\langle\phi_b(t)\bar{\phi}_f(t')\rangle &= \mathcal{D}^>(t,t') = \left(-i(n+1)e^{-i\omega_0(t-t')}\right); \\
-i\langle\phi_b(t)\bar{\phi}_f(t')\rangle &= \mathcal{D}^{\tilde{T}}(t,t') = \theta(t'-t)\mathcal{D}^>(t,t') + \theta(t-t')\mathcal{D}^<(t,t'),
\end{aligned}
\tag{8}
$$

where the angular brackets are understood as the functional integration with the action given above. The structure is general, but $\mathcal{D}^> = -i\langle|a(t)a^\dagger(t')|\rangle$ and $\mathcal{D}^< = -i\langle|a^\dagger(t')a(t)|\rangle$ are given for orientation in brackets for our simple example ($n = \rho(\omega_0)/(1-\rho(\omega_0))$ is the bosonic occupation number).

It may be surprising to see the non–zero off–diagonal correlators $\langle\phi_f\bar{\phi}_b\rangle$ and $\langle\phi_b\bar{\phi}_f\rangle$, while the action seems to be perfectly diagonal in the $f-b$ space. The trick is exactly in the presence of the upper–right element of the discretized matrix, hidden in the continuous representation. Notice, that the presence of the off–diagonal elements does not contradict to the continuous notations. Indeed, $[i\partial_t - \omega_0]\mathcal{D}^{>,<} = 0$, while $[i\partial_t - \omega_0]\mathcal{D}^{T,\tilde{T}} = \pm\delta(t-t')$. Therefore in the obvious 2×2 matrix notations $\mathcal{D}^{-1}\mathcal{D} = 1$, as it should be. The point is that the $[i\partial_t - \omega_0]^{-1}$ operator is not uniquely defined. The boundary conditions must be specified and the upper right element does exactly this.

Obviously not all four Green functions defined above are independent. Indeed, an inspection shows that

$$
\mathcal{D}^T + \mathcal{D}^{\tilde{T}} = \mathcal{D}^> + \mathcal{D}^<. \tag{9}
$$

One would therefore like to perform a linear transformation of the fields to benefit explicitly from this relation. This is achieved by the Keldysh rotation.

1.6. KELDYSH ROTATION

Let us define the new fields as

$$
\phi_{cl}(t) = \frac{1}{2}(\phi_f(t) + \phi_b(t)); \qquad \phi_q(t) = \frac{1}{2}(\phi_f(t) - \phi_b(t)) \tag{10}
$$

with the analogous transformation for the conjugated fields. The integration measure changes by unessential multiplicative constant. The subscripts *cl* and *q* stand for *classical* and *quantum* components of the fields correspondingly. The rational for such notations will become clear shortly. First, a little algebra shows that [3]

$$
-i\langle\phi_\alpha(t)\bar{\phi}_\beta(t')\rangle = \frac{1}{2}\begin{pmatrix} \mathcal{D}^K(t,t') & \mathcal{D}^R(t,t') \\ \mathcal{D}^A(t,t') & 0 \end{pmatrix}, \tag{11}
$$

[3] For fermions it is more convenient to transform ψ and $\bar{\psi}$ in the different manner (it is allowed since ψ and $\bar{\psi}$ are not actually complex conjugated, but rather notations). As a result, the standard matrix form of the fermionic propagator is

$$
\begin{pmatrix} \mathcal{G}^R(t,t') & \mathcal{G}^K(t,t') \\ 0 & \mathcal{G}^A(t,t') \end{pmatrix}.
$$

where $\alpha, \beta = cl, q$. Here indexes R, A, K stay for *retarded, advanced* and *Keldysh* components of the Green function. These three are the main objects of the Keldysh technique. They are defined as

$$\begin{aligned} \mathcal{D}^R &= \mathcal{D}^T - \mathcal{D}^< = \theta(t-t')(\mathcal{D}^> - \mathcal{D}^<)\,; \\ \mathcal{D}^A &= \mathcal{D}^T - \mathcal{D}^> = \theta(t'-t)(\mathcal{D}^< - \mathcal{D}^>)\,; \\ \mathcal{D}^K &= \mathcal{D}^> + \mathcal{D}^<\,. \end{aligned} \tag{12}$$

Notice that the retarded (advanced) component is lower (upper) triangular matrix in the time space. Moreover, our regularization of the functional integral guarantees that these functions never appear at exactly coinciding arguments, removing thus uncertainty about the value $\theta(t-t) = \theta(0)$. Fields $\bar{\phi}$ always appear at one time step δ_t *after* ϕ fields on the Keldysh contour. As a result, $\bar{\phi}_f(t)$ is after $\phi_f(t)$ and $\bar{\phi}_b(t)$ is before $\phi_b(t)$. Employing Eqs. (8) and (12), one may see that with this convention $\mathcal{D}^R(t,t) \equiv \mathcal{D}^R(t, t+\delta_t) = 0$ and $\mathcal{D}^A(t,t) \equiv \mathcal{D}^A(t+\delta_t, t) = 0$. In the continuous notations it is equivalent to the choice of the regularization, where $\theta(0) = 0$. As a result, for any number [4] of retarded or advanced functions one has

$$\begin{aligned} \mathrm{Tr}\{\mathcal{D}^R\mathcal{D}^R \ldots \mathcal{D}^R\} &= 0\,; \\ \mathrm{Tr}\{\mathcal{D}^A\mathcal{D}^A \ldots \mathcal{D}^A\} &= 0\,, \end{aligned} \tag{13}$$

where the traces are understood as time integrations. If a problem is translationally invariant in time – the same statement has the simple reinterpretation

$$\int\limits_{-\infty}^{\infty} d\omega\, \mathcal{D}^R(\omega)\mathcal{D}^R(\omega+\Omega_1)\ldots\mathcal{D}^R(\omega+\Omega_n) = 0\,, \tag{14}$$

which is an obvious consequence of retarded Green function being an analytic function in the entire upper half plane of complex energy. From Eq. (11) it is obvious that

$$\mathcal{D}^A = \left[\mathcal{D}^R\right]^\dagger\,, \tag{15}$$

where Hermit conjugation includes transposition of the time arguments.

In our simple example $\mathcal{D}^> = -i(n+1)e^{-i\omega_0(t-t')}$ and $\mathcal{D}^< = -ine^{-i\omega_0(t-t')}$, where $n = n(\omega_0) = \rho(\omega_0)/(1-\rho(\omega_0))$ is the bosonic occupation number (since the system is non–interacting the initial distribution function does not evolve). As

In this notation fermion–boson interactions take an especially simple form $S_{int} \sim \phi_\alpha \bar{\psi}\gamma^\alpha\psi$, with $\gamma^{cl} = 1$ and $\gamma^q = \sigma_1$ [15]. Since fermions are never a classical field, this form of the fermionic propagator is not related to our subsequent discussion.

[4] This is true for a single function as well. For example, if $\mathcal{D}^R = (Dq^2 - i\omega)^{-1}$, one may worry that $\int d\omega (Dq^2 - i\omega)^{-1}$ is divergent. This is not the case, however, since this integral is simply $\mathcal{D}^R(t,t) = 0$.

a result

$$\begin{aligned}\mathcal{D}^R(t,t') &= -i\theta(t-t')e^{-i\omega_0(t-t')};\\ \mathcal{D}^A(t,t') &= i\theta(t'-t)e^{-i\omega_0(t-t')};\\ \mathcal{D}^K(t,t') &= -i(2n(\omega_0)+1)e^{-i\omega_0(t-t')}.\end{aligned} \tag{16}$$

Notice that the retarded and advanced components are independent on the distribution function, whereas the Keldysh component does depend on it. This statement is much more general than our toy example. Going to the frequency representation, one finds that $\mathcal{D}^K(\omega) = -2\pi i(2n(\omega_0)+1)\delta(\omega-\omega_0) = -2\pi i(2n(\omega)+1)\delta(\omega-\omega_0)$. In the particular case of the thermal equilibrium, one has

$$\mathcal{D}^K(\omega) = \coth\frac{\beta\omega}{2}\left(\mathcal{D}^R(\omega) - \mathcal{D}^A(\omega)\right). \tag{17}$$

The last equation constitutes the statement [5] of the *fluctuation–dissipation theorem* (FDT). It implies (as we shall see below) the rigid relation between response and correlation functions at equilibrium. FDT is always satisfied in the thermal equilibrium even in the presence of the interactions. We shall see later why the interactions do not change this relation. In general case it is convenient to define a function (matrix in the time space) F via

$$\mathcal{D}^K = \mathcal{D}^R F - F\mathcal{D}^A \tag{18}$$

and refer to F (or better its Wigner transform) as the distribution function. At equilibrium $F(t-t') \to F(\omega) = \coth\frac{\beta\omega}{2} = 2n(\omega)+1$, where $n(\omega)$ is the Planck distribution.

1.7. KELDYSH ACTION AND CAUSALITY

Once we have understood the structure of the propagator in the $cl - q$ space, we may go back to the action. Inverting the matrix Eq. (11), one finds [6]

$$S[\phi_{cl},\phi_q] = 2\int\int_{-\infty}^{\infty} dt dt'\,(\bar\phi_{cl}(t),\bar\phi_q(t))\begin{pmatrix} 0 & [\mathcal{D}^A]^{-1} \\ \left[\mathcal{D}^R\right]^{-1} & [\mathcal{D}^{-1}]^K\end{pmatrix}_{(t,t')}\begin{pmatrix}\phi_{cl}(t')\\ \phi_q(t')\end{pmatrix}, \tag{19}$$

[5] The analogous statement for fermions reads as

$$\mathcal{G}^K(\epsilon) = \tanh\frac{\beta\epsilon}{2}\left(\mathcal{G}^R(\epsilon) - \mathcal{G}^A(\epsilon)\right).$$

[6] Notice that the matrix in the exponent of the functional integral, $i\mathcal{D}^{-1}$, is *anti*–Hermitian ! The convergence factors are hidden in the regularization.

where $[\mathcal{D}^{-1}]^K = [\mathcal{D}^R]^{-1}F - F[\mathcal{D}^A]^{-1} \neq [\mathcal{D}^K]^{-1}$. Of course, in our simple example $[\mathcal{D}^{R,A}]^{-1} = \delta(t-t')[i\partial_{t'} - \omega_0 \pm i0]$ and $[\mathcal{D}^{-1}]^K = 0$ and therefore the correlator matrix is basically $\delta(t-t')[i\partial_{t'} - \omega_0]\sigma_1$, as it should be. The point is that the structure of the Gaussian action given by Eq. (19) is absolutely generic and correctly encodes regularization of the functional integral. Since the Keldysh component carries information about the density matrix, there is no need any more to remember about the discreet representation. As we shall see below this structure remains intact under any perturbative renormalization. For the lack of a better terminology we shall refer to the robustness of this structure as the *causality* principle [17] (since it is the reason for the response functions being casual, see below). Let us devote some time to this structure.

Consider first the $cl - cl$ element, which is zero. This zero may be traced back to the relation (9). It has, however, the much simpler interpretation. It says that if one takes a purely classical field configuration, $\phi_q = 0$, the action is zero. But this is obvious, since in this case $\phi_f = \phi_b$ and the action on the forward contour cancels exactly that on the backward. We therefore arrive to the simple but extremely important statement:

$$S[\phi_{cl}, \phi_q = 0] = 0\,. \tag{20}$$

From this perspective the relation (9) is simply one of manifestations of the causality. Obviously Eq. (20) is not restricted to the Gaussian action. Consider e.g. the simplest interaction action $S_{int} = \frac{\lambda}{2}\int_{\mathcal{C}} dt\,(\bar\phi\phi)^2 = \frac{\lambda}{2}\int dt\,[(\bar\phi_f\phi_f)^2 - (\bar\phi_b\phi_b)^2]$. Notice that this is the continuous notation; in discreet version $\bar\phi$ fields on the Keldysh contour $\mathcal{C}$ are taken at one step δ_t *later* time with respect to ϕ fields. Transforming to the classical–quantum components, one finds

$$S_{int}[\phi_{cl}, \phi_q] = 4\lambda\,\Re\int_{-\infty}^{\infty} dt\ \bar\phi_q\bar\phi_{cl}(\phi_{cl}^2 + \phi_q^2)\,, \tag{21}$$

which obviously satisfy Eq. (20).

The fact that $cl-q$ and $q-cl$ components in Eq. (19) are advanced and retarded (means triangular in the time representation) is of the fundamental importance as well. Due to this fact Eq. (20) is intact in any order of the perturbation theory. Consider, for example, the first order correction in λ to the $cl - cl$ element of the Gaussian action (which is zero). The corresponding diagram is depicted on Fig. 1.7a. From Eq. (21) one reads out

$$\delta S_{cl-cl} = 4\lambda\,\Re\int_{-\infty}^{\infty} dt\,\bar\phi_{cl}(t)\mathcal{D}^R(t,t)\phi_{cl}(t)\,. \tag{22}$$

As was explained above, $\mathcal{D}^R(t,t) = 0$, and the $cl - cl$ element obviously remains zero. Employing Eqs. (13), (14) one may show that all higher loop corrections do

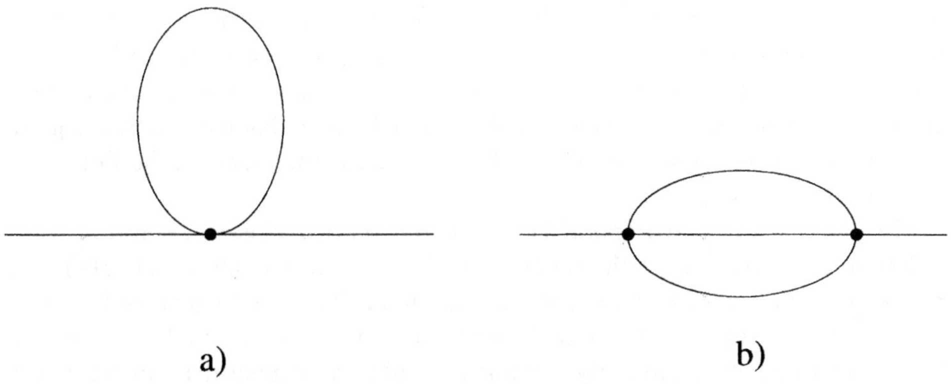

Figure 2. First and second order diagrams for the $(\bar{\phi}\phi)^2$ theory.

not change this fact. One may say that Eqs. (13), (14) are part of the causality principle. To complete the prove that causality is preserved in perturbation theory, one has to show that $cl - q$ and $q - cl$ components remain purely advanced and retarded correspondingly. A reader is advised to check that this is indeed the case in the second order, Fig. 1.7b.

1.8. SADDLE POINT EQUATIONS

Before doing the perturbation theory we have to discuss saddle points of the action. Eq. (20) states that there are no terms in the action which have zero power of both $\bar{\phi}_q$ and ϕ_q. The same is obviously true regarding $\delta S/\delta\bar{\phi}_{cl}$ and therefore the saddle point equation

$$\frac{\delta S}{\delta\bar{\phi}_{cl}} = 0 \tag{23}$$

may be always solved by

$$\Phi_q = 0\,. \tag{24}$$

The saddle point solutions are denoted by capital Φ. One may check that this is indeed the case for the action given by Eqs. (19), (21). Under the condition Eq. (24) the second saddle point equation takes the form

$$\frac{1}{2}\frac{\delta S}{\delta\bar{\phi}_q} = \mathcal{O}^R[\Phi_{cl}]\Phi_{cl} = 0\,, \tag{25}$$

where $\mathcal{O}^R[\Phi_{cl}]$ is retarded operator describing the classical dynamics of the field. In our example

$$\mathcal{O}^R[\Phi_{cl}] = i\partial_t - \omega_0 + \lambda|\Phi_{cl}|^2 \tag{26}$$

and the saddle point equation (25) is the time–dependent "Gross–Pitaevskii" equation. We have arrived to the conclusion that among possible saddle point solutions of the Keldysh action, there are always some with zero quantum component and the classical component obeying the classical equation of motion. We shall call such saddle points *classical* or *casual* (since perturbation theory on top of a classical solution preserves the causality structure). Notice, that the classical equation of motion is result of variation over the *quantum* component (in the limit where the later is zero).

In view of Eqs. (20) and (24), the action on any classical solution is zero [7]. Recall now that we do the theory for $Z = 1$ – this is the exact relation. To reproduce it in the saddle point approximation one has to show that the fluctuation determinant on top of the classical saddle point is unity. It may be a formidable task to demonstrate it explicitly, but it is helpful to remember that this must be the case.

An extremely interesting issue is an existence and role of *non–classical* saddle point solutions, which have $\Phi_q \neq 0$. Such solutions, if exist, have, in general, a non–zero action and may be loosely called *instantons*. Examples of instantons include (thermal) escape from a quantum well [20] and Wigner–Dyson statistics in metal grains [17]. Since the normalization $Z = 1$ is entirely due to classical saddle points, instantons must have strictly zero contribution to the partition function, Z. They may, however, contribute to correlation and response functions. The subject is poorly developed (unlike instantons in the imaginary time) and requires much more attention.

1.9. SEMICLASSICAL APPROXIMATION

Let me concentrate now on a classical saddle point, Eqs. (24), (25). To include quantum effects in the semiclassical approximation one has to include fluctuations of the ϕ_q field near zero. To this end we keep terms up to the second order in ϕ_q in the action. The resulting semiclassical action takes the following general form

$$S_{scl} = 2\int\!\!\int_{-\infty}^{\infty} dtdt' \, \Big[\bar\phi_q[\mathcal{D}^{-1}]^K\phi_q + (\bar\phi_q\mathcal{O}^R[\phi_{cl}]\phi_{cl} + c.c.)\Big] \, ; \tag{27}$$

$c.c.$ stays for the complex conjugation. From this point one may proceed in two directions:

(i) Since the action Eq. (27) is Gaussian in ϕ_q one may integrate it out and end up with the theory of the single filed, ϕ_{cl}. The corresponding action is

$$S_{scl}[\phi_{cl}] = 2\int\!\!\int_{-\infty}^{\infty} dtdt' \, \bar\phi_{cl}(\mathcal{O}^A[\bar\phi_{cl}]\mathcal{D}^A)[\mathcal{D}^K]^{-1}(\mathcal{D}^R\mathcal{O}^R[\phi_{cl}])\phi_{cl} \, . \tag{28}$$

[7] This statement is a close relative of the Parisi–Efetov–Wegner theorem in supersymmetry or the fact that a replica symmetric saddle point results in zero action in the replica limit.

If the nonlinearity in ϕ_{cl} is neglected, one ends up with the Gaussian theory governed by the inverse Keldysh Green function, $S \sim \bar{\phi}_{cl}[\mathcal{D}^K]^{-1}\phi_{cl}$ [8]. In general, the action Eq. (28) is pretty complicated. If it contains only a finite number of the degrees of freedom and is short–ranged in time, one may use the transfer–matrix technique to write down the corresponding "Schrödinger" equation. The result is the Focker–Planck equation.

(ii) One may perform the Hubbard–Stratonovich transformation with the auxiliary stochastic field $\xi(t)$ to decouple the quadratic term in Eq. (27). The resulting action is linear in ϕ_q. Therefore the integration over ϕ_q leads to the functional δ–function. As a result one obtains stochastic Langevin equation [9]

$$\mathcal{O}^R[\phi_{cl}]\phi_{cl}(t) = \xi(t)\,, \tag{29}$$

where $\xi(t)$ is a Gaussian noise with the correlator

$$\langle \xi(t)\bar{\xi}(t')\rangle = \frac{i}{2}[\mathcal{D}^{-1}]^K(t,t')\,. \tag{30}$$

For the diffusive dynamics, $[\mathcal{D}^{R,A}]^{-1} = D\nabla^2 \mp i\omega$, in the thermal equilibrium $i[\mathcal{D}^{-1}]^K(\omega) = i\coth\frac{\beta\omega}{2}([\mathcal{D}^R]^{-1}(\omega) - [\mathcal{D}^A]^{-1}(\omega)) \to 4T$ in the limit $\omega \ll T$. In this case one obtains the classical Langevin force with the correlator $\langle \xi(t)\bar{\xi}(t')\rangle = 2T\delta(t-t')$. The opposite limit is called sometimes the quantum Langevin equation.

Notice that the above procedure may be reversed. Start from the classical Langevin equation, then introduce functional δ–function. Finally exponentiate it with the auxiliary field called ϕ_q and integrate out the Gaussian noise. Such procedure is known in the literature as Martin–Siggia–Rose–DeDominics (MSRD) [5] technique. The result is the semi–classical Keldysh action, Eq. (27), in the high temperature limit. In the MSRD literature the field ϕ_{cl} is called ϕ, while ϕ_q is usually called $\hat{\phi}$. One may wonder if one can form their symmetric and antisymmetric combinations $\phi_f = \phi + \hat{\phi}$ and $\phi_b = \phi - \hat{\phi}$ to recover an "underlying quantum theory" with the action $S[\phi_f] - S[\phi_b]$. In general it does not work since MSRD is only the semiclassical limit of Keldysh. Therefore terms of the order $O(\hat{\phi}^3)$ may be "missing".

---

[8] If $\mathcal{D}^{R,A} = (0 \mp i\omega)^{-1}$, then in the thermal equilibrium (c.f. Eq. (17)), $[\mathcal{D}^K]^{-1} = \frac{-i}{2}\omega\tanh\frac{\beta\omega}{2}$. In the $T=0$ limit one obtains the non–local quantum dissipative action, $S \sim |\omega||\phi_{cl}(\omega)|^2$, while in the high temperature regime the action is local $S \sim \beta|\dot{\phi}_{cl}(t)|^2$.

[9] One may worry about appearance of the Jacobean upon integration of the functional δ–function over ϕ_{cl}. It does not happen with our regularization of the functional integral (Jacobean is unity). The reason is exactly the same which guarantees $Z=1$ upon integration of the fluctuations around a classical saddle point.

1.10. FDT AND KINETIC EQUATION

Having understood the structure of the saddle points, we can construct a perturbative expansion around a *classical* saddle point [10]. Let say we are about to calculate the full Green function, defined as $\mathcal{D}_{\alpha,\beta}(t,t') = -2i\langle\phi_\alpha(t)\bar{\phi}_\beta(t')\rangle$, where $\alpha,\beta = cl, q$. This is done in a standard way by expanding the exponent of the non–Gaussian action and applying the Wick theorem according to Eq. (11). One then rearranges the perturbation theory and defines the irreducible self–energy Σ through the relation:

$$\mathcal{D} = \mathcal{D}_0 + \mathcal{D}_0\Sigma\mathcal{D}_0 + \mathcal{D}_0\Sigma\mathcal{D}_0\mathcal{D}_0\Sigma\mathcal{D}_0 + \dots\,, \tag{31}$$

where multiplication is understood as the matrix one in the time and $cl-q$ spaces; $\mathcal{D}_0$ is the bare propagator, Eq. (11). The causality guarantees (this may be checked order by order) that the self–energy matrix has the structure of the correlator (i.e. $\mathcal{D}_0^{-1}$; cf. Eq. (19))

$$\Sigma_{\alpha,\beta}(t,t') = \begin{pmatrix} 0 & \Sigma^A(t,t') \\ \Sigma^R(t,t') & \Sigma^K(t,t') \end{pmatrix} \tag{32}$$

Consider e.g. the second term on the r.h.s. of Eq. (31), after a little algebra one finds

$$\begin{aligned}
&\delta\mathcal{D}cl,cl = \mathcal{D}_0^K\Sigma^A\mathcal{D}_0^A + \mathcal{D}_0^R\Sigma^R\mathcal{D}_0^K + \mathcal{D}_0^R\Sigma^K\mathcal{D}_0^A\,;\\
&\delta\mathcal{D}cl,q = \mathcal{D}_0^R\Sigma^R\mathcal{D}_0^R\,;\\
&\delta\mathcal{D}q,cl = \mathcal{D}_0^A\Sigma^A\mathcal{D}_0^A\,;\\
&\delta\mathcal{D}q,q = 0\,.
\end{aligned} \tag{33}$$

As a result the $q-q$ component remains zero, while $cl-q$ and $q-cl$ remain retarded and advanced correspondingly. Indeed, product of triangular matrices is again a triangular matrix. Obviously, the same is true in all higher orders as well. Therefore the causality structure remains intact even in the presence of nonlinearity (at least in perturbation theory). Eq. (31) may be rewritten as the matrix Dyson equation

$$\left(\mathcal{D}_0^{-1} - \Sigma\right)\mathcal{D} = 1\,. \tag{34}$$

It is a simple matter to show that the $q-cl$ component of this equation is nothing but the kinetic equation on the distribution function F. Since this procedure is well documented in the literature [1, 8, 10] we shall not describe it here [11]. Let

[10] I am not aware about any attempt of doing the perturbation theory on top of a *non–classical* saddle point.

[11] The simplest exercise is the theory of the real bosonic field with the interaction $\lambda\phi^3 \to 3\lambda\phi_{cl}^2\phi_q + \lambda\phi_q^3$. In the second order the self–energy Fig. 1.10 is given by $\Sigma^{R,A} \sim \lambda^2\mathcal{D}_0^K * \mathcal{D}_0^{R,A}$ and

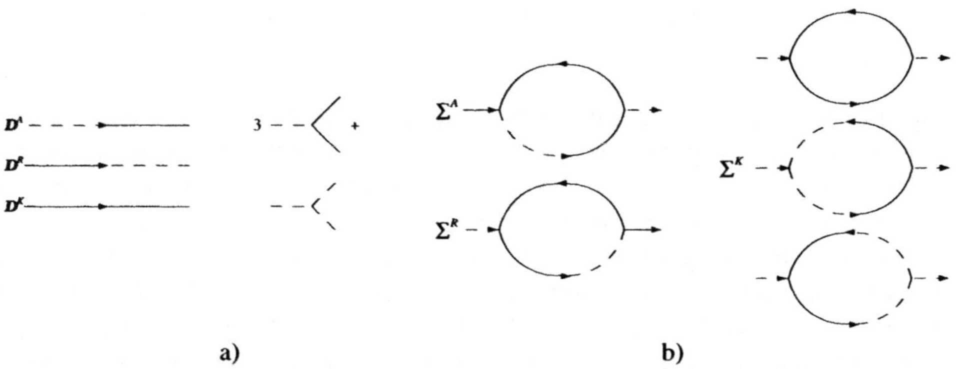

Figure 3. Example of the Keldysh diagrammatic for the ϕ^3 theory. Fields ϕ_{cl} are denoted by full lines and ϕ_q by dashed lines. There are two vertices with relative coefficient 3. There is one retarded, one advanced and three Keldysh self–energy diagrams

me only mention that to obtain the textbook (semi–classical) kinetic equation one has to assume that all external fields are slow on the scale of temperature. This assumption justifies the time–Wigner transform of the matrix F, which is a necessary step in the derivation of the kinetic equation.

The other way to obtain the kinetic equation is to introduce time non–local field $Q(t, t')$, canonically conjugated to the composite operator $\phi(t)\bar{\phi}(t')$. A small amount of disorder makes $Q(t, t')$ a dynamic field which is governed by a certain effective action (this action may be obtained by integrating out the fields ϕ and $\bar{\phi}$). The saddle point equation for this effective action $S_{eff}[Q]$ is the kinetic equation. As usually, the (semi) *classical* (kinetic) equation is obtained by variation over the *quantum* component of the Q–field. For the details of this procedure see Ref. [15]. One can possibly go beyond the saddle–point approximation by including the Gaussian fluctuations of the quantum component of Q (pretty much in the way we did above). The result of such procedure is probably the Langevin–kinetic equation of Kogan and Shulman [21]. To the best of my knowledge, it was not demonstrated explicitly yet.

$\Sigma^K \sim \lambda^2(\mathcal{D}_0^K * \mathcal{D}_0^K + \mathcal{D}_0^R * \mathcal{D}_0^A + \mathcal{D}_0^A * \mathcal{D}_0^R)$, where the star denotes convolution in the energy space. Employing the "magic" identity $\coth\epsilon\coth(\epsilon+\omega) - 1 = \coth\omega\,(\coth\epsilon - \coth(\epsilon+\omega))$, one may immediately see that if $\mathcal{D}_0$ satisfy FDT, Eq. (17), then so does Σ, namely $\Sigma^K = \coth\frac{\beta\omega}{2}(\Sigma^R - \Sigma^A)$. Then the Dyson equation (34) says that the exact propagator also satisfy FDT. The "magic" identity is directly responsible for the fact that the Planck distribution ($\coth\frac{\beta\omega}{2}$) nullifies the collision integral.

1.11. SOURCES, EXTERNAL FIELDS, RESPONSE AND CORRELATION FUNCTIONS

Up until now we have been busy with complicated representation of unity: $Z = 1$. This is an interesting, but useless job. To make the technique work we need to calculate something more informative. This something is the generating function, which is obtained by adding the source fields to the action. It is convenient to discuss it in parallel with introduction of external classical fields (such as scalar or vector potentials). Let us add the source term $S_{source} = \frac{1}{2}\int_C dt\, V\bar{\phi}\phi$ to the action. The source field $V(t)$ is defined on the Keldysh contour, therefore it is rather a couple of fields (V_f, V_b), residing on the forward and backward branches correspondingly. We decompose it now on the classical and quantum components exactly in the same way we did it with the ϕ field: $V_{cl} = (V_f + V_b)/2$ and $V_q = (V_f - V_b)/2$. The source action takes the form

$$S_{source} = \int_{-\infty}^{\infty} dt\, [V_{cl}(\bar{\phi}_{cl}\phi_q + \bar{\phi}_q\phi_{cl}) + V_q(\bar{\phi}_{cl}\phi_{cl} + \bar{\phi}_q\phi_q)] \,. \tag{35}$$

Notice that the *quantum* source is coupled to the *classical* component of the density (defined in a usual way as a half sum of the density on the forward and backward branches) and vise–versa. We define now the generating function as

$$Z[V_{cl}, V_q] = \left\langle e^{iS_{source}} \right\rangle \,. \tag{36}$$

The classical component of the source field is nothing but classical external scalar potential, which is supposed to be the same on the two branches. The presence of such field does not violate the fact that the forward and the backward evolution brings the system back to the initial state. Therefore it does not change the normalization $Z = 1$. We therefore arrive to the conclusion

$$Z[V_{cl}, V_q = 0] = 1 \,. \tag{37}$$

One can also check this statement explicitly by e.g. expanding the action in powers of V_{cl} and making contractions. All terms in this expansion vanish due to the fact that $\langle \phi_q\bar{\phi}_q \rangle = 0$. Eq. (37) is a close relative of Eq. (20) and is just another manifestation of the causality. As a result, adding to the theory classical fields alone does not do any good – we still work with unity. Therefore, one necessarily has to introduce *quantum* sources (they change sign by going from the forward to the backward branches of the contour). The presence of such source fields explicitly violates causality, and thus changes the partition (generating) function. On the other hand such fields usually do not have a physical meaning (see, however, the next section) and play only auxiliary role. In most cases one uses them only to generate observables (by appropriate differentiation) and then put them to zero, restoring the causality structure of the action. Notice that the classical component, V_{cl}, does *not* have to be taken to zero.

Let us see how it works. Suppose we are interested in the average density of bosons at time t in presence of a certain scalar potential $V_{cl}(t)$. Of course, we are looking for the classical component of the density. According to Eqs. (35) and (36) it is given by

$$n(t;V_{cl}) = -i\,\frac{\delta}{\delta V_q(t)} Z[V_{cl},V_q]\Big|_{V_q=0}. \tag{38}$$

To calculate this quantity one has to find first an appropriate (classical) saddle point of the action in presence of V_{cl} (as was mentioned above this amounts to solving the kinetic equation) and then study quantum corrections. The problem is simplified if the external field, V_{cl}, is weak in some sense. Then one may go away with the linear response

$$\chi^R(t,t') \equiv \frac{\delta}{\delta V_{cl}(t')} n(t;V_{cl})\Big|_{V_{cl}=0} = -i\,\frac{\delta^2}{\delta V_{cl}(t')\delta V_q(t)} Z[V_{cl},V_q]\Big|_{V_q=V_{cl}=0}. \tag{39}$$

I put the subscript R, anticipating on the physical ground that the response function must be *retarded*. We shall demonstrate it momentarily. First, let us introduce for esthetical reasons the matrix

$$\chi_{\alpha,\beta}(t,t') \equiv -i\,\frac{\delta^2}{\delta V_\beta(t')\delta V_\alpha(t)} Z[V]\Big|_{V=0} = \begin{pmatrix} 0 & \chi^A(t,t') \\ \chi^R(t,t') & \chi^K(t,t') \end{pmatrix}. \tag{40}$$

The fact that $\chi_{cl,cl} = 0$ is obvious from Eq. (37). To evaluate the linear response matrix χ consider first the Gaussian action, Eq. (19). Adding the source term, Eq. (35), and integrating out the boson fields $\bar{\phi}$, ϕ one obtains

$$Z[V] = e^{-\mathrm{Tr}\ln(1+\mathcal{D}\hat{V})}, \tag{41}$$

where $\hat{V} = (V_q\sigma_0 + V_{cl}\sigma_1)/2$. Notice that the normalization is just right, since $Z[0] = 1$. One may now expand $\ln(1+\mathcal{D}\hat{V})$ to the second order in $\hat{V}$ and then differentiate twice. It is a straightforward matter to see that the causality structure of $\mathcal{D}$ imposes it also on χ. E.g. $\chi^R(t,t') \sim \mathrm{Tr}\{\mathcal{D}\sigma_1\mathcal{D}\sigma_0\} = \mathcal{D}^R(t,t')\mathcal{D}^K(t',t) + \mathcal{D}^K(t,t')\mathcal{D}^A(t',t)$, which is obviously retarded. In presence of interactions, Eq. (21), one may perform Hubbard–Stratonovich transformation and then integrate out $\bar{\phi}$, ϕ fields. At this stage the causality structure of χ becomes apparent (it does not mean yet that one knows to evaluate χ).

The purpose of the above discussion was to demonstrate that even for the linear response problems (which may be always solved in Matsubara technique) the Keldysh formulation is natural and transparent. For one thing it allows one to overcome the analytical continuation procedure, which may happen to be a headache.

The other advantage is that it gives both response and correlation functions in the single framework [12].

1.12. COUNTING STATISTICS AND COUNTING FIELDS

A very interesting application of the Keldysh formalism was suggested recently by L. S. Levitov and coworkers [11]. Suppose we are interested in a certain quantum mechanical observable $\hat{O}$ (for example, a total charge which passed through a wire during a long time $t_f - t_i$). One may look for e.g. its (quantum) expectation value, $\langle\hat{O}\rangle$, or noise $\langle\hat{O}^2\rangle - (\langle\hat{O}\rangle)^2$, or even higher order moments. The most general quantity, however, is the distribution function: $P(O)$ – probability to observe the value O in a single measurement of $\hat{O}$. If one knows $P(O)$, then $\langle\hat{O}^k\rangle = \int dOP(O)O^k$. Instead of discussing $P(O)$ it is convenient to consider its Fourier transform $F(\lambda) = \int dOP(0)e^{i\lambda O}$. The variable λ - is called the counting field, and $P(O)$ (or $F(\lambda)$) is called the counting statistics.

From the above expressions one may deduce that

$$F(\lambda) = \left\langle e^{i\lambda\hat{O}} \right\rangle . \tag{42}$$

The non–trivial fact is that the angular brackets in this expression must be understood as the averaging along the Keldysh contour. The prove of this statement requires a careful consideration of the time–ordering convention for the higher moments of $\hat{O}$. We refer a reader to the original works [11, 12] for more details. If the Keldysh nature of the averaging in Eq. (42) is accepted, it should be obvious by now that the counting field λ is a purely *quantum* field. Indeed, λ was introduced to generate moments of $\hat{O}$ upon differentiation. As we have seen in the previous section – it is the quantum component of the source field (having different sign on forward and backward branches (!)), which serves for this purpose. As a result, Eq. (42) suggests that the Keldysh action of the system at hand must be modified as

$$S^{\lambda}[\phi] = S[\phi] + \int\limits_{-\infty}^{\infty} dt\, \lambda_q(t)\hat{o}_{cl}(t)\,. \tag{43}$$

We put $\hat{O} = \int_{t_i}^{t_f} dt\, \hat{o}_{cl}(t)$, where the classical observable $\hat{o}_{cl}(t)$ is expressed through the fields $\phi(t)$. We have also put $\lambda_q(t) \equiv \lambda\theta(t - t_i)\theta(t_f - t)$ – thus the name – *counting* field.

The novel feature is that the quantum source, λ_q, should *not* be taken to zero in the end of calculations. It rather remains finite and possibly even large to allow the inverse Fourier transform of $F(\lambda)$. The presence of the finite quantum source in the action spoils immediately the causality structure of the theory. For example,

---

[12] In the equilibrium response and correlation are connected by the FDT: $\chi^K(\omega) = \coth\frac{\beta\omega}{2}(\chi^R(\omega) - \chi^A(\omega))$. However, out of the equilibrium they require separate calculations.

$S^\lambda[\phi_q = 0] \sim \lambda \neq 0$. As a result, $\Phi_q = 0$ is *not* a saddle point solution any more. There are no classical saddle points in the theory ! A new non–classical saddle point has in general a non–zero action $S_{sp} = S_{sp}(\lambda) \neq 0$. Neglecting fluctuation effects, one finds for the generating function of the irreducible moments: $-i \ln F(\lambda) = S_{sp}(\lambda)$. Such program (including fluctuations) was realized by Yu. V. Nazarov [12] for the counting statistics of charge passing through a diffusive wire. See also related works on counting statistics of adiabatic pumps [22].

To conclude this section I want to mention a striking similarity between counting statistics in Keldysh theory and statistics of rare events in supersymmetric technique [23]. To my opinion this subject is far from being closed.

1.13. KELDYSH TECHNIQUE IN DISORDERED SYSTEMS

Due to the presence of the intrinsic normalization condition, $Z = 1$, the Keldysh technique is ideally suited for treatment of disordered systems, such as spin glasses or dirty metals. Notice that the response and correlation functions were defined by differentiation of $Z[V]$, Eq. (40), rather than more familiar $\ln Z[V]$ (since $Z[V_q = 0] = 1$ this is completely equivalent). Let me also mention again that the normalization constant, $\mathrm{Tr}\{\rho_0\}$, introduced in Eq. (2), is disorder–*independent*. The point is that disorder is switched on (and off) on forward (backward) parts of the contour after (before) $t = -\infty$. As a result, the "partition" or generating function Z (rather than its logarithm) is the proper object to be averaged over the quenched disorder. This circumvent the need to introduce replica [24] (or supersymmetry [25]) to take care of the logarithm. This was first realized in the context of classical spin glasses (MSRD version of Keldysh) by H. Sompolinsky [13] and substantially advanced later [14]. In the context of electrons in the random potential the corresponding theory (Keldysh nonlinear σ–model) was developed in Refs. [15, 16, 26].

Here I restrict myself to a few remarks only. On the perturbative level the replica and supersymmetry methods are supposed to eliminate "parasitic" diagrams. Such decoupled closed loop diagrams are generated if one erroneously averages Z instead of $\ln Z$. The replica trick first prescribes factor n (number of replica) to such loops and then eliminates them by taking the limit $n \to 0$. In the supersymmetry bosonic loops exactly cancel fermionic ones. The Keldysh analog of such cancellation mechanism is given by Eqs. (13) and (14). Indeed, the closed loops coupled to the rest of a diagram by the static disorder carry either only retarded or only advanced Green functions [13]. The later vanish according

[13] Indeed, the static disorder U_{dis} enters the Keldysh matrix action as $U_{dis}\sigma_1$. Integrating out of the ϕ fields one ends up with $\mathrm{Tr}\ln(1 + \mathcal{D}\sigma_1 U_{dis})$, cf. Eq. (41). Expanding the logarithm one finds terms of the form $\mathrm{Tr}\{\mathcal{D}\sigma_1 \ldots \mathcal{D}\sigma_1\}$, the later are obviously $\mathrm{Tr}\{\mathcal{D}^R \ldots \mathcal{D}^R\} + \mathrm{Tr}\{\mathcal{D}^A \ldots \mathcal{D}^A\} = 0$.

to Eqs. (13) (14). Therefore it is not surprising that all *perturbative* results of the nonlinear σ–model (including electron–electron interactions) are correctly and elegantly reproduced in the Keldysh version [15, 16].

Whether or not the Keldysh σ–model contains the *non–perturbative* effects (such as instantons) – is a hundred dollar question. I believe that the answer is positive. At least, it is possible to obtain the non–perturbative (though only asymptotic) part of the level statistics via the Keldysh σ–model [17]. To this end, one has to analyze carefully non–classical (having a non–zero action) saddle points on the massless manifold of the σ–model. In any case, a lot to be done before the Keldysh technique will become a reliable tool for the non–perturbative calculations.

2. Doi–Peliti technique for classical systems

2.1. A BIT OF HISTORY

The technique was invented by Doi in 1976 [6] and later rediscovered by Peliti [7], and others [27, 28]. J. Cardy and B. P. Lee were probably most persistent in applying the technique to various non–equilibrium statistical problems [29]. These notes are partially based on lectures of J. Cardy [30].

2.2. MOTIVATION

We are talking about extremely broad and diverse types of problems whose applicability ranges from epidemiology and dynamics of biological populations to models of chemical reactions and glasses. The problems at hand are usually determined on some d–dimensional lattice. They are defined by specifying the following entities:

- The agents, usually named as $A, B, \ldots$ These may be molecules, animal species, stock options, vehicles, *etc.*
- Rules of agent's motion on the lattice. It may be e.g. random walk, or directed ballistic motion.
- Rules of agent's interactions. E.g. the simplest one is the binary annihilation: if two agents meet at the same lattice site they annihilate with probability μ. The short way to write is

$$A + A \xrightarrow{\mu} 0\,. \tag{44}$$

This model may, for example, describe a simple chemical reaction, where two identical molecules form a stable aggregate which segregates from the solution. An other famous example is the Lotka–Volterra system [31, 32]

$$\begin{aligned} &A \xrightarrow{\mu} 0\,; \\ &B \xrightarrow{\sigma} 2B\,; \\ &A + B \xrightarrow{\lambda} 2A\,, \end{aligned} \tag{45}$$

where A designate "foxes" and B — "rabbits". Foxes die with probability μ; rabbits proliferate with probability σ, if a fox and a rabbit meet — the rabbit is eaten and another fox is produced with probability λ.

2.3. MEAN–FIELD

The simplest way to treat systems described by Eqs. (44) or (45) is to employ the *mean field* equations. To this end one introduces the mean density of agents, say A, at a lattice site r at time t – $n_A(r,t)$. For the system Eq. (44) with random walk motion (ARW – annihilating random walk) the mean field equation for the density is

$$\frac{\partial n_A}{\partial t} = D\nabla_r^2 n_A - \mu n_A^2\,, \tag{46}$$

where D is the diffusion rate. Derivation of the mean field equation is self–evident from the definition of the model, Eq. (44). We shall see later, that from the perspective of Doi–Peliti technique the mean–field is nothing but equation for a *classical* saddle point. A reader is advised to write down and investigate the mean–field equations for the Lotka–Volterra system, Eq. (45).

The mean–field ignores fluctuations. It works reasonably well, when the number of agents at each site is large. In the opposite limit where the average number of agents on a site is less then unity, the system's behavior may be solely determined by fluctuation effects (especially in low spatial dimensionality). For example, Eq. (46) predicts $n_A \sim t^{-1}$ at $t \to \infty$, while the correct result [29, 30] is $n_A \sim t^{-d/2}$ for $d < 2$ and $n_A \sim t^{-1}$ for $d > 2$. The behavior is clearly very different from the mean–field prediction for dimensionality below the critical one, $d_c = 2$. A traditional approximate way to handle fluctuations (which also stops short to obtain the above mentioned results) is by means of the Focker–Planck or Langevin equations. I shall comment below on the approximations involved in their derivation.

2.4. MASTER EQUATION

The formally exact way to handle the problem is the master equation. One first defines a many–body microscopic configuration α, determined by the (integer) number of agents at each lattice cite

$$\alpha = \{n_1, n_2, \ldots n_M\}\,, \tag{47}$$

where $M = L^d$ is the total number of cites in the d–dimensional lattice. Next one defines $P(\alpha,t)$ – probability to find the system in the many–body configuration α at time t. The master equation has the form

$$\frac{d}{dt}P(\alpha,t) = \sum_\beta \mathcal{R}_{\beta\to\alpha}P(\beta,t) - \sum_\beta \mathcal{R}_{\alpha\to\beta}P(\alpha,t)\,, \tag{48}$$

where the sums run over all many–body configurations β. The first term on the r.h.s. describes "in", while the second one "out" processes. The coefficients $\mathcal{R}_{\beta\to\alpha}$ and $\mathcal{R}_{\alpha\to\beta}$ are the transition rates. A particular case when,

$$\frac{\mathcal{R}_{\beta\to\alpha}}{\mathcal{R}_{\alpha\to\beta}} = \frac{W(\beta)}{W(\alpha)}\,, \tag{49}$$

where W is a certain positive function uniquely defined for each state, is said to satisfy the *detailed balance* condition.

As an example let us consider the binary annihilation on a single site. The microscopic state is specified by a positive integer number of agents, $\alpha = \{n\}$. According to Eq. (44) the master equation takes the form

$$\frac{d}{dt}P(n,t) = \mu\,\frac{(n+2)(n+1)}{2}\,P(n+2,t) - \mu\,\frac{n(n-1)}{2}\,P(n,t)\,. \tag{50}$$

The only non–zero rates are $\mathcal{R}_{n+2\to n}$ and $\mathcal{R}_{n\to n-2}$, which are proportional to the number of possible *pairs* in the corresponding state. These rates do not satisfy the detailed balance.

2.5. QUANTIZATION

The basic idea [6] is to draw the analogy between the master equation and the many–body Schrödinger equation. Indeed, both equations are linear in P (or Ψ) and first order in ∂_t. To make the analogy explicit one introduces the "second quantization". Define the ket–vector $|n\rangle$ as the microscopic state with n–agents. Let us also define vector

$$|\Psi(t)\rangle \equiv \sum_{n=0}^{\infty} P(n,t)\,|n\rangle\,. \tag{51}$$

Note that the weight, P, is probability rather than the amplitude. In these notations the master equation takes the form

$$\begin{aligned} \frac{d}{dt}\,|\Psi(t)\rangle &= \sum_{n=0}^{\infty}\frac{dP(n,t)}{dt}\,|n\rangle \qquad (52)\\ &= \frac{\mu}{2}\sum_{n=0}^{\infty}\Big[(n+2)(n+1)P(n+2,t) - n(n-1)P(n,t)\Big]|n\rangle\,. \end{aligned}$$

The aim now is to write the r.h.s. of this expression as $-\hat{H}|\Psi(t)\rangle$, where $\hat{H}$ is the "Hamiltonian" operator. To this end we introduce the creation and annihilation operators:

$$a^\dagger|n\rangle = |n+1\rangle\,; \tag{53}$$

$$a|n\rangle = n|n-1\rangle\,. \tag{54}$$

As a byproduct, one has $a|0\rangle = 0$ and $|n\rangle = (a^\dagger)^n|0\rangle$. One may immediately check that such operators are "bosonic" [14]

$$[a, a^\dagger] = 1\,. \tag{55}$$

As for any pair of operators satisfying Eq. (55) one may prove the identity

$$e^a f(a, a^\dagger) = f(a, a^\dagger + 1)e^a\,, \tag{56}$$

where f is an arbitrary operator–value function. With these definitions one may check that the Hamiltonian, we are looking for, has the form

$$\hat{H} = \frac{\mu}{2}((a^\dagger)^2 a^2 - a^2)\,, \tag{57}$$

where the first term on the r.h.s. is the "out" and the second one is "in" term. Generalization for the many site problem with the random walk (ARW) is straightforward, the resulting Hamiltonian is

$$\hat{H} = D \sum_{<ij>} (a_i^\dagger - a_j^\dagger)(a_i - a_j) + \frac{\mu}{2} \sum_j ((a_j^\dagger)^2 - 1)a_j^2\,, \tag{58}$$

where $< ij >$ are nearest neighbors.

2.6. OBSERVABLES

So far we succeed to put the master equation into the Schrödinger form

$$\frac{d}{dt} |\Psi(t)\rangle = -\hat{H}\, |\Psi(t)\rangle\,. \tag{59}$$

The differences with the true Schrödinger equation are:

- There is no imaginary unity, meaning that we are dealing with the imaginary time Schrödinger equation.
- $\hat{H}$ is, in general, non–Hermitian (unless the detailed balance condition is satisfied).
- The most important: the mean value of an observable $\hat{O}$ is *not* $\langle\Psi(t)|\hat{O}|\Psi(t)\rangle$. Indeed, such combination is quadratic in P, while the mean value should be linear in probability, P.

To obtain the proper mean value one defines the bra–coherent state

$$\langle\Psi_0| \equiv \langle 0|\, e^a\,. \tag{60}$$

[14] In problems where not more than one agent is allowed on a given site, (e.g. traffic flow models) one may probably employ "fermions". I am not aware, however, of such attempts.

This state has the special property: $\langle\Psi_0|n\rangle = 1$ for any n. This fact may be checked using Eq. (56). Employing this coherent state, one may show that the mean value of an observable may be written as

$$\langle\hat{O}\rangle = \langle\Psi_0|\hat{O}|\Psi(t)\rangle \,. \tag{61}$$

Indeed, let us consider density as an example, $\hat{O} = a^\dagger a$

$$\langle n(t)\rangle = \langle\Psi_0|a^\dagger a|\Psi(t)\rangle = \langle\Psi_0|a|\Psi(t)\rangle = \langle 0|e^a a^\dagger a \sum_n P(n,t)|n\rangle = \sum_n nP(n,t)\,, \tag{62}$$

in agreement with the expectations based on the meaning of $P(n,t)$. The second equality on the r.h.s. is very important; it is an immediate consequence of the coherent state being the left eigenstate of the creation operator, $\langle\Psi_0|a^\dagger = \langle\Psi_0|$.

2.7. NORMALIZATION

To have a consistent theory, one has to worry that the total probability to find the system in some state is unity

$$1 = \sum_n P(n,t) = \langle\Psi_0|\Psi(t)\rangle \,. \tag{63}$$

The last equality is just the mean value of the unit operator. One may solve formally the Schrödinger equation and write $|\Psi(t)\rangle = \exp\{-\hat{H}(a^\dagger, a)t\}|\Psi(0)\rangle$. Eq. (63) is satisfied iff $\langle\Psi_0|\hat{H}(a^\dagger, a) = 0$. Since the coherent state is an eigenstate of the creation operator, $\langle\Psi_0|a^\dagger = \langle\Psi_0|$, one arrives at the conclusion that any legitimate Hamiltonian must obey

$$\hat{H}(a^\dagger = 1, a) = 0 \,. \tag{64}$$

E.g. the Hamiltonian of ARW, Eq. (58), indeed satisfy this condition. To introduce the functional integral representation we'll need to have the normal ordering of creation and annihilation operators. Therefore it is a good idea to commute e^a through the Hamiltonian employing Eq. (56).

$$\langle 0|e^a e^{-\hat{H}(a^\dagger, a)t}|\Psi(0)\rangle = \langle 0|e^{-\hat{H}(a^\dagger+1, a)t}e^a|\Psi(0)\rangle \,. \tag{65}$$

If an observable is considered, the same commutation may be performed provided that the observable is written in terms of annihilation operators only. An example, of this procedure is provided by the second equality in Eq. (62). We shall redefine from now on the Hamiltonian as $H(a^\dagger, a) \equiv \hat{H}(a^\dagger + 1, a)$. In terms of the new Hamiltonian the fundamental normalization condition, Eq. (63), is satisfied iff

$$H(a^\dagger = 0, a) = 0 \,. \tag{66}$$

We shall see momentarily that this condition is the direct analog of the Keldysh normalization condition, Eq. (20).

2.8. FIELD THEORY

Consider the "partition function" of the theory

$$Z \equiv \langle 0|e^{-H(a^\dagger,a)t}|\tilde{\Psi}(0)\rangle \,, \tag{67}$$

where $|\tilde{\Psi}(0)\rangle = e^{a}\,|\Psi(0)\rangle$ is the renormalized initial microscopic state, its precise form is not very important for our further discussion. Due to normalization condition and according to Eqs. (63), (65), $Z = 1$. We divide now the time interval $[0, t]$ onto $N \to \infty$ slices and introduce the coherent state resolution of unity at each time slice. As a result, one obtains in the standard way [19]

$$Z = \int \mathcal{D}\phi^*\phi \, \exp\{-S[\phi^*(t), \phi(t)]\} \,, \tag{68}$$

where

$$S[\phi^*(t), \phi(t)] = \int_0^t dt \left[\phi^* \partial_t \phi + H(\phi^*, \phi)\right] . \tag{69}$$

For example, in the case of ARW one has

$$S[\phi^*(t,r), \phi(t,r)] = \int_0^t dt \int dr \left[\phi^*(\partial_t - D\nabla_r^2)\phi + \frac{\mu}{2}\left((\phi^*)^2\phi^2 + 2\phi^*\phi^2\right)\right] . \tag{70}$$

2.9. ANALOGY WITH THE KELDYSH TECHNIQUE

Notice that both in Keldysh and Doi–Peliti techniques we deal with the "partition function" $Z = 1$. (No matter that the reasons for this normalization are very different: closed time contour in Keldysh versus conservation of probability in Doi–Peliti.) In both cases we succeed to write the partition function as the coherent state functional integral, cf. Eq. (3) and Eq. (68). To this end in both cases we had to *double* number of degrees of freedom: classical and quantum in Keldysh versus ϕ and ϕ^* in Doi–Peliti [15]. The normalization requires the strong condition on an acceptable action: Eq. (20) in Keldysh versus

$$S[\phi^* = 0, \phi] = 0 \tag{71}$$

[15] Notice that since we are interested in *real* density of agents, one may naively expect to obtain a field theory of the single real bosonic field. We have to deal with both ϕ and ϕ^*, however, despite of the fact that the object is purely classic and the phase has no evident meaning.

in Doi–Peliti. This equation is a direct consequence of Eqs. (66) and (69). Comparing Eqs. (20) and (71) one may suspect the analogy

$$\phi^* \sim \phi_q\,; \qquad\qquad \phi \sim \phi_{cl}\,. \tag{72}$$

We shall see below that this analogy indeed works for all intends and purposes. Let us rewrite the quadratic part of the action of the ARW model in the following form

$$S[\phi^*,\phi] = \frac{1}{2}\int dt\,(\phi,\phi^*)\begin{pmatrix} 0 & \overbrace{-\partial_t - D\nabla_r^2}^{A} \\ \underbrace{\partial_t - D\nabla_r^2}_{R} & \underbrace{X}_{K} \end{pmatrix}\begin{pmatrix} \phi \\ \phi^* \end{pmatrix}. \tag{73}$$

Here R and A designate retarded and advanced linear classical operators. Symbol K stays for the Keldysh component of the correlator. In our simplistic example it is absent, $X = 0$. This fact is not generic by any means: in more complicated models e.g. Lotka–Volterra there is non–zero Keldysh component. Moreover, even if it is zero in the bare action – it is usually generated in the process of renormalization. On the other hand, the fact that the ϕ–ϕ element is zero is not accidental. This zero is direct a consequence of Eq. (71) and it is protected against any renormalization.

Once the structure of the quadratic action is understood, one may write down propagators to build the perturbation theory (cf. Eq. (11))

$$\begin{aligned}
\langle\phi(t)\phi^*(t')\rangle &= \mathcal{D}^R(t,t') \sim (Dq^2 - i\omega)^{-1}\,;\\
\langle\phi^*(t)\phi(t')\rangle &= \mathcal{D}^A(t,t') \sim (Dq^2 + i\omega)^{-1}\,;\\
\langle\phi^*(t)\phi^*(t')\rangle &= 0\,;\\
\langle\phi(t)\phi(t')\rangle &= \mathcal{D}^K(t,t')\,.
\end{aligned} \tag{74}$$

As in the Keldysh case, $\mathcal{D}^R(t,t) = \mathcal{D}^A(t,t) = 0$. It is exactly the same causality structure, which is responsible for the conservation of probability (normalization) and the entire internal consistency of the theory. The analogy between the two techniques goes actually much further than the perturbative expansion.

2.10. SADDLE POINT EQUATIONS

Exactly as in the case of Keldysh, the saddle point equation

$$\frac{\delta S}{\delta\phi} = 0 \tag{75}$$

may be always solved by

$$\Phi^* = 0\,. \tag{76}$$

The saddle point solutions are denoted by capital Φ. Under the condition Eq. (76) the second saddle point equation takes the form

$$\frac{\delta S}{\delta \phi^*} = \mathcal{O}^R[\Phi]\Phi = 0\,, \tag{77}$$

where $\mathcal{O}^R[\Phi]$ is the retarded operator. In the example of ARW, Eq. (70)

$$\mathcal{O}^R[\Phi] = \partial_t - D\nabla_r^2 + \mu\Phi \tag{78}$$

and the saddle point equation (77) is nothing but the mean–field equation, Eq. (46). Notice, that the classical (mean–field) equation of motion is a result of variation over the "quantum" (ϕ^*) component (in the limit where the later is zero). In view of Eqs. (71) and (76), the action on any classical solution is zero.

The existence and role of "non–classical" saddle points (with $\Phi^* \neq 0$ and $S_{sp} \neq 0$) in the context of Doi–Peliti technique, to the best of my knowledge, was not yet investigated.

2.11. SEMICLASSICAL APPROXIMATION

To include fluctuation effects in the semiclassical approximation one keeps terms up to the second order in ϕ^* in the action. For the ARW problem the entire action is semiclassical (this is not generic by any means). Let us rewrite it in the following generic semiclassical form

$$S_{scl} = \int dt\,\left[\frac{1}{2}\phi^*[\mathcal{D}^{-1}]^K\phi^* + \phi^*\mathcal{O}^R[\phi]\phi\right]. \tag{79}$$

For the ARW problem $[\mathcal{D}^{-1}]^K = \mu\phi^2 > 0$. Positive definiteness of this quantity is necessary for the convergence of the ϕ^* integral. From this point one may again proceed in two directions:

(i) Since the action Eq. (79) is Gaussian in ϕ^* one may integrate it out and end up with the theory of the single filed, ϕ. The corresponding action is practically identical to Eq. (28). Employing the transfer–matrix method, one arrives at the Focker–Planck equation.

(ii) One may perform the Hubbard–Stratonovich transformation with the auxiliary stochastic field $\xi(t)$ to decouple the term quadratic in ϕ^*. Subsequent integration over ϕ^* results in the functional δ–function. This way one obtains stochastic Langevin equation

$$\mathcal{O}^R[\phi]\phi(t) = i\xi(t)\,, \tag{80}$$

where $\xi(t)$ is Gaussian noise with the correlator

$$\langle\xi^2(t)\rangle = [\mathcal{D}^{-1}]^K\,. \tag{81}$$

Some studies start from postulating Langevin (or Focker–Planck) dynamics for problems at hand. We note that, this approach is not exact, but rather a semiclassical approximation. In some cases, like the ARW, the semiclassical approximation and hence the Langevin dynamics is indeed exact. However, even in such cases the correlator of noise is far from being trivial. Moreover, the noise term in Langevin equation, Eq. (80), contrary to any intuition, happens to be pure imaginary. It is hard to imagine how such dynamics could be postulated from the outset.

2.12. SOURCES

To calculate an observable, e.g. mean density, one need to introduce sources. As was mentioned above, mean value of any observable may be expressed through annihilation operators only, c.f. the second equality in Eq. (62). In the functional integral language it means that pre–exponential factors at hand are given in terms of the "classical" field, ϕ, only. Therefore the source term, one needs, has the form $\int dtV^*(t)\phi(t)$ and the generating functional is given by

$$Z[V^*] = \left\langle e^{\int dtV^*(t)\phi(t)} \right\rangle , \tag{82}$$

where the angular brackets stay for averaging with the action, Eq. (69). With this generating functional one may obtain, e.g., the mean density of agents

$$\langle n(t)\rangle = \langle \Psi_0|a|\Psi(t)\rangle = \int \mathcal{D}\phi^*\phi\, e^{-S[\phi^*(t),\phi(t)]}\phi(t) = \frac{\delta}{\delta V^*(t)} Z[V^*]\Big|_{V^*=0} . \tag{83}$$

Other observables are evaluated in the similar fashion. Note that as in the Keldysh technique, observables are given in terms of the "classical" field, ϕ, while the generating function is functional of the fictions "quantum" source, $V^*(t)$, which is to be taken to zero after differentiation.

2.13. SHORT SUMMARY

The aim of this chapter is not to give a detailed account of a specific calculation employing the Doi–Peliti technique. Lectures and papers of J. Cardy *et al.* [30, 29] are highly recommended for this purpose. My intention was to highlight the similarities and common concepts with the Keldysh techniques. The hope is to facilitate intermixing of the two largely non–intercepting communities. Let me stress once again common points of the two techniques:

- Internal fundamental *normalization*, $Z = 1$.
- Doubling of degrees of freedom: "*classical–quantum*".
- Causality: *retarded–advanced* structure of the action and the propagators.
- Existence of the *classical* extrema: $\Phi_q = 0\,;\ \mathcal{O}^R\Phi_{cl} = 0$.

- Langevin and Focker–Planck approaches as the *semiclassical* approximation — second order in ϕ_q.
- Generating function of *classical* observables is functional of the *quantum* source.

I am indebted to V. Lebedev, A. Andreev, A. Altland and V. Elgart for teaching me various aspects of the techniques and to L. V. Keldysh, M. Doi and L. Peliti for inventing them. This research was supported in part by the BSF grant N 9800338.

References

1. L. V. Keldysh, Zh. Eksp. Teor. Fiz. **47**, 1515 (1964); [Sov. Phys. JETP **20**, 1018 (1965)].
2. J. Schwinger, J. Math. Phys. **2**, 407 (1961).
3. R. P. Feynman and F. L Vernon Jr., Ann. Phys. **24**, 118 (1963).
4. H. W. Wyld, Ann. Phys., **14**, 143 (1961).
5. P. C. Martin, E. D. Siggia, and H. A. Rose, Phys. Rev. **A 8**, 423 (1973); DeDominics, J. Physique (Paris), **37**, C1 (1976).
6. M. Doi, J. Phys. **A 9**, 1465 (1976).
7. L. Peliti, J. Physique **46**, 1469 (1985).
8. E. M. Lifshitz, and L. P. Pitaevskii, *Statistical Physics, part II*, Pergamon Press (1980).
9. G. D. Mahan, *Many–particle physics*, Plenum Press, NY, 1990.
10. J. Rammer, and H. Smith, Rev. Mod. Phys. **58**, 323, (1986).
11. L. S. Levitov, and G. B. Lesovik, JETP Lett., **58**, 230 (1993); L. S. Levitov, H. Lee, and G. B. Lesovik, J. Math. Phys. **37**, 4845 (1996); D. A. Ivanov, H. W. Lee, and L. S. Levitov, Phys. Rev. **B 56**, 6839 (1997).
12. Yu. V. Nazarov, Ann. Phys. (Leipzig), **8**, 507 (1999); W. Belzig, Yu. V. Nazarov, Phys. Rev. Lett. **87**, 067006 (2001); Yu. V. Nazarov, M. Kindermann, cond-mat/0107133.
13. H. Sompolinsky, Phys. Rev. Lett **47**, 935 (1981); H. Sompolinsky, and A. Zippelius, Phys. Rev. **B 25**, 6860 (1982).
14. V. S. Dotsenko, M. V. Feigelman, and L. B. Ioffe, *Spin glasses and related problems*, v. 15, pt. 1 *Soviet scientific reviews* **15** (Harwood Academic, New–York 1990); L. F. Cugliandolo, and G. Lozano, cond-mat/9807138.
15. A. Kamenev, and A. Andreev, Phys. Rev. **B 60**, 2218, (1999).
16. C. Chamon, A. W. W. Ludwig, and C. Nayak, Phys. Rev. **B 60**, 2239, (1999).
17. A. Altland and A. Kamenev, Phys. Rev. Lett., **85** 5615 (2000).
18. A. A. Abrikosov, L. P. Gorkov, I. E. Dzyaloshinski, *Methods of quantum field theory in statistical physics*, Dover, NY, 1963.
19. J.W. Negele and H. Orland, *Quantum Many-Particle Systems*, Adelison-Wesley, 1988.
20. L. B. Ioffe, D. Sherrington, cond-mat/9709270; A. V. Lopatin, L. B. Ioffe, cond-mat/9904219.
21. Sh. M. Kogan, A. Ya. Shulman, Zh. Eksp. Teor. Fiz. **56**, 862 (1969) [Sov. Phys, JETP **29**, 467 (1969)].
22. A. V. Andreev and A. Kamenev, Phys. Rev. Lett. **85**, 1294 (2000); L.S. Levitov, cond-mat/0103617; A. V. Andreev, E. G. Mishchenko, cond-mat/0104211.
23. B. A. Muzykantskii and D. E. Khmelnitskii, Pisma Zh. Eksp. Teor. Fiz. **62**, 68 (1995) [JETP Lett. **62**, 76 (1995)].
24. S. F. Edwards and P. W. Andreson, J. Phys. F. **5**, 89 (1975).
25. K. B. Efetov, Adv. Phys. **32**, 53 (1983); K. B. Efetov, *Supersymmetry in Disorder and Chaos*, Cambridge University Press, 1997.
26. M. V. Feigel'man, A. I. Larkin, M. A. Skvortsov, Phys. Rev. **B 61** , 12361 (2000).

27. D. C. Mattis and M. L. Glasser, Rev. Mod. Phys. **70**, 979 (1998).
28. P. Grassberger and M. Scheunert, Fortschr. Phys. **28**, 547 (1980).
29. B. P. Lee, J. Phys. A **27**, 2633 (1994); B. P. Lee and J. Cardy. Phys. Rev. E **50**, 3287 (1994); B. P. Lee and J. Cardy, J. Stat. Phys. **80**, 971 (1995).
30. J. Cardy, http://www-thphys.physics.ox.ac.uk/users/JohnCardy/home.html
31. A. J. Lotka, Proc. Natl. Acad. Sci. USA **6**, 410 (1920); V. Volterra, *Lecon sur la Theorie Mathematique de la Lutte pour le via.* (Gauthier-Villars, Paris, 1931).
32. E. Bettelheim, O. Agam, and N. M. Shnerb, cond-mat/9908450.

NONLINEAR SIGMA MODEL FOR DISORDERED MEDIA: REPLICA TRICK FOR NON-PERTURBATIVE RESULTS AND INTERACTIONS

IGOR V. LERNER
School of Physics and Astronomy, The University of Birmingham, Edgbaston, Birmingham B15 2TT, UK

1. Introduction

Starting from the seminal papers of Wegner [1] and Efetov [2], a field-theoretical description based on the nonlinear σ model (NLσM) has become one of the main analytical approaches to various problems in disordered electronic systems. The ensemble averaging over all configurations of disorder is performed either using bosonic [1, 3] or fermionic [4] n-replicated fields and taking the $n \to 0$ limit in the results, or using supersymmetric fields [2, 5]. The main advantage of this approach lies in formulating the theory in terms of low lying excitations (diffusion modes) which greatly simplifies perturbative and renormalization group calculations, and also allows a non-perturbative treatment.

The first application of this approach was a derivation [1–4] of the renormalization-group (RG) equations of the scaling theory [6] of Anderson localization. For such a perturbative approach, generalized later for mesoscopic systems [29], both the replica and the supersymmetric methods are equally justified: they ensure the cancellation of unphysical vacuum loops in a diagrammatic expansion.

However, it soon became conventional wisdom that there exist two sets of problems, for each only one of these methods being applicable. The first set of problems can be schematically specified with the following TOE[1] model:

$$H = \sum_i \frac{\hat{p}_i^2}{2m} + \sum_i V_i + \tfrac{1}{2} \sum_{ij} V_{ij} \,. \qquad (1)$$

Here the second term represents schematically both lattice and disorder potential, while the last represents all possible two-particle interactions (including the BCS

[1] TOE stands for "Theory of Everything". In order to be really a TOE model in a condensed-matter context, the Hamiltonian (1) should also include spin terms, as later in this paper in relation to superconductivity. Naturally, either with or without spin terms, there is no hope for a rigorous approach to, let alone the exact solution of, this model for an arbitrary disorder/interaction strength.

I. V. Lerner et al. (eds.),
Strongly Correlated Fermions and Bosons in Low-Dimensional Disordered Systems, 341–373.

one where relevant). To include the interactions, at least at the perturbative and RG level, the original fermionic replica [4] has been generalized by Finkelstein [7]. The interest in this approach has been greatly enhanced by the recent discovery [8] of an apparent metal-insulator transition in 2D disordered systems in zero magnetic field. Naturally, the TOE model and its σ model implementation covers a much wider variety of systems and phenomena than in the scope of the original NLσM [1–4] describing only non-interacting disordered systems.

After a few earlier attempts [9–11], it has been recently demonstrated [12, 13] that the Keldysh technique [14] provides a viable alternative to the replica approach for interacting systems. However, the latter still remains one of the best available tools for consideration of interacting electrons in disordered systems. In the very least, it is clear that there is no simple way of applying the supersymmetry method to a many-particle *fermionic* system, so that the choice is between using Keldysh technique or replica trick.

On the other hand, there exist another set of problems for disordered electron systems where meaningful results are essentially non-perturbative (or at least look like that – see section 3 below). In the absence of interactions, many of them have been solved the help of the supersymmetry method (see, e.g., Refs. [15–20] out of many), while the viability of the replica approach was seriously questioned.

The first, and arguably most famous, of such problems was solved by Efetov [15] who used the supersymmetric NLσM to derive the two-level correlation function (TLCF) in the universal ergodic regime for electrons in disordered metallic grains. The results proved to be identical to those for eigenvalue correlations [21] in random matrix theory (RMT), as had much earlier been conjectured by Gor'kov and Eliashberg [22]. This was the first microscopic derivation of essentially a non-perturbative result for noninteracting electrons in disordered media which has opened the way to numerous new results (again, see Refs. [15–20] out of many – this paper is not a proper venue to give a comprehensive list, let alone even a short review, of them).

In many ways, this first non-trivial "supersymmetric" result [15] seemed to be an excellent illustration of why the replica trick could only be used within a perturbative approach. For the (easiest) case of the unitary symmetry (one of Dyson's symmetry classes corresponding to the absence of the time-reversal symmetry, e.g. due to an external magnetic field) the irreducible TLCF is given by

$$R_2(\omega) = -\frac{\sin^2\omega}{\omega^2}, \tag{2}$$

where ω is the distance between two levels in units of δ/π and δ is the mean level spacing. This result is valid in the ergodic regime, i.e. for $\omega \ll g$ where $g \gg 1$ is the dimensionless conductance. For $\omega \gg 1$, the TLCF averaged over fast oscillations could be readily obtained from the standard diagrammatic techniques [23] with $1/\omega$ being the perturbation parameter (or from the perturbation theory in the framework of either the supersymmetric or replica NLσM). However, the non-perturbative factor $\sin^2\omega$ cannot be restored from the perturbation series. Since the replica trick is well justified only within the perturbative approach, it might seem rather hopeless to obtain the result (2) within the replica approach. And indeed, quite involved calculations by Verbaarschot and Zirnbauer [24] have shown that a direct application of the replica trick (using either the bosonic or fermionic NLσM) has apparently not reproduced the TLCF given by Eq. (2).

In section 2, following Refs. [25–27], I will show that introducing a proper replica-symmetry breaking within the fermionic NLσM leads to the non-perturbative result (2), albeit only in the asymptotic region $\omega \gg 1$. This raises hope that one might eventually apply the replica approach for obtaining *non-perturbative results* for interacting electrons in disordered systems (see Ref. [28] for the first steps in this direction).

The replica-symmetry breaking is not the only way to get (at least seemingly) a non-perturbative result from the replica NLσM. In section 3 I will briefly summarise how quite non-perturbative calculations [18–20] (that could only be performed within the supersymmetric method) have reproduced *part* of the results for long-range asymptotics of different mesoscopic distribution obtained much earlier [29–32] by using a perturbative RG approach valid within any variant of the NLσM.

Then, I will describe in section 4 how to include interactions into a modern variant of the replica NLσM and will use this model in section 5 to describe different approaches to superconductor-insulator transitions in thin disordered films.

2. Replica-symmetry breaking: a way to non-perturbative results?

In this section we consider non-interacting electrons in disordered media – the problem described in a standard way by the first two terms in Hamiltonian (1) (this is equivalent to the Anderson model of disorder). The aim is to reproduce non-perturbative oscillatory contribution to the TLCF of Eq. (2) using the replica NLσM.

Before describing the derivation, it is worth specifying what is meant here by the replica-symmetry breaking. It is the standard fermionic NLσM which does contain correct non-perturbative oscillatory contributions to the TLCF. The replica symmetry is not broken at the level of choosing a correct saddle-point approximation but rather fully exploited by involving a set of additional (in comparison to standard considerations [4]) saddle-point submanifolds within the usual NLσM. The symmetry is broken in a rather natural way when accounting for the contribution of these submanifolds to the TLCF. For the unitary case, such a procedure gives the exact results for all ω, although a real reason for this [33] may be incidental. There is no doubt, however, that it gives the asymptotically ($\omega \gg 1$) correct results for all the three Dyson's symmetry classes.

Note an important analogy: in the supersymmetric NLσM Andreev and Altshuler [34] have reproduced the large-ω limit of Eq. (2) by the saddle-point calculation of the supersymmetric integral that involved an additional saddle-point which "breaks" the supersymmetry in a similar way. The same integral has been exactly calculated by Efetov [15] without any breaking of the supersymmetry.

This analogy will be extended by showing (with the help of a trick resembling the replica-symmetry breaking in the meaning explained above) that the $n = 0$ replica limit of the *exact* integral representation of the TLCF obtained in Ref. [24] for the fermionic replica NLσM leads to the same asymptotically correct oscillatory behavior of Eq. (2).

2.1. FORMULATION OF THE PROBLEM IN TERMS OF NLσM

Consider the two-level correlation function (TLCF) defined by

$$R_2(\omega) = \frac{1}{\nu^2}\left\langle \nu(\varepsilon+\omega)\nu(\varepsilon)\right\rangle - 1\,. \tag{3}$$

Here $\langle \dots \rangle$ stands for the ensemble averaging, i.e. averaging over all the realizations of the impurity potential $V(\mathbf{r})$ in Hamiltonian (1), $\nu(\varepsilon)$ is the electronic density of states per unit volume defined in terms of the spectrum $\{\varepsilon_\alpha\}$ for a given sample as $\nu(\varepsilon) = L^{-d}\sum_\alpha \delta(\varepsilon-\varepsilon_\alpha)$, $\nu \equiv \langle \nu(\varepsilon)\rangle = 1/L^d\delta$, and L, d are the sample size and dimensionality. All energies are measured in units of δ/π, in which the TLCF has the form of Eq. (2). In these units R_2 is expressed via the product of the retarded and advanced Green's functions [35] as follows:

$$R_2(\omega) = \frac{1}{2}\left[\Re\mathrm{e}\, S_2(\omega) - 1\right], \quad S_2(\omega) \equiv \left\langle G^r\!\left(\varepsilon+\frac{\omega}{2}\right) G^a\!\left(\varepsilon-\frac{\omega}{2}\right)\right\rangle . \tag{4}$$

The function S_2 can be expressed in the standard way [15, 36] in terms of a generating functional Z_n

$$S_2(\omega) = -\lim_{n\to 0}\frac{1}{n^2}\frac{\partial^2}{\partial\omega^2}\left\langle Z_n(\omega)\right\rangle, \quad Z_n = \int\! \mathcal{D}\bar\psi\mathcal{D}\psi\, \mathrm{e}^{iS} \tag{5}$$

$$S[\bar\psi,\psi] = \int\! \mathrm{d}\mathbf{r}\, \bar\psi(\mathbf{r})\left[\hat\xi - \left(\frac{\omega}{2}+i\eta\right)\Lambda + V(\mathbf{r})\right]\psi(\mathbf{r}), \quad \eta\to +0. \tag{6}$$

Here $\hat\xi = \hat p^2/2m - \varepsilon_{\rm F}$, $\bar\psi_\sigma$ and ψ_σ are conjugate $2n$-component fermionic (Grassmannian) fields (corresponding to n replica components and 2 retarded-advanced components – further 2 spin components are redundant in this section but will be included in section 4), $\Lambda \equiv \mathrm{diag}(1,-1)_{ra}\otimes \mathbb{1}_n$.[2] The replica trick, $n\to 0$,

[2] The same notation for Λ will be used in all extended spaces introduced later, assuming that it is a dyadic product of Pauli's σ_z in the RA space and of the unit matrices in all other sectors of the extended space.

allows one to write the above expression for the product of Green's functions in terms of derivatives of Z (rather than of $\ln Z$) which makes the ensemble averaging straightforward. Performing it in the assumption that $V(\mathbf{r})$ is a Gaussian δ-correlated random potential with the dispersion $(2\pi\nu\tau_{\rm el})^{-1}$, where $\tau_{\rm el}$ is the mean free time of elastic scattering from impurities, one obtains the following quartic term standing in the action S for $\bar{\psi}V\psi$:

$$S_{dis} = -\frac{1}{4\pi\nu\tau_{\rm el}}\int d\mathbf{r}\ [\bar{\psi}(\mathbf{r})\cdot\psi(\mathbf{r})]\ [\bar{\psi}(\mathbf{r})\cdot\psi(\mathbf{r})] \tag{7}$$

The next step is the standard Hubbard–Stratonovich decoupling that allows one to single out all "slow" modes of electron motion. In the absence of any symmetry breaking (due to an external magnetic field, or magnetic impurities, or spin-orbit coupling), there are two modes, equally contributing to quantum interference effects: the diffuson mode, corresponding to small transferred momenta, and the cooperon mode, corresponding to small sum momenta. It is instructive to perform the decoupling directly in the $\mathbf{r}$ space, thus emphasizing that the diffuson mode corresponds to the normal pairing while the cooperon – to the anomalous pairing:

$$\begin{aligned}\exp\{-S_{dis}\} = \int\frac{\mathcal{D}d}{\mathcal{N}_d}\exp\left\{-\int d\mathbf{r}\left[\frac{\pi\nu}{4\tau_{\rm el}}\mathrm{tr}\ d^2_\omega(\mathbf{r}) - \frac{i}{2\tau_{\rm el}}\bar{\psi}d(\mathbf{r})\psi\right]\right\} + \\ \int\frac{\mathcal{D}c}{\mathcal{N}_c}\exp\left\{-\int d\mathbf{r}\left[\frac{\pi\nu}{4\tau_{\rm el}}\mathrm{tr}\ c^+_\omega(\mathbf{r})c(\mathbf{r}) - \frac{i}{4\tau_{\rm el}}\left[\bar{\psi}c_\omega(\mathbf{r})\bar{\psi} - \psi c^+_\omega(\mathbf{r})\psi\right]\right]\right\}\end{aligned} \tag{8}$$

The diffuson field $d_\omega(\mathbf{r})$ is a matrix field of the same symmetry as $\psi\otimes\bar{\psi}$ while the cooperon field is of the same symmetry as $\bar{\psi}\otimes\psi$. These two fields can be combined into the single $4N\times 4N$ matrix field $\sigma(\mathbf{r})$ of the form

$$\sigma = \begin{pmatrix} d & c \\ c^+ & d^T \end{pmatrix}, \qquad \text{where} \quad d^+ = d,\ c^T = -c. \tag{9}$$

The extra 2×2 "diffuson-cooperon" space, explicit in Eq. (9), is usually referred to as the "charge conjugation" space (sometimes, as the time-reversal space): apart from being Hermitian, the matrix field σ has the charge conjugation symmetry,

$$\sigma = \bar{\sigma} \equiv C\sigma^* C^{-1}, \qquad \bar{\Psi} = (C\Psi)^T \tag{10}$$

where $\bar{\Psi} = \frac{1}{\sqrt{2}}\left(\psi^+, -\psi^T\right)$, $\Psi^T = \frac{1}{\sqrt{2}}\left(\psi^T, \psi^+\right)$, so that the second equation in (10) defines the charge-conjugation matrix C. The disorder functional (8) can be written in such an extended space as follows:

$$\exp\{-S_{dis}\} = \int\frac{\mathcal{D}\sigma}{\mathcal{N}_\sigma}\exp\left\{\int d\mathbf{r}\left[-\frac{\pi\nu}{8\tau_{\rm el}}\mathrm{tr}\,\sigma^2 + \frac{i}{2\tau_{\rm el}}\bar{\Psi}\sigma\Psi\right]\right\} \tag{11}$$

The functional integration measure in Eqs. (8) and (11) is irrelevant in the replica limit. After performing the Gaussian functional integration over the fermionic fields in Eq. (5), one reduces the generating functional Z_n to $Z_n = \left\langle e^{-\mathcal{F}(\sigma)} \right\rangle_\sigma$ with $\langle \ldots \rangle_\sigma$ standing for the functional integration over all the independent components of the field σ, and the "free energy" is

$$\mathcal{F}[\sigma] = \frac{1}{L^d} \int \mathrm{d}^d r \left\{ \frac{\pi\nu}{8\tau_{\mathrm{el}}} \mathrm{tr}\,\sigma^2 - \frac{1}{2} \mathrm{tr}\,\ln \left[\left(\frac{\omega}{2} + i\delta \right) \Lambda - \hat{\xi} + \frac{i}{2\tau_{\mathrm{el}}} \sigma \right] \right\} . \quad (12)$$

A saddle-point condition for this functional, found by varying the field σ is

$$-i\pi\nu\sigma = \langle r | \hat{G} | r \rangle , \qquad \hat{G} \equiv \left[\left(\frac{\omega}{2} + i\delta \right) \Lambda - \hat{\xi} + \frac{i}{2\tau_{\mathrm{el}}} \sigma \right]^{-1} . \quad (13)$$

In the class of spatially-homogeneous fields σ, this condition is obviously satisfied by the field $\sigma = \Lambda$ as can be easily seen by transforming Eq. (13) into the reciprocal momentum space and performing integration over all momenta in the r.h.s. of this equation in the pole approximation, justified provided that

$$\varepsilon_{\mathrm{F}} \tau_{\mathrm{el}} \gg 1 \quad (14)$$

What is usually not stressed is that the saddle-point condition (13) is satisfied in the same approximation not only by Λ but by any spatially-homogeneous diagonal matrix commuting with Λ with eigenvalues equal to ± 1. For most applications (but not for what follows) the existence of this wide class of the saddle-point solutions at $\omega \neq 0$ is irrelevant. The reason is that for $\omega = 0$ Eq. (13) is satisfied by a much wider class of matrices, $\sigma = Q$, where Q satisfies the following constraints[3]

$$Q^2 = 1\!\!1_{2n} , \qquad \mathrm{tr}\, Q = 0 . \quad (15)$$

These constraints are resolved by representing Q as follows:

$$Q = U^\dagger \Lambda U = T^\dagger \Lambda T , \qquad T = \exp \begin{pmatrix} 0_n & t \\ -t^\dagger & 0_n \end{pmatrix} . \quad (16)$$

Since Q obeys the charge conjugation condition (10), U is a symplectic matrix (a unitary matrix whose elements are real quarternions [4]), i.e. $U \in \mathsf{Sp}\,(2n)$. Then T is obtained by factorizing matrices U with respect to redundant matrices $R \in \mathsf{Sp}\,(n) \times \mathsf{Sp}\,(n)$ that commute with Λ, i.e. $U = RT$, which reduces T to the form given in Eq. (16) with t being an arbitrary $n \times n$ real-quarternionic matrix. This means [37] that Q belongs to the *compact* Grassmannian manifold (coset space), $\mathsf{Sp}\,(2n)/\mathsf{Sp}\,(n) \times \mathsf{Sp}\,(n)$. This class of symmetry corresponds to Dyson's orthogonal class in RMT. I will discuss other symmetry classes slightly later.

[3] It is worth stressing that the condition $\mathrm{tr}\, Q = 0$ arises from choosing an equal number of replicas in both retarded and advanced Green's functions G^r and G^a. With this choice, non-zero values of $\mathrm{tr}\, Q$ would correspond to "massive" modes that will not be contributing to the results.

The final step of the NLσM derivation is the expansion of tr ln in Eq. (12) in small ω and ∇Q around a spatially-homogeneous zero-frequency saddle point of Eqs. (15) and (16). To this end, one performs the similarity transformation in the tr ln-term representing it as tr ln $\left\{G_0^{-1} - U[\hat{\xi}, U^\dagger] + \frac{1}{2}\omega U \Lambda U^\dagger\right\}$, where G_0 is the Green's function of Eq. (13) at $\omega = 0$ and $\sigma = \Lambda$. After this, expanding the tr ln to the first power in ω and the second power in the gradient operator $\hat{\xi} \equiv \hat{\mathbf{p}}^2/2m - \varepsilon_{\mathrm{F}} \approx v_{\mathrm{F}} \mathbf{n} \cdot \nabla$, performing the pole integration over ξ, justified in the already used approximation $\varepsilon_{\mathrm{F}} \tau_{\mathrm{el}} \gg 1$, and neglecting in the same approximation tr σ^2 term in Eq. (12), one finally arrives at the non-linear sigma model functional

$$\mathcal{F}[Q;\omega] = \frac{1}{L^d} \int \mathrm{d}^d r \mathrm{Tr} \left[\frac{1}{8} D(\nabla Q)^2 - \frac{i\omega\alpha}{4} \Lambda Q\right], \tag{17}$$

where $D = v_{\mathrm{F}}{}^2 \tau_{\mathrm{el}}/d$ is the diffusion coefficient; at $d = 2$, the dimensionless conductance $g = 2\pi^2 \nu D$.

If time reversal invariance is broken by a magnetic field or magnetic impurities, or spin-rotation symmetry is broken by the spin-orbit interaction, the above functional remains the same but the matrix U in Eq. (16) belongs, respectively, to the unitary group U (n) (Dyson's unitary class) or to the *orthogonal* group, SO (n) (Dyson's symplectic class). Here $\alpha = 1$ for the orthogonal class and $\alpha = 2$ for the unitary and symplectic classes. This factor arises because unitary and symplectic classes have been obtained from orthogonal [4] by the suppression of massive modes corresponding to the time-reversal or spin-rotational symmetry breaking and a subsequent reduction of the Q matrix rank. The coefficient α also absorbs an extra factor in the symplectic case due to the redefinition of the mean level spacing δ in the chosen units.

For simplicity, I consider in the remaining part of this section only the unitary class, referring to the original publication [26, 27] for the other two classes. I will also limit considerations to the ergodic regime corresponding to the level separations much smaller than the Thouless energy, $E_{\mathrm{T}} \sim D/L^2$ (which in the chosen units coincides, up to a numerical factor, with the dimensionless conductance g). In this regime the gradient term in Eq. (17) may be neglected, and the NLσM functional reduces to the zero-dimensional limit: [15]

$$\mathcal{F}[Q;\omega] = -\frac{i\omega}{2} \mathrm{Tr}\ [\Lambda Q]\ , \tag{18}$$

with Q becoming a spatially homogeneous matrix.

2.2. CALCULATING THE ASYMPTOTICS OF THE TLCF

Limiting all further considerations to the ergodic regime only, $Z(\omega)$ given by Eqs. (5) and (18) can be represented as

$$Z_n(\omega) = \int \mathcal{D}Q \exp\left[-i\frac{\omega}{2} \mathrm{Tr}\, \Lambda Q\right], \tag{19}$$

where the measure is defined by

$$\mathcal{D}Q = \prod_{i,j=1}^{n} \mathrm{d}\Omega_{ij}^{ra}\, \mathrm{d}\Omega_{ij}^{ra\,*}\,, \qquad d\Omega \equiv dT \cdot T^{-1}\,. \tag{20}$$

Here T is the matrix parameterizing Q, Eq. (16), and r and a refer to the replica indices which originate from G^r and G^a, respectively. In the large-ω limit this integral is mainly contributed by the extrema of the functional which obey the standard condition $[\Lambda, Q] = 0$. This condition is satisfied by any matrix of the form $Q = \mathrm{diag}(Q^r, Q^a)$, where Q^r and Q^a are the $n \times n$ Hermitian matrices whose eigenvalues are ± 1 and $\mathrm{Tr}\,(Q^r + Q^a) = 0$. This defines a highly degenerate saddle-point manifold which consists of C_{2n}^n submanifolds specified by a particular distribution of n eigenvalues '+1' and n eigenvalues '-1' between Q^r and Q^a. These submanifolds can be divided into $n + 1$ classes of equivalence, $Q_p = \mathrm{diag}(Q_p^r, Q_p^a)$, labeled by $\mathrm{Tr}\,Q_p^r = -\mathrm{Tr}\,Q_p^a = n - 2p$, with $p = 0, 1, \ldots n$. The p-th class has weight $(C_n^p)^2$, with $C_n^p \equiv \binom{n}{p}$.

The matrix Q_p^r with $(Q_p^r)^2 = 1\!\!1_n$ and $\mathrm{Tr}\,Q_p^r = n-2p$ can be parameterized by analogy with Eq. (16) as

$$Q_p^r = (T_p^r)^\dagger \lambda_p T_p^r\,, \quad \lambda_p \equiv \mathrm{diag}(1\!\!1_{n-p}, -1\!\!1_p)\,, \quad T_p^r = \exp\begin{pmatrix} 0_{n-p} & t^r \\ -(t^r)^\dagger & 0_p \end{pmatrix} \tag{21}$$

where t^r is an arbitrary $p \times (n-p)$ matrix. This defines the coset space $\mathsf{G}_p = \mathsf{U}\,(n)/\mathsf{U}\,(n{-}p){\times}\mathsf{U}\,(p)$, i.e. Therefore, $Q_p = \mathrm{diag}(Q_p^r, Q_p^a)$ belongs to the manifold $\mathsf{G}_p \times \mathsf{G}_p$ and can be parameterized as

$$Q_p = T_p^\dagger \Lambda_p T_p\,, \qquad T_p = \mathrm{diag}(T_p^r, T_p^a)\,, \qquad \Lambda_p = \mathrm{diag}(\lambda_p, -\lambda_p) \tag{22}$$

The integer p specifies the replica-symmetry breaking, as it describes the number of the -1 eigenvalues in each Q^r block (equal to the number of the $+1$ eigenvalues in each Q^a block): in the symmetry-unbroken case, $p = 0$, and hence retarded and advanced blocks, $Q^{r,a}$, contain only positive or negative eigenvalues, respectively.

Now one needs to take into account contributions from 'massive' modes (with mass $\propto 1/\omega$, not to be confused with the massive modes $\propto 1/\varepsilon_{\mathrm{F}}\tau$ neglected upon the derivation of the NLσM) in the vicinity of each manifold (22). In the large-ω limit these contributions may be considered as independent and the partition function is then represented by the sum of all of them:

$$Z_n(\omega) = \sum_{p=0}^{n} (C_n^p)^2 \int \mathcal{D}Q \exp\left[-i\frac{\omega}{2}\mathrm{Tr}\,\Lambda Q_p\right], \tag{23}$$

This expression is somewhat symbolic, as $Q_p \equiv U\Lambda_p U^\dagger$, covers the entire symmetric manifold of the NLσM, Eq. (16), including all the massive modes: indeed,

$\mathrm{Tr}\, U\Lambda_p U^\dagger\Lambda \to \mathrm{Tr}\,\Lambda_p T^\dagger\Lambda T = \mathrm{Tr}\,\Lambda_p Q$, where we have substituted $U = RT$, as defined after Eq. (16). This can be justified only as a perturbative (in $1/\omega$) procedure: a possible overlapping of massive modes originated from different manifolds is irrelevant in the large-ω limit, and each of the integrals in the sum (23) can be calculated independently of the others.

Each term in the sum (23) contains both massive and massless modes. Indeed, we have used above the factorization $U = RT$ with R being block-diagonal matrices commuting with Λ. The matrices T in $\mathrm{Tr}\,\Lambda_p Q = \mathrm{Tr}\,\Lambda T\Lambda_p T^\dagger$ still contain the subset of matrices commuting with Λ_p that correspond to the massless modes. Therefore, we need to parameterize T in a way which enables us to factorize out these massless modes and perform the integration over the massive ones.

The most suitable parameterization of T, analogous to that used in Ref. [24] for the bosonic NLσM, can be obtained by expanding the matrix exponent in Eq. (16). By introducing matrix $B \equiv t(t^\dagger t)^{-1/2}\sin\sqrt{t^\dagger t}$, we represent T and thus $Q = T^\dagger\Lambda T$ as follows:

$$T = \begin{pmatrix} \mathcal{R}(BB^\dagger) & B \\ -B^\dagger & \mathcal{R}(B^\dagger B) \end{pmatrix}, \quad Q = \begin{pmatrix} \mathbb{1}_n - 2BB^\dagger & B\mathcal{R}(B^\dagger B) \\ B^\dagger\mathcal{R}(BB^\dagger) & -(\mathbb{1}_n - 2B^\dagger B) \end{pmatrix}, \tag{24}$$

where $\mathcal{R}(X) \equiv \sqrt{\mathrm{I}_n - X}$. The matrix B in this parameterization is not unconstrained, though. The $Q = Q^\dagger$ condition is fulfilled only when the matrices $\mathcal{R}(BB^\dagger)$ and $\mathcal{R}(B^\dagger B)$ are Hermitian. This is so only when all the eigenvalues of $BB^\dagger$ and $B^\dagger B$ do not exceed unity. Only under this constraint does Q, parameterized as in Eq.24, still belong to the coset space $\mathsf{U}(2n)/\mathsf{U}(n)\times\mathsf{U}(n)$. Nevertheless, this parameterization is very convenient. First, the corresponding Jacobian is equal to one [27] so that the measure of integration (20) can be written as

$$\mathcal{D}Q = \prod_{i,j} \mathrm{d}B_{ij}\mathrm{d}B^*_{ij} \equiv \mathcal{D}B\,. \tag{25}$$

In addition, the representation of all the exponents in the sum (23) in terms of B is also very simple, $\mathrm{Tr}\,\Lambda_p Q = 2(n-2p) - 2\mathrm{Tr}\,\lambda_p(BB^\dagger + B^\dagger B)$, so that we obtain:

$$Z_n(\omega) = \sum_{p=0}^{n} (C_n^p)^2 \cdot \mathrm{e}^{i\omega(2p-n)}\, Z_n^p(\omega)\,, \tag{26}$$

$$Z_n^p(\omega) = \int \mathcal{D}B \exp\left[i\omega \mathrm{Tr}\,\lambda_p(BB^\dagger + B^\dagger B)\right]. \tag{27}$$

The region of integration in (27) is restricted by the constraint described after Eq. (24). Last, but not least, the parameterization (24) allows one to separate out the massless modes, which obey the condition $[T, \Lambda_p] = 0$, in each integral (27). Indeed, this condition is satisfied by all matrices T constructed from B which anticommute with λ_p, i.e. have the off-diagonal block structure.

This means that in the representation of B in the block form reflecting the structure of $\lambda_p = \mathrm{diag}(1\!\!1_{n-p}, -1\!\!1_p)$,

$$B = \begin{pmatrix} B_1 & b_1 \\ b_2^\dagger & B_2 \end{pmatrix}, \tag{28}$$

the matrices $B_{1,2}$ represent the massive modes, and $b_{1,2}$ massless. When the massive modes are suppressed ($B_1 = 0$ and $B_2 = 0$), the T matrices in Eq. (24) constructed from $p \times (n-p)$ matrices $b_{1,2}$ only, parameterize the same degenerate p-the manifold, $\mathsf{G}_p \times \mathsf{G}_p$ described in Eq. (22), as one expects.

By substituting the representation (28) into Eq. (27), we reduce Z_n^p to the product of integrals over the massive and massless modes:

$$Z_n^p(\omega) = \int \mathcal{D}B_1\, \mathcal{D}B_2 \exp\left[-2i\omega \mathrm{tr}\,(B_1 B_1^\dagger - B_2 B_2^\dagger)\right] \int \mathcal{D}b_1 \mathcal{D}b_2, \tag{29}$$

Here the region of integration over $b_{1,2}$ depends on $B_{1,2}$ due to the constraint on the eigenvalues of the matrices $BB^\dagger$ and $B^\dagger B$ in the representation (28). Since the integral over $B_{1,2}$ is contributed only by the region where both $\mathrm{tr}\, B_1 B_1^\dagger$ and $\mathrm{tr}\, B_2 B_2^\dagger \lesssim 1/\omega \ll 1$, in the leading in $1/\omega$ approximation we may put both $B_{1,2}$ to 0 in the constraint of the integration region over the massless modes $b_{1,2}$. In this approximation, as we have noticed after Eq. (28), matrices $b_{1,2}$ parameterize the p-th manifold (22) so that

$$\int \mathcal{D}b \equiv \int \mathcal{D}b_1 \mathcal{D}b_2 = \int \mathcal{D}Q_p = \Omega^2(\mathsf{G}_p)\,, \tag{30}$$

where the measure of integration over $\mathcal{D}Q_p$ is defined in terms of T_p in the same way that $\mathcal{D}Q$ is defined in terms of T, Eq. (20), and $\Omega(\mathsf{G}_p)$ is the volume of the compact coset space G_p. This volume is expressed via the well-known volumes of the unitary group, $\Omega(\mathsf{U}\,(n))$, as follows:

$$\begin{aligned} \Omega(\mathsf{G}_p) &= \frac{\Omega(\mathsf{U}\,(n))}{\Omega(\mathsf{U}\,(n-p))\Omega(\mathsf{U}\,(p))} \\ &= (2\pi)^{\frac{1}{2}[n^2-(n-p)^2-p^2]} \prod_{j=1}^{p} \frac{\Gamma(1+j)}{\Gamma(n+2-j)}. \end{aligned} \tag{31}$$

In the same large-ω approximation, the variables $B_{1,2}$ parameterizing the massive modes are unconstrained. Then the Gaussian integral over the $2[(n-p)^2 + p^2]$ independent massive modes yields

$$\tilde{Z}_n^p(\omega) \equiv \int \mathcal{D}B \exp\left[2i\omega \mathrm{tr}\,(B_1 B_1^\dagger - B_2 B_2^\dagger)\right] = \left(\frac{\pi}{-i\omega}\right)^{(n-p)^2} \left(\frac{\pi}{i\omega}\right)^{p^2} \tag{32}$$

Combining Eqs. (32) and (31) and omitting an irrelevant overall factor which goes to 1 when $n \to 0$, we arrive at the following expression which is essentially the same as that derived in Ref. [26] via the Itzykson-Zuber integral:

$$Z_n(\omega) = \sum_{p=0}^{\infty} [F_n^p]^2 \cdot \frac{e^{i\omega(2p-n)}}{(2\omega)^{(n-p)^2+p^2}}, \qquad F_n^p \equiv C_n^p \prod_{j=1}^{p} \frac{\Gamma(1+j)}{\Gamma(n+2-j)}. \quad (33)$$

Here the summation over p has been extended to ∞ since $F_n^p = 0$ for all integer $n > p$. This allows one to take the replica limit, $n \to 0$, in each of the terms in Eq. (33). Due to the fact that $F_n^p \propto n^p$ as $n \to 0$, only the terms with $p = 0$ and $p = 1$ in Eq. (33) contribute to S_2 in Eq. (5).

Let us stress that the replica symmetry is broken only now, in the $n \to 0$ limit. Indeed, for any integer $n \neq 0$ contributions of the terms with p and $n - p$ are complex conjugate to each other, but for $n \to 0$ we no longer treat them on equal footing. Thus, the result for $Z_n(\omega)$ below is no longer a real function. Note, however, that in order to treat $S_2(\omega)$ for all ω one should imply the $\omega \to \omega + i\delta$ substitution which results in $Z_n(\omega)$ being no longer a real function for any n.

Omitting all the terms with $p \geq 2$, one obtains

$$Z_n(\omega) = \frac{e^{-i\omega n}}{\omega^{n^2}} + n^2 \frac{e^{i\omega(2-n)}}{4\omega^{(n-1)^2+1}}. \quad (34)$$

Substituting this into Eqs. (5) and (4) and keeping the leading in $1/\omega$ terms only, one arrives at the expression (2) for the TLCF. Although this expression coincides with the exact one (implying the above mentioned substitution $\omega \to \omega + i\delta$), it has been actually derived only in the large-ω limit, as is the case of the 'supersymmetry breaking' method of Andreev and Altshuler [34].

2.3. THE LARGE-ω LIMIT OF THE VERBAARSCHOT-ZIRNBAUER INTEGRAL

In order to obtain an explicit multiple-integral representation of the 'zero-mode' partition function, Eq. (19), one uses the following 'polar' decomposition of Q:

$$Q = \begin{pmatrix} u_1^+ & \\ & u_2^+ \end{pmatrix} \begin{pmatrix} \lambda & \sqrt{1-\lambda^2}e^{i\phi} \\ \sqrt{1-\lambda^2}e^{-i\phi} & -\lambda \end{pmatrix} \begin{pmatrix} u_1 & \\ & u_2 \end{pmatrix},$$

where $-1 \leq \lambda_i \leq 1$ and $u_{1,2}$ are unitary matrices. The appropriate measure of integration is given by

$$\mathrm{D}Q = \prod_{i<j} (\lambda_i - \lambda_j)^2 \prod_i \mathrm{d}\lambda_i \, \mathrm{d}\phi_i \, \mathrm{d}\mu(u_1) \, \mathrm{d}\mu(u_2).$$

The action in Eq. (19) depends only on λ, and the integrations over $\mathrm{d}\mu(u)$ give the volumes of the appropriate unitary group. Thus one obtains

$$Z_n(\omega) = \Omega^2(\mathsf{U}(n)) \int_{-1}^{+1} \delta^2(\lambda) \prod_{i=1}^{n} e^{-i\omega\lambda_i} \mathrm{d}\lambda_i, \qquad \delta(\lambda) \equiv \prod_{i<j} (\lambda_j - \lambda_i). \quad (35)$$

This is equivalent (with accuracy up to factors going to 1 in the $n \to 0$ limit) to the representation for S_2 given in Eq. (2.24) of the paper by Verbaarschot and Zirnbauer, [24] which has been used for the critique of the replica trick. We will show that it leads, in the very least, to the exact large-ω assymptotic behavior of $S_2(\omega)$.

The leading in $1/\omega$ contributions to this highly oscillatory integral (which does not have stationary points inside integration region) come from the end points. To single out these contributions, we must take some λ's close to $+1$ and the rest close to -1, which imitates replica symmetry breaking. Let us choose $n-p$ of λ's close to $+1$ and p of λ's close to -1. Then we can split up the Vandermonde determinant in the following way:

$$\delta^2(\lambda) = \prod_{i,j}^{n} |\lambda_j - \lambda_i| \approx 2^{2p(n-p)} \Delta_+^2 \, \Delta_-^2$$

where

$$\Delta_+^2 = \prod_{i,j=1}^{n-p} |\lambda_i - \lambda_j| \quad \Delta_-^2 = \prod_{i,j=n-p+1}^{n} |\lambda_i - \lambda_j|. \tag{36}$$

Reducing the integral (35) to the sum of such contributions, we represent it as

$$Z_n(\omega) \approx \Omega^2(\mathsf{U}\,(n)) \sum_{p=1}^{n} (C_n^p)\, 2^{2p(n-p)} \int_{-\infty}^{+1} \Delta_+^2 \prod_{j=1}^{n-p} \mathrm{e}^{-i\omega\lambda_j}\, \mathrm{d}\lambda_j$$

$$\times \int_{-1}^{+\infty} \Delta_-^2 \prod_{j=n-p+1}^{n} \mathrm{e}^{-i\omega\lambda_j}\, \mathrm{d}\lambda_j\,. \tag{37}$$

Since in each of the integrals all the variables are close to one of the limits of integration, the second limit was extended to infinity. Now we make substitutions $\lambda_i = 1 - x_i$ in the first integral, and $\lambda_i = -1 + x_i$ in the second one, reducing the above sum to the form:

$$Z_n(\omega) \approx \Omega^2(\mathsf{U}\,(n)) \sum_{p=1}^{n} (C_n^p)\, 2^{2p(n-p)} \mathrm{e}^{i\omega(2p-n)} I_{n-p} I_p \tag{38}$$

where I_p are integrals of Selberg's type: [21]

$$I_p = \int_0^\infty \Delta^2(x) \prod_{j=1}^{p} \mathrm{d}x_j \mathrm{e}^{-i\omega x_j} \tag{39}$$

Substituting the known Selberg integrals and discarding an overall factor which goes to unity in the replica limit we arrive at

$$Z_n(\omega) = \sum_{p=0}^{n} [F_n^p]^2 \cdot \frac{\mathrm{e}^{i\omega(2p-n)}}{2\omega^{(n-p)^2+p^2}}, \quad F_n^p = C_n^p \prod_{j=1}^{p} \frac{\Gamma(1+j)}{\Gamma(n+2-j)}. \tag{40}$$

This expression is exactly the same as Eq. 33 obtained in section 2.2 with the help of the replica-symmetry breaking. Therefore, the exact representation (35) *does contain* the true oscillatory asymptotic behavior of the TLCF.

The authors of Ref. [24] have also drawn attention to the fact that there is an apparent contradiction between the $\omega = 0$ limit for S_2 obtained from the replica trick and the exact supersymmetric result. Indeed, if ω is put to 0 in the expression for S_2 following from Eqs. (35) and (5) the $n \to 0$ limit is taken *after that*, one obtains $S_2(\omega=0) = -1$. This cannot be correct as $\Re e\, S_2(\omega \to 0) > 0$ as follows from the definition (4) and, moreover, it is known that $S_2(\omega \to 0) \to \delta(\omega)$. What is interesting, however, is that if one separates the singular, $S_2^{\rm sing}(\omega)$, and regular, $S_2^{\rm reg}(\omega)$, parts of the exact $S_2(\omega)$, then $S_2^{\rm reg}(\omega \to 0) = -1$. Therefore, the replica method gives $S_2^{\rm reg}(\omega=0)$ correctly, and it is just $S_2^{\rm sing}$ which is missing. However, the fact that $Z_n(\omega \to 0)$ is finite for any integer n does not necessarily implies that it is also finite (as a function of $\omega \to 0$) in the replica limit. For example, if the expansion of $S_2(\omega)$ *before* taking the $n = 0$ limit contained a term proportional to ω^{n^2-1}, it would be singular in the replica limit. In other words, if a non-trivial dependence on the order of limits $n \to 0$ and $\omega \to 0$ existed, in a spirit of the replica trick the $n \to 0$ limit should be taken first. At the moment, though, this remains only a speculation. However, the fact that the large-ω limit of this integral reproduces the correct results (2) makes it plausible that there are only technical difficulties rather than one of principle in the application of the replica method.

2.4. THE SMALL ω LIMIT

The purpose of this section was to demonstrate explicitly that non-perturbative oscillatory contributions to the TLCF of electrons in a random potential could be extracted from the standard NLσM formulated in fermionic replicas[4] many years ago. To this end, all one needs is to parameterize all the non-trivial saddle-point manifolds corresponding to the broken replica-symmetry and describing 'massless modes' of the theory, and expand the action in the vicinity of these manifolds to include 'massive modes'. The very similar approach has been used in the supersymmetric NLσM: the non-perturbative oscillations have been extracted by the expansion around two extremal points one of which breaks the supersymmetry [34]. Since the exact supersymmetric calculation of the TLCF was well known [15], it was clear that the supersymmetry breaking [34] was just a convenient method of extracting the large-ω limit (and going beyond the universal 'zero-mode' approximation). It has also shown in section 2.3 that the exact integral representation of $R_2(\omega)$ does contain the correct behavior in the large-ω limit.

The small ω limit is, arguably, more interesting as it governs a number of non-perturbative results obtained within the supersymmetric NLσM. I will briefly summarise some of them [18–20] in the following section, emphasizing that they have reproduced some of the results obtained much earlier [29–32] by using a perturbative RG approach valid within any variant of the NLσM.

3. Tails of distribution functions

All extensive physical characteristics of disordered systems fluctuate from sample to sample. However, only in mid-eighties it was understood that the scale of such fluctuations is governed by the quantum coherence length so that at $T \to 0$ they are not reduced with increasing the sample size. Immediately after the discovery [38, 39] of the universal conductance fluctuations (UCF), it was shown [29] that their distribution function is Gaussian in its bulk, but has long lognormal tails. The existence of such tails appears to be a common feature of distribution functions of many observable quantities [29–32] such as global and local density of states, different relaxation times, etc. With increasing the disorder, the part of these tails in distribution functions was increasing so that the entire distribution of any local quantity was becoming lognormal in the critical regime in the vicinity of the metal-insulator transition [31]. The characteristic feature of the lognormal distribution is that the logarithm of its sth moment is proportional to $s(s-1)\ln L$ where L is the coherence length (at zero T, this is just a sample size). Therefore, all the moments (or in the limit of weak disorder, where only the tails are lognormal, all the high moments) scale with different exponents, i.e. system shows multifractal behaviour.

The results for the tails of the distribution functions that had been originally obtained [29–32] by the RG treatment of an extended (in a way described below) NLσM, have later been reproduced [18–20] in a much more elegant way in the framework of the standard supersymmetric σ model. I do not intend to describe this approach here but only want to stress that some of the steps seem to be absolutely impossible within the replica treatment. The Q field in the supersymmetric model belongs to a "supermanifold" that includes both compact and a noncompact sectors. The existence of the noncompact sector is absolutely crucial for finding the distribution tails. With properly imposed boundary conditions required for finding the tails, $Q = \Lambda$ does no longer represents the saddle-point. A correct saddle-point equation reduces, e.g., to the Liouville equation [19] for a single scalar parameter that parameterises the (only relevant) noncompact sector of the space of the field Q (instead of, e.g., the parameterisation (24) in terms of the unconstrained matrix B or any other matrix parameterisation required in the replica σ model). A non-perturbative solution to this equation provides for the existence of lognormal tails of the local density of states distribution [20] or, equivalently, for the multifractality of the wave function [19]. Had such a solution been found before the RG solution of Refs. [29–32], it would be taken as yet another 'example' of the incapacity of the replica method. However, although the replica method works perfectly within the RG approach, there is still no clear understanding of why the results obtained within the non-perturbative approach outlined above are exactly the same as those within the RG approach to the extended NLσM (in the limit of the weak disorder, to which the validity of the non-perturbative supersymmetric approach is limited).

3.1. LOCAL DENSITY OF STATES IN OPEN SYSTEMS

The two-level correlation function considered above is not well defined for $\omega \lesssim g$ for an open system, as level widths become of the order of Thouless energy, $E_T = \hbar/\tau_{erg}$ which is much bigger than the level mean spacing δ (their ratio is of order $g \gg 1$ in the metallic limit). Its direct analog in this regime is the variance of the density of states (DoS) at the Fermi energy. It is useful to consider also the entire distribution of DoS which can be found in terms of its irreducible moments (cumulants). The relative values of the DoS cumulants can be expressed via the effective field-theoretical functional in a way similar to that in Eqs. (5)–(6):

$$R_s \equiv \frac{\langle\langle \nu^s \rangle\rangle}{\langle \nu \rangle^s} = \lim_{n\to 0} \frac{(i)^{2s}}{(2n^2)^s} \frac{\partial^{2s}}{\partial \omega^{2s}} \langle\langle Z_n(\omega) \rangle\rangle \ . \tag{41}$$

The cumulants of local DoS are given by the same expression with partial derivatives substituted by variational ones, ω being considered in this case as a spatial-dependent source field in Eq. (6). As the higher derivatives with respect to ω are involved in the above expression, the Tr ln in Eq. (12) should now be expanded to the higher powers of ω. This leads to the following additional contribution [32] to the σ model functional (17):

$$\mathcal{F}_{\rm add}[Q;\omega] = -\frac{i\alpha}{4} \sum_{s=1}^{\infty} \Gamma_s {\rm Tr}\, [\omega \Lambda Q]^{s+1} \ , \quad \Gamma_{s0} = \left(\frac{i\tau_{\rm el}\delta}{2\pi}\right)^s \frac{(2s-1)!!}{(s-1)!} \ . \tag{42}$$

As the bare value, Γ_{s0}, of the additional "charge" is proportional to the $s^{\rm th}$ power of the small parameter $\tau_{\rm el}\delta \sim (\lambda_F)^{d-1}\ell/L^d$, their direct contribution to the DoS cumulants is negligible compared to that obtained by the repeated differentiation (41) of the ω term in the standard σ model.[4]

However, the charges Γ_s sharply increase under RG transformations [32]. While "normal" contribution to the DoS cumulants, i.e. those from the standard σ model, Eq. (17), scale with the system size as appropriate powers of the DoS variance, the additional contribution from the vertices (42) show the multifractal scaling due to this RG increase of Γ_s. For Dyson's orthogonal class, it reads

$$R_s^{\rm add} \propto g_0^{1-s} \Gamma_{s0}\, {\rm e}^{us(s-1)} \ , \quad u \equiv \ln \frac{\sigma_0}{\sigma} \approx g_0^{-1} \ln \frac{L}{\ell} \ , \tag{43}$$

[4] In calculating the local DoS cumulants, this small parameter is compensated by the appearance of powers of $\delta(\mathbf{r}=0) \to \pi\nu/\tau_{\rm el}$ so that the contribution of the additional terms to the variance and higher cumulants of the local DoS is comparable [31] to that of the standard σ model.

where σ is the conductivity at the scale of the system size L and σ_0 is its bare value, that is the conductivity at the scale of the mean free path ℓ. As the small parameter Γ_{s0} scale as a linear power of s, for large enough s the $e^{us(s-1)}$ factor becomes dominant.

The additional contribution to the sth cumulant of the local DoS differs from that in Eq. (43) by the absence of the small parameter Γ_{s0}. Such a contribution becomes dominant for $s \gtrsim u^{-1}$ (or for $s \gtrsim g_0$ in the weak disorder limit). The approximate equality in Eq. (43) refers to the weak disorder limit. In this limit, the result for the multifractal cumulants (43) has been reproduced within the supersymmetric approach [19, 20] outlined above. Such a behaviour of the high cumulants leads to the lognormal tails of the distribution function. With increasing the disorder, u becomes of order 1 in the vicinity of the metal-insulator transition in $d > 2$, or for the region of strong localization in $d = 2$ dimensions. In this region, the entire distribution becomes lognormal. The physical meaning of this is that there exist rare realisations of disorder which gives values of the local DoS (or, equivalently, local values of the wave function amplitude $|\Psi(\mathbf{r})|^2$) which are much higher than the average value. The importance of this untypical realisations greatly increases with increasing the disorder so that they correspond to prelocalized states.

3.2. IS THERE A PARADOX?

The properties of tails of different distributions governed by the existence of the prelocalized states are described in detail in reviews [32, 40]. The point of this presentation is to underline certain peculiarities of the replica method. To this end, let me first outline how the RG results which lead to Eq. (43) have been obtained. The composite vertices (42) are not closed under the RG transformations of the σ model given by Eqs. (17) and (42). The index-permutation symmetry lacking in Eq. (42) is restored under RG transformations via generating the following additional vertices in each sth order:

$$\omega^{s+1}\prod_{l=1}^{\infty}\left[\mathrm{Tr}\,(\Lambda Q)^l\right]^{s_l}\,,\qquad \sum_{l=1}^{\infty}s_l = s+1\,. \tag{44}$$

Thus in each sth order the RG equations become matrix equations, the matrix rank being equal to the number of partitions of the integer s into the sum of positive integers. These equations are exactly solvable [31] for any integer n, and keeping only the largest eigenvalue leads in the replica limit $n \to 0$ to the result (43). As the RG approach is purely perturbative,[5] the applicability of the replica method is beyond doubt. However, there exist some apparent contradiction.

[5] Eq. (43) has been obtained as a result of the one-loop RG analysis. The higher-loop contributions limit its validity to $s \lesssim g_0^{3/2}$. Such a limitation ensures that no moment will be proportional to a positive power of the system volume, L^d, i.e. all the multifractal dimensionalities remain

The point is that one can choose different number of replicas for to represent the advanced and retarded sectors. In the simplest unitary case the Q matrix would belong to the coset space $\mathsf{U}(n)/\mathsf{U}(n-m)\times\mathsf{U}(m)$ where n is the total number of replicas, and m is the difference in their numbers in the advanced and retarded sectors. In the replica limit, $n\to 0$ and $m\to 0$, the result is naturally the same as before. However, the largest eigenvalue of the RG equations happens to depend only on n but not on m. Then, instead of taking the replica limit, one could fix $m = 1$ and take an arbitrary integer n. The resulting model is defined on the coset space $\mathsf{SU}(n)/\mathsf{U}(n-1)$ isomorphic to sphere, i.e. in this limit the model becomes equivalent to the n-vector model. The composite vertices (42) reduce in this case simply to $\cos^s\varphi$ where φ is the angle between the n-field and the direction of an external (magnetic) field. It is well known (see, e.g., Ref. [41]) that such operators are irrelevant in the n-field theory, i.e. they only decrease under the RG transformations, in contrast to those of Eq. (42). So here is the contradiction as they should correspond, as described above, to a particular case of the matrix operators. It is clear, though, that a direct correspondence is anyway impossible, as the rank of the Q-matrices in Eq. (42) and thus the number of the eigenvalues in the RG equations depends on m and in the symmetric case $m=0$ it is equal to the number of partitions of the integer s into the sum of positive integers, i.e. much larger than in the case $m=1$ when it is just equal to s.

The resolution of the apparent paradox is in the fact that one should look only at correlation functions (observables) rather than at the RG dimensions of the composite vertices. In particular, only vertices $[\operatorname{Tr}\Lambda Q]^s$ out of all the variety in Eq. (44) contribute to the cumulants (41), while all the rest serve to give correct eigenvalues. In calculating such a contribution, each eigenvalue enters with a coefficient of proportionality which is a polynomial of n and m. It appears that all the polynomials attached to "wrong" (in the case $m = 1$) eigenvalues vanish at all integer values of n thus restoring the correspondence between the matrix and the standard treatments of the n-model. Simultaneously, the $n = m = 0$ replica limit gives the results (43) discussed above. On the face of it, it looks similar to the "replica symmetry breaking" discussed in the previous section: there exists a set of coefficients vanishing at all integer n and giving a nonzero contribution at $n = 0$. However, all the results discussed here have been obtained within the perturbative RG approach, and thus no trick was involved – the replica limit here is just to cancel unwanted vacuum loops in the RG diagrams. All these RG results can be equally easy obtained within the supersymmetric σ model.

One question that remains, though, is the following. The multifractal cumulants (43) have been obtained here within the *extended* σ model, Eqs. (17) and (42), while later they have been reproduced within the standard SUSY σ model. So, does one really need the additional vertices (42)?

---

positive. A similar restriction in the supersymmetric method of Ref. [19] follows directly from the inapplicability of the standard σ model description at ballistic scales.

3.3. HIGH GRADIENT VERTICES: RG OR NOT RG?

The answer to this question is in getting – intermediately – even more vertices. Those in Eq. (42) have been obtained in expanding the Tr ln term in Eq. (12) in higher powers of ω. But it could be also expanded in higher powers of ∇Q which yields the following set of vertices [42]:

$$\mathcal{F}_s[Q] \equiv \gamma_s \varepsilon_{\alpha_1 \ldots \alpha_{2s}} \int \operatorname{Tr} \left(\prod_{i=1}^{2s} \partial_{\alpha_i} Q \right) \mathrm{d}^d r \,. \tag{45}$$

The renormalization of these vertices results in the same increase of the largest eigenvalue as in Eq. (43), and actually, taken together with the ω-vertex in the standard σ model of Eq. (17), they contribute to the cumulants, Eq. (43), in the same way (but for unknown – in any method – pre-exponential factors) as the composite vertices (42). Therefore, the latter are not even necessary to obtain the multifractality.

An important point now is that, although the higher-gradient vertices (45) have been obtained from the higher-order expansion of the Tr ln term in Eq. (12), they are also generated in performing the RG transformation of the standard NLσM. Usually they are omitted as being naïvely irrelevant. However, they become relevant at the one-loop level and actually govern long-tail asymptotics of different distributions. Note that in contrast to the composite vertices (42), the high-gradient operators remain relevant (in the same sense) also for the n vector model [43].

Now the possible relation between the RG approach outlined here and the supersymmetric approach of Refs. [19, 20] is the following. The boundary conditions imposed to find correctly the tails of the distributions in the latter approach makes the saddle-point solution spatially inhomogeneous, while in the former approach the relevance of the high gradients may indicate that such a spatial inhomogeneity develops spontaneously when starting from the standard homogeneous saddle-point.

Finally, there is a possibility that the role of the high-gradient operators (45) is considerably more crucial. In the perturbative RG approach, they have only contributed to the higher moments of conductance, DoS, etc, since the structure of the RG equations turns out to be triangular, i.e. the higher order gradient terms do not contribute to the renormalization of the lower order terms. However, in contrast to the composite operators (42), the operators (45) do not break any symmetry of the NLσM. Therefore, one cannot exclude a possibility of a non-perturbative contribution from these operators to, say, the average conductance. Had such a contribution existed, it would totally ruin the one-parameter RG for the average conductance, and this the one-parameter scaling description of the Anderson localization. At the moment, such a possibility remains purely speculative.

4. Coulomb and pairing interactions in the sigma model

The main point of the previous sections is that the replica σ model works perfectly for perturbative problems and – in certain cases – does reasonably well even for certain non-perturbative problems within the non-interacting model. Since the inclusion of interactions beyond the mean-field approach is impossible within the supersymmetric method, the replica model the most reliable tool for models with interactions. The original fermionic replica σ model [4] has been generalized to include the interactions by Finkelstein [7]. It allowed him to reproduce earlier perturbative results (reviewed, e.g., in Ref. [44]), to account where necessary for the Landau Fermi-liquid constants, and to derive the renormalization group equations describing (at least at a qualitative level) a metal–insulator transition in disordered interacting systems. This model has also been successfully extended [45] to allow for effects of Coulomb interaction in superconducting systems (like lowering the transition temperature T_c by disorder – in an apparent, but well understood, deviation from the Anderson theorem [46]). The interest in this approach has been greatly enhanced by the recent discovery [8] of an apparent metal-insulator transition in 2D disordered systems in zero magnetic field. Although it is not at all clear whether the observed effects are, indeed, due the transition [47, 48], and if they are – whether such a transition is, indeed, driven by interactions, a possibility of having the transition in a 2D disordered interacting system (schematically described by the TOE model (1)), in contrast to a disordered non-interacting system, is intriguing by itself. Such a possibility is, undoubtedly, a driving force in a considerable revival of interest in Finkelstein's σ model (see, e.g., Refs. [12, 13, 49–51]). At the moment, there is no clear evidence whether Keldysh techniques employed in Refs. [12, 13] would give any edge over the replica method used throughout this chapter. Below I first describe how to include the Coulomb interaction (for simplicity, in a singlet channel only, as the inclusion of a triplet channel is technically almost identical) and the BSC pairing interaction into the derivation of NLσM given in section 2. Then I will show (section 5) how to use such a model for describing the superconductor-insulator transition.

4.1. HUBBARD–STRATONOVICH DECOUPLING

Since the pair interaction does not conserve single-particle energy, in contrast to the elastic scattering which was the only mechanism included up to now, the effective interaction functional should be dynamical. As usual, it is convenient to introduce imaginary time τ implying everywhere the thermodynamic Gibbs averaging (together with the averaging over quenched disorder where applicable). Thus one considers thermodynamic Green's functions related (in the Matsubara frequency representation) to the retarded and advanced Green's functions of the previous sections by the standard procedure of analytical continuation [35]. Then the effective functional corresponding to the interaction term in Eq. (1) can be written as

$$\mathcal{S}_{\text{int}} = \frac{1}{2}\int \mathrm{d}x\mathrm{d}x'\bar{\psi}_s(x)\bar{\psi}_{s'}(x')V_{xx'}\psi_{s'}(x')\psi_s(x)\,, \tag{46}$$

where $x \equiv \mathbf{r}, \tau$, and all the fermionic fields are anti-periodic in imaginary time τ with period $1/T$, $s = (\uparrow, \downarrow)$ is the spin index, $V_{xx'} \equiv \delta(\tau - \tau')V_c(\mathbf{r} - \mathbf{r}')$, and V_c represents the Coulomb interaction. After the replication, all the fields in Eq. (46) naturally have the same replica index. Similarly, the BCS functional describing the attraction in the Cooper channel can be written as

$$\mathcal{S}_{\text{sc}} = \lambda_0 \int \mathrm{d}x\, \bar{\psi}_\uparrow(x)\bar{\psi}_\downarrow(x)\psi_\downarrow(x)\psi_\uparrow(x), \tag{47}$$

where λ_0 is the BCS coupling constant.

The Hubbard–Stratonovich decoupling of the functional (46)-(47) is similar to that for the disorder functional (7). To allow for all slow modes in the interaction functional(46) one should introduce three matrix fields: the fields Φ and $\hat{f}$ to take account of a small-angle scattering (a singlet channel) and large-angle scattering, and yet another one that corresponds to the Coulomb repulsion in the Cooper channel. The last one would lead to the standard renormalization of the BCS attraction (see, e.g., Ref. [52]) and is not considered here; further, it is assumed that systems under considerations still have an effective attraction in the Cooper channel after such a renormalization. Under these conditions, the decoupling has the following form:

$$\begin{aligned}
\mathrm{e}^{-\mathcal{S}_{\text{int}}} &= \int \mathcal{D}\Phi \exp\left\{-\frac{1}{2}\int \mathrm{d}x\mathrm{d}x'\, \Phi(x)V^{-1}_{xx'}\Phi(x') + i\int \mathrm{d}x\, \bar{\psi}_s(x)\Phi(x)\psi_s(x)\right\} \\
&+ \int \mathcal{D}\hat{f} \exp\left\{-\frac{1}{2}\int \mathrm{d}x\mathrm{d}x'\, \mathrm{tr}\left[\hat{f}(x)V^{-1}_{xx'}\hat{f}(x')\right] + i\int \mathrm{d}x\, \bar{\psi}_s(x)f_{ss'}(x)\psi_{s'}(x)\right\}, \\
\mathrm{e}^{-\mathcal{S}_{sc}} &= \int \mathcal{D}\Delta \exp\Big\{-\frac{1}{\lambda}\int \mathrm{d}x\, |\Delta(x)|^2 \\
&\quad + i\int \mathrm{d}x\, \left[\Delta(x)\bar{\psi}_\uparrow(x)\bar{\psi}_\downarrow(x) - \bar{\Delta}(x)\psi_\downarrow(x)\psi_\uparrow(x)\right]\Big\}
\end{aligned} \tag{48}$$

Now one doubles the number of components of the fields ψ and $\bar{\psi}$ as in Eq. (10) and performs the Gaussian integration to obtain

$$\mathsf{S} = \mathsf{S}_{\text{fields}} + \frac{\pi\nu}{8\tau}\mathrm{Tr}\,\sigma^2 - \frac{1}{2}\mathrm{Tr}\,\ln\left[\hat{\xi} - i\left(\frac{1}{2\tau}\sigma + \Phi + \Delta + \hat{\varepsilon}\right) - \hat{f}\right]. \tag{49}$$

Here the operator $\hat{\varepsilon}$ equals $i\hat{\tau}_3\partial_\tau$ in imaginary time representation and becomes the diagonal matrix of fermionic Matsubara frequencies ($\varepsilon_n = \pi(2n+1)T$) in frequency representation; Tr refers to a summation over all the matrix indices and to an integration over $\mathbf{r}$ and τ (or summation over Matsubara frequencies in the frequency representation); $\mathsf{S}_{\text{fields}}$ includes the ψ-independent part of the action

(48) quadratic in the fields Φ, $\hat{f}$ and $\hat{\Delta}$. The triplet channel, described by the field $\hat{f}$, is quite important: in particular, it can lead to the delocalization in the presence of disorder [7, 49]; however, such effects will not be considered here and this term will be ignored from now on as it is not relevant for the application of the model in section 5. The field σ has the same structure as in Eqs. (9) and (10), apart from explicitly including the 2×2 spin sector and replacing the 2×2 retarded-advanced sector by the dependence on the imaginary time τ (since it is diagonal in τ, it becomes a matrix field in the Matsubara frequencies). In the absence of the fields Φ and Δ, the saddle-point solution for the action (49) is formally the same as in the zero-temperature techniques of section 2 but the matrix Λ has a non-unit structure in the Matsubara (instead of advanced-retarded) sector:

$$Q = U^{\dagger}\Lambda U\,, \qquad \Lambda = \mathrm{diag}\,\{\mathrm{sgn}\,\hat{\varepsilon}\}\,. \tag{50}$$

The field Φ, corresponding to the singlet part of the Coulomb interaction, is diagonal in all the sectors. The "order-parameter" field $\hat{\Delta}$ is Hermitian and self-charge-conjugate, diagonal in the replica indices and coordinates $\mathbf{r}$ and τ, and has the following structure in the spin and time-reversal space:

$$\hat{\Delta}(x) = |\Delta(x)| e^{\frac{i}{2}\chi(x)\hat{\tau}_3} \hat{\tau}_2^{sp} \otimes \hat{\tau}_2 e^{-\frac{i}{2}\chi(x)\hat{\tau}_3}\,, \tag{51}$$

where $|\Delta|$ and χ are the amplitude and the phase of the pairing field $\Delta(\mathbf{r}, \tau)$, $\hat{\tau}_i$ and $\hat{\tau}_i^{\mathrm{sp}}$ are Pauli matrices ($i = 0, 1, 2, 3$ with $\hat{\tau}_0 = 1$) that span the charge-conjugate and spin sectors, respectively.

In the presence of the fields Φ and Δ, the saddle-point equation for the effective functional (49) can be formally written similarly to Eq. (13) for the non-interacting zero-temperature case:

$$-i\pi\nu\sigma(\mathbf{r}) = \left\langle \mathbf{r} \left| \left[-\hat{\xi} + \frac{i}{2\tau_{\mathrm{el}}}\sigma + i\left(\hat{\varepsilon} + \hat{\Delta} + \Phi\right)\right]^{-1} \right| \mathbf{r} \right\rangle \tag{52}$$

Now $\sigma = \Lambda$ does no longer represent the saddle point solution for $\varepsilon \neq 0$. Still, one can derive an effective functional by expanding the above Tr ln within the manifold (50) in the symmetry-breaking field $\varepsilon + \Delta + \Phi$ and in gradients of Q, as has been done in the original works by Finkelstein [7]. An alternative is to make first a similarity transformation around Λ within the manifold (50) to find the saddle point solution σ_{sp} in the presence of the fields. This can be formally done with the help of matrix U_0 that diagonalizes the Hermitian field $\hat{\varepsilon} + \hat{\Delta} + \Phi$:

$$\hat{\varepsilon} + \hat{\Delta} + \Phi = U_0^{\dagger}\lambda U_0\,, \qquad \sigma_{\mathrm{sp}} = U_0^{\dagger}\Lambda U_0\,. \tag{53}$$

Here U_0 belongs to the same symmetry group that defines the manifold (16) in the absence of the interaction fields. By substituting Eq. (53) into the saddle-point equation (52), one can easily verify that this is, indeed, a spatially-homogeneous solution, provided that $\lambda\tau_{\rm el} \gg 1$ which will always be the case for a dirty superconductor ($\Delta\tau_{\rm el} \gg 1$) or a dirty metal (dimensionless conductance $g \gg 1$).

Now one can perform the expansion of Tr ln, Eq. (49), in gradients of Q and in the symmetry-breaking fields represented by the eigenvalues λ and the matrix U_0, Eq. (53). It is convenient to employ the following parameterization:

$$\sigma = U_0^\dagger Q \, U_0 \,, \qquad Q = U^\dagger \Lambda U \,, \tag{54}$$

where Q represents the saddle-point manifold in the metallic phase and σ is obtained from Q by the same rotation (53) as $\sigma_{\rm sp}$ is obtained from the metallic saddle point Λ. Therefore, Q is defined, as in the metallic phase, on the coset space $S(2N)/S(N) \otimes S(N)$ where, depending on the symmetry, S represents the unitary, orthogonal or symplectic group as in the noninteracting case of section 2. The parameterization (54) simplifies considerably all the subsequent derivations and leads to a new variant of the nonlinear σ model which can be more convenient for many applications than the original Finkelstein's NLσM. After substituting $\sigma = U_0^\dagger U^\dagger \Lambda \, U_0 U$, Eq. (54), into Eq. (49), one obtains the following representation for the Tr ln term:

$$\delta S = -\tfrac{1}{2} {\rm Tr} \ln \{ \hat{G}_0^{-1} + U_0 [\hat{\xi}, U_0^\dagger] - i (U \lambda U^\dagger) \} \,,$$

where one can also include an external magnetic field with the vector potential $\mathbf{A}$:

$$\hat{G}_0 \equiv \left(\hat{\xi} - \frac{i}{2\tau_{\rm el}} \Lambda \right)^{-1} , \qquad \hat{\xi} \equiv \frac{1}{2m} (\mathbf{p} - e\mathbf{A})^2 - \varepsilon_{\rm F} \,.$$

The expansion to the lowest powers of gradients and λ is now straightforward and similar to that for the noninteracting case. It results after some calculations in the following effective action:

$$S = \frac{1}{2} {\rm Tr} \left[\Phi V_{\mathbf{rr}'}^{-1} \Phi \right] + \frac{1}{T\lambda_0} \sum_\omega \int \! d\mathbf{r} \, |\Delta_\omega|^2 + \frac{\pi\nu}{2} {\rm Tr} \left[\frac{D}{4} (\partial Q)^2 - \lambda Q \right] . \tag{55}$$

The long derivative in Eq. (55) is defined as

$$\partial Q \equiv \nabla Q + \left[\mathbf{A}_0 - ie\mathbf{A}\hat{\tau}_3 , \, Q \right] \equiv \partial_0 Q + [\mathbf{A}_0, Q] \,, \tag{56}$$

where the matrix $\mathbf{A}_0$ is given by

$$\mathbf{A}_0 = U_0 \partial_0 U_0^\dagger \,, \tag{57}$$

and $\partial_0 \equiv \nabla - [ie\mathbf{A}\hat{\tau}_3, \ldots]$ is the long derivative (56) in the absence of the pairing field Δ. Both U_0 and λ should be found from the diagonalization of $\epsilon + \Delta + \Phi$, Eq. (53). Although such a diagonalization cannot be done in general, it will be straightforward in many important limiting cases. For $\Delta = \Phi = 0$, the field $\mathbf{A}_0$

vanishes, $\partial \to \partial_0$, and $\lambda \to \varepsilon$, so that the functional (55) goes over to that of the standard nonlinear σ model for non-interacting electrons.

The action (55) is most general in the current context. It allows one to develop a fully self-consistent approach to superconductivity of dirty metals in the presence of Coulomb interaction. However, any application requires a set of further simplifications. As a simple illustration, I will show below how to use the model in a dirty superconductor near the metal-superconductor transition in the absence of Coulomb interaction. Then, I will show in section 5 how to use this model to describe the insulator-superconductor transition as a result of combined effects of disorder and interactions.

4.2. GINZBURG-LANDAU FUNCTIONAL

In the vicinity of the metal-superconductor transition one can expand the action (55) (in the absence of the Coulomb field Φ) in the pairing field Δ. A further simplification is possible in the weak disorder limit, $g \gg 1$, when the Q-field can be integrated out to obtain an effective action for the Δ-field only. In the quadratic in Δ approximation, the kernel of this action will give, with due account for the disorder, an effective matrix propagator of the pairing field which governs properties of a disordered superconducting sample near the transition.

To integrate over the Q-field, one splits the action (55) into $\mathcal{S} \equiv \mathcal{S}_0 + \mathcal{S}_\Delta$ where $\mathcal{S}_0$ is the standard nonlinear σ model functional as in the metallic phase. Then one makes a cumulant expansion, i.e. first expands $\mathrm{e}^{-(\mathcal{S}_0+\mathcal{S}_\Delta)}$ in powers of $\mathcal{S}_\Delta$, then performs the functional averaging with $\mathrm{e}^{-\mathcal{S}_0}$ (denoted below by $\langle\ldots\rangle_Q$) and finally re-exponentiates the results. The expansion involves only the first and second order cumulants since the higher order cumulants generate terms of higher order in Δ. Then the only contributions to the action quadratic in Δ are

$$\begin{aligned}\mathcal{S}_{\text{eff}}[\Delta] &= \frac{1}{\lambda_0 T}\sum_\omega \int \mathrm{d}\mathbf{r}\, |\Delta_\omega|^2 - \frac{\pi\nu}{2}\Big\langle \mathrm{Tr}\,(\lambda - \epsilon) Q\Big\rangle_Q \\ &-\Big\langle \frac{\pi\nu D}{8}\mathrm{Tr}\;[\mathbf{A}_0, Q]^2 + \frac{(\pi\nu D)^2}{8}\Big(\mathrm{Tr}\, Q\partial_0 Q\mathbf{A}_0\Big)^2\Big\rangle_Q .\end{aligned} \tag{58}$$

Expanding λ and $\mathbf{A}_0$ to the lowest power in Δ and performing the functional averaging one finds [51] the action quadratic in Δ as follows:

$$\mathcal{S}_{\text{eff}}[\Delta] = \frac{\nu}{T}\sum_\omega \int \mathrm{d}\mathbf{r}\, \Delta^*_\omega(\mathbf{r})\langle \mathbf{r}|\, \hat{\mathcal{K}}_\omega\, |\mathbf{r}'\rangle \Delta_\omega(\mathbf{r}')\,, \tag{59}$$

with the operator $\hat{\mathcal{K}}_\omega$ given by

$$\hat{\mathcal{K}}_\omega = \frac{1}{\lambda_0 \nu} - 2\pi T \sum_{\epsilon(\omega-\epsilon)<0} \left\{ \hat{\Pi}^c_\omega + \frac{1}{\pi\nu}\frac{\Pi^d_{|2\epsilon-\omega|}(0)\hat{\mathcal{C}}}{(2\epsilon-\omega)^2}\right\}. \tag{60}$$

Here $\Pi^{c,d}_{|\omega|}(\mathbf{r},\mathbf{r}') = \langle \mathbf{r}|\hat{\Pi}^{c,d}|\mathbf{r}'\rangle$ are the cooperon and diffuson propagators, with

$$\hat{\Pi}^c_{|\omega|} = \left(\hat{\mathcal{C}} + |\omega|\right)^{-1}, \tag{61}$$

where $\hat{\mathcal{C}} \equiv -D\left(\nabla - 2ie\mathbf{A}\right)^2$ defines the cooperon modes; $\hat{\Pi}^d$ is obtained from $\hat{\Pi}^c$ by putting the external vector potential $\mathbf{A} = \mathbf{0}$. In the last term in Eq. (60), $\Pi^d_{|\omega|}(0) \equiv \Pi^d_{|\omega|}(\mathbf{r},\mathbf{r})$; this term may be obtained by expanding in g^{-1} the cooperon propagator with the renormalized diffusion coefficient,

$$\hat{\mathcal{C}} \to \left[1 - \frac{1}{\pi\nu}\Pi^d_{|\omega|}(0)\right]\hat{\mathcal{C}},$$

which is a weak localisation correction to free cooperon propagator $\Pi^c_{|\omega|}(\mathbf{r},\mathbf{r}')$.

The summation over Matsubara frequencies in Eq. (60) yields

$$\hat{\mathcal{K}}_\omega = \ln\frac{T}{T_0} + \psi\left(\frac{1}{2} + \frac{|\omega| - \hat{\mathcal{C}}}{4\pi T}\right) - \psi\left(\frac{1}{2}\right) - \frac{a_\omega \hat{\mathcal{C}}}{4\pi T}, \tag{62}$$

where $T_0 \equiv T_{c0}(B=0)$ is the transition temperature of the clean superconductor in the absence of a magnetic field and ψ is the digamma function. The weak localisation correction is proportional to the coefficient a_ω given by

$$a_\omega(T) = \frac{1}{\pi\nu V}\sum_{\mathbf{q}}\frac{1}{Dq^2}\left\{\psi'\left(\frac{1}{2} + \frac{|\omega|}{4\pi T}\right)\right.$$

$$\left. - \frac{4\pi T}{Dq^2}\left[\psi\left(\frac{1}{2} + \frac{|\omega| + Dq^2}{4\pi T}\right) - \psi\left(\frac{1}{2} + \frac{|\omega|}{4\pi T}\right)\right]\right\}.$$

For $\omega = 0$ the coefficient $a_0 \equiv a_{\omega=0}(T)$ can be simplified in the two limits:

$$a_0 = \begin{cases} \dfrac{\psi'(1/2)}{\pi\nu L^d}\displaystyle\sum_{L_T^{-1}<q<\ell^{-1}}\frac{1}{Dq^2}, & L \gg L_T, \\[2ex] -\dfrac{\psi''(1/2)}{8\pi^2\nu L^d T}, & L \ll L_T, \end{cases} \tag{63}$$

where $L_T \equiv \sqrt{D/T}$ is the thermal smearing length.

The instability of the normal state (i.e. a transition into the superconducting state) occurs when the lowest eigenvalue of the operator $\hat{\mathcal{K}}_\omega$ becomes negative. The eigenfunctions of this operator coincide with the eigenfunctions of the cooperon operator $\hat{\mathcal{C}}$. The lowest eigenvalue of $\hat{\mathcal{C}}$ is known to be $\mathcal{C}_0 = DB/\phi_0$, where ϕ_0 is the flux quanta. This ground state cooperon eigenfunction corresponds to the lowest eigenvalue $\mathcal{K}_0$ of the operator $\hat{\mathcal{K}}_\omega$. The condition $\mathcal{K}_0 = 0$ implicitly defines the line $T_c(B)$ in the (T, B)-plane where the transition occurs:

$$\ln\frac{T_c}{T_0} + \psi\left(\frac{1}{2} + \frac{\mathcal{C}_0}{4\pi T_c}\right) - \psi\left(\frac{1}{2}\right) = \frac{a_0\mathcal{C}_0}{4\pi T_c}. \tag{64}$$

The term in the r.h.s. of Eq. (64) describes a $1/g$-correction to the main result. This weak localisation is linear in the magnetic field B and vanishes as $B \to 0$ as expected (the Anderson theorem [46]). In a nonzero magnetic field the weak localisation correction to B_c is positive which has a very simple explanation. The superconductivity is destroyed by the magnetic field when $\Phi(\xi) \gtrsim \Phi_0$, where $\Phi(\xi)$ is the flux over the area with linear size of the order of the coherence length $\xi \sim \sqrt{D/T}$ and Φ_0 is the flux quanta. The weak localisation corrections reduce D and thus ξ. Therefore, one needs a stronger field to destroy the quantum coherence. The same reasoning explains the growth of T_c in a fixed magnetic field.

Note finally that it would be straightforward to include the leading weak-localization corrections in all orders of g^{-1} ln by calculating the Q - averages in Eq. (58) via the renormalization group. This would lead to the renormalization of D in the cooperon propagator (61), thus changing the shape of the $T_c(B)$ curve. However, the value of $T_c(0)$ will again remain unaffected, since the superconducting instability is defined by the onset of the homogeneous zero mode in the operator $\hat{\mathcal{K}}$, Eq. (62), which does not depend on the value of the diffusion coefficient in the cooperon propagator.

It is worth stressing that the Anderson theorem reflects certain properties of a model rather than those of real superconductors. If one allows for Coulomb interaction, then the critical temperature of the superconducting transition is no longer independent of disorder. Combined effects of the interaction and disorder lead to corrections to the transition temperature proportional (in a slightly simplified way) to $g^{-1} \ln^3(T_c\tau_{\rm el})$ [53, 54, 45]. For a relatively weak disorder, the system still remains superconducting at $T \to 0$ while for a sufficiently strong disorder the above corrections would suppress the superconducting pairing at any temperature and make a 2D system insulating at $T \to 0$. Such a mechanism of suppressing Δ by disorder gives a possible scenario for a widely observed superconductor–insulator (SI) transition in two-dimensional structures [55–58] which is most adequately described within the NLσM [45, 7] similar to those developed in this section. However, a widely accepted scenario for such a transition is based on very different, both in origin and in implementation, models [59–63] where $|\Delta|$ remains finite and the superconductivity is suppressed by the loss of the phase coherence due to the fluctuations of the phase χ of the order parameter. I will show in the following section that these "phase-only" models can also be derived under certain parametrically controlled assumptions from the NLσM developed in this section.

5. Superconductor–insulator transition in 2D systems: phase and amplitude fluctuations of the order parameter

A wide variety of experimental data indicate the existence of a superconductor–insulator (SI) transition in two-dimensional films [55–58] (see, in particular, [58] for earlier references). The transition can be tuned by either disorder (changing with the thickness of a superconducting film) or magnetic field, thus being one of the most intensely studied examples of quantum phase transitions [64].

One of the most accepted ways to understand the problem of the SI transition is based on the so-called Bose-Hubbard (or "dirty-boson") models [63]. In this class of models exists a duality between charge-$2e$ bosons (preformed Cooper pairs) and vortices. The superconducting phase is due to the bose-condensation of the charged bosons with localized vortices while the insulating phase is due to the bose-condensation of the vortices with the localized charged bosons. Another approach which captures the basic physics of granular superconductors is based on dissipative models[65] of resistively shunted charged Josephson arrays[59–62], with the emphasis on the role of dissipation and Coulomb interaction. In both group of models [63, 65, 59–62], the transition is driven by fluctuations of the phase of the order parameter, i.e. the superconductivity is destroyed in spite of the existence of nonvanishing $|\Delta|$ locally. This approach seems to be rather different from that mentioned in the previous section where the disorder and interaction destroy the superconductivity by destroying $|\Delta|$ everywhere in a homogeneously disordered system. An experimental distinction between homogeneous and granular systems is not that strict[55] as it seemed a few years ago, and recent experimental observations [55, 56] strongly suggest that the amplitude fluctuations in the vicinity of the SI transition are no less important than the phase fluctuations.

The purpose of this section is to modify a general NLσM action for the description of granular systems with the BCS and Coulomb interactions. Such a description takes account of fluctuations of both amplitude and phase of the order parameter Δ, thus encompassing all the above described approaches. Both the Bose-Hubbard model [63] and the dissipative models [59–62] will be derived from this action via certain controlled simplifications made within the NLσM. The latter model is more general and allows one to go beyond certain limitations necessary in the derivation of the phase-only action.

The starting point is a coarse-grained version of the TOE model Hamiltonian of Eq. (1) for a granular superconductor, where the kinetic energy includes terms $t_{ij}a_i^\dagger a_j$ corresponding to tunnelling hops between the grains. The derivation of the NLσM from such Hamiltonian follows the procedure described in the previous section. Thus a 'half-baked' effective functional obtained by integrating out the fermionic fields is similar to that in Eq. (49):

$$\begin{aligned} S[\hat\sigma,\hat\Delta,\Phi] = \frac{\pi\nu}{8\tau_{\rm el}}\,Tr\,\sigma^2 + \frac{1}{4\lambda_0}{\rm T}r\,|\hat\Delta|^2 + \frac{1}{2}{\rm T}r\,\Phi U^{-1}\Phi \\ -\frac{1}{2}{\rm T}r\ln\left[-\hat\xi - \hat t + \frac{i}{2\tau_{\rm el}}\,\hat\sigma + i\left(\hat\Delta + \Phi + \hat\epsilon\right)\right]. \end{aligned} \tag{65}$$

The difference here is that $\hat{\xi}$ is the operator of the intra-grain kinetic energy (counted from the chemical potential), while $\hat{t}$ is the tunneling amplitude matrix (i.e. the inter-grain kinetic energy). Note that in the granular case the singlet field Φ is sufficient to decouple the Coulomb interaction term so that the triplet field f does not enter Eq. (65). All the bosonic fields have the same structure as before but for the addition of the $m \times m$ grain sector. The symbol Tr refers both to a summation over all these matrix indices and to an integration over intra-grain position $\mathbf{r}$ and the imaginary time τ.

The principal simplification for granular systems is that all the fields are spatially homogeneous inside each grain when the grains are zero-dimensional, i.e. their sizes $L \lesssim \xi, L_T$ (ξ and L_T are the superconducting and thermal coherence lengths) which is equivalent to $|\Delta|, T \lesssim 1/\tau_{\rm erg}$. For a diffusive grain $\tau_{\rm erg} = L^2/D \equiv \hbar/E_{\rm T}$ while for a ballistic (chaotic) grain $\tau_{\rm erg} \sim L/v_{\rm F}$. The tunneling matrix $\hat{t} = \{t_{ij}\}$ depends only on grain indices, and the Coulomb interaction thus reduces to the capacitance matrix, $U^{-1} \to C_{ij}/e^2$.

To follow the procedure described in the previous section one needs, first, to solve (formally) the saddle point conditions, Eqs. (53) and (54) separately for each grain. As all the fields are spatially homogeneous inside each grain, $U_0 \equiv U_0(i)$ commutes with the operator $\hat{\xi}$. Then one only needs to expand the Tr ln to the first nonvanishing orders in t_{ij} and λ_i, this expansion being justified when $|t|, |\Delta|, T \ll 1/\tau_{\rm el} \ll \varepsilon_{\rm F}$. Thus one arrives at the following effective action:

$$
\begin{aligned}
S[Q,\Delta,\Phi] = \int_0^\beta \mathrm{d}\tau \Big\{ \sum_i \frac{|\Delta_i|^2}{\nu\lambda_0\delta_i} + \sum_{ij} \frac{C_{ij}}{2e^2}\,\Phi_i\Phi_j \Big\} \\
- \sum_i \frac{\pi}{2\delta_i} \mathrm{Tr}\,\lambda_i Q_i - \frac{g^{\rm t}_{ij}}{2} \sum_{ij} \mathrm{Tr}\,Q_i S_{ij} Q_j S_{ji}\,,
\end{aligned}
\tag{66}
$$

where $S_{ij} \equiv U_0(i)U_0^\dagger(j)$, all the fields depend on τ, Tr refers to all indices except i, j numerating grains, δ_i is mean level spacing in the i-th grain, and the tunneling conductance is defined by $g^{\rm t}_{ij} \equiv 2\pi^2|t_{ij}|^2/\delta_i\delta_j$ (which is nonzero only for neighboring grains). Both λ_i and $U_0(i)$ (and thus S_{ij}) should be found from the diagonalization procedure in Eq. (53), while Q is given by Eq. (54).

The next step is to represent $U_0(i)$ as

$$
U_0(i) = V_i\,\mathrm{e}^{-\frac{i}{2}\chi_i(\tau)\hat{\tau}_3}\,. \tag{67}
$$

This is similar to the gauge transformation suggested in Refs. [66] and used in Ref. [67] to gauge out the Coulomb field. However, one cannot gauge out two independent fields, Δ and Φ. Substituting the transformation (67) into the diagonalization condition (53), one reduces it to

$$
\hat{\epsilon} + \widetilde{\Phi}_i + \hat{\Delta}^{\rm o}_i = V_i^{+}\lambda_i V_i\,, \tag{68}
$$

where $\hat{\Delta}_i^{\mathrm{o}}$ is the field (51) taken at $\chi = 0$ and the field $\widetilde{\Phi}$ is given in the τ representation by $\widetilde{\Phi}_i \equiv \Phi_i - \frac{1}{2}\partial_\tau \chi_i$.

Both $\widetilde{\Phi}$ and $|\Delta|$ are massive fields whose fluctuations are strongly suppressed. It is straightforward to show that the fluctuations of $\widetilde{\Phi}$ are of order δ which is much smaller than both T and $|\Delta|$. Therefore, in the mean field approximation in $\widetilde{\Phi}$ this field can be neglected, $\widetilde{\Phi} = 0$. This condition is nothing more than the Josephson relation in imaginary time [60]. This locks the fluctuations of the Coulomb field Φ with the phase-fluctuations of the pairing field $\hat{\Delta}$,

$$\Phi = \tfrac{1}{2}\partial_\tau \chi \,, \tag{69}$$

thus reducing the action (66) to one depending only on the fields Q and Δ.

The mean field approximation in $|\Delta_i|$ is valid for $|\Delta_i| \gg \delta$. It is equivalent to the standard self-consistency equation which formally follows from the variation of the action (66) with respect to $|\Delta|$. In this approximation one finds $|\Delta_i|$ to be independent of i and τ. Thus the first term in Eq. (66) becomes a trivial constant, so that the action depends on Q and χ only:

$$\mathcal{S}[Q,\chi] = \sum_{ij}\frac{C_{ij}}{8e^2}\int_0^\beta \mathrm{d}\tau\, \partial_\tau\chi_i\,\partial_\tau\chi_j - \sum_i \frac{\pi}{2\delta_i}\mathrm{Tr}\lambda_i Q_i - \frac{g_{ij}^{\mathrm{t}}}{2}\sum_{ij}\mathrm{Tr}\, Q_i S_{ij} Q_j S_{ji} . \tag{70}$$

The field χ in this action obeys the boundary condition $\chi(\tau+\beta) = \chi(\tau) \bmod 2\pi$. In calculating the partition function, one should in general allow for different topological sectors corresponding to different winding numbers in χ.

The "phase-only" action (70) includes neither fluctuations of $|\Delta|$, nor fluctuations of Φ beyond the Josephson relation, Eq. (69). Below this action will be reduced to the AES action. Still, we stress that the action (70) is more general than the AES action. Thus, in the absence of superconductivity, $\Delta \equiv 0$, it was shown [67] that the former contains a correct screening of the Coulomb interaction at low T, in contrast to the latter. This may also be important in the case when Δ is much smaller than the charging energy.

To further simplify the action (70) we note that the diagonalization conditions (68), in the absence of the $|\Delta|$ fluctuations, are the same for each grain and reduced to those solved in Ref. [51]:

$$\begin{aligned} V_{\varepsilon\varepsilon'} &= \cos\frac{\theta_\varepsilon}{2}\,\delta_{\varepsilon,\varepsilon'} + \hat{\tau}_2^{sp}\otimes\hat{\tau}_2 \sin\frac{\theta_\varepsilon}{2}\,\mathrm{sgn}\,\varepsilon\,\delta_{\varepsilon,-\varepsilon'} \\ \lambda &= \mathrm{diag}\sqrt{\varepsilon^2+|\Delta|^2}\,\mathrm{sgn}\,\varepsilon\,, \quad \cos\theta_\varepsilon \equiv \frac{|\varepsilon|}{\sqrt{\varepsilon^2+|\Delta|^2}}\,. \end{aligned} \tag{71}$$

Then S_{ij} in Eq. (70) can be expressed in terms of V as

$$S_{ij} \equiv V \mathrm{e}^{-\frac{i}{2}\chi_{ij}\hat{\tau}_3} V^\dagger\,, \qquad \chi_{ij} \equiv \chi_i - \chi_j\,. \tag{72}$$

Finally note that large-$|\varepsilon|$ contributions to the action (13) are strongly suppressed, while for $|\varepsilon| \ll |\Delta|$ one has $\lambda = |\Delta|\Lambda$ which suppresses fluctuations of Q in each grain imposing $Q=\Lambda$. Then, all matrices in the action (70) are diagonal in the replica indices so that these indices become redundant. It finally reduces the action (70) to that depending only on one *scalar* bosonic field $\chi_i(\tau)$, the phase of the order parameter, which is a function of only the imaginary time and the grain number. Indeed, the second term in Eq. (70) reduces to a trivial constant; evaluating the tunneling term with the help of Eqs. (71) and (72), one obtains

$$\begin{aligned} S[\chi] &= \sum_{ij}\Bigg\{\frac{C_{ij}}{8e^2}\int_0^\beta d\tau\, \partial_\tau\chi_i\, \partial_\tau\chi_j \\ &\quad -2g^{\mathrm{t}}_{ij}\int_0^\beta d\tau\int_0^\beta d\tau' g^2_{\mathrm{n}}(\tau-\tau')\cos\chi^-_{ij} + g^2_{\mathrm{a}}(\tau-\tau')\cos\chi^+_{ij}\Bigg\}, \end{aligned} \tag{73}$$

where $\chi_{ij}(\tau) \equiv \chi_i(\tau) - \chi_j(\tau)$, $\chi^{\pm}_{ij} \equiv \frac{1}{2}\Big[\chi_{ij}(\tau) \pm \chi_{ij}(\tau')\Big]$, and the normal and anomalous Green's functions $g_{n,a}$ (integrated over all momenta) are given by

$$g_n(\tau)=T\sum_{\varepsilon}\frac{\varepsilon\sin\varepsilon\tau}{\sqrt{\varepsilon^2+|\Delta|^2}}\,,\quad g_a(\tau)=T\sum_{\varepsilon}\frac{|\Delta|\cos\varepsilon\tau}{\sqrt{\varepsilon^2+|\Delta|^2}}\,. \tag{74}$$

The action (73) is exactly the AES action introduced in Refs. [59–61]. Further simplifications are possible in two limiting cases.

First, in the normal case ($\Delta=0$ one has in Eq. (74)

$$g_a = 0\,, \qquad g_n^2(\tau) = \frac{T^2}{\sin^2 \pi T\tau}\,. \tag{75}$$

Then the field χ should be substituted, according to Eq. (69), by $2\int^\tau d\tau'\Phi(\tau')$. This corresponds to using the action (73), (75) for a set of normal tunnel junctions [68], as has been recently shown in Ref. [67]; the functional (70) in the limit $\Delta = 0$ is equivalent to that of Ref. [67]. Including disorder-induced fluctuations (i.e. going beyond the $Q = \Lambda$ approximation) allows one to obtain [67] a correct low-T limit for the phase correlation function missing in the action (73).

The action (70) is more general than that considered in Ref. [67]: although under the mode locking condition (69) it depends only on the fields χ and Q, the matrix S_{ij}, Eqs. (71) and (72), reduces to a simple $U(1)$ gauge transformation as in in Ref. [67] only in the limit $\Delta = 0$.

The second limiting case, $T \ll |\Delta|$, is just the limit relevant in the context of the SI transition in granular superconductors. For $T = 0$, the summation in Eq. (74) can be substituted by integration which yields

$$g_n(\tau) = \frac{|\Delta|}{\pi}K_1(|\Delta|\tau), \quad g_a(\tau) = \frac{|\Delta|}{\pi}K_0(|\Delta|\tau)$$

This is also a good approximation for a low-T case; substituting this into Eq. (73) gives the action for the dissipative model[60, 61]. Note that for $|\tau-\tau'|\ll|\Delta|^{-1}$, the main contribution in the tunneling action (73) is given by the normal term with the corresponding kernel proportional to $|\tau-\tau'|^{-2}$. The Fourier transform of this would give a term of the Caldeira-Leggett type[65] proportional to $|\omega|$.

The tunneling action (73) is non-local in τ. As has been noted in Ref. [60] for the case of one Josephson junction, for sufficiently large capacitance the phase χ_{ij} changes slowly in comparison with $|\Delta|^{-1}$, and in the adiabatic approximation $\chi(\tau')$ is changed by $\chi(\tau)+(\tau'-\tau)\partial_\tau\chi(\tau)$. Making such an expansion, one obtains from Eq. (73) the following local action:

$$\mathcal{S}[\chi]=\int_0^\beta \mathrm{d}\tau\left\{\sum_{ij}\tfrac{1}{2}u_{ij}^{-1}\dot{\chi}_i\dot{\chi}_j-|\Delta|g^{\mathrm{t}}_{ij}\cos\chi_{ij}\right\}, \tag{76}$$

where $\dot{\chi}_i\equiv\partial_\tau\chi_i$ and

$$\frac{1}{u_{ii}}\equiv\frac{C_{ii}}{4e^2}+\sum_j\frac{g^{\mathrm{t}}_{ij}}{|\Delta|}\frac{3+\cos\chi_{ij}}{8},$$

$$u_{ij}^{-1}\equiv\frac{C_{ij}}{4e^2}-\frac{g^{\mathrm{t}}_{ij}}{|\Delta|}\frac{3+\cos\chi_{ij}}{8}.$$

If all the self-capacitances are equal to C with $E_c\propto e^2/C$ being the charging energy, and all $g^{\mathrm{t}}_{ij}=g^{\mathrm{t}}$, then $u_{ii}\equiv U$ has the meaning of the renormalized charging energy. Ignoring a weak dependence of u on $\cos\chi_{ij}$ in the above relations, one obtains the renormalized charging energy:

$$U=\frac{E_c}{1+\#E_c g^{\mathrm{t}}/|\Delta|}. \tag{77}$$

Here the coefficient # depends on the number of next neighbors for each grain, etc. A similar renormalization takes place for the next-neighbor off-diagonal energy u_{ij}. Now one can see that on the face of it the adiabatic approximation employed to obtain Eq. (76) is valid for $U\ll|\Delta|$. However, in the region $g^{\mathrm{t}}\gg|\Delta|/E_c$, where the charging energy (77) is strongly renormalized, the instanton-like solutions [69] may be important. This may further reduce the region of applicability for the local in τ action (76).

Finally, by introducing the operator $\hat{n}$ canonically conjugate to the phase χ, one finds the Hamiltonian that corresponds to the action (76):

$$\hat{H}=\sum_{ij}\tfrac{1}{2}u_{ij}\hat{n}_i\hat{n}_j-|\Delta|g^{\mathrm{t}}_{ij}\cos\left(\chi_i-\chi_j\right). \tag{78}$$

This is just the Hamiltonian of the Bose-Hubbard model [63] which was first microscopically derived by Efetov [70] in the context of granular superconductors.

To conclude, in this section the effective σ model-type action (66) has been derived for a granular system with zero-dimensional grains in the presence of the Coulomb and pairing interactions. This is the most general (in the present context) action which takes into account fluctuations of both amplitude and phase of the order parameter Δ. Neglecting fluctuations of $|\Delta|$ and fluctuations of the Coulomb field Φ beyond the Josephson relation (69) reduces this action to the "phase-only" action (70) which still contains intra-granular disorder important for the correct screening for $|\Delta|$ small compared to the charging energy. Neglecting this disorder further reduces the action (70) to that of the AES model, Eq. (73). When the renormalized charging energy, Eq. (77), is much smaller than $|\Delta|$, the action (73) finally goes over to that of the Bose-Hubbard model, Eq. (78), which is widely used for the description of the superconductor-insulator transition[63]. However, the above estimations show that this reduction is parametrically justified only for the region $E_c \ll |\Delta|$ where the transition happens at $g^{\rm t} \ll E_c/|\Delta| \ll 1$ which corresponds to a strongly granular system. Note finally that the most general (in the present context) action (66) describes both amplitude and phase fluctuations of the order parameter, being still considerably different from the NLσM action for homogeneous systems.

Acknowledgments. This work has been supported by the Leverhulme Trust under the contract F/94/BY and by the EPSRC grant GR/R33311.

References

1. F. Wegner, Z. Phys. B **35**, 207 (1979); L. Schäfer and F. Wegner, Z. Phys. B **38**, 113 (1980).
2. K. B. Efetov, Sov. Phys. JETP **55**, 514 (1982).
3. A. J. McKane and M. Stone, Ann. Phys. (N. Y.) **131**, 36 (1981).
4. K. B. Efetov, A. I. Larkin, and D. E. Khmel'nitskii, Sov. Phys. JETP **52**, 568 (1980).
5. J. J. M. Verbaarschot, H. A. Weidenmüller, and M. R. Zirnbauer, Phys. Rep. **129**, 367 (1985).
6. E. Abrahams, P. W. Anderson, D. C. Licciardello, and T. V. Ramakrishnan, Phys. Rev. Lett. **42**, 673 (1979).
7. A. M. Finkelshtein, Sov. Phys. JETP **57**, 97 (1983); Sov. Sci. Rev. 2 **14**, 1 (1990).
8. S. V. Kravchenko et al., J. Phys.–Cond. Matter **7**, L41 (1995); Phys. Rev. B **51**, 7038 (1995).
9. A. G. Aronov and A. S. Ioselevich, JETP Letters **42**, 84 (1985).
10. V. S. Babichenko and A. N. Kozlov, Sol. State Commun. **59**, 39 (1986).
11. M. Horbach and G. Schön, Physica **A167**, 93 (1990).
12. A. Kamenev and A. Andreev, Phys. Rev. B **60**, 3944 (1999).
13. C. Chamon, A. W. W. Ludwig, and C. Nayak, Phys. Rev. B **60**, 2239 (1999).
14. L. Keldysh, Sov. Phys. JETP **20**, 1018 (1964).
15. K. B. Efetov, Sov. Phys. JETP **56**, 467 (1982).
16. K. B. Efetov, *Sypersymmetry in Disorder and Chaos*, CUP, Cambridge (1997).
17. V. E. Kravtsov and A. D. Mirlin, JETP Letters **60**, 656 (1994).
18. B. A. Muzykantskii and D. E. Khmelnitskii, Phys. Rev. B **51**, 5480 (1995).
19. K. B. Efetov and V. I. Fal'ko, Europhys. Lett. **32**, 627 (1995).
20. A. D. Mirlin, Phys. Rev. B **53**, 1186 (1996).
21. M. L. Mehta, *Random matrices*, Academic Press, Boston (1991).
22. L. P. Gor'kov and G. M. Eliashberg, Sov. Phys. JETP **21**, 940 (1965).

23. B. L. Altshuler and B. I. Shklovskii, Sov. Phys. JETP **64**, 127 (1986).
24. J. J. M. Verbaarschot and M. R. Zirnbauer, J. Phys. A **17**, 1093 (1985).
25. A. Kamenev and M. Mezard, J. Phys. A **32**, 4373 (1999).
26. A. Kamenev and M. Mezard, Phys. Rev. B **60**, 3944 (1999).
27. I. V. Yurkevich and I. V. Lerner, Phys. Rev. B **60**, 3955 (1999).
28. A. Kamenev, Phys. Rev. Lett. **85**, 4160 (2000).
29. B. L. Altshuler, V. E. Kravtsov, and I. V. Lerner, JETP Letters **43**, 441 (1986); Sov. Phys. JETP **64**, 1352 (1986).
30. B. L. Altshuler, V. E. Kravtsov, and I. V. Lerner, Sov. Phys. JETP **67**, 795 (1988).
31. I. V. Lerner, Phys. Lett. **133A**, 253 (1988); I. V. Lerner and F. Wegner, Z. Phys. B **81**, 95 (1990).
32. B. L. Altshuler, V. E. Kravtsov, and I. V. Lerner, in *Mesoscopic Phenomena in Solids*, edited by B. L. Altshuler, P. A. Lee, and R. A. Webb, North-Holland, Amsterdam, 449 (1991).
33. M. Zirnbauer, cond-mat/9903338 (1999).
34. A. V. Andreev and B. L. Altshuler, Phys. Rev. Lett. **75**, 902 (1995).
35. A. A. Abrikosov, L. P. Gor'kov, and I. E. Dzyaloshinskii, *Methods of Quantum Field Theory in Statistical Physics*, Pergamon Press, New York (1965).
36. R. A. Smith, I. V. Lerner, and B. L. Altshuler, Phys. Rev. B **58**, 10343 (1998).
37. B. A. Dubrovin, A. T. Fomenko, and S. P. Novikov, *Modern Geometry, Part II*, Springer-Verlag, Berlin (1985).
38. B. L. Altshuler, JETP Letters **41**, 648 (1985).
39. P. A. Lee and A. D. Stone, Phys. Rev. Lett. **55**, 1622 (1985).
40. A. D. Mirlin, Phys. Rep. **326**, 260 (2000).
41. A. Z. Patashinskii and V. I. Pokrovskii, *Fluctuation Theory of Phase Transitions*, Pergamon Press, Oxford (1979).
42. V. E. Kravtsov, I. V. Lerner, and V. I. Yudson, Sov. Phys. JETP **67**, 1441 (1988).
43. F. Wegner, Z. Phys. B **78**, 33 (1990).
44. B. L. Altshuler and A. G. Aronov, in *Electron-Electron Interactions in Disordered Systems*, edited by A. L. Efros and M. Pollak, North-Holland, Amsterdam, 1 (1985).
45. A. M. Finkelshtein, JETP Letters **45**, 46 (1987); Physica B **197**, 636 (1994).
46. P. W. Anderson, J. Phys. Chem. Sol. **11**, 26 (1959).
47. M. Y. Simmons, A. R. Hamilton, M. Pepper, E. H. Linfield, P. D. Rose, D. A. Ritchie, A. K. Savchenko, and T. G. Griffiths, Phys. Rev. Lett. **80**, 1292 (1998).
48. A. R. Hamilton, M. Y. Simmons, M. Pepper, E. H. Linfield, P. D. Rose, and D. A. Ritchie, Phys. Rev. Lett. **82**, 1542 (1999); *ibid* **84**, 2489 (2000).
49. C. Castellani, C. DiCastro, and P. A. Lee, Phys. Rev. B **57**, R9381 (1998).
50. M. V. Feigel'man, A. I. Larkin, and M. A. Skvortsov, Phys. Rev. B **61**, 12361 (2000).
51. I. V. Yurkevich and I. V. Lerner, Phys. Rev. B **63**, 4522 (2001); *ibid* 13043 (2001).
52. N. Nagaosa, *Quantum Field Theory in Condensed Matter Physics*, Springer, Heidelberg (1999).
53. Y. N. Ovchinnikov, Sov. Phys. JETP **37**, 366 (1973).
54. S. Maekawa, H. Ebisawa, and H. Fukuyama, J. Phys. Soc. Jpn. **53**, 2681 (1984).
55. N. Mason and A. Kapitulnik, Phys. Rev. Lett. **82**, 5341 (1999).
56. J. A. Chervenak and J. M. Valles Jr., Phys. Rev. B **59**, 11209 (1999).
57. N. Markovic, C. Christiansen, A. M. Mack, W. H. Huber, and A. M. Goldman, Phys. Rev. B **60**, 4320 (1999).
58. A. M. Goldman and N. Markovic, Physics Today **51**, 39 (1998).
59. V. Ambegaokar, U. Eckern, and G. Schön, Phys. Rev. Lett. **48**, 1745 (1982).
60. A. I. Larkin and Y. N. Ovchinnikov, Phys. Rev. B **28**, 6281 (1983).
61. U. Eckern, G. Schön, and V. Ambegaokar, Phys. Rev. B **30**, 6419 (1984).

62. S. Chakravarty et al., *Phys. Rev. Lett.* **56**, 2303 (1986); M. P. A. Fisher, *ibid* **57**, 885 (1986); D. Dalidovich and P. Phillips, *ibid* **84**, 737 (2000); A. Cuccolu, A. Fibini and V. Tognetti, *Phys. Rev.* B **61**, 11289 (2000).
63. M. P. A. Fisher, G. Grinstein, and S. M. Girvin, Phys. Rev. Lett. **64**, 587 (1990); M. P. A. Fisher, *ibid* **65**, 923 (1990); E. S. Sorensen et al., *ibid* **69**, 828 (1992); C. Bruder, R. Fazio, and G. Schön, *Phys. Rev.* B **47**, 342 (1993); M. Wallin et al., *ibid* **49**, 12115 (1994); J. Lidmar et al., *ibid* **58**, 2827 (1998); D. Das and D. Doniach, *ibid* **60**, 1261 (1999).
64. S. L. Sondhi, S. M. Girvin, J. P. Carini, and D. Shahar, Rev. Mod. Phys. **69**, 315 (1997).
65. A. O. Caldeira and A. J. Leggett, Phys. Rev. Lett. **46**, 211 (1981).
66. A. Kamenev and Y. Gefen, Phys. Rev. B **54**, 5428 (1996).
67. I. S. Beloborodov et al., Phys. Rev. B **63**, 115109 (2001).
68. G. Schön and A. D. Zaikin, Phys. Rep. **198**, 237 (1990).
69. S. E. Korshunov, JETP Letters **45**, 434 (1987).
70. K. B. Efetov, Sov. Phys. JETP **51**, 1015 (1980).

EXACT FUNCTIONALS, EFFECTIVE ACTIONS AND (DYNAMICAL) MEAN-FIELD THEORIES: SOME REMARKS

ANTOINE GEORGES
Laboratoire de Physique Théorique de l'Ecole Normale Supérieure, 24 rue Lhomond, 75231 Paris Cedex 05, France

1. Introduction

In these lecture notes, I would like to discuss (hopefully in a pedagogical manner), a general approach to strongly interacting systems which is based on the construction of a functional of some local quantity (effective action) by the Legendre transform method. This method has a wide range of applicability [6], ranging from field theory, the statistical mechanics of a classical magnet to interacting fermion systems. Though exact in principle, it requires in practice that the exact functional is approximated in some manner.

The examples considered in this lecture will be: the Weiss mean-field theory (MFT) of a classical magnet, the density functional theory (DFT) of the inhomogeneous electron gas in solids and the dynamical mean-field theory (DMFT) of strongly correlated electron systems.

There are common concepts underlying all these constructions (cf. Table), as will become clear after reading each specific section below, namely:

- i) These theories focus on a specific *local quantity*: the local magnetization in MFT, the local electronic density in DFT, the local Green's function (or spectral density) in DMFT.
- ii) They replace the original system of interest by an *equivalent system*, which is used to provide a representation of the selected quantity: a single spin in an effective field for a classical magnet, free electrons in an effective one-body potential in DFT, a single impurity Anderson model within DMFT. The effective parameters entering this equivalent problem define *generalized Weiss fields* (the Kohn-Sham potential in DFT, the effective hybridization within DMFT), which are self-consistently adjusted. I note that the associated equivalent system can be a non-interacting (one-body) problem, as in MFT and DFT, or a fully interacting many-body problem (albeit simpler than the original system) such as in DMFT and its extensions.

I.V. Lerner et al. (eds.),
Strongly Correlated Fermions and Bosons in Low-Dimensional Disordered Systems, 375–394.

– iii) In order to pave the way between the real problem of interest and the equivalent model, the method of coupling constant integration will prove to be very useful in constructing (formally) the desired functional using the Legendre transform method. The coupling constant can be either the coefficient of the interacting part of the hamiltonian (which leads to a non-interacting equivalent problem, as in DFT), or in front of the non-local part of the hamiltonian (which leads in general to a local, but interacting, equivalent problem such as in DMFT).

Some issues and questions are associated with each of these points:

– i) While the theory and associated functional primarily aims at calculating the selected local quantity, it always come with the possibility of determining some more general object. For example, classical MFT leads to a specific (Ornstein-Zernike) form for the spatial dependence of correlation functions. DFT comes with Kohn-Sham orbitals, which raises the question of their interpretation as one-electron excitations. DMFT produces a local self-energy which one may interpret as the lattice self-energy from which the full k-dependent Green's function can be reconstructed. In each of these cases, the precise status and interpretation of these additional quantities can be questioned.
– ii) I emphasize that the choice of an equivalent representation of the local quantity has nothing to do with subsequent approximations made on the functional. The proposed equivalent system is in fact an exact representation of the problem under consideration (for the sake of calculating the selected local quantity). It does raise a *representability* issue, however: is it always possible to find values of the generalised Weiss field which will lead to a specified form of the local quantity, and in particular to the exact form associated with the specific system of interest? For example: given the local electronic density $n(x)$ of a specific solid, can one always find a Kohn-Sham effective potential such that the one-electron local density obtained by solving the Schrödinger equation in that potential coincides with $n(x)$? Or, in the context of DMFT: given the local Green's function of a specific model, can one find a hybridisation function such that it can be viewed as the local Green's function of the specified impurity problem ?
– iii) There is also a stability issue of the exact functional: is the equilibrium value of the local quantity a minimum ? More precisely, one would like to show that negative eigenvalues of the stability matrix correspond to true physical instabilities of the system. I will not seriously investigate this issue in this lecture (for a discussion within DMFT, where it is still quite open, see [4]).

TABLE I. Comparison of various functional-based theories

Theory	MFT	DFT	DMFT
Quantity	Local magnetization m_i	Local density $n(x)$	Local GF $G_{ii}(\omega)$
Equivalent system	Spin in effective field	Electrons in effective potential	Quantum impurity model
Generalised Weiss field	Effective local field	Kohn-Sham potential	Effective hybridisation

2. The example of a classical magnet

For the sake of pedagogy, I will consider in this section the simplest example on which the above ideas can be made concrete: that of a classical Ising magnet with hamiltonian

$$H = -\sum_{ij} J_{ij} S_i S_j \tag{1}$$

2.1. CONSTRUCTION OF THE EFFECTIVE ACTION

We want to construct a functional $\Gamma[m_i]$ of a *preassigned* set of local magnetizations m_i, such that minimizing this functional yields the equilibrium state of the system. This functional is of course the Legendre transform of the free-energy with respect to a set of local magnetic fields. It is also called the *effective action* in the field-theory literature[1].

I will give a formal construction of this functional, following a method due to Plefka [19] and Yedidia and myself [9]. Let us introduce a varying coupling constant $\alpha \in [0,1]$, and define:

$$H_\alpha \equiv \alpha H = \sum_{ij} \alpha J_{ij}\, S_i S_j \tag{2}$$

Introducing local Lagrange multipliers λ_i, we consider the functional:

$$\Omega[m_i, \lambda_i; \alpha] \equiv -\frac{1}{\beta} \ln \mathrm{Tr}\, e^{-\beta H_\alpha + \beta \sum_i \lambda_i (S_i - m_i)} = F[\lambda_i] + \sum_i \lambda_i m_i \tag{3}$$

Requesting stationarity of this functional with respect to the λ_i's amounts to impose that, *for all values of* α, $\langle S_i \rangle$ coincides with the preassigned local magnetization m_i. The equations $m_i = \langle S_i \rangle$ which expresses the magnetization as a

[1] The effective action is actually $S = \beta\,\Gamma$ in our notations

function of the sources λ_i can then be inverted to yield the λ_i's as functions of the m_j's and of α:

$$\langle S_i \rangle_{\lambda,\alpha} = m_i \to \lambda_i = \lambda_i[m_j;\alpha] \tag{4}$$

(The average $\langle \cdots \rangle_{\lambda,\alpha}$ in this equation is with respect to the Boltzmann weight appearing in the above definition of Ω, including λ_i's and α). The Lagrange parameters can then be substituted into Ω to obtain the α-dependent Legendre transformed functional:

$$\Gamma_\alpha[m_i] = \Omega[m_i, \lambda_i[m_j,\alpha]] = F\left[\lambda_i[m]\right] + \sum_i \lambda_i[m] m_i \tag{5}$$

Of course, the functional we are really interested in is that of the original system with $\alpha = 1$, namely:

$$\Gamma[m_i] \equiv \Gamma_{\alpha=1}[m_i] \tag{6}$$

Let us first look at the non-interacting limit $\alpha = 0$ for which the explicit expression of Ω is easily obtained as:

$$\Omega_0 = \sum_i \left(-\frac{1}{\beta} \ln \cosh \beta\lambda_i + m_i \lambda_i \right) \tag{7}$$

Varying in the λ's yields:

$$\tanh \beta \lambda_i^{(\alpha=0)} = m_i \tag{8}$$

and finally:

$$\Gamma_{\alpha=0}[m_i] = \frac{1}{\beta} \sum_i \left(\frac{1+m_i}{2} \ln \frac{1+m_i}{2} + \frac{1-m_i}{2} \ln \frac{1-m_i}{2} \right) \tag{9}$$

The $\alpha = 0$ theory defines the *equivalent problem* that we want to use in order to deal with the original system. Here, it is just a theory of *independent spins in a local effective field.* The expression (9) is simply the entropy term corresponding to independent Ising spins for a given values of the local magnetizations.

The value taken by the Lagrange multiplier in the equivalent system, $\lambda_i^{\alpha=0}$ (denoted λ_i^0 in the following), must be interpreted as the *Weiss effective field.* We note that, in this simple example, there is an explicit and very simple relation (8) between the Weiss field and m_i, so that one can work *equivalently* in terms of either quantities. Also, because of the simple form of (8), *representability* is trivially satisfied: given the true equilibrium local magnetization m_i's ($\in [-1,1]$) of the model under consideration, one can always represent them by the Weiss fields $\beta h_i^{eff} = \text{arctanh}\, m_i$.

To proceed in the construction of Γ, we use a coupling constant integration and write:

$$\Gamma[m_i;\alpha=1] = \Gamma_0[m_i] + \int_0^1 d\alpha \frac{d\Gamma_\alpha}{d\alpha}[m_i] \tag{10}$$

It is immediate that, because of the constraint $\langle (S_i - m_i)\rangle = 0$:

$$\frac{d\Gamma_\alpha}{d\alpha} = \langle H\rangle_{\alpha,\lambda[\alpha]} = -\sum_{ij} J_{ij}\langle S_i S_j\rangle_{\alpha,\lambda[\alpha,m]} \tag{11}$$

In this expression, the correlation must be viewed as a functional of the local magnetizations (thanks to the inversion formula (4)). Introducing the connected correlation function:

$$g^c_{ij}[\{m_k\};\alpha] \equiv \langle (S_i - m_i)(S_j - m_j)\rangle_{\alpha,\lambda[\alpha,m]} \tag{12}$$

we obtain:

$$\frac{d\Gamma_\alpha}{d\alpha} = -\sum_{ij} J_{ij} m_i m_j - \sum_{ij} J_{ij}\, g^c_{ij}[\{m_k\};\alpha] \tag{13}$$

So that finally, one obtains the formal expression for $\Gamma[m_i] \equiv \Gamma_{\alpha=1}[m_i]$:

$$\Gamma[m_i] = \Gamma_0[m_i] - \sum_{ij} J_{ij} m_i m_j - \sum_{ij} J_{ij} \int_0^1 d\alpha g^c_{ij}[m_k;\alpha] \equiv \Gamma_0 + E_{MF} + \Gamma_{corr} \tag{14}$$

In this expression, g^c denotes the connected correlation function for a given value of the coupling constant, *expressed as a functional of the local magnetisations.*

Hence, the *exact functional* Γ appears as a sum of three contributions:

- The part associated with the equivalent system (corresponding here to the entropy of constrained but otherwise free spins)
- The mean-field energy $\sum_{ij} J_{ij} m_i m_j$
- A contribution from correlations which contains all corrections beyond the mean-field

As explained in the next section, there is a direct analogy between this and the various contributions to the density functional within DFT (kinetic energy, Hartree energy and exchange-correlation).

2.2. AN EXACT EQUATION FOR THE EFFECTIVE ACTION

At this stage, I would like to note that by combining differentiation with respect to the coupling constant (13) with the fluctuation-dissipation theorem, one can obtain a closed equation satisfied by the exact functional. The key observation is that the connected correlation g^c is directly related to the matrix of second derivatives of Γ. Indeed, since: $\frac{\delta\Gamma}{\delta m_i} = \lambda_i[m]$, we have: $\frac{\delta^2\Gamma}{\delta m_i \delta m_j} = \frac{\delta\lambda_i}{\delta m_j}$. Differentiation of $m_i = \langle S_i\rangle$ viewed as a functional of the λ's yields:

$$\frac{\delta m_i}{\delta\lambda_j} = \beta\langle (S_i - m_i)(S_j - m_j)\rangle \equiv \beta g^c_{ij} \tag{15}$$

Thus, one can relate the correlation function to Γ as:

$$\beta g^c_{ij}[\{m_k\};\alpha] = \left[\frac{\delta^2\Gamma_\alpha}{\delta m_k \delta m_l}\right]^{-1}_{ij} \tag{16}$$

in which the r.h.s involves a matrix inversion. This expresses the fluctuation - dissipation relation between correlation and response (generalised susceptibility). Inserting this into (13) yields:

$$\frac{d\Gamma_\alpha}{d\alpha} = -\sum_{ij} J_{ij} m_i m_j - \frac{1}{\beta}\sum_{ij} J_{ij}\left[\frac{\delta^2\Gamma_\alpha}{\delta m_k \delta m_l}\right]^{-1}_{ij} \tag{17}$$

which can also be written in integral form:

$$\Gamma_\alpha[m_i] = \Gamma_0[m_i] - \alpha\sum_{ij} J_{ij} m_i m_j - \frac{1}{\beta}\sum_{ij} J_{ij}\int_0^\alpha d\alpha' \left[\frac{\delta^2\Gamma_{\alpha'}}{\delta m_k \delta m_l}\right]^{-1}_{ij} \tag{18}$$

This can also be turned into an equation for the temperature dependence of the effective action $\mathcal{S}[\beta, m_i] \equiv \beta\,\Gamma_{\alpha=1}[m_i]$:

$$\mathcal{S}[\beta, m_i] = \mathcal{S}[\beta = 0, m_i] - \beta\sum_{ij} J_{ij} m_i m_j - \sum_{ij} J_{ij}\int_0^\beta d\beta' \left[\frac{\delta^2\mathcal{S}_{\beta'}}{\delta m_k \delta m_l}\right]^{-1}_{ij} \tag{19}$$

This equation fully determines in principle the effective action functional. However, in order to use it in practice, one generally has to start from a limit in which the functional is known explicitly, and expand around that limit. For example, as described below, an expansion around the high-temperature limit yields systematic corrections to mean-field theory [9].

I finally note that a slightly more general form of (17) is obtained by considering an arbitrary variation of the exchange couplings:

$$-\frac{1}{\beta}\frac{\delta}{\delta J_{ij}}\mathcal{S} = m_i m_j + \left[\frac{\delta^2\mathcal{S}}{\delta m_k \delta m_l}\right]^{-1}_{ij} \tag{20}$$

This equation is analogous to the Wilson-Polchinsky equation [20] for the effective action (after a Legendre transformation: see also [21]), which can be taken as a starting point for a renormalisation group analysis by starting from the local limit and expanding in the "locality" (see e.g [2, 21].

2.3. EQUILIBRIUM CONDITION AND STABILITY

The physical values of the magnetisations at equilibrium are obtained by minimising Γ, which yields:

$$m_i^* = \tanh\left(\beta\sum_j J_{ij}m_j^* - \beta\frac{\delta\Gamma_{corr}}{\delta m_i}\right) \tag{21}$$

and the Weiss field takes the following value:

$$h_i^{eff} \equiv (\lambda_i^0)^* = \sum_j J_{ij}\, m_j^* - \frac{\delta\Gamma_{corr}}{\delta m_i}|^* \tag{22}$$

This equation is a *self-consistency condition* which determines the Weiss field in terms of the local magnetizations on all other sites. Its physical interpretation is clear: h_i^{eff} is the true (average) local field seen by site i. It is equal to the sum of two terms: one in which all spins are treated as independent, and a correction due to correlations.

The stability of the functional around equilibrium is controlled by the fluctuation matrix:

$$\frac{\delta^2\Gamma}{\delta m_i\delta m_j} = \frac{\delta\lambda_i^0}{\delta m_j} - J_{ij} + \frac{\delta^2\Gamma_{corr}}{\delta m_i\delta m_j} \tag{23}$$

At equilibrium, this is nothing else than the inverse of the susceptibility (or correlation function) matrix:

$$\frac{\delta^2\Gamma}{\delta m_i\delta m_j} \equiv (\chi^{-1})_{ij} = (\chi_0^{-1})_{ij} - J_{ij} + \frac{\delta^2 A_{corr}}{\delta m_i\delta m_j} \tag{24}$$

with:

$$(\chi_0^{-1})_{ij} = \frac{1}{\beta(1-m_i^2)}\delta_{ij} \tag{25}$$

Hence, our functional does satisfy a stability criterion as defined in the introduction: a negative eigenvalue of this matrix (i.e of $\chi(\vec{q})$) would correspond to a physical instability of the system.

2.4. MEAN-FIELD APPROXIMATION AND BEYOND

Obviously, this construction of the *exact* Legendre transformed free energy, and the exact equilibrium condition (21) has formal value, but concrete applications require some further approximations to be made on the correlation term Γ_{corr}. The simplest such approximation is just to neglect Γ_{corr} altogether. This is the familiar Weiss mean-field theory:

$$\Gamma_{MFT} = \frac{1}{\beta}\sum_i\left(\frac{1+m_i}{2}\ln\frac{1+m_i}{2} + \frac{1-m_i}{2}\ln\frac{1-m_i}{2}\right) - \sum_{ij} J_{ij}m_im_j \tag{26}$$

For a ferromagnet (uniform positive J_{ij}'s), this approximation becomes *exact in the limit of infinite coordination* of the lattice.

The formal construction above is a useful guideline when trying to improve on the mean-field approximation. I emphasize that, within the present approach, *it is the self-consistency condition (22) (relating the Weiss field to the environment) that needs to be corrected*, while the equation $m_i = \tanh \beta h_i^{eff}$ is attached to our choice of equivalent system and will be always valid. For example, in [19, 9] it was shown how to construct Γ_{corr} by a systematic high-temperature expansion in β. This expansion can be conveniently generated by iterating the exact equation (19). It can also be turned into an expansion around the limit of infinite coordination [9]. The first contribution to Γ_{corr} in this expansion appears at order β (or α^2) and reads:

$$\Gamma_{corr}^{(1)} = -\frac{\beta}{2} \sum_{ij} J_{ij}^2 (1 - m_i^2)(1 - m_j^2) \tag{27}$$

This is a rather famous correction to mean-field theory, known as the "Onsager reaction term". For spin glass models (J_{ij}'s of random sign), it is crucial to include this term even in the large connectivity limit. The corresponding equations for the equilibrium magnetizations are those derived by Thouless, Anderson and Palmer [25].

One can think of other, more sophisticated ways, of improving around the mean-field approximation by inserting into (14) an approximate form of the correlation function(e.g that of a Gaussian model, or of a spherical model), hence yielding an improved correlation functional Γ_{corr}. As an example, I note that Eq.(18) suggests a natural way of doing this, by using the correlation function calculated from Γ_{MFT} itself. We have:

$$\frac{\delta^2 \Gamma_\alpha^{MFT}}{\delta m_i \delta m_j} = \frac{1}{\beta(1 - m_i^2)} \delta_{ij} - \alpha J_{ij} \tag{28}$$

Let us evaluate Γ for a set of magnetisation such that $m_i^2 = m^2$ is independent of the site. Then, in Fourier space:

$$\left[\frac{\delta^2 \Gamma_\alpha^{MFT}}{\delta m_i \delta m_j}\right]^{-1} = \frac{\beta(1 - m^2)}{1 - \alpha\beta(1 - m^2) J_{\vec{q}}} \tag{29}$$

Inserting this into (18) and performing the integral over α, we obtain the improved functional:

$$\Gamma = \Gamma_0 - \sum_{ij} J_{ij} m_i m_j + \frac{1}{\beta} \sum_{\vec{q}} \ln \left[1 - \beta(1 - m^2) J_{\vec{q}}\right] \tag{30}$$

This amounts to an RPA summation of the interaction beyond the mean-field approximation.

3. Density functional theory

In this section, I explain how density-functional theory [2] (DFT) [11, 14] can be derived along very similar lines. This section borrows from the work of Fukuda et al. [5, 6] and of Valiev and Fernando [26]. For a recent pedagogical review emphasizing this point of view, see [1].

Let us consider the inhomogeneous electron gas of a solid, with hamiltonian:

$$H_\alpha = \hat{T} + \int dx v(x)\hat{n}(x) + \frac{\alpha}{2}\int dx dx' \hat{n}(x)U(x-x')\hat{n}(x') \tag{31}$$

where $\hat{T}$ is the kinetic energy, $v(x)$ is the external potential due to the nuclei and $U(x-x') = e^2/|x-x'|$ is the electron-electron interaction. I will use conventions in which $\hbar = m = 1$ in this section. Again, the coupling-constant parameter α was introduced for the sake of mere convenience, and the physical case is $\alpha = 1$.

We want to construct the free energy functional of the system while constraining the average density to be equal to some specified function $n(x)$. In complete analogy with the previous section, we introduce a Lagrange multiplier function $\lambda(x)$, and consider [3]:

$$\Omega_\alpha[n(x), \lambda(x)] \equiv -\frac{1}{\beta}\ln \mathrm{Tr} \exp\left(-\beta H_\alpha + \beta\int dx \lambda(x)(n(x) - \hat{n}(x))\right) \tag{32}$$

A functional of *both* $n(x)$ and $\lambda(x)$. As before, stationarity in λ insures that:

$$\langle \hat{n}(x)\rangle_{\lambda,\alpha} = n(x) \quad \rightarrow \quad \lambda(x) = \lambda_\alpha[n(x)] \tag{33}$$

This will be used to eliminate $\lambda(x)$ in terms of $n(x)$ and construct the functional of $n(x)$ only:

$$\Gamma_\alpha[n(x)] \equiv \Omega_\alpha[n(x), \lambda_\alpha[n(x)]] \tag{34}$$

3.1. EQUIVALENT SYSTEM: NON-INTERACTING ELECTRONS IN AN EFFECTIVE POTENTIAL

Again, I first look at the non-interacting case $\alpha = 0$. Then we have to solve a one-particle problem in an x-dependent external potential. This yields:

$$\Omega_0[n[x], \lambda[x]] = -\mathrm{tr} \ln[i\omega_n - \hat{t} - \hat{v} - \hat{\lambda}] - \int dx \lambda(x) n(x) \tag{35}$$

In this equation, tr denotes the trace over the degrees of freedom of a single electron, $i\omega_n$ is the usual Matsubara frequency, and $\hat{t} \equiv -\nabla^2/2$, $\hat{v}$, $\hat{\lambda}$ are the

[2] I actually consider the finite-temperature extension of DFT [17]

[3] Note that I chose in this expression a different sign convention for λ than in the previous section, and also that Tr denotes the full many-body trace over all N-electrons degrees of freedom.

one-body operators corresponding to the kinetic energy, external potential and $\lambda(x)$ respectively. The identity $\ln det = \mathrm{tr} \ln$ has been used.

Minimisation with respect to $\lambda(x)$ yields the following relation between λ^0 and $n(x)$:

$$\frac{1}{\beta}\sum_n \langle x|\frac{1}{i\omega_n - \hat{t} - \hat{v} - \hat{\lambda}_0}|x\rangle = n(x) \tag{36}$$

This defines the functional $\lambda_0[n(x)]$, albeit in a somewhat implicit manner. This is directly analogous to Eq.(8) defining the Weiss field in the Ising case (but in that case, this equation was easily invertible). If we want to be more explicit, what we have to do is solve the one-particle Schrodinger equation:

$$\left(-\frac{1}{2}\Delta + v_{eff}(x)\right)\phi_l(x) = \varepsilon_l \phi_l(x) \tag{37}$$

where the *effective one-body potential* is *defined* as:

$$v_{eff}(x) \equiv v(x) + \lambda^0(x) \tag{38}$$

It is convenient to construct the associated resolvent:

$$R(x, x'; i\omega_n) = \sum_l \frac{\phi_l(x)\phi_l^*(x')}{i\omega_n - \varepsilon_l} \tag{39}$$

and the relation (36) now reads:

$$\sum_l |\phi_l(x)|^2 f_{FD}(\varepsilon_l) = n(x) \tag{40}$$

in which f_{FD} is the Fermi-Dirac distribution.

This relation expresses the local density in an interacting many-particle system as that of a one-electron problem *in an effective potential* defined by (36) (Kohn-Sham potential). In so doing, the effective one-particle wave functions and energies (Kohn-Sham orbitals) have been introduced, whose relation to the original system (and in particular their interpretation as excitation energies) is far from obvious (see e.g [12]). There is, for example, no fundamental justification in identifying the resolvent (39) with the true one-electron Green's function of the interacting system. The issue of *representability* (i.e whether an effective potential can always be found given a density profile $n(x)$ is also not obvious, in contrast to the previous section.

To summarize, the non-interacting functional $\Gamma_0[n(x)]$ reads:

$$\Gamma_0[n(x)] = -\mathrm{tr}\ln[i\omega_n - \hat{t} - \hat{v} - \hat{\lambda}_0[n]] - \int dx \lambda^0[x;n]n(x)$$

$$= -\mathrm{tr}\ln[i\omega_n - \hat{t} - \hat{v}_{eff}[n]] - \int dx v_{eff}(x)n(x) + \int dx v(x)n(x)$$

$$= -\frac{1}{\beta}\sum_l \ln\left[1 + e^{-\beta\varepsilon_l[n]}\right] - \int dx v_{eff}(x)n(x) + \int dx v(x)n(x) \tag{41}$$

in which λ_0 and v_{eff} are viewed as a functional of $n(x)$, as detailed above. As often deplored in the DFT litterature, this is not a very explicit functional of $n(x)$. It is however a completely explicit functional of $\lambda_0(x)$ (or equivalently of the KS effective potential $v_{eff}(x)$) so that it is sometimes more convenient to think in that manner.

3.2. THE EXCHANGE-CORRELATION FUNCTIONAL

We turn to the interacting theory, and use the coupling constant integration method (see [10] for its use in DFT):

$$\Gamma[n(x)] = \Gamma[n(x); \alpha = 0] + \int_0^1 d\alpha \frac{d\Gamma_\alpha}{d\alpha} \tag{42}$$

Similarly as before:

$$\frac{d\Gamma_\alpha}{d\alpha} = \langle \hat{U} \rangle_{\lambda,\alpha} = \frac{1}{2} \int dx dx' U(x - x') \langle \hat{n}(x) \hat{n}(x') \rangle_{\lambda,\alpha} \tag{43}$$

Separating again a Hartree (mean-field) term, we get:

$$\Gamma[n(x)] = \Gamma_0[n(x)] + E_{Hartree}[n(x)] + \Gamma_{xc}[n(x)] \tag{44}$$

with:

$$E_{Hartree}[n(x)] = \frac{1}{2} \int dx dx' U(x - x') n(x) n(x') \tag{45}$$

and Γ_{xc} is the correction-to mean field term (the exchange-correlation functional):

$$\Gamma_{xc}[n(x)] = \frac{1}{2} \int dx dx' U(x - x') \int_0^1 d\alpha g_\alpha^c[n; x, x'] \tag{46}$$

In which:

$$g_\alpha^c[n; x, x'] \equiv \langle (\hat{n}(x) - n(x))(\hat{n}(x') - n(x')) \rangle_{\lambda_\alpha[n],\alpha} \tag{47}$$

is the (connected) density-density correlation function, expressed as a functional of the local density, for a given value of the coupling α.

It should be emphasized that the exchange-correlation functional Γ_{xc} is *independent* of the specific form of the crystal potential $v(x)$: it is a *universal functional* which depends only on the form of the inter-particle interaction $U(x - x')$! This is clear from the fact that Γ can also be rewritten, using (41):

$$\Gamma[n(x)] = \Gamma_{HEG}[n(x)] + \int dx v(x) n(x) \tag{48}$$

in which Γ_{HEG} is the density functional for the homogeneous problem (i.e the electron gas without the periodic crystal potential $v(x)$).

I finally note that an exact relation can again be derived for the density functional (or alternatively the exchange-correlation functional) by noting that:

$$\beta g^c_\alpha[n; x, x'] = \left[\frac{\delta\Gamma_\alpha}{\delta n(x)\delta n(y)}\right]^{-1}_{xx'} \tag{49}$$

Inserting this relation into (44,46), one obtains:

$$\Gamma_\alpha[n] = \Gamma_0[n] + \alpha E_H[n] + \frac{1}{2}\int dx dx' U(x-x') \int_0^\alpha d\alpha' \left[\frac{\delta\Gamma_\alpha}{\delta n(x)\delta n(y)}\right]^{-1}_{xx'} \tag{50}$$

in complete analogy with (18). I am not aware of practical applications of this exact equation, which could perhaps serve as a starting point for treatments of the inhomogenous electron gas in the spirit of exact RG.

3.3. THE KOHN-SHAM EQUATIONS

Let us now look at the condition for equilibrium. We vary $\Gamma[n(x)]$, and we note that, as before, the terms originating from the variation $\delta\lambda^0/\delta n(x)$ cancel because of the relation (36). We thus get:

$$\frac{\delta\Gamma}{\delta n(x)} = -\lambda_0(x) + \int dx' U(x-x')n(x') + \frac{\delta\Gamma_{xc}}{\delta n(x)} \tag{51}$$

so that the equilibrium density $n^*(x)$ is determined by:

$$\lambda^0(x)^* = \int dx' U(x-x')n^*(x') + \frac{\delta\Gamma_{xc}}{\delta n(x)}|_{n=n^*} \tag{52}$$

which equivalently specifies the KS potential at equilibrium as:

$$v_{eff}(x) = v(x) + \int dx' U(x-x')n^*(x') + \frac{\delta\Gamma_{xc}}{\delta n(x)}|_{n=n^*} \tag{53}$$

Equation (52) is the precise analog of Eq.(22) determining the Weiss field in the Ising case, and v_{eff} is the true effective potential seen by an electron at equilibrium, in a one-electron picture. Together with (37), it forms the fundamental (Kohn-Sham) equations of the DFT approach.

Concrete applications of the DFT formalism require an approximation to be made on the exchange-correlation term. The celebrated *local density approximation* (LDA) reads:

$$\Gamma_{xc}\left[n(x)\right]|_{LDA} = \int dx\, n(x)\, \epsilon_{xc}^{HEG}[n(x)] \tag{54}$$

in which $\epsilon_{xc}^{HEG}(n)$ is the exchange-correlation energy density of the *homogeneous* electron gas, for an electron density n. Discussing the reasons for the successes of this approximation (as well as its limitations) is quite beyond the scope of these lectures. The interested reader is referred e.g to [12, 1].

Finally, we observe that DFT satisfies the stability properties discussed in the introduction, since $\delta^2\Gamma/\delta n(x)\delta n(x')$ is the inverse of the density-density response function ($\vec{q}$-dependent compressibility). A negative eigenvalue would correspond to a charge ordering instability.

4. Exact functional of the local Green's function, and the Dynamical Mean-Field Theory approximation

In this section, I would like to explain how the concepts of the previous sections can provide a broader perspective on the dynamical mean field approach to strongly correlated fermion systems. In contrast to DFT which focuses on ground-state properties (or thermodynamics), the goal of this approach (see [8] for an early review) is to address excited states by focusing on the *local Green's function* (or the *local spectral density*). Thus, it is natural to formulate this approach in terms of a functional of the local Green's function. This point of view has been recently emphasized by Chitra and Kotliar [3].

I describe below how such an *exact functional* can be formally constructed for a correlated electron model (irrespective, e.g of dimensionality), hence leading to a *local Green's function (or local spectral density) functional theory*. I will adopt a somewhat different viewpoint than in [3], by taking the *atomic limit* (instead of the non-interacting limit) as a reference system. This leads naturally to represent the exact local Green's function as that of a quantum impurity model, with a suitably chosen hybridisation function. There is no approximation involved in this mapping (only a representability assumption). This gives a general value to the construction of Ref.[7].

Dynamical mean field theory can then be viewed as a subsequent approximation made on the non-local contributions to the exact functional (e.g. the kinetic energy).

For the sake of simplicity, I will take the Hubbard model as an example throughout this section. The hamiltonian is decomposed as:

$$H_\alpha = U\sum_i n_{i\uparrow}n_{i\downarrow} - \alpha\sum_{ij,\sigma} t_{ij}\, c^+_{i\sigma}c_{j\sigma} \tag{55}$$

I emphasize that the varying coupling constant $\alpha \in [0,1]$ has been introduced in front of the hopping term, which is the non-local term of this hamiltonian, and *not* in front of the interaction. When dealing with a more general hamiltonian, we would similarly decompose $H = H_{\text{loc}} + \alpha H_{\text{non-loc}}$.

4.1. REPRESENTING THE LOCAL GREEN'S FUNCTION BY A QUANTUM IMPURITY MODEL

In order to constrain the local Green's function $\langle c_i(\tau)c_i^+(\tau')\rangle$ to take a specified value $G(\tau-\tau')$, we introduce conjugate sources (or Lagrange multipliers) $\Delta(\tau-\tau')$ and consider [4]:

$$\Omega_\alpha[G(\omega),\Delta(\omega)] \equiv$$
$$-\frac{1}{N_s\beta}\ln\int DcDc^+ \exp\{\int_0^\beta d\tau(\sum_{i\sigma} c_{i\sigma}^+(-\partial_\tau+\mu)c_{i\sigma} - H_\alpha[c,c^+]) +$$
$$+\int_0^\beta\int_0^\beta d\tau d\tau' \sum_{i\sigma}\Delta(\tau-\tau')[G(\tau-\tau')-c_{i\sigma}^+(\tau)c_{i\sigma}(\tau')]\} \tag{56}$$

Inverting the relation $G = G_\alpha[\Delta]$ yields $\Delta = \Delta_\alpha[G]$, and a functional of the local Green's function is obtained as $\Gamma_\alpha[G] = \Omega_\alpha[G,\Delta_\alpha[G]]$.

I would like to emphasize that this construction is quite different from the Baym-Kadanoff formalism, which considers a functional of the complete lattice Green's function G_{ij}, not only of its local part G_{ii}. The Baym-Kadanoff approach also gives interesting insights into the DMFT construction [4], but will not be considered further in these lectures.

Consider first the $\alpha = 0$ case, in which the hamiltonian is purely local. Then, we have to consider a local problem defined by the action:

$$S_{imp} = -\int_0^\beta d\tau \int_0^\beta d\tau' \sum_\sigma c_\sigma^+(\tau)\left[(-\partial_\tau+\mu)\delta(\tau-\tau')-\Delta_0(\tau-\tau')\right]c_\sigma(\tau')$$
$$+U\int_0^\beta d\tau\, n_\uparrow(\tau)n_\downarrow(\tau) \tag{57}$$

Hence, the local Green's function $G(i\omega_n)$ is represented as that of a quantum impurity problem (an Anderson impurity problem in the context of the Hubbard model):

$$G = G_{imp}[\Delta_0] \tag{58}$$

As before, Δ_0 plays the role of a Weiss field (analogous to the effective field for a magnet, or to the KS effective potential in DFT). Formally, this Weiss field specifies [7] the effective bare Green's function of the impurity action (57):

$$\mathcal{G}_0^{-1}(i\omega_n) = i\omega_n + \mu - \Delta_0(i\omega_n) \tag{59}$$

There are however two important new aspects here:

[4] In this section, I will divide the free energy functional by the number N_s of lattice sites (restricting myself for simplicity to an homogeneous system)

- i) The Weiss function Δ_0 is a *dynamical* (i.e frequency dependent) object. As a result the local equivalent problem (57) is not in Hamiltonian form but involves retardation
- ii) The equivalent local problem is not a one-body problem, but involves local interactions.

We note that, as in DFT, the explicit inversion of (58) is not possible in general. In practice, one needs a (numerical or approximate) technique to solve the quantum impurity problem (an *"impurity solver"*), and one can use an iterative procedure. Starting from some initial condition for Δ_0 (or $\mathcal{G}_0$) , one computes the interacting Green's function G_{imp}, and the associated self-energy $\Sigma_{imp} \equiv \mathcal{G}_0^{-1} - G_{imp}^{-1}$. One then updates $\mathcal{G}_0$ as: $\mathcal{G}_0^{new} = [\Sigma_{imp} + G^{-1}]^{-1}$, where G is the specified value of the local Green's function.

4.2. EXACT FUNCTIONAL OF THE LOCAL GREEN'S FUNCTION

We proceed with the construction of the exact functional of the local Green's function, by coupling constant integration (starting from the atomic limit).

At $\alpha = 0$ (decoupled sites, or infinitely separated atoms), we have [5] :

$$\Omega_0[\Delta_0, G] = F_{imp}[\Delta_0] - \mathrm{Tr}\,(G\Delta_0) \tag{60}$$

where F_{imp} is the free energy of the local quantum impurity model viewed as a functional of the hybridisation function. By formal inversion $\Delta_0 = \Delta_0[G]$:

$$\Gamma_0[G] = F_{imp}\,[\Delta_0[G]] - \mathrm{Tr}\ (G\Delta_0[G]) \tag{61}$$

We then observe that (since the α-derivatives of the Lagrange multipliers do not contribute because of the stationarity of Ω):

$$\frac{d\Gamma_\alpha}{d\alpha} = \langle H_{\text{non-loc}} \rangle \tag{62}$$

which, for the Hubbard model, reduces to the kinetic energy:

$$\frac{d\Gamma_\alpha}{d\alpha} = \langle \hat{T} \rangle = -\frac{1}{N_s}\sum_{ij} t_{ij}\langle c_i^+ c_j\rangle|_G = \mathrm{Tr}\,\frac{1}{N_s}\sum_{\vec{k}} \varepsilon_{\vec{k}}\, G_\alpha(\vec{k}, i\omega_n)|_G \tag{63}$$

In this expression, the lattice Green's function $G_\alpha(\vec{k}, i\omega_n)$ should be expressed, for a given α, as a functional of the local Green's function G.

---

[5] In this formula and everywhere below, Tr denotes $\frac{1}{\beta}\sum_n$, with possibly a convergence factor $e^{i\omega_n 0+}$.

This leads to the following formal expression for the exact functional $\Gamma[G] = \Gamma_{\alpha=1}[G]$:

$$\Gamma[G] = F_{imp}[\Delta_0[G]] - \text{Tr}\,(G\Delta_0[G]) - \int_0^1 d\alpha \frac{1}{N_s}\sum_{ij} t_{ij}\langle c_i^+ c_j\rangle|_G \tag{64}$$
$$= F_{imp}[\Delta_0[G]] - \text{Tr}\,(G\Delta_0[G]) + \int_0^1 d\alpha \text{Tr}\frac{1}{N_s}\sum_{\vec{k}} \varepsilon_{\vec{k}} G_\alpha(\vec{k}, i\omega_n)|_G$$

4.3. A SIMPLE CASE: THE INFINITE CONNECTIVITY BETHE LATTICE

It is simple to see from this formal expression that $\Gamma[G]$ can be obtained in closed form in the limit of infinite lattice connectivity. This is particularly straightforward for a Bethe lattice with connectivity z, and the hopping scaled as: $t_{ij} = t/\sqrt{z}$ [18], with $z \to \infty$. Expanding the kinetic energy functional in (64) in powers of α, one sees that only the term of order α remains in the $z = \infty$ limit thanks to the tree-like geometry, namely:

$$\alpha\sum_{ijkl} t_{ij}t_{kl}\langle c_i^+ c_j c_k^+ c_l\rangle_{\alpha=0} = \alpha\sum_{ij} t_{ij}^2 \text{Tr}\, G^2 = \alpha(zN_s)\frac{t^2}{z}\text{Tr}\, G^2 \tag{65}$$

So that, integrating over α:

$$\Gamma_{Bethe}[G] = F_{imp}[\Delta_0[G]] - \text{Tr}\,(G\Delta_0[G]) + \frac{t^2}{2}\text{Tr}\, G^2 \tag{66}$$

This functional is similar (although different in details) to the one recently used by Kotliar [15] in a Landau analysis of the Mott transition within DMFT.

The self-consistency condition that finally determines both the Weiss field and the local Green's function is obtained from $\delta\Gamma/\delta G = 0$ (note that $G = \delta F_{imp}/\delta\Delta_0$). This yields:

$$\Delta_0[G, i\omega_n] = t^2\, G(i\omega_n) \tag{67}$$

One has to solve the single-impurity Anderson model (57) subject to this self-consistency condition [8].

4.4. DMFT AS AN APPROXIMATION TO THE KINETIC ENERGY FUNCTIONAL.

Now, I will show that the usual form of DMFT [8] (for a general non-interacting dispersion ε_k) corresponds to a very simple approximation made on the exact functional above. We need to evaluate the kinetic energy term in this functional. The DMFT approximation assumes that this can be done by using the self-energy of the impurity model (i.e at $\alpha = 0$) in order to construct the lattice Green's function $G_\alpha(\vec{k}, i\omega_n)$. Namely:

$$G_\alpha(\vec{k}, i\omega_n)|_{DMFT} = \frac{1}{i\omega_n + \mu - \Delta_\alpha[i\omega_n; G] - \alpha\varepsilon_{\vec{k}} - \Sigma_{\alpha=0}[i\omega_n, G]} \tag{68}$$

With:

$$\Sigma_{\alpha=0}[G; i\omega_n] \equiv \mathcal{G}_0^{-1} - G^{-1} = i\omega_n + \mu - \Delta_0[i\omega_n, G] - G^{-1} \tag{69}$$

Summing over $\vec{k}$, one then expresses the local Green's function in terms of the hybridisation as:

$$G(i\omega_n) = \int d\varepsilon \frac{D(\varepsilon)}{\zeta - \alpha\varepsilon} = \frac{1}{\alpha}\widetilde{D}\left(\frac{\zeta}{\alpha}\right) \tag{70}$$

With $\zeta \equiv i\omega_n + \mu - \Delta_\alpha - \Sigma_0 = \Delta_0 - \Delta_a + G^{-1}$. In this expression, $D(\varepsilon) = \frac{1}{N_s}\sum_{\vec{k}} \delta(\varepsilon - \varepsilon_{\vec{k}})$ is the non-interacting density of states, and $\widetilde{D}(z) = \int d\varepsilon \frac{D(\varepsilon)}{z-\varepsilon}$ its Hilbert transform. Introducing the inverse function such that $\widetilde{D}[R(g)] = g$, we can invert the relation above to obtain the hybridisation function as a functional of the local G for $U = 0$:

$$\Delta_\alpha[i\omega_n; G] = G^{-1} + \Delta_0[G] - \alpha R\left[\alpha G\right] \tag{71}$$

So that the lattice Green's function is expressed as a functional of G as:

$$G_\alpha(\vec{k}, i\omega_n) = \frac{1}{\alpha R(\alpha G) - \alpha\varepsilon_{\vec{k}}} \tag{72}$$

Inserting this into (64), we can evaluate the kinetic energy:

$$\frac{1}{N_s}\sum_{\vec{k}} \varepsilon_{\vec{k}} G_\alpha(\vec{k}) = \frac{1}{\alpha}\int d\varepsilon \frac{\varepsilon\, D(\varepsilon)}{R(\alpha G) - \varepsilon} = \frac{1}{\alpha}\left[-1 + \alpha G R(\alpha G)\right] \tag{73}$$

This leads to the following form of the functional, in the DMFT approximation:

$$\begin{aligned}\Gamma_{DMFT}[G] = \; & F_{imp}\left[\Delta_0[G]\right] - \mathrm{Tr}\left(G\Delta_0[G]\right) + \\ & + \int_0^1 d\alpha \mathrm{Tr}\left[G(i\omega_n)R\left(\alpha G(i\omega_n)\right) - \frac{1}{\alpha}\right]\end{aligned} \tag{74}$$

In the case of an infinite-connectivity Bethe lattice, corresponding to a semi-circular d.o.s of width $4t$, one has: $R[g] = t^2 g + 1/g$, so that the result (66) is recovered from this general expression.

I note that the DMFT approximation to the kinetic energy part of the functional is completely independent of the interaction strength U.

The condition for equilibrium $\delta\Gamma/\delta G = 0$ reads [6]:

$$\Delta_0[i\omega_n, G] = R[G(i\omega_n)] - \frac{1}{G(i\omega_n)} \qquad (75)$$

This is the self-consistency condition relating the Weiss field Δ_0 to the local Green's function, within the DMFT approximation [8]. It can be rewritten in the more familiar form, using (69):

$$G(i\omega_n) = \int d\varepsilon \frac{D(\varepsilon)}{i\omega_n + \mu - \Sigma_{imp}(i\omega_n)} \quad , \quad \text{with: } \Sigma_{imp} = \mathcal{G}_0^{-1} - G^{-1} \qquad (76)$$

which expresses that the local Green's function coincides with the impurity one.

The self-consistency condition is equivalent to the condition $\Delta_{\alpha=1}[G] = 0$, as expected fm the fact that $\Delta_{\alpha=1} = \delta\Gamma/\delta G$. Hence, within the DMFT approximation, the lattice Green's function is obtained by setting $\alpha = 1$ into (68):

$$G(\vec{k}, i\omega_n)|_{DMFT} = \frac{1}{i\omega_n + \mu - \varepsilon_{\vec{k}} - \Sigma_{imp}(i\omega_n)} \qquad (77)$$

4.5. BEYOND THE DMFT APPROXIMATION

The construction of an exact functional of the local Green's function described here suggests some routes for going beyond the DMFT approximation. Among the possible strategies:

- The most naive extension (while not in fact the simplest technically, or the most desirable on physical grounds[7] !) is to attempt an expansion around the limit of large lattice coordination. The above formalism lends itself very naturally to such an expansion (particularly in the case of the Bethe lattice, Eq.(66)), by extending the methods developed in [9] for classical magnets. In the present context however, one quickly realizes that higher order local correlation functions of the impurity model enter this expansion. For this reason, it is actually advisable to extend the present formalism from the start to a functional of both the local Green's function and of higher order local correlation functions. This is the central idea of extended DMFT (E-DMFT) [23, 13, 22, 24, 4].
- One may consider a cluster covering of the lattice, and construct a functional of the intra-cluster components of the lattice Green's function. Following the same lines as above, one can construct this functional by starting from the limit of decoupled clusters, i.e consider the model with inter-cluster hoppings

[6] When deriving this equation, it is useful to note that $R(\alpha G) + \alpha G R'(\alpha G) = \partial_\alpha[\alpha R(\alpha G)]$.

[7] The expansion around large dimensions is expected to display some singularities associated in particular with the physics of superexchange, as discussed by several authors.

scaled by α, and choose $\alpha = 0$ as the reference system. The unknown part of the functional corresponds to the inter-cluster kinetic energy. A straightforward generalization of the DMFT approximation as presented above will then naturally lead to the "free cluster" (or C-DMFT [16]) scheme in which the intra-cluster self-energy is used into the self-consistency condition.

Acknowledgments: I am grateful to R. Chitra, G. Kotliar and P. Le Doussal for discussions on some of the issues presented here.

References

1. Argaman, N. and G. Makov: 2000, 'Density functional theory: an introduction'. *Am. J. Phys.* **68**, 69. preprint physics/9806013.
2. Chauve, P. and P. L. Doussal: 2000, 'Exact renormalization group and application to disordered problems'. preprint cond-mat/0006057.
3. Chitra, R. and G. Kotliar: 2000, 'Dynamical mean-field theory and electronic structure calculations'. *Phys. Rev. B* **62**, 12715.
4. Chitra, R. and G. Kotliar: 2001, 'Effective action approach to strongly correlated fermion systems'. *Phys. Rev. B* **63**, 115110.
5. Fukuda, R., T. Kotani, Y. Suzuki, and S. Yokojima: 1994, 'Density functional theory through Legendre transformation'. *Prog. Theor. Phys.* **92**, 833.
6. Fukuda, R., T. Kotani, and S. Yokojima: 1996, 'Novel use of Legendre transformation in field theory and many-particle systems'. *Prog. Theor. Phys. Suppl.* **121**, 1.
7. Georges, A. and G. Kotliar: 1992, 'Hubbard model in infinite dimensions'. *Phys. Rev. B* **45**, 6479.
8. Georges, A., G. Kotliar, M. Rozenberg, and W. Krauth: 1996, 'Dynamical mean-field theory of strongly correlated fermion systems and the limit of infinite dimensions'. *Rev. Mod. Phys.* **68**, 13.
9. Georges, A. and J. Yedidia: 1991, 'How to expand around mean-field theory using high-temperature expansions'. *J. Phys. A* **24**, 2173.
10. Harris, J.: 1984, 'Adiabatic connection approach to Kohn-Sham theory'. *Phys. Rev. A* **29**, 1648.
11. Hohenberg, P. and W. Kohn: 1964, 'Inhomogeneous electron gas'. *Phys. Rev.* **136**, B864.
12. Jones, R. and O. Gunnarsson: 1989, 'The density functional formalism, its application and prospects'. *Rev. Mod. Phys.* **61**, 689.
13. Kajueter, H.: 1996, 'Interpolating perturbation scheme for correlated electrons systems'. *PhD thesis, Rutgers University.*
14. Kohn, W. and L. Sham: 1965, 'Self-consistent equations including exchange and correlation effects'. *Phys. Rev.* **140**, A1133.
15. Kotliar, G.: 1999, 'Landau theory of the Mott transition in the fully frustrated Hubbard model in infinite dimensions'. *Eur. J. Phys. B* **27**, 11.
16. Kotliar, G., S. Savrasov, G. Palsson, and G. Biroli: 2001, 'Cellular Dynamical Mean Field Approach to Strongly Correlated Systems'. *Phys. Rev. Lett.* **87**, 186401.
17. Mermin, N.: 1965, 'Thermal properties of the inhomogeneous electron gas'. *Phys. Rev.* **137**, A1441.
18. Metzner, W. and D. Vollhardt: 1989, 'Correlated Lattice Fermions in $d = \infty$ Dimensions'. *Phys. Rev. Lett.* **62**, 324.
19. Plefka, T.: 1982, 'Convergence condition of the TAP equation for the infinite-ranged Ising spin-glass model'. *J. Phys. A* **15**, 1971.

20. Polchinsky, J.: 1984, 'Renormalization and effective Lagrangian'. *Nucl. Phys. B* **231**, 269.
21. Schehr, G. and P. L. Doussal, 'Exact multilocal renormalization on the effective action'. *preprint.*
22. Sengupta, A. and A. Georges: 1995, 'Non-Fermi liquid behaviour near a T = 0 spin-glass transition'. *Phys. Rev. B* **52**, 10295.
23. Si, Q. and J. L. Smith: 1996. *Phys. Rev. Lett.* **77**, 3391.
24. Smith, J. L. and Q. Si: 2000, 'Spatial correlations in dynamical mean-field theory'. *Phys. Rev. B* **61**, 5184.
25. Thouless, D., P. Anderson, and R. Palmer: 1977, 'Solution of 'solvable model of a spin-glass''. *Phil. Mag.* **35**, 593.
26. Valiev, M. and G. Fernando: 1997, 'Density functional theory of one-electron propagators'. *Phys. Lett. A* **227**, 265.

Index

Batch number: 10755284

Printed by Printforce, the Netherlands